# The Biology of Nematocysts

# The Biology of Nematocysts

Edited by

**David A. Hessinger**
*Department of Physiology and Pharmacology*
*School of Medicine*
*Loma Linda University*
*Loma Linda, California*

**Howard M. Lenhoff**
*Department of Developmental and Cell Biology*
*University of California*
*Irvine, California*

**Academic Press, Inc.**
*Harcourt Brace Jovanovich, Publishers*
San Diego New York Berkeley Boston
London Sydney Tokyo Toronto

ACADEMIC PRESS, INC.
San Diego, California 92101

*United Kingdom Edition published by*
ACADEMIC PRESS LIMITED
24-28 Oval Road, London NW1 7DX

**Library of Congress Cataloging-in-Publication Data**

The Biology of nematocysts / edited by David A. Hessinger and Howard M. Lenhoff.
p. cm.
Based on a symposium held in Aug. 1986 at the University of California, Irvine under the co-sponsorship of the Dept. of Physiology and Pharmacology of Loma Linda University School of Medicine and the Dept. of Developmental and Cell Biology of the University of California, Irvine.
Includes index.
ISBN 0-12-345320-8 (alk. paper)
1. Nematocysts—Congresses. I. Hessinger, David A. II. Lenhoff, Howard M. III. Loma Linda University. Dept. of Physiology and Pharmacology. IV. University of California, Irvine. Dept. of Developmental and Cell Biology
[DNLM: 1. Coelenterata—congresses. QL 375 B615 1986]
QL378.B56 1989
593'.5—dc19
DNLM/DLC
for Library of Congress 88-7751
CIP

PRINTED IN THE UNITED STATES OF AMERICA
88 89 90 91 9 8 7 6 5 4 3 2 1

# Contents

## II. Development



## III. Control and Morphodynamics of Discharge



## IV. Biochemistry

# Contributors

*Numbers in parentheses indicate the pages on which the authors' contributions begin.*

Peter A. V. Anderson, C. V. Whitney Laboratory, University of Florida, St. Augustine, Florida 32084 (273)

Charles H. Bigger, Department of Biological Sciences, Florida International University, Miami, Florida 33199 (295)

Richard S. Blanquet, Department of Biology, Georgetown University, Washington, D.C. 20057 (407)

Hans Bode, Developmental Biology Center, University of California, Irvine, California 92717 (209)

Joseph W. Burnett, Division of Dermatology, University of Maryland School of Medicine, Baltimore, Maryland 21201 (369)

Gary J. Calton, Division of Dermatology, University of Maryland School of Medicine, Baltimore, Maryland 21201 (369)

Richard D. Campbell, Department of Developmental and Cell Biology, University of California, Irvine, California 92717 (115,123)

M. G. Denaro, Institute of General Physiology, University of Messina, Messina, Italy (551)

G. Falzea, Institute of General Physilolgy, University of Messina, Messina, Italy (551)

Daphne Gail Fautin, Department of Invertebrate Zoology, California Academy of Sciences, San Francisco, California 94118-9961 (487)

Mark T. Ford, Department of Physiology and Pharmacology, Loma Linda University, Loma Linda, California 92350 (75)

Paul G. Greenwood, Department of Biological Science, Florida State University, Tallahassee, Florida 32306 (445)

Klaus Hausman, Institut fur Allgemeine Zoologie, Freie Universität Berlin, D-1000 Berlin, West Germany (53)

David A. Hessinger, Department of Physiology and Pharmacology, Loma Linda University, Loma Linda, California 92350 (75, 233, 255, 333)

Thomas Holstein, Zoologisches Institut, Universität München, D-8000 Munich 2, West Germany (53)

Thomas G. Honegger, Department of Zoology, University of Zürich, Zürich, Switzerland (567)

L. A. Hufnagel, Department of Microbiology, University of Rhode Island, Kingston, Rhode Island 02881 (519)

G. Kass-Simon, Department of Zoology, University of Rhode Island, Kingston, Rhode Island 02881 (519, 531)

William R. Kem, Department of Pharmacology and Therapeutics, University of Florida, Gainesville, Florida 32610 (375)

Marianne Klug, Zoological Institute, University of Zürich-Irchel, 8057 Zürich, Switzerland (427, 543)

G. La Spada, Institute of General Physiology, University of Messina, Messina, Italy (551)

Howard M. Lenhoff, School of Biological Sciences, University of California, Irvine, California 92717 (1)

Sylvia G. Lenhoff, School of Biological Sciences, University of California, Irvine, California 92717 (1)

Richard N. Mariscal, Department of Biological Science, Florida State University, Tallahassee, Florida 32306 (95)

M. Craig McKay, C. V. Whitney Laboratory, University of Florida, St. Augustine, Florida 32084 (273)

Claudia E. Mills, Friday Harbor Laboratories, University of Washington, Friday Harbor, Washington 98250 (463)

Carina Östman, Department of Zoology, Uppsala University, S-751-22 Uppsala, Sweden (501)

Jennifer E. Purcell, Horn Point Environmental Laboratories, University of Maryland, Cambridge, Maryland 21613 (463)

Elaine A. Robson, Department of Pure and Applied Zoology, The University, Reading RG6 2AJ, England (179)

A. Salleo, Institute of General Physiology, University of Messina, Messina, Italy (551)

R. J. Skaer, Peterhouse, Cambridge, England (165)

Robert P. Stidwill, Department of Zoology, University of Zürich, Zürich, Switzerland (567)

Pierre Tardent, Zoological Institute, University of Zürich-Irchel, Zürich, Switzerland (309, 427, 543, 567)

J. C. Thomason, Department of Zoology, University of Newcastle upon Tyne, Newcastle upon Tyne, Tyne & Wear, United Kingdom NE1 7RU (575)

Glyne U. Thorington, Department of Physiology and Pharmacology, Loma Linda University, Loma Linda, California 92350 (233)

Glen M. Watson, Department of Physiology and Pharmacology, Loma Linda University, Loma Linda, California 92350 (21, 143, 255)

Jakob Weber, Zoological Institute, University of Zürich-Irchel, 8057 Zürich, Switzerland (427, 543)

Jane A. Westfall, Department of Anatomy and Physiology, Kansas State University, Manhattan, Kansas 66506 (41)

Richard L. Wood, Department of Anatomy and Cell Biology, University of Southern California, Los Angeles, California 90033 (21, 25)

# Preface

One of the most complex and enigmatic of all eukaryotic cells is the cnidocyte, with its unique secretory product, the cnida, known more generally as the nematocyst. Cnidocytes and nematocysts are found in all members of the phylum Cnidaria and are diagnostic of the phylum.

The cnidocyte originates from an undifferentiated, multipotent stem cell, the interstitial cell. Within its lifespan, the cnidocyte passes through two functional phases: Initially, each cnidocyte functions as an ectodermal secretory cell which synthesizes and assembles a single organelle, the nematocyst, within a secretory vesicle. In its second phase, the cnidocyte functions as both a specialized sensory cell and an effector cell. Between those two phases, however, the cnidocyte, with its developing nematocyst, migrates through the interstitial spaces of the ectoderm to its final functional site, usually in the tentacles. There the cnidocyte matures and, when properly stimulated, initiates discharge of the nematocyst with the accompanying eversion of its tubule. The nematocyst is used primarily in capturing prey or in turning away predators.

To fully understand this unique intracellular structure and the cell that makes, transports, and controls it requires a multidisciplinary approach incorporating virtually all the basic biological sciences, from natural history and taxonomy to ultrastructure, developmental biology, and biochemistry.

Spurred by recollections of the multidisciplinary nature of the 1961 symposium on the biology of hydra, where the mysteries of the nematocyst were first brought to the attention of scientists normally not familiar with the literature of invertebrate biologists, the organizers decided to hold a similar kind of conference.

Scientists trained in many different disciplines were brought together for the first time in a symposium devoted exclusively to the biology of nematocysts and cnidocytes. The symposium was held on the campus of the University of California, Irvine, in August 1986 under the co-sponsorship of the Department of Physiology and Pharmacology of Loma Linda University School of Medicine and the Department of Developmental and Cell Biology of the University of California,

Irvine. Special tribute was paid to Professor Cadet Hand for his life's contribution to furthering the study of nematocysts through his work and that of his students, a number of whom are contributors to this volume.

After the symposium, many of the contributors were asked to submit up-to-date reviews of the results obtained by various approaches to the investigation of nematocysts and cnidocytes. Those reviews, along with a few of the research proceedings presented at the symposium, make up the contents of this volume.

Nematocysts have long intrigued marine biologists, zoologists, and naturalists, but this modern compilation of multidisciplinary reviews should excite the interest as well of developmental and cell biologists, sensory physiologists, neurophysiologists, toxicologists, and biochemists. In addition, because virtually every college course in biology involves a laboratory exercise using hydra as a representative cnidarian, this book should be of special interest to teachers of biology, invertebrate zoology, and marine biology. In fact, immediately after the scientific sessions of the symposium, a day-long workshop for high school and community college instructors was held to highlight the main subject areas covered by the research symposium.

This book is organized into six sections: (1 ) ultrastructure; (2) development; (3) control of discharge; (4) biochemistry; (5) natural history; and (6) some contributed specialized papers of work in progress. An introductory historical paper discusses how the animal nature of the so-called "zoophytes" came to be recognized and the structure and function of nematocysts determined. The book also includes the proceedings of a colloquium on the nomenclature of nematocysts that was part of the symposium. The colloquium constituted an attempt to standardize the conflicting terminology of the past, as modern research on the nematocyst and cnidocyte begins to unfold.

The editors believe that as the current reductionist trends of research in molecular biology, cell biology, neurobiology, immunology, pharmacology, and biochemistry inevitably peak, the study of the cellular, intracellular, and developmental mysteries of the nematocyst and cnidocyte may offer an exciting interdisciplinary and holistic challenge to young biologists of the twenty-first century.

## Acknowledgments

We thank the following individuals and institutions for their support: Loma Linda University (Lyn Behrens, Dean of the School of Medicine; Maurice Hodgen, Dean of the Graduate School; Ian Fraser, Chairman of the Department of Physiology and Pharmacology); University of California, Irvine (Stuart Krasner, Acting Dean of the Graduate Division; Grover Stephens, Dean of the School of Biological Sciences; Albert Bennett, Chairman of the Department of Developmental and Cell Biology); and the National Science Foundation (grant DCB-8519265).

# Introduction

## HOW THE ANIMAL NATURE OF MARINE CNIDARIANS WAS RECOGNIZED AND THE NEMATOCYST DISCOVERED

Howard M. Lenhoff
Sylvia G. Lenhoff

School of Biological Sciences
University of California
Irvine, California 92717

SUMMARY

Such aquatic organisms as bryozoans, and corals and hydrozoans of the Cnidaria, were once termed "zoophytes." The designation connoted organisms which had properties of both plants and animals. Through the early eighteenth century, views that the so-called zoophytes were animals were not taken seriously, and beliefs that zoophytes were of plant origin prevailed. In 1727, the French physician and naturalist, Jean-André Peysonnel, presented strong reasons for refuting the dominant belief that the polyps of corals were plants. He was not believed, however, until René-Antoine Ferchault de Réaumur, a leading figure in science of the time, became influenced by the work of Abraham Trembley on hydra and on the bryozoan *Lophopus*. At Réaumur's request, his colleague, Bernard de Jussieu, extended Peysonnel's observations, affirming that a wide variety of organisms said to be zoophytes were animals. Nonetheless, objections to recognizing zoophytes as animals persisted even in the 1750s when John Ellis published his thorough works in which he examined large numbers of cnidarians and bryozoans. We describe this saga of the "zoophytes" in which authority and belief long prevailed over evidence from observation and experiment.

Although nematocysts are considered the diagnostic feature of cnidarians, they were not discovered until nearly 100 years after cnidarians were finally recognized as animals. Functional and morphological clues as to the existence of nematocysts were observed in the mid-eighteenth century by Trembley, Martin Folkes, and Henry Baker. Not until the techniques of histology were developed in the mid-nineteenth century, however, were nematocysts actually detected. By the 1860s, the structure, role, and overall function of nematocysts were relatively well worked out.

## I. ANIMAL NATURE OF MARINE CNIDARIANS

> *I felt keenly that nature was too vast and too little known for us to decide without temerity that such and such a property was not to be found in such and such a class of organisms.* (A. Trembley, 1744)

In the mid-eighteenth century, Abraham Trembley (1744), using hydras or "freshwater polyps" as he called them, startled the scientific world with his revolutionary discoveries of regeneration, asexual reproduction by budding, grafting of animal tissues, and phototaxis in eyeless animals. It also took Trembley's breakthrough experimentalism and refusal to accept unquestioned theories of long standing to set the stage for widespread recognition of large groups of marine cnidarians as animals. Trembley's careful studies of his freshwater polyps convinced him early on that, despite certain plant-like characteristics, they were definitely animals, not sensitive plants and not "zoophytes." At the time that Trembley published his first revolutionary findings in the early 1740s the term "zoophytes" was applied to such aquatic organisms as corals (Fig. 1) and hydrozoans of the Cnidaria, and bryozoans; the designation connoted organisms which had properties of both plants and animals.

*Fig. 1. "Coral Flowers" as depicted in the early eighteenth century (from Hughes, 1744).*

To some, the presumed zoophytes represented the connecting link between the plant and animal kingdoms. They were thus an important construct in the venerable and broadly accepted concept of a Great Chain or Ladder of Being, or Scale of Nature, that included all living things, and sometimes in addition, angelic creatures at the top and mineral matter at the bottom (see Lovejoy, 1936).

The animal nature of the so-called zoophytes, especially corals, was first postulated by Ferrante Imperato in his 1599 volume, *Historia Naturale*. His views, however, were not taken seriously, and the beliefs that zoophytes were of plant origin continued to prevail for almost 150 years longer. This belief was further promulgated by the Count de Marsigli. In 1711 he published his *Histoire Physique de la Mer* in which he referred to the polyps of corals as "flowers and blossoms." His definition was accepted even by the reknowned French scientist and model for Trembley, René-Antoine Ferchault de Réaumur, who stated, "this discovery of the 'flowers' of the coral made a great noise in the world of naturalists" (see Pennington, 1885, p. 3).

Sixteen years later, Jean-André Peysonnel, a physician from Marseilles, came up with strong reasons for refuting the dominant belief that the polyps of corals were plants and the mineralized (calcareous) coral productions were the result of plant or inanimate activity. Peysonnel entrusted his notes to Réaumur, asking him to present them to the Academy at Paris. Réaumur, still influenced by the prestigious savant, Marsigli, did not want to embarass his young friend. Although he transmitted Peysonnel's findings to the Academy of Sciences in Paris, he did not reveal their authorship, including them instead in "Observations upon the formation of Coral and other productions called Stony Plants," (Réaumur, 1729) In this paper, Réaumur does not support Peysonnel's views.

Almost another fifteen years passed and the views of the recognized "authorities" regarding the "coral flowers" still held sway. At that point, however, the work of Réaumur's young follower, Trembley, on hydra (the "polyp with arms shaped like horns") and on the bryozoan (polyzoon) *Lophopus* (the "plumed polyp,") began to raise doubts in Réaumur's mind as to the plant-like nature of "marine flowers."

At Réaumur's request, his colleagues, Bernard de Jussieu and Jean-Étienne Guettard, went to the French coasts and to the rivers around Paris to investigate some of the organisms in question. Their research led them to conclude that Peysonnel was correct in his conclusions regarding the animal nature of the so-called "zoophytes" (see Réaumur, 1742). De Jussieu extended Peysonnel's observations, affirming that a wide variety of organisms said to be zoophytes were animals.

In the preface to his sixth volume on the *Insectes* (1742) in which he first describes Trembley's discoveries on hydra, Réaumur makes amends for his mishandling of Peysonnel's work and admits that Peysonnel's original interpretation was the correct one. As he wrote, "The care taken by Monsieur Peysonnel in making his observations should have convinced me sooner that the flowers [of] ... Count de Marsigli ... were really animals" (translated by Baker, 1952).

Réaumur also salutes Trembley's findings for leading the way to "these new knowledges" which had escaped so many astute observers. In a letter to Trembley (see M. Trembley, 1943) he wrote, "The vast and curious country of the polyps has devolved upon you ... ," and he urged Trembley to investigate marine polyps. Trembley, however, continued to devote those years to the study of freshwater organisms; somewhat later he did study the hydroid *Sertularia* that he obtained on the English coast (see Baker, 1952, p.127).

After the revelations of de Jussieu, Guettard, and Réaumur, Peysonnel returned to the attack. In 1751 he wrote a 400 page manuscript and transmitted it to the Royal Society of London rather than to the Academy in Paris. In it he restated the results of his researches of the previous 30 years. [According to Johnston (1847), the manuscript was never published as such and is in the library of the Paris Museum.]

Johnston pointed out that at first Peysonnel's treatise "excited considerable attention among members of the Royal Society." Eventually, however, he met opposition in England, chiefly from Dr. James Parsons and Henry Baker. Baker, the microscopist, who was intrigued with the crystallization of salts, stated his belief based on reason and analogy , but no evidence, that stony corals were nothing more than mineral productions. In Baker's (1753) words:

> Great Disputes have arisen among *Naturalists* concerning the formation of *Corals* and *Corallines*... .Such as believe these corals to be the *Work* of *Insects*, ground their Opinions ... on the sole Supposition that each of their *starry Cells* or *Hollows* is inhabited by *an Insect*, who therefore was the Maker of it. ...I say, to conclude thus, is somewhat over-hasty and unphilosophical. ...[Regarding] the *stony Corals* ... they are certainly formed by an *Apposition* of Particles some how or other brought together ... [to] ... be produced with all their beautiful Regularity and Variety ... [as] concreted by Frost. ... I have seen, in Winter Time, formed this Way, upon Windows ... very elegant Representations in Miniature of Landscapes, with Groves of Trees, and a seeming Variety of Shrubs and Herbs ... .

As for the horny and more pliant coverings of some forms, Baker thought them to be of a vegetable nature:

> Though, living far from the Sea, I have never been able to examine *Corallines* just taken from thence...[Their] Cases are of a membraneous Texture ... entirely different from the *stony Corals* ... . I believe that the most accurate Examiner, *void of any Hypothesis*, would declare it to be vegetable. (Baker, 1753)

Parsons, a physician and naturalist, also based his objections on belief. Without performing any experiments or making any observations, he said that he could not believe "that so fine an arrangement of parts, such regular ramifications, and such well-contrived organs to serve for vegetation, should be the operations of little, poor, jelly-like animals." Instead, Parsons believed that the inanimate "productions" of zoophytes were made by plants (see Pennington, 1885, p. 7). A popular view at that time held that once these non-animal productions had been made, then small polyp-like animals would inhabit them.

Ellis's book on the corallines (1755) and his subsequent volume with Solander (1786) would seem sufficient to have resolved the issue once and for all. Using a strong single-lens microscope (Ellis, 1755, Plate XXXVIII), Ellis examined a great many live and preserved organisms; he also carried out a few crude chemical tests. His evidence regarding the animal nature of the many organisms he studied, complete with accurate drawings, was most convincing in confirming Peysonnel's original contention.

Dr. John Baker, the author of the biography of Trembley (1952), wrote (1945) that: "Ellis, more than any other single man, caused scientists to accept into the animal kingdom the great and diverse groups of hydroids, sea-fans and their allies, and Polyzoa. No other man has ever made so great an enlargement of the subject matter of zoology."

Nonetheless, even after Ellis' convincing work, there were still those who clung to their beliefs and maintained that zoophytes were either partly or wholly "vegetable in nature." Linnaeus, in a letter to Ellis, wrote "Zoophyta ... are therefore vegetables, with flowers like small animals, which you have most beautifully delineated." He suggested that zoophytes were somewhere between animal and plant, with the plant portion building the hardened stem. In a letter to Linnaeus, Ellis responded to a conjecture that zoophytes became more plant-like as the stem material increased, by wryly asserting that "Artful people may puzzle at the vulgar, and tell us that the more hairy a man is, and the longer his nails grow, he is more of a vegetable than a man who shaves his hair or cuts his nails... ." In another letter to Linnaeus, Ellis wrote "I can not reconcile myself to vegetating animals ... . We have not sufficient proof to determine it; and I am averse to hypothesis." (See Johnston, 1847, pp. 424-429.) Clearly, Ellis was not intimidated by the views of Linnaeus, this giant figure among biologists, and his followers. In strong and certain terms, Ellis (1768) wrote that "zoophytes are true animals, and in no way part vegetable..."

To summarize, in the eighteenth century, it was Peysonnel who first proposed that the "sea-flowers" of corals were animals, but his findings were ignored; it was Trembley's research on hydra and on the bryozoan *Lophopus* that stimulated students of natural history to reopen the question regarding the possible animal nature of the so-called zoophytes; and it was the work of Ellis that finally provided conclusive evidence to put to rest the question of the animal or vegetable nature of the "zoophytes," in particular the marine cnidarians.

These, we now know, were the animals which in the 19th century were to become the objects of research for an incredible number of distinguished biologists, among them Dujardin, Morgan, Loeb, Wilson, Metchnikoff, Nussbaum, Agassiz, Haeckel, Allman, Hertwig, and Huxley.

It seems fitting to close the subject of how the animal nature of large groups of cnidarians was recognized with a cautionary comment by Abraham Trembley (1744):

> [Until we] know precisely all the attributes of plants and animals ... it is much more natural to consider the polyps and various other animals which have been given the name of zoophytes as animals which show more noteworthy similarities to plants than do other animals. ...It is too dangerous in the subject of natural history to abandon experience and allow the imagination to lead us. In following such a path, one risks arriving at uncertain hypotheses which can become detrimental to the progress of this science should one have the misfortune to become prejudiced in their favor. Instead of clarifying phenomena through new experiments, we have recourse to a hyphothesis, or rather a prejudgment, which not only spares us the trouble of observing, but which also serves to compound our errors.

## II. STUDIES LEADING TO THE DISCOVERY OF NEMATOCYSTS

> *Thus we cannot do better to explain the facts we know than by trying to discover new ones. Nature must be explained by nature and not by our own views.* (A. Trembley, 1744)

It was Antony van Leeuwenhoek in 1703, in his one paper mentioning hydra, who first described "knot-like lumps" on the "horns" of the hydra that he observed (see Dobell, 1960, pp. 282-283). From his drawings it is obvious that he is referring to the batteries on the tentacles. Leeuwenhoek even goes one step further to state "These lumps look to me as though they were made up of seven round globules; to wit, one in the middle, which sticks up a bit above the others, and the rest lying round it in a rosette." The "round globules" probably refer to the nematocysts in the battery, and the "one in the middle," to a stenotele.

It was not until Trembley's investigations of hydra (1744) stimulated thorough studies by others, however, that we began to gain our first notions of the function of nematocysts. Despite his careful observation, ingenious experimentation, and amazing findings, Trembley was not destined to become the discoverer of the structure and function of these complex cellular secretions. He does start the process of investigating these organelles by publishing certain "facts he knows." For example, in his *Mémoires* (see Lenhoff and Lenhoff, 1986, p. 34) Trembley describes "beads" on the "arms" of hydra:

> Imagine an arm as it extends. Little by little one sees these surface granules, which touched or nearly touched each other when the arm was strongly contracted, become spaced wider apart .... When the arm reaches a certain degree of elongation, its surface appears merely sprinkled with beads .... Drawing further away from each other as the arm continues to elongate, the beads finally turn out to be strung in a single row and separated by a transparent thread .... These beads are formed by a clustering of a number of granules.

We now know that the "beads" described by Trembley were the nematocyst batteries and the "granules" the nematocysts. But because Trembley also saw "granules" in the endoderm (i.e. food granules), he mistakenly thought that they were the same and functioned in some manner to aid the nutrition of the hydra.

Trembley also described hairlike structures (Fig. 2) along the arms of the polyps (see Lenhoff and Lenhoff, 1986, p. 35):

> The hairlike structures ... become noticeable in the elongated arm of a polyp when it is examined under the strong lens of a microscope. They appear to be transparent.

*Fig. 2. Portion of a tentacle of a hydra showing "clustering[s] . . . of granules" and "hairlike structures" (from Trembley, 1744, Plate 5).*

Some modern authors have thought the hairs depicted in Plate 5 of the *Mémoires* to be cnidocils. Weill (1934) was not sure whether they were cnidocils or discharged threads of nematocysts. From their relatively long size as drawn by Lyonet, we presume the hairlike structures described by Trembley to be the tubules of discharged stenotele nematocysts.

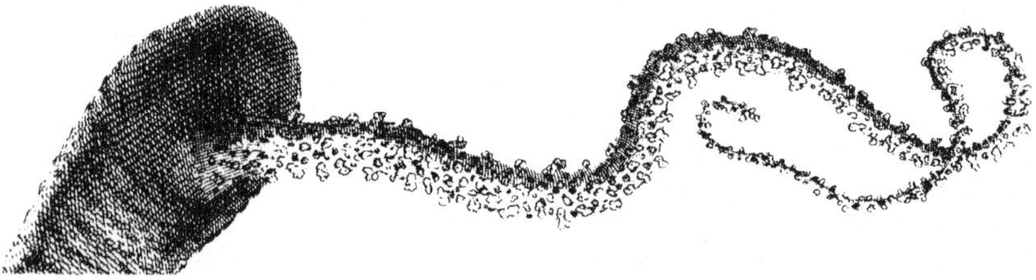

*Fig. 3. Drawing of tentacle, emerging from head of a hydra, showing the "little* Papillae*" (from Folkes, 1744).*

While observing the capture of prey, Trembley deduced that the arms of the polyps were able to hold the prey fast by some unexplained means (see Lenhoff and Lenhoff, 1986, pp. 54-55):

> [The millepede] need but touch the arm for it to be seized. As soon as the millepede feels itself captured, it struggles vigorously, making great efforts to free itself... Rarely do they free themselves from the polyp at all..

The first mention of the possibility of "hooks" being present on the arms of the polyps was made by a non-biologist, Martin Folkes (1744), President of the Royal Society of London, who took an avid interest in Trembley's experiments.

> [It] appears to me, that the little *Papillae* [Fig. 3] described on the Surface of the Arms assist them like so many Hooks or Tenters to hold their Worms barely by touching them; for I have more than once seen a *Polypus* draw a Worm to him, and nimbly turn it about with a single Arm, only laid over it, without folding or clasping it; which last method, however, he makes use of also, when the Worm comes to struggle and strive hard to be disengaged.

Actually, the microscopist Henry Baker, whose erroneous views about the marine cnidarians we have noted, was first to show the tubules of discharged nematocysts along the tentacles of hydra. Baker (1744) thought that if he could fix and then dry a polyp, he might be able to see some features not otherwise visible in the live animal. After applying a variety of fixatives - vinegar, dissolved salts, and spirits of wine - he comments:

> By observing the Arms thus dried, we obtain a clear Idea of the Means whereby this Creature catches fast hold of its Prey, the Moment of its touching it, and before it can bring its Arms to clasp about it: For we plainly see here, that the Arms are thick beset with Hairs

[Fig. 4], or rather sharp Hooks, which possibly are moveable, and can strike easily into the Body of a tender Worm. But these Hooks or Hairs perhaps, some-how or other generally drawn in, or laid flat and close along the Sides of the Arms, as I have seen them in some sorts of Star-fish.

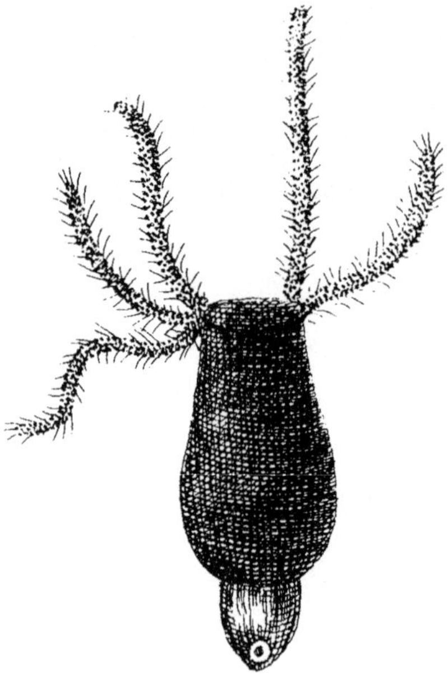

*Fig. 4. Drawing of dried hydra showing tentacles "beset with Hairs" after the animal was exposed to vinegar, salts, and wine (from Baker, 1744).*

Baker got carried away by his observations, and again was too quick to draw conclusions. For example in his book on the polyp (1743, p.64) he wrote:

> [The Polyp's] Arm's ... sense of Feeling is so delicate, that if a Worm touches even the utmost Extremity of these slender Arms, they immediately lay hold of it ... [The Arms] invelop and fetter it in so many Places, and to such a Degree, that notwithstanding it be much larger and seemingly stronger than they, it is soon rendered uncapable of struggling to any Purpose. In this Condition the *Polype* lengthens out his Snout, and bites the Worm violently on one Side, which not only gives it a great deal of Pain, as its wriggling motion testifies, but likewise makes the Blood run out: and it continues sucking at the Wound till the internal Juices are so far extracted, that the Body of the Worm is reduced to a size not over-large for the *Polype*'s Mouth ... .

Despite his erroneous speculations, Baker was a keen observer. For example, he suggested that the same structures present in the arms that capture prey, might also be present along the body of the hydra (Baker,1743, p. 30), a fact we now know to be true.

> Although I do not pretend to have yet been able perfectly to distinguish such, I have Reason to believe there may be certain minute Scales, or Protuberances, or some Sort of little *Spiculae* or Hooks all over the *Polype*'s Body, which can lengthen and protrude ... [If] a Worm happens to touch any Part of the Body, one shall plainly see it catch'd and held fast, one knows not how ... .[The] worm too instantly begins to struggle, and shews great Sense of Pain, but can very seldom get away; which almost proves it must be held by some invisible Hooks or Claspers [along the body] running into it, and which are probably of the same Sort as those along the Arms.

Baker also was the first to suggest that a poisonous substance might be transmitted from the polyp to the prey leading to a "sudden death." But he goes a little too far when he suggests that the poison may come from the "bite" of the "*Polype*'s snout." (Baker, 1743, pp. 32-33):

> We shall never, perhaps, be able to discover, certainly, by ocular Demonstration, whether the Mouth of this Creature is really armed with Teeth; but we may conjecture it so to be, from the Ease wherewith it bites or breaks the Skin of a Worm, in order to suck the Blood and Juices .... There is, likewise, a farther Probability of its having such offensive Weapons, from the violent and painful Agonies a Worm expresses the Moment it is taken hold on by the *Polype*'s Snout, and from the sudden Death that follows ... [One] would almost incline ... to imagine there must be something poisonous in the Bite; and that the *Polype*, as well as the Viper, does not only bite, but even inject a Venom into the Wound it gives, for the more speedy Destruction of its Prey.

Thus Trembley, and much more so Baker, approached the concept of the nematocyst, but did not describe it because histology was not developed as an art or science during their time. It took nearly 100 years more before nematocysts were clearly identified both as to their structure and function.

Ehrenberg (1836) carried out the first systematic investigation of nematocysts, studying primarily the stenotele of hydra, which he called "anglers." He made one serious error based upon one observation in which the tip of a discharged stenotele was found imbedded in the operculum of an undischarged stenotele. Ehrenberg concluded that the discharged tube was contractile, that the nematocyst could invaginate after it was discharged, and that at the end of the discharged tube was a capsule (Fig. 5).

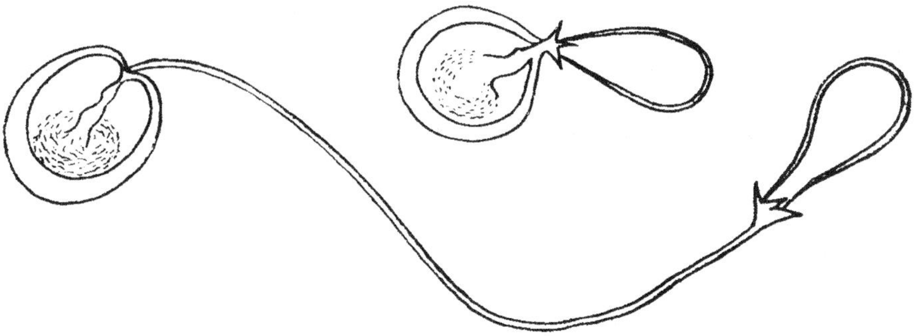

*Fig. 5. Drawings of first nematocysts (stenoteles) seen; these show the tip of a discharged one imbedded in the operculum of an undischarged one (from Weill's, 1934, redrawing of Ehrenberg, 1836).*

Wagner (1835) made an even more serious error while observing nematocysts in acontia. He concluded that the acontia were testicles, and the tadpole shaped discharged nematocysts were sperm, with the capsule being the head of the sperm and the thread the tail; however, he did draw the barbed armature at the base of the tube. In 1841 he corrected his errors. From his studies of *Pelagia noctiluca* he postulated that a stinging filament innoculated the prey with a venomous liquid which was contained in a capsule.

Corda in 1837 described a structure resembling a stenotele, but said that it contained muscle and that the barb was able to go in and out of the capsule. He also postulated that the structure was capable of innoculating prey with a venom. Finally it was Erdle (1841) who correctly described the stenoteles of hydra as well as nematocysts of some actinians.

Doyère (1842) accurately described hydra nematocysts. He correctly pointed out that the tube is invaginated and coiled at the bottom of the undischarged stenotele, and that it everts on discharge. He also was the first to point out the three points on the calyx, the glutinant nematocysts, and cnidocils. Dujardin (1845) made some exact drawings of the nematocysts that he observed, and suggested that nematocysts might be used for taxonomic purposes.

By the middle of the nineteenth century, however, the facts began to be sorted out. Haime (1854) said that he "would call them nematocysts." Gosse (1858, 1860), known to coin numerous new terms, was the first to use the term "cnidae." Huxley's drawings of nematocysts (1859) are quite accurate and sophisticated.

Weill (1934) credits the first major large work on nematocysts as that of Moebius (1866). Moebius described the fine structure, function, and development of nematocysts (see Fig. 1 in Tardent, this volume). He said that there was general agreement that the nematocyst is an intracellularly-produced body, formed from a capsule open at one end, with a tube that is long and complicated which invaginates into its capsule, and which could be made to evaginate under certain conditions.

## III. CONCLUSION

> *It seemed to me from the start of my observations that knowledge of the remarkable properties of the polyps could bring pleasure to the inquisitive and contribute something to the progress of natural history.* (Trembley, 1744)

From Trembley to the nineteenth century researchers studying the nematocyst stretched a hundred years of trial and error and of "authoritative" theories dethroned only with great difficulty. Gradually advances were made in the understanding of the marine cnidarians and their enigmatic nematocysts, organoids that still fascinate and intrigue us today. Trembley's pioneering studies stimulated much of the ensuing work on the "polyps," both freshwater and marine.

As Réaumur had predicted in 1742, later naturalists have indeed sought out additional species of marine polyps and studied differences in their morphology, feeding, growth, and reproduction. As we now begin a symposium to stimulate the unraveling of the mysteries still locked up in this complex cellular secretion, we continue to fulfill Réaumur's cherished hope that: "At last, a part of natural history that is so interesting and new and that has been sketched only roughly, will be studied thoroughly as it deserves."

REFERENCES

Baker, Henry. 1743. *An Attempt towards a Natural History of the Polype*. R. Dodsley, London.

Baker, Henry. 1744. Some Observations on a *Polype* dried. Phil. Trans. 42:616-619.

Baker, Henry. 1753. *Employment for the Microscope*. R. Dodsley, London.

Baker, John R. 1945. *Science and Planned State*. MacMillan. New York.

Baker, John R. 1952. *Abraham Trembley of Geneva: Scientist and Philosopher 1710 - 1784*. Edward Arnold. London.

Corda, A. J. 1837. Anatome *Hydrae fuscae*. Ann. sc. nat. Zool. 8:363-366.

Dobell, C. 1960. *Antony van Leeuwenhoek and his "Little Animals."* Dover, New York.

Doyère. 1832. Note sur quelques points de l'anatomie des Hydres d'eau douce. C. R. Ac. Sc. 15:429-432.

Dujardin, F. 1845. Mémoires sur le developpement des Polypes hydraires et Meduses. Ann. sc. nat. Zool. 4:257-281.

Ehrenberg, C. G. 1836. Beobachtungen uber die organisation der Armpolypen. Mitt. Naturf. Ges. Berlin. 2$^{0}$ Quartal:27-29.

Ellis, John 1755. *An Essay Towards a Natural History of the Corallines, and other Marine Productions of the like Kind*. Millar, Rivington, and Dodsley, London.

Ellis, John 1768. An Account of the Actinia Sociata, or Clustered Animal-flower, lately found on the Sea-Coasts of the new-ceded Islands. Phil. Trans. 57:428-437.

Ellis, J. and D. Solander. 1786. *The Natural History of Many Curious and Uncommon Zoophytes, Collected from Various Parts of the Globe*. B. White & Son, London.

Erdle, P. 1841. Über Organisation der Fangarme der Polypen. Arch. Anat. Physiol. 423-434.

Folkes, Martin. 1744. Some Account of the Insect called the Fresh-water *Polypus*, before-mentioned in these *Transactions*. Phil. Trans. 42:422-436.

Gosse, P. H. 1858. Researches on the poison apparatus in the Actiniidae. Proc. Roy. Soc. 9:125-128.

Gosse, P. H. 1860. *A History of the British Sea-Anemones and Corals*. Van Vorst, London.

Haime, J. 1854. Memoire sur le Cerianthe. Ann. Sc. nat. Zool. 1:341-389.

Hughes, G. 1744. Concerning a *Zoophyton*, somewhat resembling the Flower of the *Marigold*. Phil. Trans. 42:590-593.

Huxley, T. H. 1859. *Oceanic Hydrozoa; Calycophoridae and Physophoridae*. Ray Society, London.

Johnston, George. 1847. *British Zoophytes*, Vol. 1, John Van Voorst, London.

Lovejoy, A. O. 1936. *The Great Chain of Being*. Harvard, Cambridge, MA.

Moebius, K. 1866. Uber den Bau, den Mechanismus und die Abhdln. naturw. Ver Hamburg. 5:1-22.

Pennington, A. S. 1885. *British Zoophytes: An introduction to the Hydroida, Actinozoa, and Polyzoa found in Great Britain, Ireland, and the Channel Islands*. L. Reeve, London.

Réaumur, R. A. F. 1729. Observations sur la formation du corail , & des autres productions appelées plantes pierreuses. Hist. Acad. Roy. Sci. (Mém. Math. Phys.) pp. 226.

Réaumur, R. A. F. 1742. *Mémoires pour servir à l'histoire des insectes*. Vol 6. Imprimerie Royale, Paris.

Trembley, A. 1744. *Mémoires, pour servir à l'histoire d'un genre de polypes d'eau douce, à bras en forme de cornes*. Verbeek, Leiden.

Trembley, M. 1943. *Correspondance inédite de Réaumur et d'Abraham Trembley comprenant 113 lettres recueilles et annotées par Maurice Trembley*. Introduction by É. Guyénot. Georg, Geneva.

Tardent, P. 1988. History and current state of knowledge concerning discharge of cnidae (this volume).

Wagner, R. 1835. Entdeckung männlicher Geschlectsteile bei den Aktinien. Arch. Naturg. 2:215-219.

Weill, R. 1934. *Contribution a l'etude des cnidaires et de leurs nematocystes*. Vols. 1 and 2. University Press of France, Paris

# 1

# COLLOQUIUM ON TERMINOLOGY

Glen M. Watson

Department of Physiology and Pharmacology
Loma Linda University
Loma Linda, California 92350

Richard L. Wood

Department of Anatomy and Cell Biology
University of Southern California
School of Medicine
Los Angeles, California 90033

A colloquium was held to attempt to standardize some of the terminology used to describe cnidae and cnidocytes, and to agree on a terminology that would be simple, precise, and that would not conflict with terminology in other areas of biology. Criteria used to judge specific terms were clarity, general applicability to all cnidae and cnidocytes, ease of understanding and usefulness for biologists specializing in other areas, and historical precedence. The terms and definitions below represent the views of the majority of those participating in the colloquium.

1. cnida (cnidae) - the intracellular secretory product diagnostic of all members of the phylum Cnidaria. It consists of a capsule containing a highly-folded, eversible tubule. Cnidae are used to capture prey, to defend against predators, to attack (or to defend against attack by) unrelated, neighboring cnidarians, and to attach to suitable substrates. Nematocysts, spirocysts and ptychocysts are the three major categories of cnida. In the special case of the ptychocyst, discharged cnidae

provide the structural framework for protective layers around the body column of certain burrowing anemones.

2. cnidocyte - the cell that synthesizes the cnida, and from which the cnida discharges. The immature cnidocyte is referred to as a cnidoblast. The stage of development at which a cnidoblast becomes a cnidocyte was not agreed upon. The alternate views were, (1) once the cell reaches its definitive location in the epithelium and the cnida is positioned with the capsule apex oriented outward and, (2) once the cell begins to migrate from its site of origin to its definitive location in the animal. If the type of cnida is known, the more specific term, nematocyte (nematoblast), spirocyte (spiroblast) or, ptychocyte (ptychoblast) should be used.

3. tubule - the portion of the cnida that everts during discharge. The tubule is a cylindrical structure that is continuous with, or attached to, the apex of the capsule. Basally enlarged regions of the tubule are called shafts rather than butts. No separate term was agreed upon to describe the distal region of the tubule. The tubule can exist in several different states. Prior to discharge, the inverted tubule is located within the capsule. During discharge, the tubule is everting and after discharge, the tubule is everted. During the development of the cnida, the tubule is synthesized in the cytoplasm outside the capsule. Because all tubules now are thought to invert into the capsule after synthesis, tubules in the cytoplasm are referred to as preinverted tubules or, when appropriate, as inverting tubules. Because all cnida tubules are hollow and evert during discharge, such terms as thread and filament were considered to be inaccurate and inappropriate. The term, external tube, used to describe tubules in the cytoplasm of cells synthesizing cnidae, was considered to be confusing because it does not clearly indicate that the tube is external to the capsule but is not external to the cell.

4. spine - the armature decorating the surface of an everted tubule of a nematocyst. The shape of spines varies from one type of nematocyst to another. Generally speaking, spines are barb-shaped, although they may be shaped as spades (in many mastigophores) or even as anvils (in certain holotrichs). The term, spine is intended to be the general term, while other terms that more adequately describe the shape of a particular spine are encouraged. For example, the initial penetrating spines of the stenotele of hydra are accurately described by the term stylet. The spines of nematocysts of *Physalia* are barb-shaped, and the term barb may be used to describe them.

The term, barb also may be used to describe small, secondary extensions of a particular spine.

5. cnida membrane - the limiting membrane that lies outside of the cnida capsule, and which separates the outer wall of the capsule from the cytoplasm of the cnidocyte.
6. wall of the capsule - in nematocysts, walls are composed of two or more layers that can be distinguished in the electron microscope. The outer layer of the capsule wall is in contact with the cnida membrane; the inner layer is in contact with the contents of the capsule. Any additional layers that may be distinguishable within capsule walls of certain cnidae are designated as intermediate layers. Such terms as inner layer and outer layer are preferable to the terms sclera and propria because the former are descriptive and less confusing.
7. wall of the tubule - walls exist in as many as five configurations: preinverted, inverting, inverted, everting and everted (see "tubule" above). Thus, authors are encouraged to provide complete descriptions of tubule walls such as, "outer layer of the wall of the preinverted tubule", etc. in order to avoid confusion.
8. matrix contents - the material within the capsule lumen or within the lumen of the inverted tubule that is of an undefined chemical composition (such as that visible in a thin-section of a cnida viewed in the electron microscope). The term soluble contents refers to the material obtained from cnidae that is intended for chemical analysis. The soluble contents may or may not be equivalent to the venom, depending on the procedures used to obtain the soluble contents.
9. venom - the contents of a nematocyst that are injected into a target following discharge. The venom can include toxins, materials having a known negative influence on biological systems, and other materials such as salts, inert fluids, and remnants of nematocyst assembly, etc.

Further recommendations:

For terms not considered in the colloquium, the participants recommended that descriptive terms be used until definite functions of structures are known. They also suggested that any term coined by an author during the preparation of a manuscript be checked against published material for possible conflict or ambiguity.

# 2

# Survey of the Ultrastructure of Cnidocytes

Richard L. Wood

Department of Anatomy and Cell Biology
University of Southern California
Los Angeles, California 90033

## SUMMARY

Cnidae or nematocysts are specialized secretory products of cnidoblasts. Exocytosis is a two-stage process; there is an explosive discharge of the contents of the storage capsule, and after a finite delay, a release of the entire capsule from the cell. Ultrastructural studies have aided in the classification of nematocysts, provided a better understanding of capsule formation, confirmed that capsular discharge involves the eversion of an internal tubule, and clarified the nature of the cnidocil-associated apparatus. Major questions remain regarding the mechanism(s) of mechano- and chemosensory transduction, the mechanism of capsular activation, and the role of innervation or other factors in the control of discharge. Freeze-fracture preparations of hydra reveal intramembrane particles aggregated in the opercular region of the cnidocyte. This specialization could represent transport sites involved in sensory transduction or water permeation, but also could reflect attachment of the capsular vacuole membrane (VM) to the apical membrane of the cnidocyte. Gap junctional profiles have not been detected on cnidocytes. However, mature cnidocytes are attached apically to adjacent epithelial cells by septate junctions, and basally to the mesoglea by hemidesmosomes. Intracellular filaments of the actin type ( 7 nm diameter, phallocidin-binding), and the intermediate type (12 nm diameter, phallocidin-negative) occur in the cnidocyte. The two filament types have distinct distribution patterns and both appear to be involved primarily in structural support.

# I. INTRODUCTION

The cnidae or nematocysts are highly specialized intracellular organelles characteristic of the phylum Cnidaria. Between twenty-five and thirty different types of nematocysts have been described, and they are known to be involved in a variety of functions ranging from simple adhesion to the hypodermic injection of toxins into unwary prey. These remarkable structures are actually a form of secretory storage for an unusual mode of secretion. The final morphology of nematocysts is achieved by a sequence of synthesis reminiscent of typical secretory products, but the process of capsule assembly continues to elude full comprehension despite decades of attention by many different investigators.

A notable feature of this special type of secretion is that the secretory product is stored as a complex capsule that is capable of at least semi-autonomous activity. The biological function of nematocysts depends on ejection of part or all of the capsular content. The final release of the capsule itself, the process most akin to typical secretion, is not the biologically important event. It is not surprising then that the structure, mode of formation and mechanism(s) of discharge of nematocysts continue to intrigue cell biologists.

Fig. 1. Oblique section through the cnidocyte apex showing the cnidocil (CN), stereocilia (ST), supporting rods (SR), septate junction (SJ), and condensed cuticle (C) associated with the opercular region. Bar = 0.5 uM

Fig. 2. Longitudinal section of a cnidocil in hydra. The dense core is cross-striated and appears unrelated morphologically to its microtubular origin. Peripheral microtubules are still evident (MT). Bar = 0.5 uM

Fig. 3a. Tentacle of hydra treated with NBD-phallocidin showing the distribution of f-actin. Ectodermal myonemes (MY) and the cores of stereocilia (arrow) are brightly fluorescent. 3b. Phase micrograph of the same field as 3a. Bar = 20 uM

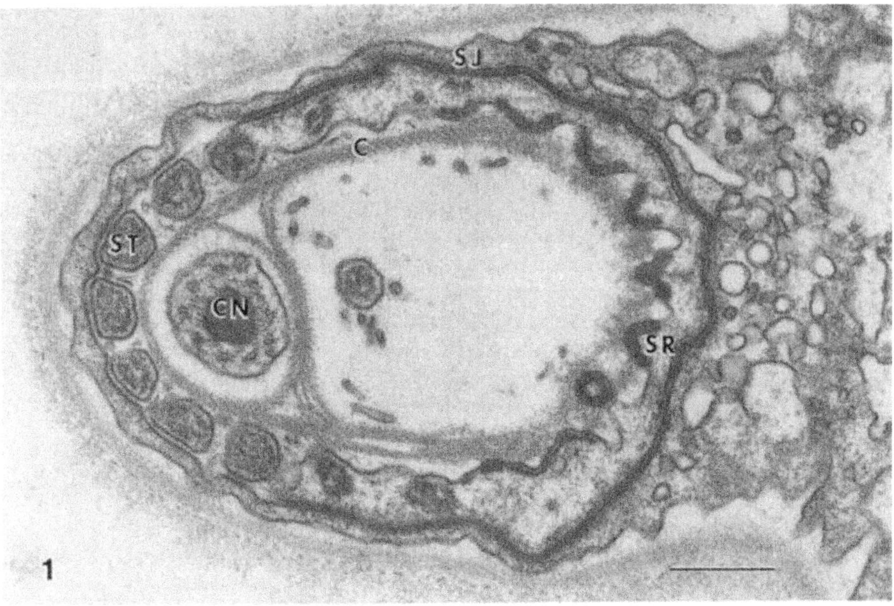
SJ
C
ST
CN
SR
1

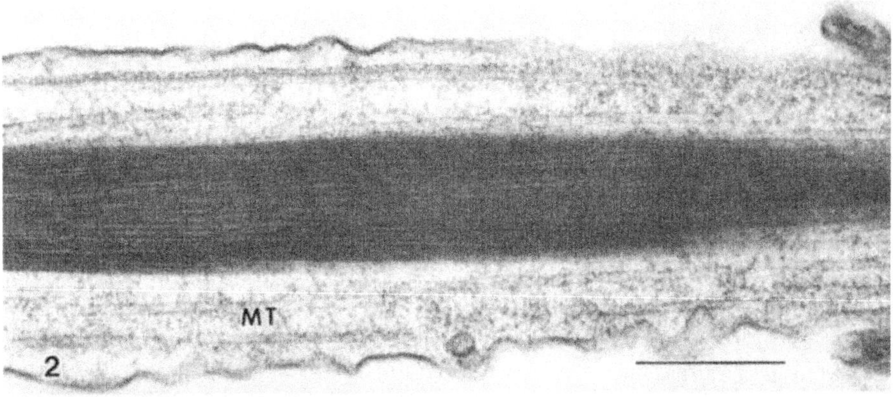
MT
2

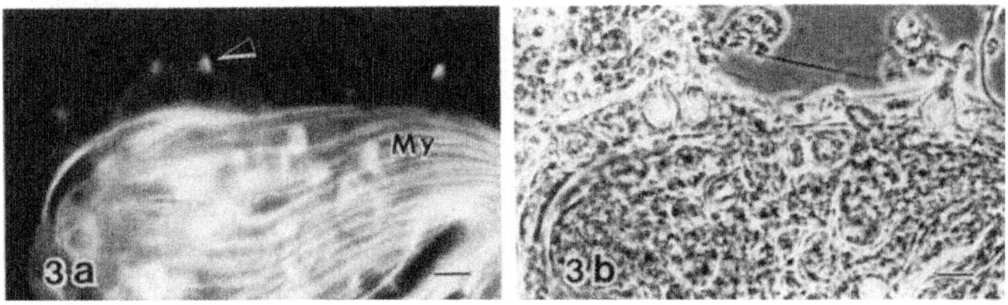
My
3a
3b

## II. CONVENTIONAL EM OF THE CNIDOCYTE

Recent ultrastructural investigations have provided a wealth of new factual information about cnidocytes. Nevertheless, the physiological significance of much of the ultrastructural information is poorly understood. For example, the functional significance of different forms of the cnidocil/cnidocyte complex, of neuro-cnidocyte synapses and of cnidocyte/mesogleal attachments are still largely speculative. Similarly, the morphological basis for cnidocyte mechano- and chemosensory transduction remains obscure. It seems obvious that careful documentation of structural relationships will contribute to the ultimate resolution of these quandaries but it is equally obvious that detailed knowledge of the ultrastructure will not be sufficient by itself.

Until recently, greater attention has been directed at describing the structure of nematocysts proper than to the cells which produce and house them. Cnidoblasts and cnidocytes have a number of unusual, if not unique, features. The apical end of the cell becomes specialized for reception and transduction of environmental stimuli appropriate for the elicitation of a specific response by each nematocyst type. The basic mechanism of capsular discharge may be similar for the various categories of nematocysts, but how selective reception and response are achieved is obscure. There are differences in the morphology of mechanosensory devices in different cnidarian groups, but many organisms possessing multiple types of nematocysts with different functions show no distinct morphological differences in the cnidocil/cnidocyte complex.

The cnidocil is a flagellum (cilium) or flagellar derivative (Figs. 1,2), but the axoneme structure may be so extensively modified that the flagellar origin is masked (Fig. 2). A complex of stereocilia (long microvilli), short microvilli and accompanying "supporting rods" are common adjuncts to the cnidocil proper. Such modifications appear to be related primarily to mechanical stiffening so as to make the apex of the cell more efficient in detecting environmental perturbations and transmitting them to the opercular region of the nematocyst capsule. The cores of the stereocilia and "supporting rods" of short microvilli ensheath the apical portion of the capsule in hydra but may extend around the entire capsule in some species. Various kinds of interconnections among these structures have been described. The precise function of these structures remains conjectural. It is logical to assume that they are supportive, yet the cores of the stereocilia of cnidocytes in hydra bind NBD-phallocidin, a specific marker for f-actin (Fig. 3) (Wood and Novak, 1982). However, actin filaments are known to serve a supportive function in many cell types (Pollard, 1976), and they are known to be the basis for a non-contractile elongation of the acrosomal filament in sperm cells of certain echinoderms (Tilney, et al., 1973). As yet there has been no definitive published evidence demonstrating a contractile role for the actin filament cores of stereocilia, but the possibility of a dual supportive and contractile role cannot be discounted.

The basal part of cnidocytes contains the nucleus and large numbers of filaments 10-12 nm in diameter. Reports by light microscopists and from early electron microscopy (Slautterback, 1967), that filaments associated with the stereocilia and short microvilli are directly continuous with the basal filaments have not been corroborated by more recent studies. The larger diameter of the basal filaments and the fact that they do not bind phallocidin suggest strongly that they correspond to a type of intermediate filament (Wood and Novak, 1982). This conclusion is also supported by the fact that in most hydrozoans and scyphozoans the basal filaments insert at the cell surface at fascial desmosomes which anchor the cnidocyte to processes of epithelial cells. The latter in turn are anchored to the underlying mesoglea by fascial hemi-desmosomes, and the intervening epithelial cytoplasm is reinforced with short microtubules (Figs. 4,5), (Slautterback, 1967; Westfall, 1973; Wood and Novak, 1982). In siphonophores and the Anthozoa, cnidocytes are anchored to the mesoglea by 10-12 nm filaments inserting directly into hemidesmosomes. In higher organisms desmosomes and hemidesmosomes are associated with intermediate filaments (tonofilaments), not actin-type microfilaments.

## III. FREEZE-FRACTURE OF THE CNIDOCYTE

One of the principal advantages of the freeze-fracture technique is that it reveals unique information about membrane organization. It also provides images that demonstrate three-dimensionality because the fracture planes tend to follow membrane contours and replication preserves those contours. Figure 6 illustrates the latter feature well. It is a low magnification view of a battery of nematocysts in a tentacle of hydra. This replica actually showed two additional nematocysts, so the total number visualized was approximately one-third of the average number contained in a tentacle battery. Of course, the fracture plane is achieved in large part by chance so different aspects of the battery cell and cnidocytes are revealed in adjacent areas. For example, in this field the fracture deflected around the upper surfaces of four cnidocytes (that is, the capsules remained in the tissue under the replica surface), and around the lower surface of another cnidocyte (the capsule was pulled away during fracture). The cnidocil was not revealed for any of these cnidocytes, but is shown in a different replica in Figure 7. The membrane fluting at the base of the cnidocil is similar to that seen at the base of endodermal flagella in hydra (Wood, 1979), further substantiating the flagellar origin of the cnidocil.

The opercular region of the nematocyst is a crucial area in capsular discharge, yet it does not display remarkable ultrastructural features by conventional methods. However, freeze-fracture replicas have provided new details of potential significance for understanding the mechanism of capsular discharge. The apical plasma membrane of the cnidocyte, and the immediately underlying cnida membrane surrounding the nematocyst capsule (apical cap) have eccentrically located clusters of 10-14 nm intramembrane particles (Figs. 8,9) or a well defined particle-depleted zone (Fig. 10). The specific configuration depends on which fracture face of which membrane is being viewed. In figure 8 the inner aspect (P face) of the cnidocyte apical plasma membrane is replicated, whereas figure 9 shows the outer aspect (E face) of the same membrane. Adjacent to the prominent particle enriched areas there is a region which displays a different population of particles (Fig. 8) or an overlying patch of vacuolar membrane, (VM) (Fig. 9). The P-face image of the VM has an eccentrically located region depleted of particles that is outlined by a single row of particles (Fig. 10). A smooth membrane profile of the E face of the VM appears through a break in the apical plasma membrane in Figure 8. Intramembrane particles revealed by freeze-fracture are known in many instances to represent

Fig. 4. Low magnification TEM of hydra tentacle showing mounted nematocytes and basally located mesogleal attachments (arrows). ME, mesoglea. Bar = 1.0 uM

Fig. 5. Higher magnification of nematocyte/mesogleal attachment. 12 nm filaments insert at a fascial desmosome (DE) with a narrow process of the battery cell which is traversed by short microtubules (MT). The battery cell process is attached to the mesoglea (ME) by a fascial hemidesmosome (HD). Bar = 0.5 uM

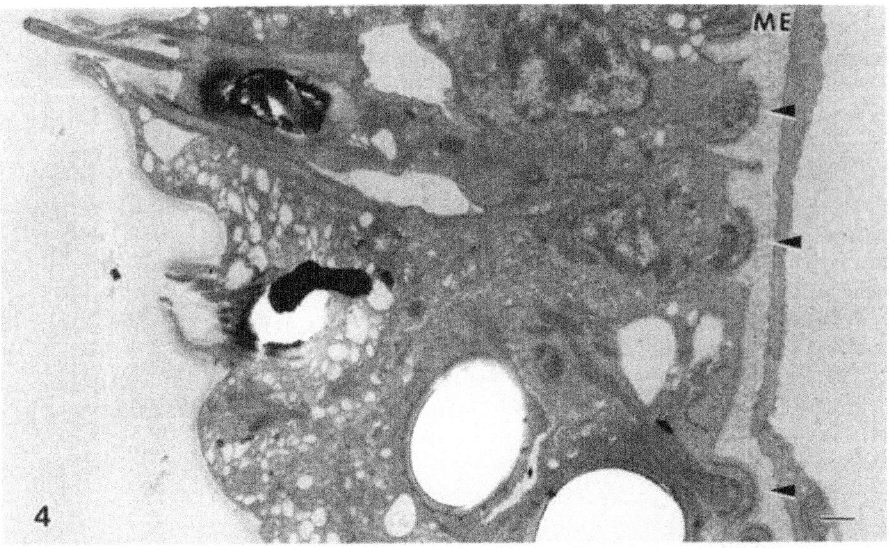
ME
4

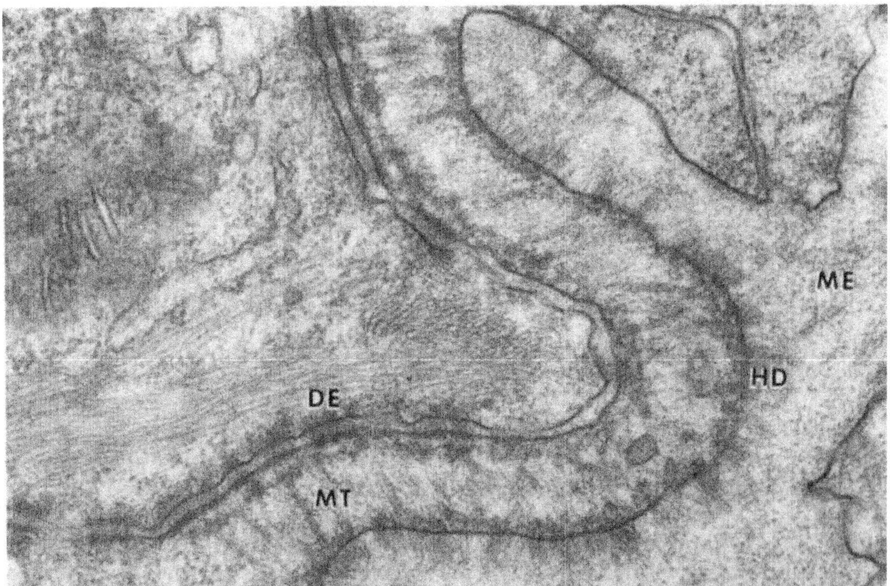
ME
HD
DE
MT
5

complexes of transport proteins (Alberts et al., 1983), and an opercular location of specialized configurations is suggestive of function in chemoreception, $Ca^{2+}$ transport or water permeability, all of which have been implicated in nematocyst discharge. It is of particular interest in this regard that specialized arrays of apically located intramembrane particles in toad bladder epithelium have been shown to be involved with antidiuretic hormone-sensitive water permeation (Wade, 1985). It would be of interest to know whether ADH affects nematocyst discharge or the packing of intramembrane particles in the apex of the cnidocyte.

Intramembrane particle aggregates are also associated with cellular attachments. A zone of possible attachment between the VM and the apical plasma membrane of the cnidocyte has been described for an anemone (Watson and Mariscal, 1985). Such attachment could occur in hydra as well, but it has not been detected in conventional preparations.

Two other features of the cnidocytes of hydra have been revealed by freeze-fracture replication methodology. First, mature stenoteles (and possibly desmonemes) have a radial disposition of minute grooves and ridges associated with the VM surrounding the apical half of the capsule (Figs. 8,9,10). This configuration probably corresponds to delicate fluting of the membrane seen in thin sections, but its regularity is particularly evident in freeze-fracture preparations. There are also linear patterns of intramembrane particles on the P fracture face, perhaps reflecting sites of binding of cytosolic constituents to the membrane. Microtubular arrays are known to be associated

Fig. 6. Freeze-fracture replica of a nematocyst battery in hydra. Bar = 1.0 uM

Fig. 7. Replica of a nematocyte apex showing the cnidocil (CN) and one stereocilium (ST). Bar = 1.0 uM

Fig. 9. Replica of cnidocyte apex showing complememtary view to that seen in figure 8. The E face of the cnidocyte membrane has depressions corresponding to the elevated region with prominent particles, and a patch of P face of the vacuolar membrane (PVM) adheres separately from the remaining vacuolar membrane (VM). The main fragment of vacuolar membrane has linear arrays of particles corresponding to the grooves seen figure 8. Bar = 0.5 uM

Fig. 10. Replica of cnidocyte showing the P face of the vacuolar membrane at the nematocyst operculum. A particle-depleted zone is delineated by a single row of particles (arrow). Bar = 1.0 uM

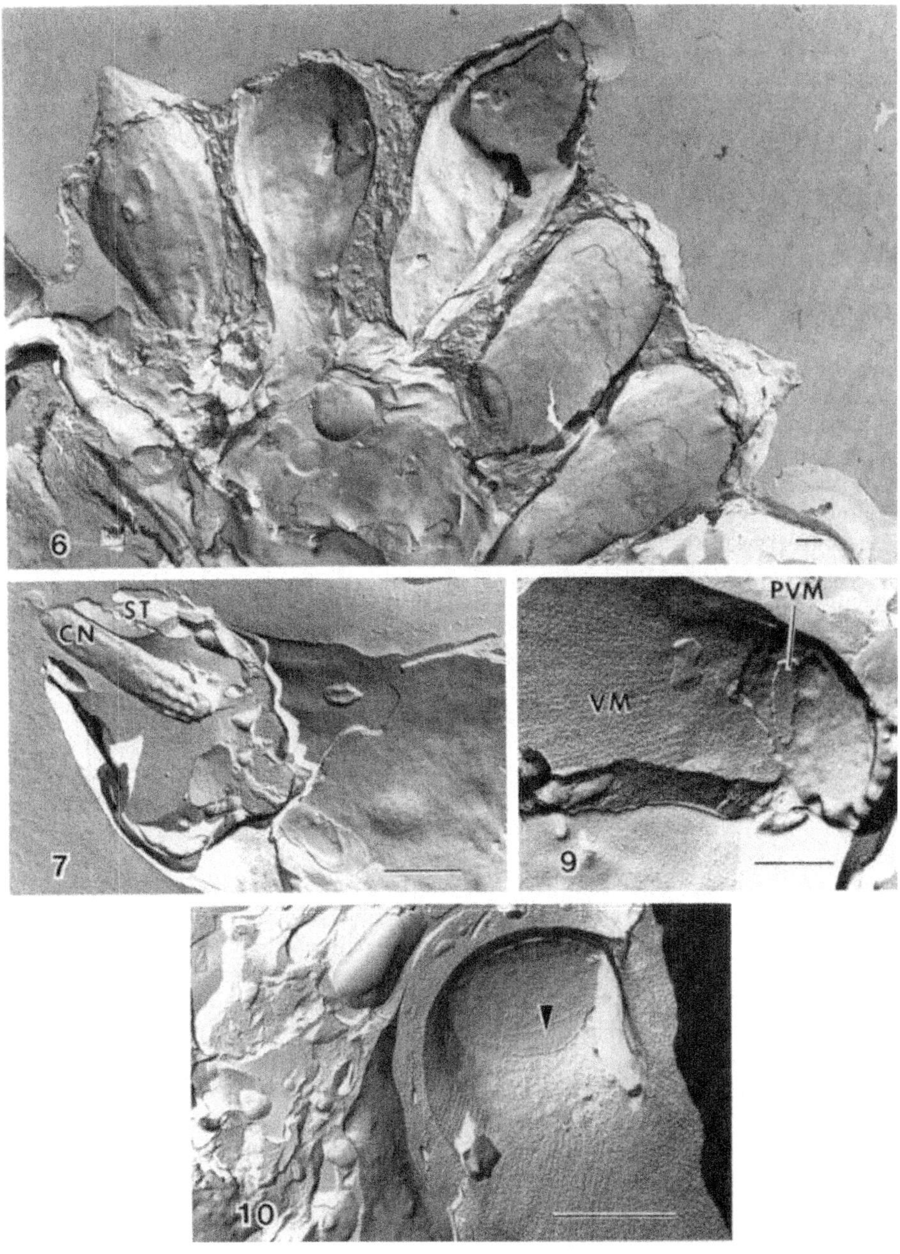
6
ST
CN
7
PVM
VM
9
10

with developing nematocysts, but generally they disappear or disperse with maturation (Westfall, 1966; Holstein, 1981). The freeze-fracture image could represent a remnant of previous microtubular attachment to the VM.

Second, the cnidocyte apparantly has no gap junctions with surrounding epithelial cells. Freeze-fracture is the most sensitive available method for detecting gap junctions because of the large membrane areas exposed to en face view (Fig. 6), and the distinctiveness of gap junctional profiles, yet none of these junctions have been seen on cnidocytes. Adjacent epithelial cells show numerous gap junctions. This observation confirms what has been observed in conventional thin sections, and indicates that nematocyst batteries probably are not coordinated to discharge by means of gap junctions. No intramembrane specializations indicative of synaptic contact have been detected on cnidocytes either, but neither have such specializations been seen on smooth muscle and glandular cells of higher organisms. The absence of gap junctions or of obvious synaptic specializations provide further evidence that nervous influence on nematocyst discharge in hydra is probably primarily for modulation and establishing thresholds of response rather than being stimulatory. It seems highly unlikely that rapid, simultaneous discharge of an entire battery of nematocysts could be mediated by a diffuse innervation resembling the autonomic system of vertebrates. Stimulation of the battery cell, and secondary activation of all associated cnidocytes would require a means of simultaneous signal transmission, for which there is at present no morphological evidence.

Fig. 8. Replica of the cnidocyte apex showing the opercular region. The P face of the cnidocyte membrane shows two kinds of particle aggregations (AG 1 and AG 2). An irregular patch of membrane has pulled away to reveal the E face of the closely apposed vacuolar membrane of the nematocyst (arrow). The E face of the vacuolar membrane has a chevron pattern of fine grooves below the level of the operculum (G). SJ, septate junction between cnidocyte and battery cell. Bar = 0.5 uM

G
G
SJ
AG1
AG2
8

## IV. NEMATOCYST MORPHOLOGY

A discussion of cnidocyte ultrastructure would hardly be complete without at least mentioning the nematocyst itself. The descriptive classification of nematocyst types by Weill (1934) has been confirmed, refined and expanded with the use of electron microscopy. Some exceptions to Weill's classification are now recognized: 1) if true atrichous isorhizas exist, they are rare; 2) the existence of amastigophores is questionable; and 3) the spirocyst, because of its peculiar morphology and chemistry, probably should be regarded as a separate class of cnida, not a nemotocyst.

Scanning electron microscopy has been particularly useful for elucidating the armature structure of discharged nematocysts (Mariscal, 1974; Mariscal and Bigger, 1976), but the internal morphology of mature undischarged capsules has been difficult to obtain. Much of the current knowledge arises from studies of immature cnidoblasts because the mature capsules do not infiltrate well with the epoxy monomers used in embedding for thin sectioning (Skaer, 1973). The toughness of the capsular wall also affects fracture planes in frozen tissue, so the freeze-fracture technique has contributed little additional information on the structure of nematocysts. One thing is abundantly clear, however. The capsule and its internal components do not have a lipid bilayer composition. Fracture planes may occasionally break across a nematocyst (Fig. 11), but when they do, there is never a deviation along the capsular wall or internal components as occurs with phospholipid bilayers. The ultrastructure of nematocysts gives no other direct clues as to specific chemical composition. Nevertheless, it is clear that there are differences in the chemistry of both capsules and content in the different categories of nematocysts. X-ray microanalysis and cytochemistry have the potential to demonstrate distribution patterns of specific elements and compounds, but the application of these methods

Fig. 11. Cross fracture of a stenotele capsule. The capsular wall (CW) fractures differently from the contents. The profiles of the shaft (S) and the tubule (T) are evident, but they do not fracture in the manner of phospholipid membranes. Bar = 1.0 uM

Fig. 12. SEM of a discharged stenotele. The cnidocil (CN) is still evident along the base of the everted shaft. Bar = 1.0 uM

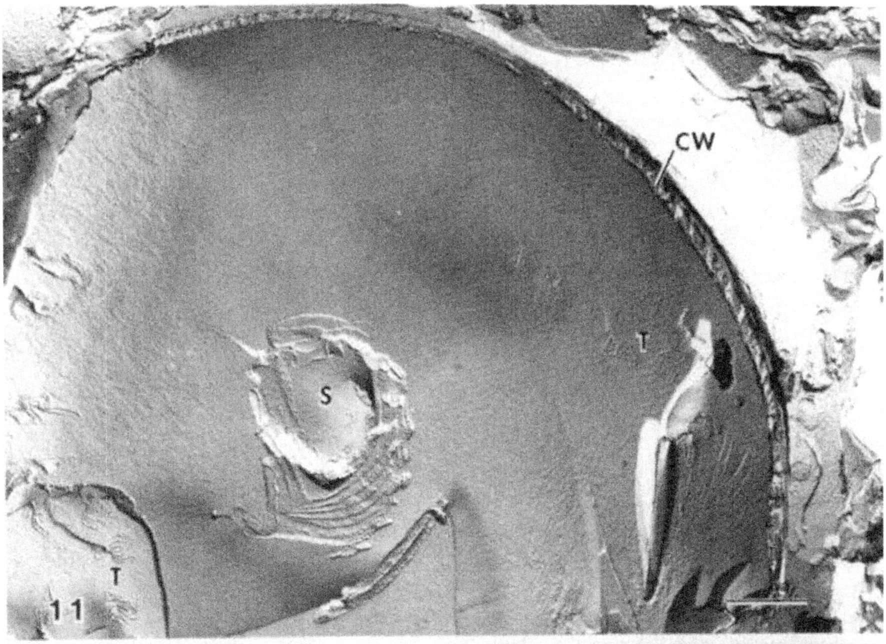
CW
T
S
T
11

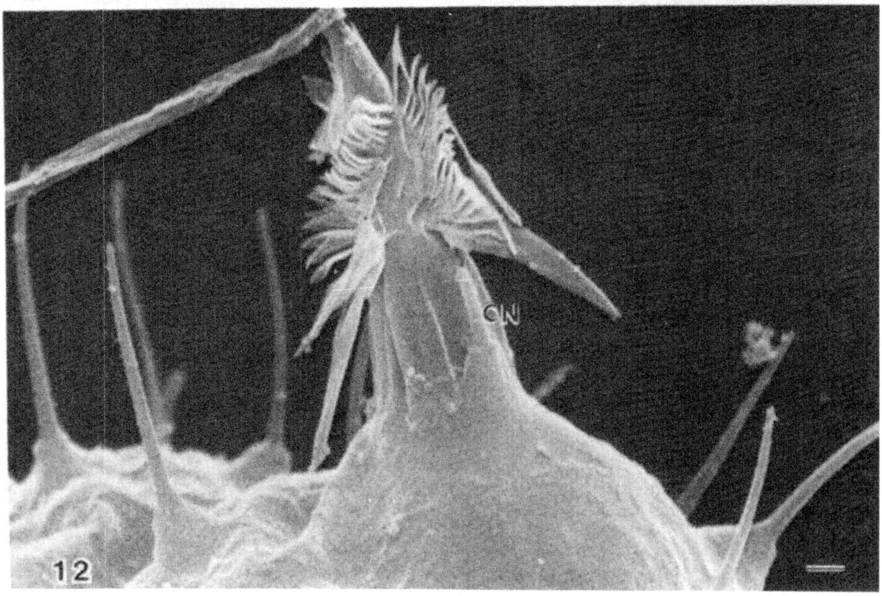
CN
12

to the study of nematocysts is still in its infancy, and has had only limited success to date (Mariscal, 1980; Gupta and Hall, 1984; Watson and Mariscal, 1985).

The morphodynamics of the discharge of stenotele nematocysts of hydra have been analyzed in detail (Tardent and Holstein, 1982; Holstein and Tardent, 1984). These elegant studies provided enlightenment on the process of nematocyst discharge, but were unable to resolve the mechanism of receptor transduction or determine the forces which cause the rapid eversion of the internal tubule. Under the conditions used in those experiments, it appeared that the stenotele capsules were catapulted from their cnidocytes at the onset or during the process of discharge. However, this is by no means a universal phenomenon in the discharge of stenoteles (Figs. 3,12), and it is unlikely that it could be an essential aspect of nematocyst discharge under normal physiological conditions. Certainly, premature ejection of the capsules of adhesive and entangling types of nematocysts would be counterproductive for their function of preventing prey escape before injected paralyzing toxins can take effect. Furthermore, organisms such as small annelids which do not have numerous bristles for desmonemes to entangle are readily captured by hydra. This implies that nematocysts of the penetrant type are involved in physical as well as paralytic immobilization during such capture.

## V. CONCLUSIONS

Observations by electron microscopy have been useful in elucidating the morphology of cnidae and cnidocytes. Nevertheless, major questions remain regarding the structural bases of sensory transduction and discharge of the capsule. Freeze-fracture replication has revealed a specialized intramembrane particle array at the cnidocyte apex. This configuration could represent receptor sites, ionic channels, water permeation channels, or an attachment site, any of which could be involved with nematocyst discharge. Clearly, a more extended freeze-fracture study would be warrented. Electron microscopy has failed to demonstrate gap junctions on the cnidocyte. Although small gap junctions could be missed in conventional thin sections, it is unlikely that they would be overlooked in freeze-fracture preparations which expose large areas of cnidocyte membrane in each replica. This indicates that coordination of nematocyst discharge probably is not through direct ionic communication with adjacent epithelial cells. Freeze-fracture has not revealed focal membrane specializations that can be correlated with chemical synapses either; thus, the effect of innervation on nematocyst discharge may be diffuse and may be associated more with modulation than with coordination of simultaneous activation.

## ACKNOWLEDGEMENTS

I thank Aileen Kuda-Cooke and Patricia Novak for contributions to the research from my laboratory reported in this paper.

## REFERENCES

Alberts, b., D. Bray, J. Lewis, M. Raff, K. Roberts, and J.D. Watson. 1983. Molecular Biology of the Cell, Garland Publishing Inc.: New York, p. 289.

Gupta, B.L. and T.A. Hall. 1984. Role of high concentrations of Ca, Cu, and Zn in the maturation and discharge in situ of sea anemone nematocysts as shown by X-ray microanalysis of cryosections. In L. Bolis, et al., (eds.), Toxins, Drugs and Pollutants in Marine Animals. Springer-Verlag: Berlin, p. 77.

Holstein, T. 1981. The morphogenesis of nematocytes in Hydra and Forskalia: An ultrastructural study. J. Ultrastr. Res. 75:276-290.

Holstein, T. and P. Tardent. 1984. An ultrahigh-speed analysis of exocytosis: Nematocyst discharge. Science 223:830-832.

Mariscal, R.N. 1974. Nematocysts. In L. Muscatine and H.M. Lenhoff (eds.), Coelenterate Biology. Academic Press: New York, pp. 129-178.

Mariscal, R.N. 1980. The elemental composition of nematocysts as determined by X-ray microanalysis. In P. Tardent, and R. Tardent (eds.), Developmental and Cellular Biology of Coelenterates. Elsevier/North Holland, Amsterdam, p. 337.

Mariscal, R.N. and C.H. Bigger. 1976. A comparison of putative sensory receptors associated with nematocysts in an anthozoan and a scyphozoan. In G.O. Mackie (ed.), Coelenterate Ecology and Behavior. Plenum Publishing Corp.: New York, pp. 559-568.

Pollard, T.D. 1976. Cytoskeletal functions of cytoplasmic contractile proteins. J. Supramol. Struct. 5:317-334.
Skaer, R.J. 1973. The secretion and development of nematocysts in a siphonophore. J. Cell Sci. 13:371-393.
Slautterback, D.B. 1967. The cnidoblast-musculoepithelial cell complex in the tentacles of *Hydra*. Zeit. Zellforsch. 79:296-318.
Tardent, P. and T. Holstein. 1982. Morphology and morphogenesis of the stenotele nematocyst of *Hydra attenuata*, Pall. (Hydrozoa, Cnidaria). Cell Tiss. Res. 224:269-290.
Tilney, L.G., S. Hatano, H. Ishikawa, and M. Mooseker. 1973. The polymerization of actin: its role in the generation of the acrosomal process of certain echinoderm sperm. J. Cell Biol. 59:109-126.
Wade, J.B. 1985. Membrane structural studies of the action of vasopressin. Fed. Proc. 44:2687-2692.
Watson, G.M. and R.N. Mariscal. 1985. Ultrastructure of nematocyst discharge in catch tentacles of the sea anemone *Haliplanella luciae* (Cnidaria: Anthozoa). Tissue and Cell 17:199-213.
Weill, R. 1934. Contribution a l'etude des cnidaires et de leurs nematocystes. Trav. Sta. Zool Wimereux 10 & 11.
Westfall, J.A. 1966. The differentiation of nematocysts and associated structures in the Cnidaria. Zeit. Zellforsch. 75:381-403.
Westfall, J.A. 1970. The nematocyte complex in a hydromedusan, *Gonionemus vertens*. Zeit. Zellforsch. 110:457-470.
Wood, R.L. 1979. The fine structure of the hypostome and mouth of *Hydra*. II. Transmission electron microscopy. Cell Tiss. Res. 199:319-338.
Wood, R.L. and P.L. Novak. 1982. The anchoring of nematocysts and nematocytes in the tentacles of *Hydra*. J. Ultrastr. Res. 81:104-116.

# 3

# PRESUMED NEURONEMATOCYTE SYNAPSES AND POSSIBLE PATHWAYS CONTROLLING DISCHARGE OF A BATTERY OF NEMATOCYSTS IN HYDRA

Jane A. Westfall

Department of Anatomy and Physiology
Kansas State University
Manhattan, Kansas, 66506, U.S.A.

## SUMMARY

Nematocytes are nematocyst-containing receptor cells with an apical cnidocil ("trigger hair") surrounded basally by stereocilia in the Hydrozoa. In the battery cell complex of hydra, each nematocyte is connected basally by a series of prominent desmosomes to microtubular cushions associated with basal myonemes of the enveloping epitheliomuscular cell (battery cell). Sensory cells and ganglion cells, also enveloped by the battery cell, form apparent synaptic contacts on one or more nematocytes in one to several batteries on the tentacle of hydra. Nematocyst discharge may involve an integration of responses by: (1) the cnidocil as a trigger hair; (2) the sensory cell as a modulator of excitatory or inhibitory input either directly or via the ganglion cell; (3) the ganglion cell as an integrator of battery response; and, (4) the myonemes as contractile elements to bend the tentacle towards the prey. Thus, both anatomically wired neural pathways and independently triggered nematocytes play a role in the synchronized firing of different types of nematocysts to capture food organisms.

---

Contribution No. 86-129-A from the Kansas Agricultural Experiment Station, supported in part by United States PHS Grant No. NS-10264 from the National Institute of Neurological and Communicative Disorders and Stroke.

## INTRODUCTION

Since G. H. Parker (1916, 1919) defined the nematocyte as an independent effector, i.e., a cell not under neural control, there have been two schools of thought on the control of nematocyst discharge. One school believes that the nematocyte responds directly to specific chemical and mechanical stimuli in the aqueous environment without internal control by the animal (see review by Mariscal, 1974). The other school believes that the animal itself can control nematocyte response. Evidence for this comes from both behavioral physiology and morphology. For example, there is physiological evidence that nematocyst discharge is reduced or inhibited in fully fed animals (Burnett *et al.*, 1960; Bouchet, 1961; Sandberg *et al.*, 1971, Sandberg, 1972; Mariscal, 1973; Smith *et al.*, 1974; Ruch and Cook, 1984). Also, there is ultrastructural evidence for presumed neuronematocyte synapses in coelenterates, which suggests a neural pathway of control (Westfall, 1969, 1970*a*, *b*; Westfall *et al.*, 1971*b*; Westfall, 1973; Westfall and Kinnamon, 1978, 1984; Satterlie and Case, 1978).

In this paper, I use data from these different lines of research to develop a hypothesis concerning possible pathways controlling discharge of nematocysts from both single cells and batteries of cells in tentacles of hydra. For example, nematocytes, sensory cells, ganglion cells, and epitheliomuscular cells may play a complex coordinated role in mechano- and chemoreception, neurotransmission, neuromodulation, and behavioral response to capture of prey. I propose that the nematocyst batteries in the tentacles of hydra are highly specialized units of structure and function in which individual nematocytes may have independent effector capabilities and also be part of a complex coordinated system under neural control of the animal.

## THE BATTERY CELL COMPLEX

Nematocysts of hydra are contained mainly in the tentacles and function in the capture of food. Four types of nematocyst are present in hydra: stenotele (penetrant), desmoneme (volvent), and holotrichous and atrichous isorhizas (glutinants) (Weill, 1934). The stenoteles and desmonemes respond to discharge by mechanical and chemical stimulation; holotrichous isorhizas appear to function in defense and atrichous isorhizas in attachment of tentacles during locomotion (Ewer, 1947). The orderly arrangement of nematocysts in batteries lying within a bulge of an epitheliomuscular cell in tentacles of hydra was well described by von Gelei (1927). At the top of the bulge, there are one or occasionally two stenoteles, which function to penetrate the prey and inject a paralyzing poison into it. Near the penetrant are one or two isorhizas, and around the slope of the bulge are numerous desmonemes. The desmonemes have the longest cnidocils and, thus, may contact the prey first (Kanaev, 1952). They entrap it with their lassolike threads, so that the stenotele can strike the prey without missing. The arrangement of the nematocytes in a battery housed by the large epitheliomuscular cell (EMC) or battery cell has been confirmed by electron microscopy (Slautterback, 1967; Westfall et al., 1971a; Hufnagel et al., 1985).

With the advent of electron microscopy, the cnidocil and supporting structures could be studied in detail. It was found that cells containing mature nematocysts had: (1) a cnidocil composed of a dense core; (2) a semi-circle of nine supporting structures at the base of the cnidocil; and (3) another 20 supporting structures with interconnections around the nematocyst capsule (Chapman and Tilney, 1959). The nonmotile cnidocil arises from a centriole-like body, which suggests that it is a modified flagellum (Fawcett, 1961). Although Hadži (1909) and von Gelei (1927) portrayed nematocyte processes extending towards the muscular base of the enveloping EMC, it was only with the electron microscope that the desmosomal contacts between the fibrillar region of the nematocyte and specialized microtubular regions of the muscular processes could be seen (Slautterback, 1967; Wood and Novak, 1982; Hufnagel et al., 1985). Slautterback (1967) proposed that the EMC serves an integrative function for the activation of its associated cnidocytes and perhaps for those of other batteries. He observed nine peripheral doublets of microtubules in the cnidocil and microfilaments in the nine stereocilia of the cnidocyte. He also noted fibrous connections between stereocilia and the supporting rods, which he called rootlets, around the capsule. Although he believed that the bundles of basally directed filaments extending from the 20 supporting rods of the cnidocyte originated in the stereocilia, I have observed that they are extensions of the supporting rods of the capsule, which are not true rootlets (Fig. 1). The rods are interconnected by a fibrillar collar, which may serve to hold the capsule in position near the cnidocil. The filaments end in desmosomes at a nematocyte-battery epithelial cell-mesolamellar complex, or NBM complex, described in detail by Wood and Novak (1982). The

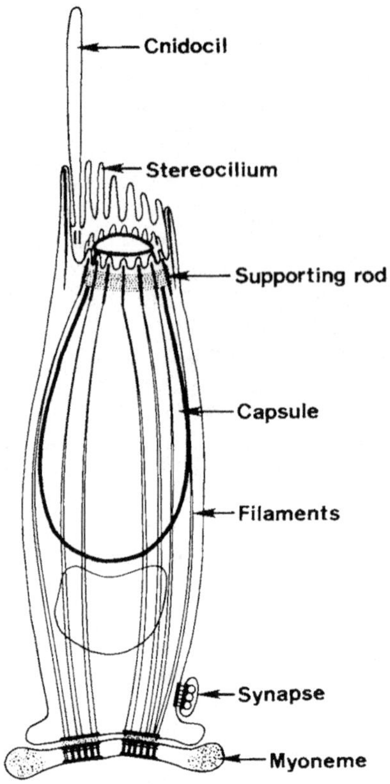


Fig. 1. A diagrammatic representation of a nematocyte with fine filaments extending from supporting rods around the upper portion of the capsule and ending in desmosomal attachments to microtubular extensions of myonemes at the base of an enveloping epitheliomuscular cell. An apical cnidocil acts as a trigger hair, and a basally located synapse provides neural input to nematocyst discharge.

functional organization of the battery cell complexes in hydra was further clarified by Hufnagel *et al.* (1985) from serial sections; these demonstrated that some nematocyte processes insert between myonemes and contact the mesoglea, whereas other processes insert into the microtubular NBM complexes associated with the myonemes (Fig. 2). I observed several patterns of nematocyte attachment to the NBM complexes. In some instances, two different nematocytes appeared to attach to the NBM complexes of one specialized

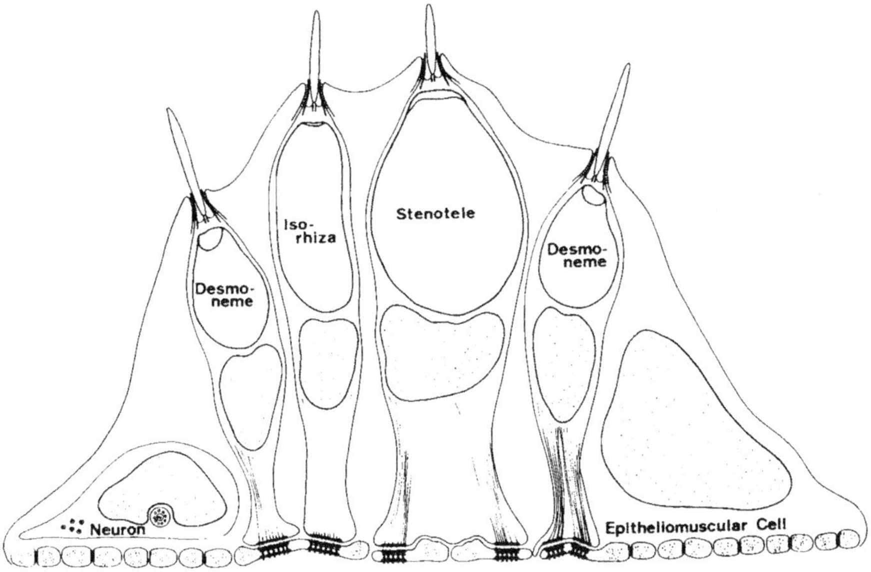


Fig. 2. Schematic drawing of a cross section through an epidermal epitheliomuscular cell of hydra illustrating variations in attachment of the nematocytes to specialized microtubular cushions (NBM complexes) associated with the basal myonemes. Two types of nematocysts (desmoneme and isorhiza) may be associated with the same specialized region; alternatively, one type (stenotele) may have filamentous attachments to specialized regions of two different myonemes, or the same nematocyst (desmoneme) may contact two or more specialized regions of the same myoneme.

region; in other cases, the filaments in a single nematocyte branched into two bundles with attachments to either separated or continuous NBM complexes. The filaments do not bind NBD-phallacidin, which suggests that they are noncontractile and serve as anchoring devices for the nematocytes (Wood and Novak, 1982). Perhaps they maintain the relative positions of the nematocytes as the battery cell becomes compressed during contraction of the longitudinal muscle of the tentacle. After nematocyst discharge, the fibrillar collar holding the unfired capsule in place may loosen to allow release of the fired capsule.

A neural unit for a typical battery cell complex may consist of a slender sensory cell oriented perpendicularly

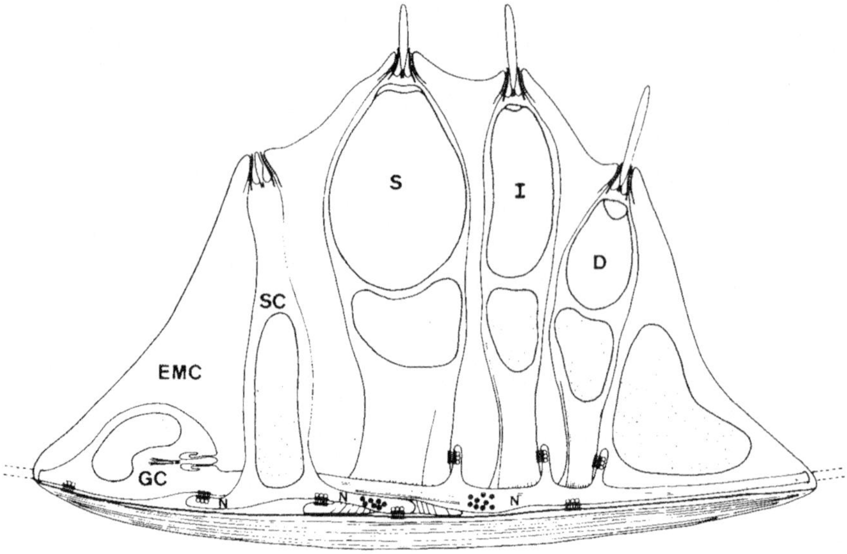


Fig. 3. Diagram of a longitudinal section of a battery cell complex composed of (1) three types of nematocysts: stenotele (S), isorhiza (I), and desmoneme (D); (2) two types of neurons: sensory cell (SC) and ganglion cell (GC) with neurites (N); and, (3) basal myonemes (horizontal lines) of the enveloping epitheliomuscular cell (EMC). Note that the sensory cell forms synapses on the ganglion cell, myoneme and nematocyte; the ganglion cell in turn synapses with the myoneme and three types of nematocyst-containing cells.

to the basal myonemes and a basal ganglion cell with major neurites lying in parallel with the myonemes (Fig. 3). The sensory cell contains an apical cilium-stereociliary complex, which is enveloped by margins of the surrounding epitheliomuscular cell (Westfall and Kinnamon, 1978; Westfall and Sims, 1978). The sensory cell may narrow below the elongate nucleus and form a single axon, which bifurcates basally, or it may appear as an elongate triangle with two or more processes extending from the base of the nucleus. Dense-cored vesicles are present in the axonal processes, although the neuronematocyte synaptic foci usually have clear vesicles. Presumed sensory cell synapses have 2-4 clear or dense-cored vesicles at paired electron dense membranes, separated by a 20 nm wide cleft containing transverse filaments. The ganglion cell is characterized by its position near the myonemes and an internal cilium surrounded by a ring of inverted stereocilia. Ganglion cells form apparent synapses with one or more nematocytes and myonemes (Westfall, 1973). The neuronematocyte synapses may have several mixed, clear and dense-cored vesicles; also, the membrane thickenings are more prominent than those of the sensory cell.

In 1883, Jickeli observed that a ganglion cell could innervate one to several nematocysts in both hydra and the hydroid Eudendrium. Spangenberg and Ham (1960), using methylene blue staining, demonstrated nerve cell contacts on nematocytes containing a desmoneme and a stenotele in Hydra littoralis. More recently, we have demonstrated by immunochemistry using a monoclonal antibody, JD1, that long, ganglion cell neurites contact stenoteles and desmonemes in up to five different batteries of nematocysts (Yu et al., 1985). This suggests that ganglion cells probably are the major control cells for battery discharge.

## CONTROL OF NEMATOCYST DISCHARGE

Ewer (1947) believed that in hydra the threshold for discharge in response to mechanical stimulation of the cnidocils of stenoteles and desmonemes was lowered by food extracts. Lentz (1966) demonstrated that either a combination of mechanical and chemical stimuli or mechanical stimuli alone, in the presence of such neurotransmitter substances as acetylcholine or certain biogenic amines, could trigger excessive nematocyst discharge in hydra. Therefore, he concluded that a combination of mechanical and chemical stimulation is essential for nematocyst discharge but postulated that the threshold to discharge may be lowered by neural influences. Although he proposed that neural influences may act through ganglion or sensory cells, he found no morphological evidence of specialized synaptic contacts in hydra.

Westfall and coworkers were the first to provide ultrastructural evidence of neuronematocyte synapses in coelenterates. In the trachymedusan Gonionemus, Westfall (1970a) observed serial neuro-neuro-nematocyte synaptic foci at the base of the nematocyte near its desmosomal junction with the modified microtubular region of the epitheliomuscular cell. The suspected synapses were characterized by short, electron-dense foci composed of a pair of uniformly apposed thickened membranes and one or more rows of mixed, clear and dense-cored vesicles associated with the presynaptic membrane. The serial nature of the axo-axo-effector cell contacts suggested that presynaptic inhibition could occur. The nerve cells involved, however, were not identified. Later, working with hydra, I found that both ganglion cells and sensory cells form apparent synaptic foci on nematocytes (Westfall *et al.*, 1971*b*; Westfall, 1973; Westfall and Kinnamon 1978, 1984). In addition, both types of neurons have similar synaptic foci on each other and on the epitheliomuscular cell in the region of the basal myonemes (Fig. 3). Because both types of neurons have a cilium and neurosecretory granules as well as apparent interneuronal and effector cell synapses, I proposed that these phylogenetically primitive neurons function as neurosecretory-sensory-motor-interneurons (Westfall 1973; Westfall and Kinnamon, 1978). Ganglion cells have been traced in serial thin sections to simultaneous synaptic-like contacts on three or more nematocyst-containing cells and on the muscular bases of the EMC (Westfall, 1973). Using immunocytochemical techniques, we have traced processes from ganglion cells to contacts with nematocyst batteries in several adjacent EMC (Yu *et al.*, 1985). Thus, we can now generate a hypothetical circuit diagram for neural control

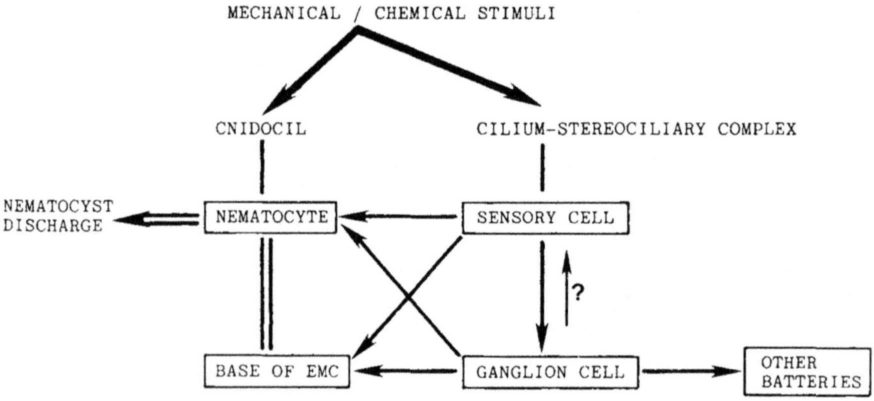


Fig. 4. Hypothetical circuit diagram for neural control of nematocyst discharge in hydra. The functional morphology of battery cell units in tentacles suggests a complex, coordinated system for nematocyst discharge involving chemo-mechano-receptivity (thick arrows), synaptic pathways (thin arrows), and muscular contraction near the desmosomal contacts (double line) of the nematocyte.

of nematocyst discharge in hydra (Fig. 4). Probably, chemical stimuli lead to a lowering of the threshold of the nematocyte to mechanical stimuli by the prey either directly or via the sensory cell and the ganglion cell as an intermediate modulator. Because ganglion cell "synapses" are more numerous and ultrastructurally more prominent than sensory cell "synapses", I believe the ganglion cell plays a major role in synchrony of nematocyst discharge and coordination of muscular contraction. Perhaps the sensory cell influences the local response of the myonemes to ganglion cell excitation, so that the tentacle is bent when contacting the prey. The prey may mechanically trigger the cnidocils of the desmonemes, which fire to entangle the food organism, and the cnidocil of the stenotele, which fires into the prey and injects a paralyzing poison into it. Subsequently, the prey is carried to the mouth by coordinated activity of the tentacles.

ACKNOWLEDGMENT

I thank Mallory R. Hoover for the artwork.

REFERENCES

Burnett, A. L., T. L. Lentz, and M. Warren. 1960. The nematocyst of hydra. I. The question of control of the nematocyst discharge reaction by fully fed hydra. Ann. Soc. Roy. Zool. Belgique 90:247-268.

Bouchet, C. 1961. Le controle de la décharge nématocystique chez l'Hydre. C.R. Acad. Sci. 252:327-328.

Chapman, G. B. and L. G. Tilney. 1959. Cytological studies of the nematocysts of hydra. I. Desmonemes, isorhizas, cnidocils and supporting structures. J. Biophys. Biochem. Cytol. 5:69-78.

Ewer, R. F. 1947. On the functions and mode of action of the nematocysts of hydra. Proc. Zool. Soc. London 117:365-376.

Fawcett, D. 1961. Cilia and flagella. _In_ J. Brachet and A. Mirsky (eds). _The Cell_, Vol. II, pp. 217-297. Academic Press, New York.

Gelei, J. von 1927. Das Rätsel der Nesselzellen. Biol. Zbl. 47:327-343.

Hadži, J. 1909. Über das Nervensystem von Hydra. Arbeiten Zool. Instit. der Universitat, Wien 17:225-268.

Hufnagel, L. A., G. Kass-Simon, and M. K. Lyon. 1985. Functional organization of battery cell complexes in tentacles of Hydra attenuata. J. Morph. 184:323-341.

Jickeli, C.F. 1883. Der Bau der Hydroidpolypen. Morph. Jahrbuch. 8:373-416.

Kanaev, I.I. 1952. Hydra: Essays on the biology of fresh water polyps. English translation edited by H.M. Lenhoff, 1969, Irvine, California.

Lentz, T. L. 1966. The Cell Biology of Hydra. Wiley, New York.

Mariscal, R. N. 1973. The control of nematocyst discharge during feeding by sea anemones. Publ. Seto. Mar. Biol. Lab., Proc. Second Internat. Symp. Cnidaria 20:695-702.

Mariscal, R. N. 1974. Nematocysts. In L. Muscatine and H.M. Lenhoff (eds). Coelenterate Biology. Reviews and New Perspectives, pp. 129-178. Academic Press, New York.

Parker, G. H. 1916. The effector systems of actinians. J. Exp. Zool. 21:461-484.

Parker, G. H. 1919. The Elementary Nervous System. Lippincott, Philadelphia.

Ruch, R. J. and C. B. Cook. 1984. Nematocyst inactivation during feeding in Hydra littoralis. J. exp. Biol. 111:31-42.

Sandberg, D. M., 1972. The influence of feeding on behavior and nematocyst discharge of the sea anemone Calliactis tricolor. Mar. Behav. Physiol. 1:219-238.

Sandberg, D. M., P. Kanciruk, and R. N. Mariscal. 1971. Inhibition of nematocyst discharge correlated with feeding in a sea anemone, Calliactis tricolor (Leseur). Nature 232:263-264.

Satterlie, R. A. and J. F. Case. 1978. Neurobiology of the gorgonian coelenterates, Muricea californica and Lophogorgia chilensis. II. Morphology. Cell Tissue Res. 187:379-396.

Slautterback, D. B. 1967. The cnidoblast-musculoepithelial cell complex in the tentacles of Hydra. Z. Zellforsch. Mikrosk. Anat. 79:296-318.

Smith, S., J. Oshida, and H. Bode. 1974. Inhibition of nematocyst discharge in hydra fed to repletion. Biol. Bull. 147:186-202.

Spangenberg, D.B. and R.G. Ham. 1960. The epidermal nerve net of hydra. J. Exp. Zool. 143:195-201.

Weill, R. 1934. Contribution à l'étude des cnidaires et de leurs nematocystes. Trav. Stat. Zool. Wimereux 10,11, 1-701.

Westfall, J. A. 1969. Nervous control of nematocyst discharge: chemical synapses. Am. Zool. 9:1141.

Westfall, J. A. 1970a. Ultrastructure of synapses in a primitive coelenterate. J. Ultrastruct. Res. 32:237-246.
Westfall, J. A. 1970b. The nematocyte complex in a hydromedusan, Gonionemus vertens. Z. Zellforsch. Mikrosk. Anat. 110:457-470.
Westfall, J. A. 1973. Ultrastructural evidence for a granule-containing sensory-motor-interneuron in Hydra littoralis. J. Ultrastruct. Res. 42:268-282.
Westfall, J. A. and J. C. Kinnamon. 1978. A second sensory-motor-interneuron with neurosecretory granules in Hydra. J. Neurocytol. 7:365-379.
Westfall, J. A. and J. C. Kinnamon. 1984. Perioral synaptic connections and their possible role in the feeding behavior of Hydra. Tissue Cell 16:355-365.
Westfall, J. A. and D. E. Sims. 1978. Stereo SEM of sensory cell papillae on tentacles of Hydra. Scanning Electron Microscopy/1978 II, 671-676.
Westfall, J. A., S. Yamataka, and P. D. Enos. 1971a. Scanning and transmission microscopy of nematocyst batteries in epitheliomuscular cells of Hydra. 29th Ann. Proc. Electron Microscopy Soc. Amer., pp. 410-411.
Westfall, J.A., S. Yamataka, and P.D. Enos. 1971b. Ultrastructural evidence of polarized synapses in the nerve net of Hydra. J. Cell Biol. 51:318-323.
Wood, R.L. and P.L. Novak. 1982. The anchoring of nematocysts and nematocytes in the tentacles of Hydra. J. Ultrastruct. Res. 81:104-116.
Yu, S-M., J. A. Westfall, and J. F. Dunne. 1985. Light and electron microscopic localization of a monoclonal antibody in neurons in situ in the head region of Hydra. J. Morph. 184:183-193.

# 4

# THE CNIDOCIL APPARATUS OF HYDROZOANS: A PROGENITOR OF HIGHER METAZOAN MECHANORECEPTORS?

Thomas Holstein

Zoologisches Institut, Universität München,
D-8000 Munich 2, West Germany

Klaus Hausmann

Institut für Allgemeine Zoologie, Freie Universität Berlin,
D-1000 Berlin, West Germany

SUMMARY

The sensory receptor of the nematocyte in hydrozoans is the cnidocil appparatus. Reinvestigations of its fine structure in the hydrozoan *Craspedacusta sowerbii* shows that the various elements of the cnidocil apparatus are intimately connected and form a highly polarized, bilaterally symmetrical structure. We discuss the physiological significance of the various elements of the cnidocil apparatus and demonstrate that it displays striking similarities with the tubular body of mechanoreceptors in arthropods and with the sensory receptor of hair cells in the lateral line and the inner ear of vertebrates.

## I. INTRODUCTION

Cnidarian nematocytes are secretory and sensory cells. During their differentiation they elaborate a secretory product which discharges in a spectacular exocytosis within a few milliseconds and is used for killing prey (Holstein and Tardent, 1984; Tardent and Holstein, 1982). To trigger its release the sensory receptor on the surface of the cell must be

The Biology of Nematocysts

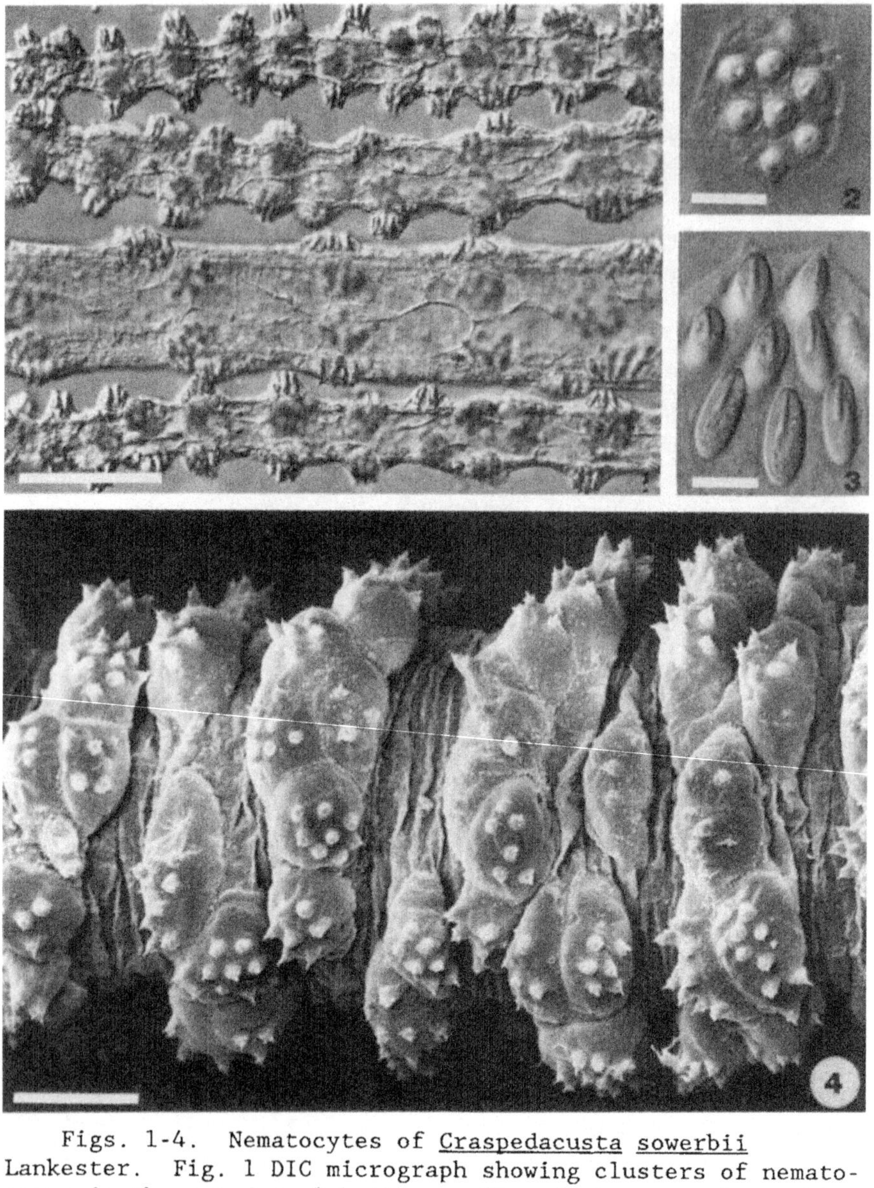

Figs. 1-4. Nematocytes of Craspedacusta sowerbii Lankester. Fig. 1 DIC micrograph showing clusters of nematocytes in the ectodermal battery cells of the tentacles. Fig. 2 DIC micrograph showing an optical cross section of a battery cell with eight nematocytes. Fig. 3 DIC micrograph of an optical longitudinal section of a battery cell with nematocytes. Fig. 4 Scanning electron micrograph of a tentacle. The cnidocil apparatus of the nematocyte protrudes from the free surface of the battery cells. Bar = 10 μm (Figs. 2,3), 20 μm (Fig. 4), and 50 μm (Fig. 1).

stimulated. The receptor of the nematocyte is a modified ciliary structure and goes by th name of the ciliary cone in anthozoans, the flagellum stereociliary complex in scyphozoans, and the cnidocil apparatus in hydrozoans (Mariscal, 1974).

In the present review we give an overview of the design and function of the various elements of the cnidocil apparatus. We conclude that the morphological specializations are due to nematocyst anchorage and to a selective reception of mechanical stimuli. In addition, there is an increasing complexity of the cnidocil apparatus from the anthozoans to the hydrozoans.

The cnidocil apparatus shows striking similarities to the mechanoreceptor of hair cells in the accousticolateralis system of vertrebrates and also to the mechanoreceptors in arthropods. Even though ciliary structures are ubiquitous throughout the animal kingdom, it is not clear how the mechanoreceptors in higher metazoans evolved this function.

## II. SPATIAL ORGANISATION OF THE NEMATOCYTE

Although electrophysiological data on nematocytes are still lacking, it is obvious that the organization of nematocytes fits well with the basic organization of an epithelial sensory cell: The distal sensory region consists of a cilium surrounded by a circle of microvilli; the inner proximal region contains numerous mitochondria and the nematocyst. The nematocyte is linked with the epithelial battery cell by a band of septate junctions (Hausmann and Holstein, 1985; Novak and Wood, 1983; Wood and Novak, 1982; Hand and Gobel, 1972; Westfall 1970a). These commence at the level of the nematocyst's operculum at the region were the stereocilia insert and they extend basally 1-3 µm (Fig.5) (Wood and Novak, 1982; Hausmann and Holstein, 1985). The septate junctions resemble septate junctions between epithelial cells (Figs. 7, 11), but they are more regular and no interdigitations can be found (Figs. 6, 10). The septae exhibit a 11-nm periodicity (Figs. 6, 8-10) (Wood and Kuda, 1980).

## III. CNIDOCIL APPARATUS OF HYDROZOANS

The Hydrozoan cnidocil apparatus consists of (i) a long cnidocil (a modified cilium), (ii) an outer row of long microvilli, known as stereocilia, (iii) an inner ring of short microvilli and (iv) a complex system of rods and fibrils, the fibrillar collar. The cnidocil and the stereocilia stick out of a collar perforation of the battery cell (accessory cell). Depending on nematocyst type and species the cnidocil extends 5-33 μm from the cellular surface. The stereocilia form a bilaterally symmetrical pallisade, which encircles the cnidocil (Beams et al., 1973; Cormier and Hessinger, 1980a, 1980b; Hausmann and Holstein, 1985).

The entire cnidocil apparatus has a characteristic bilateral symmetry which is generated by the two rows of microvilli arranged around an axis from the cnidocil to the nematocyst. The inner series of microvilli is always arranged around both cnidocil and in general their number is smaller than the number of outer microvilli (Table 1).

## A. The Cnidocil

All cnidocils which have been investigated lack dynein-arms as well as the classical 9+2 pattern of motile cilia (Table I). In the basal region 9 doublets are arranged around a variable number of microtubules. In the distal region the number of singlet microtubules is increased (up to 300 in Physalia, Cormier and Hessinger, 1980a), as are the peripheral doublet microtubules in several species. In Craspedacusta (Figs. 5, 14, 21-24) and Gonionemus single fibers (15 nm) run parallel to the microtubules (Hausmann and Holstein, 1985; Westfall, 1970a). In hydra a prominent dense core is situated in the center of the cnidocil running up to the top of the cnidocil (Slautterback, 1967); in Physalia electron dense material fills the entire cnidocil (Cormier and Hessinger, 1980a; Carre and Carre, 1980). The nature of this amorphous core material is unknown.

At the basal plate, the microtubules show the usual pattern of nine peripheral doublets. The microtubules below the basal plate are arranged as nine peripheral triplets forming a typical basal body. From the innermost microtubules of each

Fig. 5 Longitudinal section of the cnidocil apparatus (TEM). The cnidocil (CN), one of the inner microvilli (IM), and one stereocilium (SC) are sectioned. Note the bundle of longotudinally arranged microfilaments anchoring the stereocilium in the cytoplasm of the nematocyte (NC). The nematocyte is linked to the battery cell (BC) by septate junctions. Elements of the fibrillar collare are cross sectioned (*). Bar = 2.5 μm.

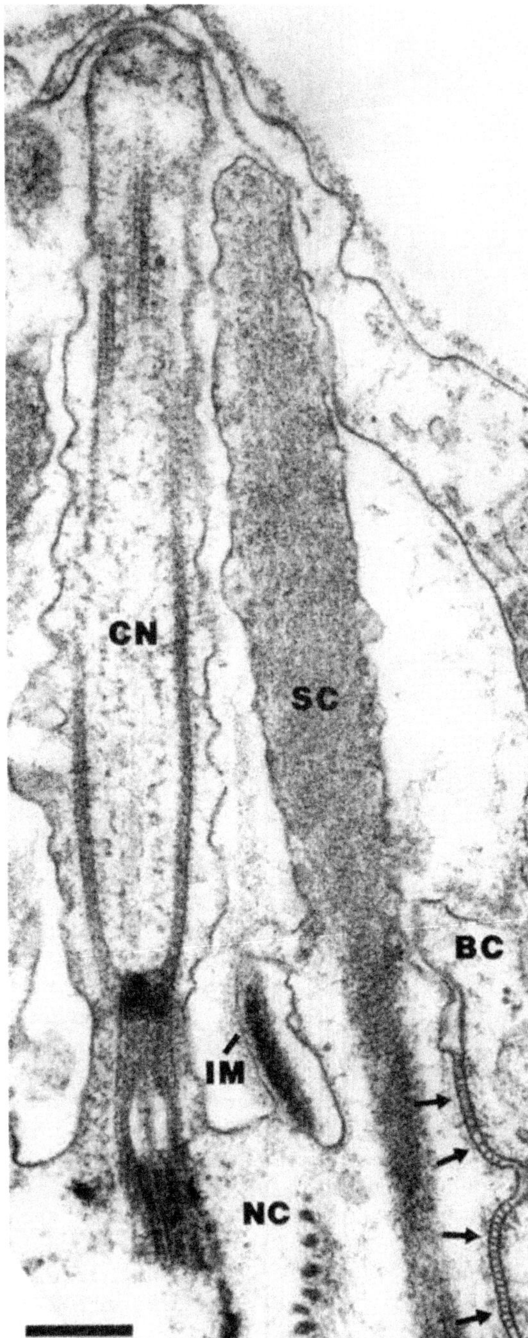
CN
SC
BC
IM
NC
5

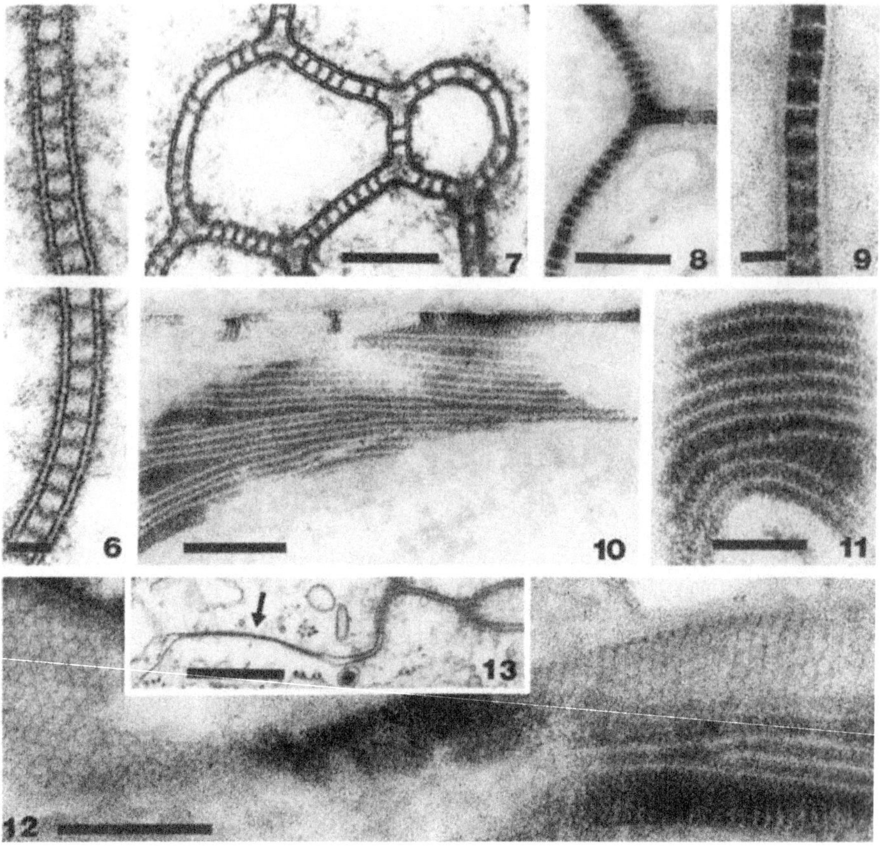


Figs. 6-13 Cell junctions. Fig. 6 Nematocyte battery cell septate junction. Note the regular arrangement of the septa. Fig. 7 Interdigitating processes of epidermal cells are linked by septate junctions. Figs. 8-12 Negative staining of cell junction by _en block_ lanthanum impregnation. Fig. 8 Epidermal septate junction; the septae are negatively stained. Fig. 9 Nematocyte battery cell septate junction. Fig. 10 Tangential view on the septate junctions between a nematocyte and a battery cell. Note the broad junctional complex. Fig. 11 _En face_ view on a septate junction showing a 1.5 nm periodicity of the septae. Fig. 12 Tangential section of gap junctions (nexus) between battery cells. Note the hexagonal arrays of polygonal subunits. Fig. 13 A thin section view on septate and gap (arrow) junctions between battery cells. Bar = 20 nm (Figs. 6, 9); 0.1 μm (Figs. 7, 8, 11, 12), 0.2 μm (Fig. 10), and 0.5 μm (Fig. 13).

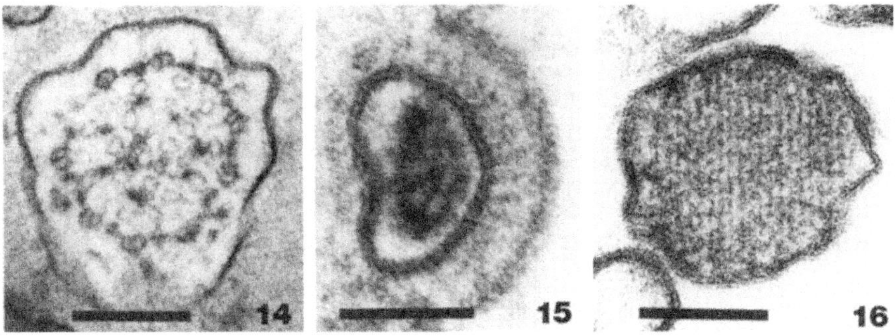


Fig. 14 Cross section through the cnidocil showing 9 doublet microtubules, numerous singlet microtubules and fibers in the center. Fig. 15 Cross section through an inner microvillus of the cnidocil apparatus. The inner microvilli have a reniform shape and are covered with the paracrystalline cuticle layer. Note the electron lucent hallow between the electron dense core and the membrane of the microvillus. Fig. 16 Cross section of a stereocilium showing numerous fine microfilaments arranged in a paracrystalline pattern. Bar = 0.1 µm (Fig. 15), 0.2 µm (Fig. 5, 14, 16).

triplet pericentriolar structures radiate out and form a circle of nine electron dense spokes, which anchors the basal body to the cellular membrane. Satellite structures and ciliary rootlets are reduced in the mature nematocyte (Slautterback, 1967; Westfall, 1970a; Hausmann and Holstein, 1985), but they are clear during morphogenesis (Holstein, 1981). In Craspedacusta and Gonionemus a basal lateral process at the kinetosome has been described (Westfall, 1970a; Hausmann and Holstein, 1985). A centriole is associated with the basal body at an angle of 90 degrees (Westfall, 1970a).

The facts that single microtubules are embedded in a central dense core, that fibrils occur between the doublets, and that dynein arms are absent clearly suggests that the generally accepted ciliary motility mechanism is not present in the cnidocil. The additional elements in the cnidocil may be responsible for the observed stiffness of the cnidocil. The increased number of microtubules in some cnidocils (approx. 400 in Physalia) generates the characteristic shape of the cnidocil, like a pencil with the sharpened point resting on the basal plate (see Slautterback, 1967, Fig. 24; Cormier and Hessinger, 1980a, Fig. 3).

B. The Inner Microvilli

The microvilli of the inner row (synonyms: supportive rods, striated rootlets) are short cellular extensions with a length of 0.5 μm - 1.0 μm and a diameter of 0.06 μm - 0.12 μm (Figs. 5, 15, 25). They have a reniform cross section and contain a homogeneous electron dense core, separated from the microvillar membrane by a 12 nm electron lucent zone. The core material shows a periodicity of approximately 20 nm (Slautterback, 1967; Westfall, 1970; Cormier and Hessinger, 1980a; Hausmann and Holstein, 1985), which looks similar to ciliary rootlets, although there is no evidence that they are ciliary structures. The electron dense core of the inner microvilli can be traced into the cytoplasm of the nematocyte, where it forms a ring of supportive rods encompassing the nematocyst (Bouillon and Levi, 1967; Slautterback, 1967; Westfall, 1970; Cormier and Hessinger, 1980a; Hausmann and Holstein, 1985; Weiss *et al*., 1985).

Consecutive cross sections in *Craspedacusta* have revealed that the tip of the inner microvillus is contiguous with the tip of the apical nematocyte membrane over the nematocyst (Hausmann and Holstein, 1985). The length of the rod cores is dependent on the cell (nematocyst) type. In *Gonionemus* and *Physalia* the rod runs toward the basal end of the encircled cyst, where it frays out into numerous fine fibrils (ca. 6 nm) (Westfall, 1970a; Cormier and Hessinger, 1980a; Hausmann and Holstein, 1985). These basal fibrillar bundles can form a long stalk in small nematocytes of *Physalia* (Cormier and Hessinger, 1980b; Will, 1911), of *Gonionemus* (Westfall, 1970), and in desmonemes of hydra (Holstein, unpublished observations); or it may be reduced to a basal pad in larger nematocytes of *Physalia* (Cormier and Hessinger, 1980) and in stenoteles of hydra (Holstein, unpublished observations).

Figs. 17-20 Glycocalyx covering epithelial battery cells. Fig. 17 Glycocalyx (GC) on a battery cell and numerous secretory vesicles are filled with fuzzy material. Figs. 18-20 Membrane retrieval by the formation of coated pits (arrows). Fig. 21-25 Cuticle layer in the cnidocil apparatus. Fig. 21 Cross section through the proximal region of the cnidocil apparatus showing the mucous material (*) surrounding inner microvilli, cnidocil, and capsular region of the nematocyte. Fig. 22 Cross section through the cnidocil apparatus immediately above the inner microvilli.

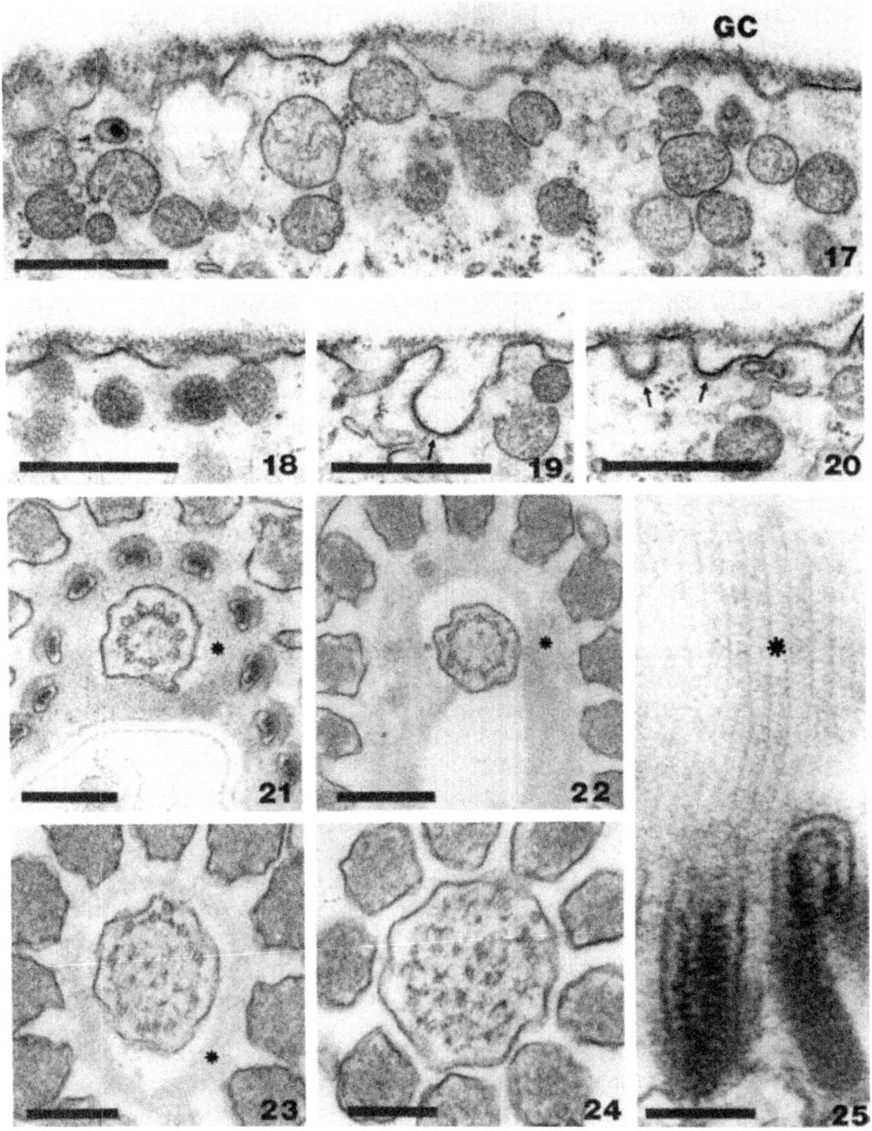


Fig. 23 Cross section through the midregion of the cnidocil apparatus. Note the semicircular arrangement of the stereocilia and the cuticle layer (*). Fig. 24 Cross section through the distal region of the cnidocil apparatus showing a circular arrangement of the stereocilia around the cnidocil. Fig. 25 Longitudinal section of the inner microvillae. Note the prominent paracrystalline cuticle layer (*) covering the microvilli. Bar = 0.5 μm (Figs. 17-22), 0.3 μm (Figs. 23, 24), and 0.2 μm (Fig. 25).

## C. The Outer Microvilli (Stereocilia, Stereovilli)

The outer row of microvilli consists of long microvilli, also known as stereocilia (Figs. 5, 16, 21-24). From as few as about 9 (*Hydractinia*) to as many as 30 (*Gonionemus*) stereocilia protrude almost perpendicularly from the apical surface of the nematocyte with relatively constant center to center spacing. The cross sectional shape of this bundle of stereocilia varies from an elliptical (*Physalia*) to a crescent like pattern (*Craspedacusta, Gonionemus, Obelia*). However, regardless of the gross figuration of the bundle, it displays a morphological polarization: The stereocilia are shortest at one edge of the bundle and grow monotonically toward the opposite edge. The longest stereocilia are located at the site of the cnidocil (Slautterback, 1967; Hausmann and Holstein, 1985).

Stereocilia are 0.2 µm in diameter and contain a prominent, longitudinally oriented bundle of up to 400 parallel microfilaments (Slautterback, 1967). The individual filaments have a diameter of 5nm and have been identified as actin filaments based on their crossreactivity with f-actin antibodies and phalloidin (Hausmann and Holstein, 1985; Novak and Wood, 1982). The paracrystalline bundles of individual stereocilia are anchored in the cytoplasm of the cell and can be traced as well shaped circular bundles up to the distal part of the nematocyte (Slautterback, 1967; Bouillon and Levi, 1967; Hausmann and Holstein, 1985).

The actin bundles in the stereocilia are highly ordered. Thin transverse sections of the stereocilia (Slautterback 1967) show the pattern in the organization of the filaments: Connections between neighbouring filaments follow arcs that are stacked upon each other in rows to give a festooned pattern (Tilney *et al.*, 1980). In longitudinal sections a transverese periodicity can be seen which runs across the filaments and creates a paracrystralline pattern. The separation of adjacent filaments in the bundle is 100 Å (Slautterback, 1967). The lateral distance from the membrane to the filaments measures 100 Å, which is a very close association of the filaments to the membrane compared with the distance of 300 Å in normal microvilli (Fig. 16).

## D. The Fibrillar Collar

In the distal region of the nematocyte a complex fibrous network, called the fibrous collar, surrounds the nematocyst and the base of the cnidocil at a length of about 0.5 µm to 0.75 µm (Figs. 5, 27. The fibrous collar links (i) the cytoplasmic processes of stereocilia and inner microvilli

(=supportive rods), (ii) the supportive rods with each other and (iii) the supportive rods with the wall of the nematocyst capsule.

In the cytoplasm the fibrillar bundles of the stereocilia are linked to the supportive rods by a bridge of filamentous electron dense material (Slautterback, 1967; Hausmann and Holstein, 1985). The supportive rods are embedded in an interlaced girdle, which encircles the neck region of nematocyst and kinetosome (Hausmann and Holstein, 1985). Between each pair of supportive rods there are three thick fibrils, which run equidistantly to each other and parallel to the rods. The individual fibrils of the cross- connecting network pass each other at an angle of about 30 degrees. Their points of intersection overlap with the inner fibrils adjacent to the rods, but never with the middle fibrills (Hausmann and Holstein, 1985). The supportive rods are separated from the capsular wall by a distance of ca. 30 nm. Groups of two to three microtubules are located between the rods and the capsule (Hausmann and Holstein, 1985). In addition Cormier and Hessinger (1980) described an outer ring of microfilaments surrounding the microfilaments of the stereocilia in a plane perpendicular to the long axis of the cell.

## D. The Cuticle Layer

The extracellular cavity enclosed by the stereocilia is filled with a prominent paracrystalline extracellular layer which covers the proximal cnidocil, the inner microvilli and the apical membrane of the nematocyte (Fig.21-25). The cuticle layer is not a simple mucous mat covering the surface of the nematocyte. Battery (accessary) cells are covered by a 50-60 nm thick glycocalycal layer, which shows no specific structural organization and is continuously produced by the discharge of vesicles situated beneath the epithelial membrane (Figs. 17-18). Membrane retrieval is mediated by coated pits (Figs. 18-20).

The cuticular layer of the cnidocil apparatus, however, radiates from the surface of the inner microvilli with a periodicity of 15 nm (*Craspedacusta*, Figs. 22, 25). The layer has a thickness of 35 nm in cross sections and it shows the same outer shape the villi have. Longitutinal sections indicate that the layer has distal thickness of 1000 nm (Fig. 25, 26). The cnidocil apparatus of nematocytes in larvae of *Hydractinia* show in addition small vesicles, which are situated around the inner microvilli. The function of this highly ordered extracellular structure is totally unclear, but it can be found in freshwater as well as in seawater hydrozoans.

## IV. FUNCTIONAL PROPERTIES OF THE CNIDOCIL-APPARATUS.

Three hypothetical functions have been suggested to explain the sophisticated structure of the cnidocil apparatus: (1) induction of nematocyst discharge, (2) anchorage of the nematocyst, and (3) sensory perception (and transduction) of stimuli controlling nematocyst discharge.

(1) Several authors have suggested that the cnidocil causes the fibrous collar around the nematocyst to constrict, forcing the operculum open, or increasing the intracapsular pressure on the capsule and thereby eliciting discharge ("toothpaste-tube model"). However there is evidence demonstrating that the network of fibrils in the fibrous collar is completely different from any known contractile system. It has also been shown that the first visible reaction of the nematocyst to an electrical stimulus is a rapid volume increase. The nematocyst subsequently decreases in volume , but at a rate which is about 1000 X too fast to be attributed to contractile filaments (Holstein and Tardent, 1984).

(2) Since a direct function in discharge is improbable, we propose that the supportive rods and the interconnecting fibrous collar are essential in order to stabilize and anchor the nematocyst within the nematocyte. Such an anchorage is mechanically necessary: It is obvious that captured prey would get lost, if the nematocysts were not be anchored in an effective way. One could also imagine that a solid attachment is required to avoid recoil of the cyst during discharge, which could injure the host tissue and vitiate the effect of discharge.

(3) There is good evidence that the nematocyte is sensitive to mechanical stimuli (Conklin and Mariscal, 1976; Jones, 1947; Lentz, 1968; Carré and Carré, 1980), although direct electrophysiological data are still lacking. The cnidocil apparatus is bilaterally symmetrical and in some hydrozoans highly polarized. Structural similarities of the cnidocil apparatus to the hair cell sensory apparatus suggest that there may be also resemblances at the functional and micromechanical level. A passing prey which displaces the protruding cnidocil's tip causes a gross displacement of the cnidocil apparatus, including the stereocilia. The stereocilia behave as rigid rods anchored in the cytoplasm by their core of longitudinally oriented filaments.  Horizontal mechanical displacement of the cnidocil's tip would result in a mainly vertical shear between stereocilia along the axis of stimulation. (This region of shear, we suggest, is a candidate for the location of transduction molecules; Hausmann and Holstein, 1985).

The sensitivity of the cnidocil apparatus to chemical stimuli is still in question. It is an old observation that cnidocyst discharge is correlated with the nutritional condition of the animal (Lasker et al., 1982; Mariscal 1973; Smith et al. 1974; Rushforth 1973). In hydra and *Craspedacusta* ectocommensals can be often found in close contact with the

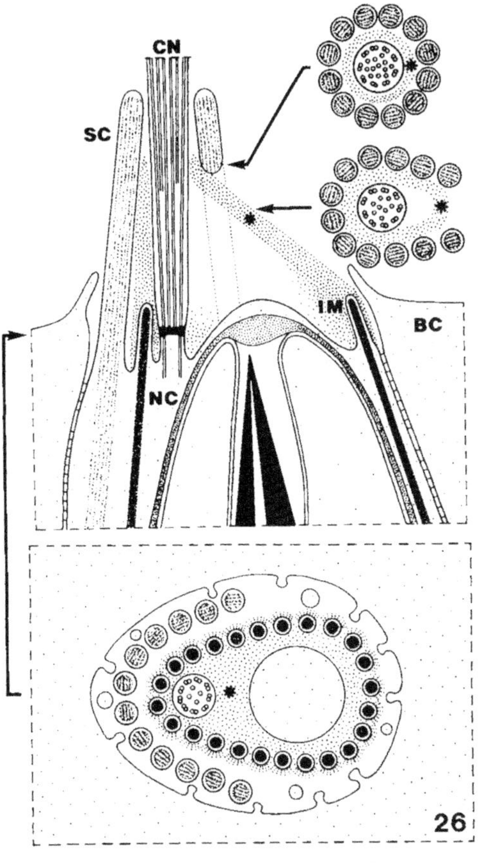


Fig.26 Schematic reconstruction of the cnidocil apparatus in *Craspedacusta*. The longitudinal section shows the stereocilia which are anchored in the cytoplasm of the nematocyte (NC) and which surround the cnidocil (CN) converging at its tip. The inner microvilli (IM) surround both, the cnidocil and the nematocyte. The nematocyte is connected with battery cell (BC). Note the prominent cuticle layer (*) covering the inner microvilli.

tentacular surface deflecting the cnidocils; however this stimulation fails to induce discharge. These observations may suggest that the cnidocil apparatus is also sensitive to chemical stimuli. However, ultrastructural and immunological studies clearly show the presence of synaptic contacts to neurites in the vicinity of nematocytes (Westfall, 1970b; Chapman, 1974). In hydra in each battery cell there is one sensory cell (Westfall and Kinnamon, 1978). Therefore, it cannot be excluded that the observed chemical sensitivity of nematocyst discharge is simply caused by an inhibitory / excitatory effect of the surrounding nervous system.

## EVOLUTION OF THE CNIDOCIL APPARATUS

The highly ordered organization of the hydrozoan cnidocil apparatus raises the question of the origin of this structure in cnidarian evolution. Much less research has been done on the organization of the sensory receptor of nematocytes in anthozoans and scyphozoans (Table 1). Nearly no data are available on the cnidocil apparatus in scyphozoans. The figures given by Chapman (1974) and Westfall (1966) on nematocytes in *Aurelia aurita* indicate that the basic elements of the hydrozoan cnidocil apparatus (i.e stereocilia, inner microvilli, fibrous collar) are present in the scyphozoa.

With the exception of spirocysts (Mariscal et al., 1976) all anthozoan cnidocytes have a ciliary sensory apparatus, which shows an organization similar to that in the basic cnidarian sensory cell (Lentz, 1966, 1968; Lyons, 1973; Mariscal, 1974a; Mariscal and Bigger, 1976; Mariscal *et al.*, 1978; Peteya, 1975; Watson and Mariscal, 1983b, ). There is a normal (9X2+2 MT pattern) motile flagellum with well developed basal rootlet complex which is surrounded by a pallisade of microvilli (Westfall, 1965; Peteya, 1975; Watson and Mariscal, 1983). Depending on the nematocyst type, the diameter of the microvilli varies from 0.05-0.3 µm and shows an increased number of microfilaments (Westfall, 1966; Peteya, 1975; Schmidt and Moraw, 1982). At present it is not clear whether these enlarged microvilli also exhibit the characteristic festooned pattern of paracrystalline actin filaments found in stereocilia of hydrozoans. Inner microvilli and a fibrous collar have not been described in anthozoan nematocytes.

The anchoring apparatus of the nematocyst is also simpler in anthozoans (Mariscal, 1974b; Mariscal and Bigger, 1976; Mariscal *et al.*, 1976 Westfall 1965). In *Calliactis parasitica* the filaments of the stereocilia (diameter 0.13 µm) fray out into single filaments in the upper capsular region

| Class and Species | Cilium | | | Microvilli | | FC | CL | References |
|---|---|---|---|---|---|---|---|---|
| | dMT | sMT | DC | MV1 | MV2 | | | |
| Hydrozoa | | | | | | | | |
| *Clava squamata* | 9 | 8 | + | 12 | 35 | + | + | Bouillon & Levi 1967 |
| *Craspedacusta sowerbii* | 9 | 8 | + | 9 | 21 | + | + | Hausmann & Holstein 1985 |
| *Gonionemus sp.* | 9 | 10 | + | 27(+) | 36 | + | + | Westfall 1970 |
| *Hydra oligactis* | 9 | 100 | + | 21(+) | 21 | + | + | Slautterback 1967 |
| *Hydractinia echinata* | 9 | 10 | + | 14(+) | 24 | + | + | Weiss *et al.* 1985 |
| *Muggiacea kochi* | 9 | 400 | + | ? | 15 | + | ? | Carré & Carré 1980 |
| *Physalia physalis* | 27 | 500 | + | 15 | 15 | + | + | Cormier & Hessinger 1980 |
| Scyphozoa | | | | | | | | |
| *Aurelia aurita* | 9 | 2 | - | ? | ? | + | ? | Chapman 1974 |
| Anthozoa | | | | | | | | |
| *Alcyonium digitatum* | 9 | 2 | - | - | 7 | (+) | - | Schmidt & Moraw 1982 |
| *Haliplanella luciae* | 9 | 2 | - | - | 9 | - | - | Watson & Mariscal 1983 |
| *Metridium senile* | 9 | 2 | - | - | ? | - | + | Westfall 1966 |

sMT=singlet microtubules, dMT= doublet microtubules, MV1=outer Microvilli (Stereocilia, Stereovilli), MV2 inner Microvilli, FC=fibrous collar, CL=cuticular layer, DC= dense core. The numbers indicate the number of structural elements present in the sensory receptor of nematocytes; MV1 brackets: the stereocilia surrounding the cnidocil.

and intermingle with other filaments (Chapman 1974). Considering that anthozoans are sometimes viewed as the most primitive group of the cnidarians (Hyman, 1940), this is a remarkable pattern. It suggests that the fibrillar collar of scyphozoan and hydrozoan nematocytes may have evolved from a diffuse fibrillar network of intermingled actin filaments in anthozoans. Such a hypothesis would be consistent with the observation that the anlage of the outer (stereocilia) and inner microvilli during hydrozoan nematocyte morphogenesis is a uniform structure (Holstein, 1981). We suspect the main difference between both cores to be in the different packing of actin filaments. Accordingly, the filament bundles of the rigid inner microvilli may have a higher degree of order than the filament bundles of the stiff but flexible stereocilia.

## VI. IS THE CNIDOCIL APPARATUS A PROGENITOR OF HIGHER METAZOAN MECHANORECEPTORS?

The cnidocil apparatus exhibits two structural features which correspond to structures proposed to be involved in the perception of mechanical stimuli in higher metazoan mechanoreceptors: (i) the multiple ciliary microtubules which are embedded in a core correspond to the tubular body in arthropods and (ii) the stereocilia (outer microvilli) correspond to the stereocilia in the hairbundle of hair cells in vertebrates (Fig. 27).

Cnidarian stereocilia show striking similarities with the stereocilia in the hair cells of the accoustico-lateralis system in vertebrates: (1) a drastically increased number of actin filaments (up to 500), (2) ordered packing of the actin filaments which creates a special festooned pattern, (3) tight contacts to the membrane of a stereocilium, and (4) cytoplasmic core of the actin bundles. These common features indicate that the main structural features characterizing the stereocilia of hair cells are not unique to vertebrates (DeRosier *et al.*, 1980; Tilney *et al.*, 1980; Hudspeth, 1983; Neugebauer and Thurm, 1984). We propose that these are basic features of mechanoreceptors which developed very early in metazoan evolution. It is interesting to note that in protists the mechanosensitive channels are situated in the membrane of the cell body and not of the ciliary membrane (Machemer and Machemer-Röhnisch, 1984).

In addition there are structural features which could be interpreted as functional adaptations in the different systems: (1) The stereocilia are anchored at a basal cuticular plate in the hair cells, whereas the stereocilia are anchored

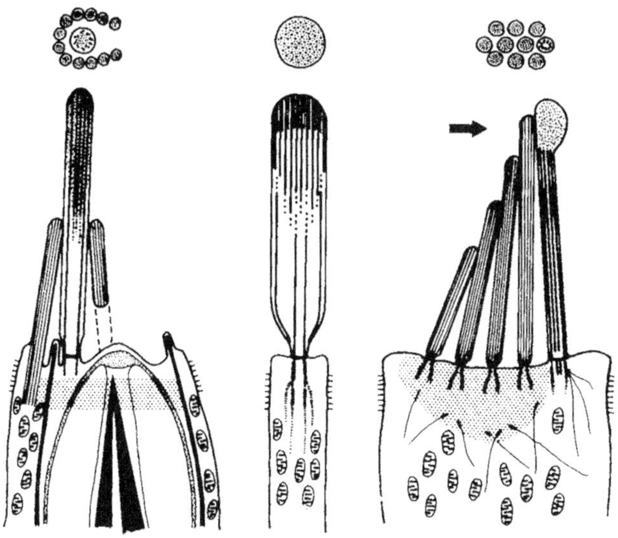


Fig.27 Schematic reconstruction of a hydrozoan cnidocil apparatus (A), an insect mechanoreceptor (B) and the sensory region of a hair cell in vertebrates (C). The cnidocil apparatus and the ciliary region of insect mechanoreceptors show a increased number of microtubules which are embedded in an electron dense core material (dotted). The cnidocil apparatus and the haircells are characterized by stereocilia (= cilium associated microvilli with an increased number of actin filaments). The stereocilia in hair cells are anchored in the cell by a cuticular plate (shaded), the stereocilia in the cnidocil apparatus an anchored with the fibrillar collar (shaded).

at the fibrillar collar in the nematocytes. (2) Starting from the cilium there develops a bilateral symmetry of the encircling stereocilia.

The fine structure of the cnidocil in some hydrozoans is also similar to the specialized distal ciliary (outer dendritic segment) region of mechanoreceptors in arthropods, the tubular body (Fig. 27). Here up to 300 microtubules are embedded in an electron dense material and linked with the ciliary membrane (Thurm, 1981). The contact region in the membrane is thought to be the site of transduction (Thurm, 1981). Since there are no electrophysiological data available we cannot exclude a mechanosensitive function of the cnidocil. Another explanation for the complicated cnidocil structure is that it increases the stiffness of the cnidocil.

## VII. CONCLUSION

The sensory receptor of nematocytes exhibits structural and functional similarities with the stereocilia of vertebrate hair cells, which indicate that this sensory type is not unique to vertebrates but is already present in the cnidarians. Therefore, cnidarians may be a suitable system in which to study the basic phenomena of mechanotransduction.

## VIII. REFERENCES

Beams, H.W., Kessel, R.G and C.-Y. Shih. 1973. The surface features of hydra as revealed by scanning electron microscopy. Trans.Amer.Microsc.Soc. 92:161-175.

Bouillon, J. and C. Levi. 1967. Ultrastructure du cnidocil, de l'appareil peri-nematocystique et du cnidopode des nematocystes d'hydroides. Ann.Sci.Nat.Zool.Biochem.Anim. 12 Ser 9e:425-456.

Carré, D. and C. Carré. 1980. On triggering and control of cnidocyst discharge. Mar.Biol.Physiol. 7:109-117.

Chapman, D.M. 1974. Cnidarian histology. In: Muscatine, L. and H.M. Lenhoff (eds.), *Coelenterate Biology*, pp.1-92. Academic Press, New York.

Conklin, J. and R.N. Mariscal. 1976. Increase in nematocyst and spirocyst discharge in a sea anemone in response to mechanical stimulation. In: Mackie, G.O. (ed.), *Coelenterate Ecology and Behaviour*, pp.549-558, Plenum Publishing Cooperation, New York.

Cormier, S.M. and D.A. Hessinger. 1980a. Cnidocil apparatus: sensory receptor of *Physalia* nematocytes. J.Ultrastr.Res. 72:13- 19.

Cormier, S.M. and D.A. Hessinger. 1980b. Cellular basis for tentacle adherence in the portuguese man-of-war (*Physalia physalis*). Tissue & Cell 12:713-721.

DeRosier, D.J., Tilney, L.G., and E.Egelman. 1980. Actin in the inner ear: the remarkable structure of the stereocilium. Nature 287:291-296.

Hand, A.R. and S. Gobel. 1972. The structural organization of septate and gap junctions of hydra. J.Cell Biol. 52:397-408.

Hausmann, K. and Th. Holstein. 1985. Bilateral symmetry in the cnidocil-nematocyst complex of the freshwater medusa *Craspedacusta sowerbii* Lankester (Hydrozoa, Limnomedusae). J. Ultrastruct.Res.90:89-104.

Holstein, Th. 1981. The Morphogenesis of Nematocytes in *Hydra* and *Forskålia*: an ultrastructural study. J.Ultrastr.Res 75:276-290.
Holstein, Th. and P. Tardent. 1984 An ultrahigh-speed analysis of exocytosis:nematocyst discharge. Science 223:830-833.
Horridge, G.A. 1969. Statocysts of Medusae and evolution of stereocilia. Tissue & Cell 1:341-353.
Hyman, L.H. 1940. The invertebrates: protozoa through Ctenophora. McGraw Hill, New York.
Hudspeth, A.J. 1983. Mechanoelectrical transduction by hair cells in the accousticolateralis sensory system. Ann.Rev.Neurosci. 6:187-215
Jones, C.S. 1947. The control of nematocyst discharge in hydra. J.Exp.Zool. 6:361-383.
Lasker, H.R., Syron, J.A., and W.S. Clayton Jr. 1982. The Feeding response of *Hydra viridis*: Effects of prey density on capture rates. Biol.Bull. 162:290-298.
Lentz, Th.L. 1966. The cell biology of hydra. Wiley, New York.
Lentz, Th.L. 1968. Primitive nervous systems. Yale Uni. Press, New Haven, Conneticut.
Machemer, H. and S. Machemer-Roehnisch. 1984. Mechanichal and electric correlates of mechanoreceptor activation of the ciliated tail in *Paramecium*. J.Comp.Physiol.A 154:273-278.
Mariscal, R.N. 1973. The control of nematocyst discharge during feeding by seaanemones. Pub.Seto Mar.Biol.Lab. 20:695-702.
Mariscal, R.N. 1974. Nematocysts. In: Muscatine, L. and H.M. Lenhoff (eds.), *Coelenterate Biology*, pp. 129-178. Academic Press, New York.
Mariscal, R.N. 1974a. Scanning electron microscopy of the sensory surface of the tentacles of sea anemones and corals. Cell Tiss.Res. 147:149-156.
Mariscal, R.N. 1974b. Scanning electron microscopy of the sensory epithelia and nematocysts of corals and a corallimorphian sea anemone. Proc.Sec.Internat.Coral Reef Symp. 1:519-532.
Mariscal, R.N. and C.H. Bigger. 1976. A comparison of putative sensory receptors associated with nematocysts in an anthozoan and a scyphozoan. In: Mackie, G.O. (ed.), Coelenterate ecology and behaviour, pp.559-568, Plenum Publishing Cooperation, New York.
Mariscal, R.N., Bigger, C.H., and R,B. McLean. 1976. The form and function of cnidarian spirocysts. 1. Ultrastructure of the capsule exterior and relationship to the tentacle sensory surface. Cell Tiss.Res. 168:465-474.

Mariscal, R.N., Conklin, E.J., and C.H. Bigger. 1978 The putative sensory receptors associated with the cnidae of cnidarians. Scanning electron Microscopy 2:959-966.

Neugebauer, D.-C. and U. Thurm 1984. Chemical dissection of sterovilli from fish inner ear reveals differences from intestinal microvilli. J.Neurocytol. 13:797-808.

Novak, P.L. and R.L. Wood. 1983. Development of the nematocyte junctional complex in hydra tentacles in relation to cellular recognition and positioning. J.Ultrastruct.Res. 83:111-121.

Peteya, D.J. 1975. The ciliary cone sensory cell of anemones and cerianthids. Tissue & Cell 7:243-252.

Rushforth, N.B. 1973. Behavioral modifications in coelenterates. In: Corning, W.C., Dyal, J.A., and A.O.D. Willows (eds.) *Invertebrate Learning* Vol.1, pp.123-169. Plenum Press, New York.

Slautterback, D.B. 1967. The Cnidoblast-muscoepithelial complex in the tentacles of hydra. Cell Tiss.Res. 79:296-318.

Schmid, H. and B. Moraw. 1982. Die Cnidogenese der Octocoralia (Anthozoa, Cnidaria): II.Reifung, Wanderung und Zerfall von Cnidoblast und Nesselkapsel. Helgolaender Meeresuntersuchungen 35:97-118.

Smith S., Oshida, J. amd H. Bode. 1974. Inhibition of nematocyst discharge in hydra fed to repletion. Biol.Bull. 147:186-202.

Tardent, P. and V. Schmid. 1972. Ultrastructure of mechanoreceptors of the polyp *Coryne pintneri* (Hydrozoa, Athecata). Exp.Cell Res. 72:265-275.

Tardent, P. and Th. Holstein. 1982. Morphology and morphodynamics of the stenotele nematocyst of *Hydra attenuata* Pall. (Hydrozoa, Cnidaria). Cell Tiss.Res. 224:269-290.

Tilney, L.G. DeRosier, D.J., and M.J. Mulroy. 1980. The Organization of actin filaments in the stereocilia of cochlea hair cells. J.Cell Biol. 86:244-259.

Thurm, U. 1981. Mechano electric transduction. Biophysics of Structure and Mechanism 7:245-246.

Watson, G.M. and R.M. Mariscal. 1983. Comparative ultrastructure of catch tentacles and feeding tentacles in the sea anemone *Haliplanella*. Tissue & Cell 15: 939-953.

Weis, V.M., Keene, D.R., and L.W.Buss. 1985. Biology of hydractiniid hydroids. 4.Ultrastructure of the planula of *Hydractinia echinata*. Biol.Bull. 168:103-418.

Westfall, J.A. 1965. Nematocysts of the sea anemone *Metridium*. Amer.Zool. 5:377-393.

Westfall, J.A. 1966. Fine structure and evolution of nematocytes. Sixth International Congress for Electron Microscopy, Kyoto: 235- 236.

Westfall, J.A. 1970a. The nematocyte complex in a hydromedusa *Gonionemus vertens*. Cell Tiss.Res. 110:457-470.
Westfall, J.A. 1970b. Ultrastructure of synapses in a primitive coelenterate. J.Ultrastruct.Res. 32:237-246.
Westfall, J.A. and J.C. Kinnamon. 1978. A second sensory-motor- interneuron with neurosecretory granules in hydra. J.Neurocytol. 7:365-379.
Wood, R.L. and A.M. Kuda. 1980 Formation of junctions in regeneration hydra: septate junctions. J.Ultrastruct.Res. 70:104-117.
Wood, R.L. and P.L. Novak. 1982. The anchoring of nematocysts and nematocytes in the tentacles of hydra. J.Ultrastruct.Res. 81:104-116.

5

# ULTRASTRUCTURE OF THE SMALL CNIDOCYTE OF THE PORTUGUESE MAN-OF-WAR (Physalia physalis) TENTACLE

David A. Hessinger and Mark T. Ford

Department of Physiology and Pharmacology
School of Medicine
Loma Linda University
Loma Linda, California 92350

SUMMARY

The nematocysts of the Portuguese Man-of-war (Physalia physalis) occur in two sizes, large and small. This paper describes the extraordinarily complicated ultrastructure of the Man-of war cnidocyte, with particular emphasis on the small cnidocyte. The nematocyst is located immediately below the apical region of the cnidocyte and is enclosed within a prominent cytoskeletal structure referred to as the fibrillar basket. The base of the fibrillar basket is anchored to the underlying mesoglea via hemidesmosomes. Capping the basal half of the nematocyst and in close association with the fibrillar basket are numerous elongate vesicles of unknown function. The apical region of the cnidocyte, all but the tip of the cnidocil apparatus which contacts the ambient medium, is enclosed by an adjacent epithelial "neighbor" cell. The cnidocil apparatus of the cnidocyte consists of a radially symmetrical ring of 15 to 21 stereocilia surrounding a modified cilium called the cnidocil. At the bases of the stereocilia, the cnidocyte invaginates to form a cavity. The cavity is continuous with the ambient medium and is divided into two interconnected chambers, one on top of the other. These chambers are separated from each other by a narrowing of the cavity wall formed by vertical septate-like processes that project medially from the base of each stereocilium. A spatulate, longitudinally oriented, striated rod originates within each of these vertical processes and terminates basally on the fibrillar basket. The apical ends of the rods are interconnected by a circular, filamentous band that is

The Biology of Nematocysts

situated in a plane perpendicular to the long axis of the cell. The assembly of interconnected striated rods is referred to as the fibrillar collar and is believed to support the cnidocil apparatus. A circular bundle of horizontally-oriented thin filaments called the annular ring surrounds the fibrillar collar. Evidence indicates that extracellular bridges form between the surface membranes of the stereocilia and cnidocil upon chemosensitization. We suggest that this occurs by medial movement of the stereocilia into contact with the cnidocil caused by a contraction of the annular ring against the supportive fibrillar collar.

## I. INTRODUCTION

The Portuguese Man-of-war is a free-floating, colonial hydrozoan having separate polyps adapted for the capture of prey, digestion of food, and sexual reproduction (Mackie, 1960).

Long fishing tentacles hang from a gas-filled float and drift passively in the ocean's currents. The tentacles consist of thin tubes of contractile tissue which can reach up to 30 meters in length when fully extended (Lane, 1960). They can contract more than twenty-fold (Parker, 1932) and draw captured prey to the digestive gastrozooids located immediately below the float. There, stimulated by reduced glutathione released from the wounded prey, the gastrozooids envelop the prey (Lenhoff and Schneiderman, 1959), which are then digested.

### A. Morphology of the fishing tentacle

The "tentacular" canal, which is continuous with the common gastrovascular cavity of the colony, extends the length of the tentacle. Bulbous evaginations of the tentacle called batteries measure approximately 1.5 mm by 0.7 mm and enclose individual fluid-filled chambers connected by narrow channels to the tentacular canal. When the tentacle is fully extended adjacent batteries occur at up to 5 mm intervals along the entire length. When the tentacle is contracted, however, adjacent batteries are in contact with each other. An endoderm lines the chamber of the battery and is separated from the ectoderm by a basement membrane called the mesoglea. The ectoderm is in direct contact with the ambient seawater and is composed of at least three cell types: gland cells, neighbor cells, and cnidocytes (Hyman, 1940). Gland cells

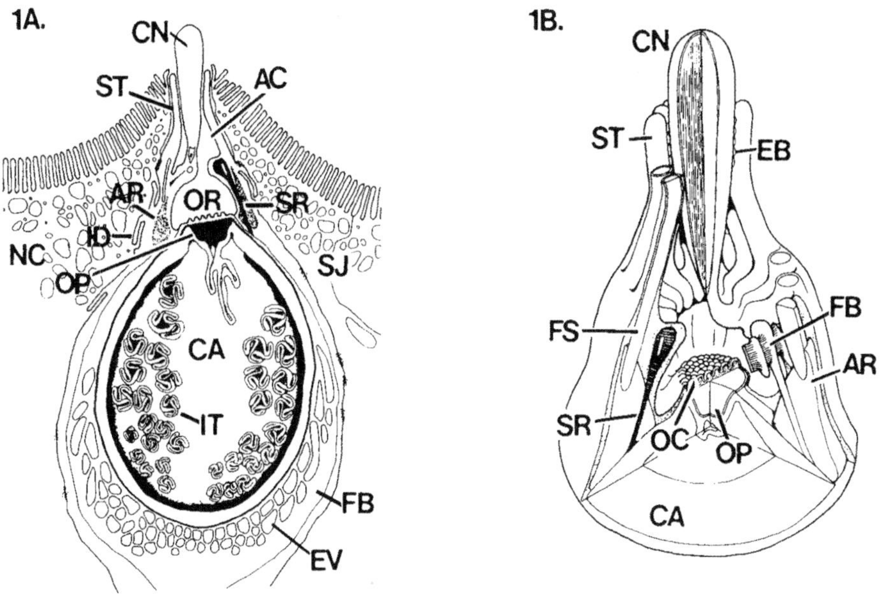


Fig. 1A. Diagram of an idealized longitudinal section of small cnidocyte.

Fig. 1B. Three-dimensional diagrammatic representation of the apical region depicting the relationship between various structural elements. Abbreviations used in these and/or in other figures: AC, apical chamber; AR, annular ring; BB, basal body; BP, basal plate; BR, barb; CA, capsule; CN, cnidocil; CR, ciliary rootlet; CV, crescent-shaped vesicle; DT, double microtubule; EB, extracellular bridge; EV, elongate vesicle; FB, fibrous band or fibrillar basket; FS, filament bundle of the stereocilia; ID, interdigitations; IT, inverted tubule; IV, invaginations; MI, mitochondrion; MT, microtubules; NC, neighbor cell; OC, opercular cap; OP, operculum; OR, opercular chamber; SJ, septate junctions; SR, striated rod; ST, stereocilia.

are columnar and contain large, electron-dense, polymorphic vesicles presumably filled with mucus which these cells secrete to the external surface. The other cells, the neighbor cell and the cnidocytes, are described in detail below.

B. Neighbor cells

The neighbor cells contain pigment granules which impart to the tentacle its characteristic blue color. These cells possess at least three morphologically distinct vesicles (Fig. 2) but it is not known in which of these, if any, the pigment granules are contained. One type of vesicle averages 0.4 μm in diameter and appears electron-lucent. The second vesicle type is 0.3 μm in diameter and contains a moderately electron-dense material. The third vesicle type measures 0.2 μm in diameter and encloses an electron-dense granular material. The large, electron-lucent vesicles are located near the external surface of the neighbor cell, while the remaining two, electron-dense types of vesicle appear to be distributed more evenly throughout the cell. These three types of vesicles may represent distinct functional compartments or developmental stages of the same type of vesicle.

The external surface of the neighbor cell is covered with numerous microvilli measuring approximately 1.0 μm in length and 80 nm in diameter. A mucous layer extends above the tips of the microvilli and covers the external surface of the battery.

Just below the external surface of the battery, the neighbor cell forms septate junctions with the adjacent cnidocyte (Figs. 1, 4). Beginning below the region of the septate junctions extensive, flap-like interdigitations occur between the neighbor cell and the cnidocyte (Fig. 5). These interdigitations appear in cross section as lamellar processes involving both cells and occur at many locations where these cells are in contact. Such interdigitations occur in tissues subject to mechanical stress. This is consistent with the observation that Man-of-War captures relatively large, swimming prey (Purcell, 1984). Such capture may produce significant local stress on involved cnidocytes.

Neighbor cells are perforated by the cone-shaped, apical

Fig. 2. Longitudinal section of small cnidocyte and adjacent neighbor cell (NC). Scale bar = 2.0 μm.

Fig. 3. Cross section of cnidocil apparatus. Extracellular bridges (EB) occur between the cnidocil (CN) and each stereocilium (ST). Scale bar = 0.2 μm.

Fig. 4. Longitudinal section of cnidocil apparatus (Fig. 2). Cnidocil (CN) attaches to cnidocyte below basal body (BB). Septate junctions (SJ) occur between the cnidocyte and neighbor cell (NC). Scale bar = 0.5 μm.

Fig. 5. Cross section of apical region of cnidocyte at level of opercular cap (OC). Striated rods (SR) occur between horizontal fine filaments of annular ring (AR). Scale bar = 0.6 μm.

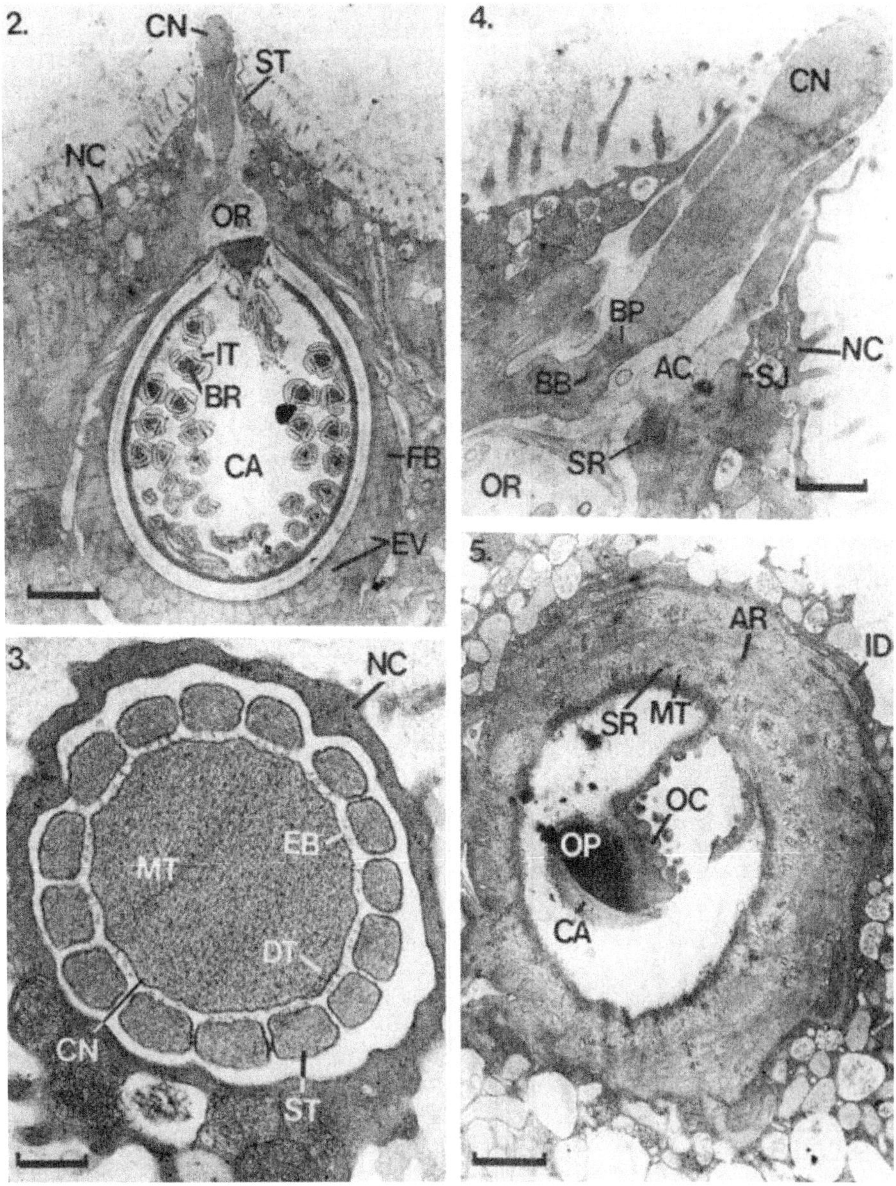
2.
CN
ST
NC
OR
IT
BR
CA
FB
EV
3.
NC
EB
MT
DT
CN
ST
4.
CN
BP
NC
AC
SJ
BB
SR
OR
5.
AR
ID
SR
MT
OC
OP
CA

region of the cnidocyte such that all cnidocytes are surrounded by and extend through an adjacent neighbor cell to make contact with the ambient seawater (Figs. 2-4). The apical margins of the "perforated" neighbor cell almost completely cover the tips of the stereocilia that originate from the cnidocyte (Figs. 1a, 2). This association is quite similar to the relationship between an accessory cell and the mechanoreceptor cell of filiform tentacles of Coryne (Tardent and Schmid, 1972).

Three types of associations between cnidocytes and adjacent accessory cells have been reported. First, in hydra several kinds of cnidocyte occur together within a single battery cell such that the only contact the cnidocytes have with the tentacle is via the battery cell, which in turn is attached to the mesoglea (Wood and Novak, 1982). The battery cell also participates in relatively complicated cellular interactions with elements of the nervous system (Hufnagel and Kass-Simon, 1988). Second, in sea anemone tentacles each cnidocyte is surrounded by two or more supporting cells (G. Watson, unpublished). The supporting cells provide the stereocilia to the sensory cone of the cnidocyte (Westfall, 1965) and carry at least one class of chemoreceptor involved in sensitizing cnidocytes to mechanical stimuli that trigger nematocysts to discharge (Watson and Hessinger, 1988). Third, the cnidocyte in the Man-of-War is anchored directly to the mesoglea (Cormier and Hessinger, 1980a) and is only surrounded at its apical region by a single neighbor cell.

Because of its apparent interaction with many types of cells including more than one type of cnidocyte it appears that the battery cell association in hydra is structurally the most complex of the three types of accessory cells, thus prompting the suggestion that it may also represent the more complicated functional association with cnidocytes. At the other extreme of structural complexity is the situation in the Man-of-War in which the neighbor cell only associates with the apical region of the cnidcyte while the cnidocyte possesses both an independent attachment to the mesoglea (Cormier and Hessinger, 1980a) and provides the full complement of the stereocilia to the cnidocil apparatus (Cormier and Hessinger, 1980b). Thus, the Man-of-War cnidocyte may perform some of the functions provided by accessory cells in the other types of accessory cell associations.

## II. THE CNIDOCYTE

The cnidocyte of the Man-of-war is an extraordinarily complicated cell (Cormier and Hessinger, 1980a, b). Cnidocytes from tentacles fixed with glutaraldehyde and osmium fixed are not easily studied by transmission electron microscopy (TEM) because of difficulty in attaining complete infiltration of the nematocyst capsule by embedding media. We have found that post-fixation of samples of tentacles for three hours in ice-cold 1% acrolein in 0.1 M PIPES buffer, pH 7.3 (Baur and Stacey, 1977) prior to preparation for TEM improves infiltration of Spurr's resin into the nematocyst capsule and provides suitable material to study the ultrastructure of the small cnidocyte. The large nematocyst, however, does not adequately infiltrate under these conditions. Indeed, these differences in infiltration may be explained by the fact that capsules of the two size classes of nematocyst have very different permeabilities. For instance, we have found that isolated large nematocysts are impermeable to solutes of molecular weights of at least 300 daltons, whereas the small nematocysts are permeable to solutes of greater than 960 daltons.

In this section we report two novel observations: elongate vesicles cap the basal half of both large and small nematocysts; and the internal structure of the nematocyst barbs is hollow.

### A. The base of the cnidocyte

The small cnidocyte of the Man-of-war possesses specialized cytoskeletal structures located both at the base and at the apex of the cell (Figs. 1, 2). At the base of the cnidocyte filaments of the fibrillar basket, which surround the nematocyst (Schultze, 1922; Mackie, 1960;), form a thickened stalk that terminates in hemidesmosomal junctions on the mesoglea (Cormier and Hessinger, 1980a). These junctions, together with the fibrillar basket anchor the nematocyst to the tentacle so that captured prey cannot escape (Cormier and Hessinger, 1980a).

### B. Elongate vesicles

Numerous elongate vesicles occur at the basal half of the nematocyst (Figs. 1, 2) where they are densely packed into a layer two to three vesicles across and interspersed in parallel fashion with elements of the fibrillar basket (Fig. 6). The contents of these vesicles in the small cnidocyte are moderately electron-dense, whereas in the large cnidocyte they appear electron-lucent. Because the vesicles are quite long and conform to the curvature of the nematocyst capsule, we have not been able to measure the full length of the vesicles from thin-sections. Nevertheless, in .the large cnidocytes we can measure them for at least 5.6 μm (Fig. 7). In both the small and large cnidocytes the vesicles measure approximately 0.25 μm in diameter.

These vesicles, not previously described in any cnidocyte, are never seen in preparations of isolated nematocysts although the fibrillar baskets still commonly surround these nematocysts (Cormier and Hessinger 1980a). Either these vesicles are fragile and rupture upon disruption of the cell or else they are easily displaced from the fibrillar basket. The presence of mitochondria in association with the basket has been cited as evidence that the fibrillar elements of the basket may be contractile and used by the cnidocyte to initiate discharge (Cormier and Hessinger, 1980a). If true, these vesicles may function in a way similar to that of the sarcoplasmic reticulum in vertebrate striated muscle by storing and releasing divalent cations. This suggestion, however, is not supported by our failure to demonstrate the presence of divalent cations in these vesicles using pyroantimonate (G. Watson, unpublished). Alternatively, these vesicles may contain water and hydrated monovalent ions which may be released directly to the nematocyst capsule to facilitate or to initiate discharge. In addition, these vesicles may also contain proteolytic enzymes which function to digest the fibrillar basket after the nematocyst has been discharged. Such a situation would

Fig. 6. Cross section of elongate vesicles (EV). Microtubules (MT) and septate junctions (SJ) occur at outer periphery of fibrillar basket (FB). Scale bar = 0.25 μm.

Fig. 7. Longitudinal section of elongate vesicles (EV) and elements of fibrillar basket (FB). Scale bar = 0.5 μm.

Fig. 8. Cross section of cnidocil (CN) above stereocilia. Only singlet microtubules (MT) occur at this level. Scale bar = 0.2 μm.

Fig. 9. Cross section of cnidocil apparatus near bases of stereocilia. Doublet microtubules (DT) persist in cnidocil and fine filaments (FS) fill stereocilia (ST). Scale bar = 0.2 μm.

Fig. 10. Cross section of apical region of cnidocyte at basal plate (BP) of cnidocil. Electron-dense striated rods (SR) occur within medial projections of stereocilia bases. Scale bar = 0.4 μm.

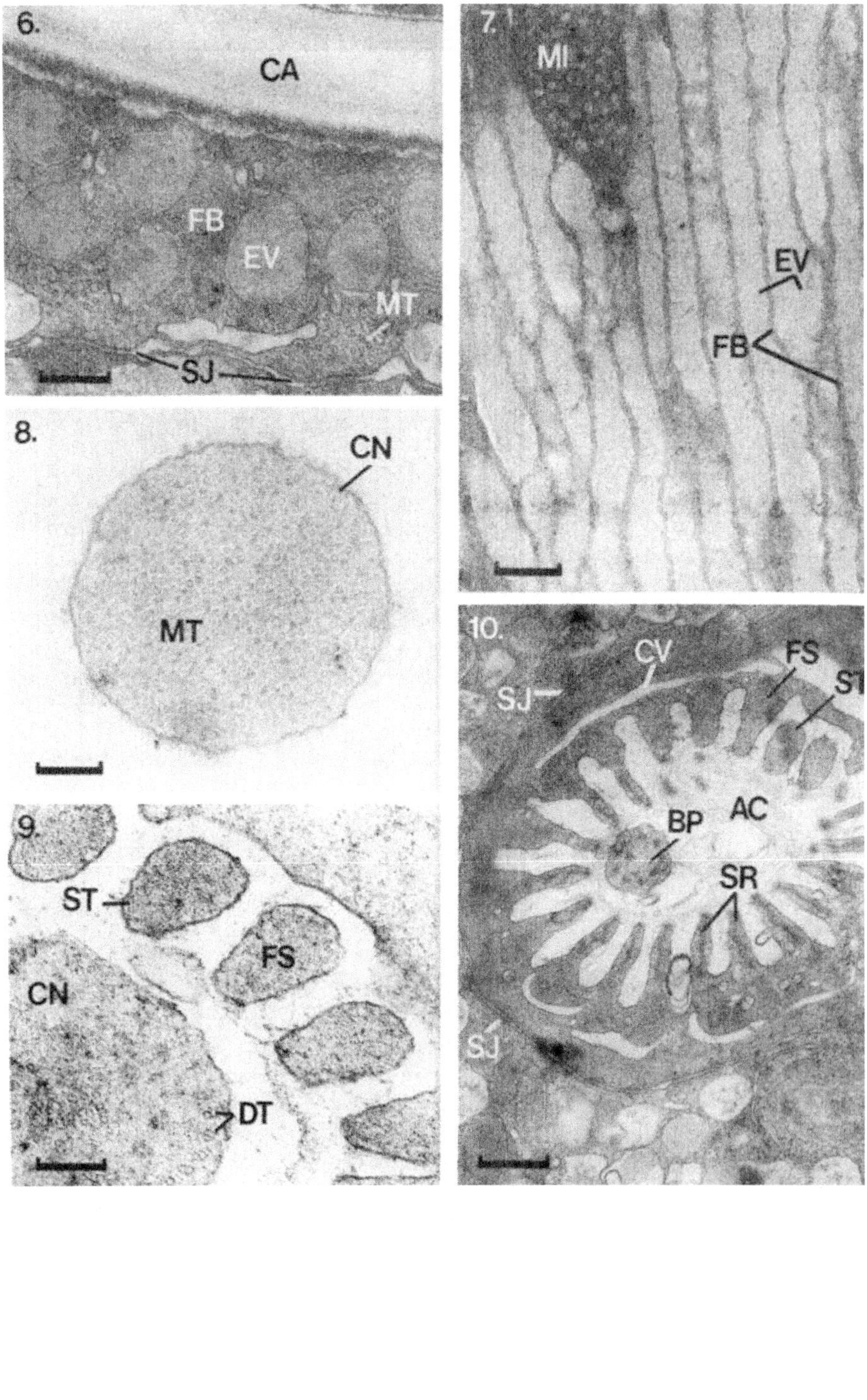
6.
CA
FB
EV
MT
SJ
7.
MI
EV
FB
8.
CN
MT
9.
ST
FS
CN
DT
10.
CV
FS
ST
SJ
BP
AC
SR
SJ

allow the tentacles to release captured prey and the attached nematocysts to the digestive gastrozooids. The presence of such a stored source of proteases could explain why the ectoderm of the Man-of-War tentacles autolyses so easily (Lane, 1958) and why many laboratories working with Man-of-War venom from isolated nematocysts report both significant protease activity and low toxicities for the nematocyst venom (Hessinger, 1988).

C. The nematocyst

Man-of-War nematocysts are capable of injecting a potent venom (Tamkum and Hessinger, 1981) into prey. Tentacle nematocysts have been classified both as atrichous isorhizas (Weill, 1930) and as holotrichous isorhizas (Cormier and Hessinger, 1980a) and occur in two sizes, large (20-28 $\mu$m dia) and small (6-8 $\mu$m).

A plug-like operculum (Figs. 2, 13), which appears to be composed of a granular material, is situated in the apical opening of the capsule wall at the junction of the capsule with the inverted tubule. The operculum seals the capsule and hollow tubule from the external environment (Figs. 1, 2, 13). At the point where the tubule is in contact with the operculum, the tubule is cylindrical in cross section. Extending beyond the operculum region, the inverted tubule becomes triply-pleated (Figs. 2, 15). In the small nematocyst the tubule gradually tapers from a diameter of 1.3 $\mu$m at its widest aspect at the apical opening of the capsule to 0.45 $\mu$m near its tip (Fig. 2) while executing 16-18 concentric coils in planes perpendicular to the polar axis in the small nematocyst.

The proximal two-thirds of the inverted tubule (representing approximately 10 coils) contains electron-dense barbs packed in a staggered, tripartite array within the tubule lumen (Fig. 15).

The distal one-third of the tubule, representing approximately 7 smaller coils, contains no detectable barbs. Such a tubular structure is difficult to classify according to conventional classification schemes (Mariscal, 1974) but the nematocyst is probably best referred to as a holotrichous isorhiza.

At high magnification, the barbs appear to be hollow, with a honeycomb-like internal structure containing crypts that parallel the base-to-tip axis of the barb (Fig. 16). The matrix of the hollow crypts are occupied by electron-dense granules. We suggest this substructure of the spination of the nematocyst tubule may be of general occurrence. In the Man-of-war this may represent a modification of the barb to minimize its total mass and/or to contain venom proteins, thereby making each barb a tiny hypodermic syringe for delivery of the nematocyst toxins.

## III. APICAL REGION OF THE SMALL CNIDOCYTE

The small cnidocyte is approximately 40 μm long and 9.8 μm in diameter at its widest aspect. The apical region of the cnidocyte is cone-shaped and houses the cnidocil apparatus and the fibrillar collar (Cormier and Hessinger, 1980b) which are situated immediately above the operculum of the nematocyst (Figs. 1, 2). The apical region of the cell is believed to be receptive to chemical and mechanical stimuli (Parker and van Alstyne, 1932). We describe below (i) the cytoskeleton of the apical region of the small cnidocyte, (ii) the reversible formation of extracellular bridges between the plasma membranes of the stereocilia and cnidocil under conditions of chemosensitization, (iii) a centrally located, double-chambered cavity in apex of the cell, and (iv) the morphodynamic nature of the apical region of the small cnidocyte.

### A. Cnidocil apparatus

The cnidocil apparatus includes the club-shaped cnidocil and a single circlet of stereocilia (Figs. 1, 3). The cnidocil is about 5 μm long, extending into the ambient medium about 1 μm beyond the tips of the microvilli at the surface of the neighbor cell. The tip of the cnidocil contains tightly packed singlet microtubules (Fig. 8; Cormier and Hessinger, 1980b). Directly below the tip of the cnidocil doublet microtubules first appear (Fig. 3). The doublets are restricted, however, to a single peripheral row just within the cnidocil membrane (Cormier and Hessinger, 1980b) where they continue to the base. The cnidocil narrows from 0.9 μm diameter at its widest aspect to 0.3 μm at its base (Fig. 1, 4). Near the base of the cnidocil only a few singlet microtubules and an incomplete ring of doublets remain (Fig.9). At the base, a typical ciliary basal plate and basal body occur (Fig. 4, 10, 11; Cormier and Hessinger, 1980b).

The cnidocil is surrounded by 15 to 21 stereocilia, each measuring about 2.8 μm in length and 0.2 μm in diameter (Figs. 1, 3, 4). Each stereocilium contains a bundle of approximately 125 fine filaments running lengthwise. The diameter of the stereocilia increase from 0.18 μm at the tips to 0.28 μm at the bases.

In non-treated tentacles the extracellular space between the stereocilia and cnidocil is approximately 0.1 μm (Cormier and Hessinger, 1980b). In tentacles treated with mucin from the surface of prey fish the space between each stereocilium and cnidocil decreases and it appears that each stereocilium becomes interconnected with the cnidocil membrane by moderately electron-dense, extracellular "bridges" (Fig. 3) measuring 34 nm in diameter. In cross-section an average of three bridges occur in the space between the plasma membranes of the cnidocil and the stereocilia. Bridges are observed only in mucin-exposed tentacles in regions where doublet microtubules are present. Bridges are not observed between adjacent stereocilia nor between the stereocilia and the neighbor cell.

The mechanoreceptors of vertebrate hair cells exhibit extracellular bridges between individual stereocilia (Comis et al., 1985; Neugebauer and Thurm, 1985). The cnidocil apparatus is purported to be the mechanoreceptor of the cnidocyte (Parker and van Alstyne, 1932; Pantin, 1942; Cormier and Hessinger, 1980b). We believe such extracellular bridges in Man-of-War cnidocytes (Fig. 3) reversibly "couple" the membrane of the cnidocil with the membranes of the stereocilia under conditions of prey-derived chemosensitization to facilitate sensory transduction of mechanical stimuli in triggering the discharge of nematocysts.

## B. Apical Chambers

Beginning at the bases of the stereocilia a deep, central cavity extends nearly to the operculum of the nematocyst (Figs. 1, 2). The apical margin of this cavity appears as a narrow rim on which each stereocilium sits (Fig. 10). Narrow vertical processes, or "flutes," project medially from the bases of the stereocilia into the central cavity (Fig. 10). Basally these processes broaden and fuse (Fig. 11) to form a circular ledge which divides the cavity into two interconnected chambers (Figs. 1, 2). The base of the cnidocil, including the basal body, originates from one side of this ledge (Figs. 1, 2, 4, 11).

Of the chambers formed by the circular ledge, the outer chamber is defined basally by the ledge and laterally by the surrounding neighbor cell (Figs. 1a, 2, 4). It is open to the ambient medium through the interstices between the cnidocil, stereocilia, and the perforation of the neighbor cell.

The narrow opening in the circular ledge connects the outer chamber with the lower, so-called "opercular" chamber (Figs. 1a, 2, 4). The opercular chamber is defined apically by the ledge, laterally by the region of the fibrillar collar (see section III. C., below), and basally by the opercular cap of the nematocyst.

The opercular cap is situated on top of the operculum of the nematocyst (Figs. 1b, 13, 14). It consists of an electron-dense layer 90 nm thick sandwiched between the plasma and nematocyst membranes and is characterized by a total of 45-50 raised, electron-dense bumps, each measuring about 0.13 μm in diameter and 90-95 nm high. At its lateral edge, the electron-dense material of the cap thins to 35 nm and appears to fuse with the electron-dense fibrillar basket (Figs. 5, 13).

Our description of two chambers formed by the apical invagination of the cnidocyte represents a unique finding. The outer chamber is similar to the heart-shaped chamber described in the perioral sensory cells of hydra (Kinnamon and Westfall, 1984), but the lower, opercular chamber of the Man-of-War cnidocyte, with the opercular cap, has not been described for other cnidarian cells. Functionally the opercular chamber likely reflects the capacity of the apical region to undergo morphodynamic changes in response to appropriate chemical and mechanical stimuli.

## C. Fibrillar collar

The fibrillar collar is composed of a ring of 15 to 21 vertically-oriented striated rods that surround the opercular chamber. In longitudinal section, each rod is alternately electron-dense and electron-lucent with a periodicity of 19 nm (Fig. 13). Each rod originates within a flute just above the ledge and extends basally for about 3 μm to the fibrillar basket on the "shoulder" of the nematocyst (Fig. 1, 13 insert).

In cross section, each striated rod appears as an electron-dense bar above the circular ledge (Fig. 10, 11) and as a disk of similar density below the ledge (Fig. 14). The diameter of the rod decreases from 0.25 μm at the ledge to less than 80 nm at the fibrillar basket, indicating that the rod tapers along its length.

Within the circular ledge, oblate-shaped, fibrous plates occur equidistant between each rod (Figs. 1b, 11, 12). The plates are 0.15 μm wide, 25 nm thick, and at least 0.32 μm long. Each plate is connected to the adjacent rods by fine filaments. This arrangement of striated rod-filaments-plate-filaments-striated rod, etc. forms a continuous, circular band within the ledge.

Near the circular band, each striated rod appears hollow. The hollow "space" within each rod is at least 0.32 $\mu$m long. In cross section, the hollow space appears elliptical, measuring approximately 0.21 $\mu$m wide by 27 nm thick at its largest aspect.

Cormier and Hessinger (1980b) described the striated rods of the fibrillar collar as "striated rootlets, one for each stereocilium." The periodicity of the striations seen in the ciliary rootlet associated with the cnidocil in Figure 8 of that report measures 75 nm, as do the striations of the cnidocyte ciliary rootlets in Aiptasia (G. Watson, unpublished). However, the periodicity of the rod striations seen in Figure 9 of Cormier and Hessinger's report and in Figures 13 and 14 of the present report measure 19 nm. Westfall (1970) observed a periodicity of 20 nm in electron-dense, structural "stiff rods" of the Gonionemus cnidocyte. Furthermore, the striated rods in Physalia appear to have no attachment to or insertion into any of the ciliary structures of the cnidocyte. From these comparisons, it appears that the striated rods in the apex of the Man-of-war cnidocyte are not ciliary rootlets but rather are more similar in structure to the "stiff rods" of Gonionemus.

Furthermore, no difference in the periodicity of rod striations occurs between those contained within untreated and mucin-treated Man-of-war cnidocytes, as described above (section III. A.), or between those contained within untreated cnidocytes and cnidocytes electrically stimulated to discharge (G. Watson, unpublished). These observations suggest that the striated rods are primarily supportive

Fig. 11. Cross section of apical region of cnidocyte at supporting ledge. Circular fibrous band (FB) links striated rods (SR) which, at this level, appear hollow. Scale bar = 0.5 $\mu$m.

Fig. 12. Cross section of apical region of cnidocyte immediately below supporting ledge. Filament bundles of stereocilia (FS) and a ciliary rootlet (CR) occur next to striated rods (SR). Horizontally oriented fine filaments of annular ring (AR) first appear at this level. Scale bar = 0.25 $\mu$m.

Fig. 13. Longitudinal section of apical region of cnidocyte. Scale bar = 1.0 $\mu$m. Insert: striated rod (SR) terminating on fibrillar basket (FB). Scale bar = 0.5 $\mu$m.

Fig. 14. Cross section of apical region of cnidocyte at opercular cap (OC). Each striated rod (SR) is surrounded by an array of microtubules (MT). Scale bar = 0.3 $\mu$m.

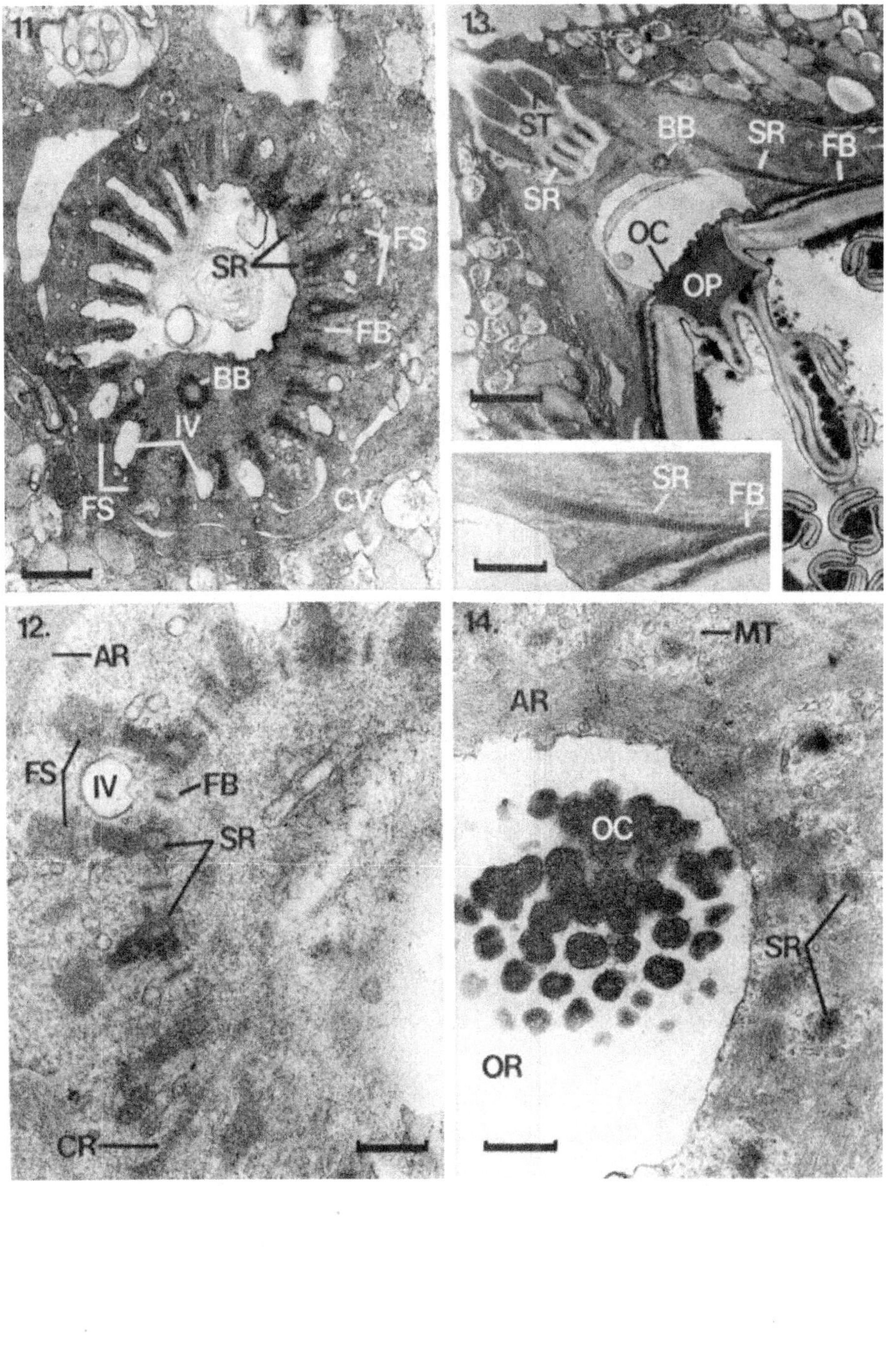
11.
SR
FS
FB
BB
IV
FS
CV
13.
ST
SR
BB
SR
FB
OC
OP
SR
FB
12.
AR
FS
IV
FB
SR
CR
14.
MT
AR
OC
SR
OR

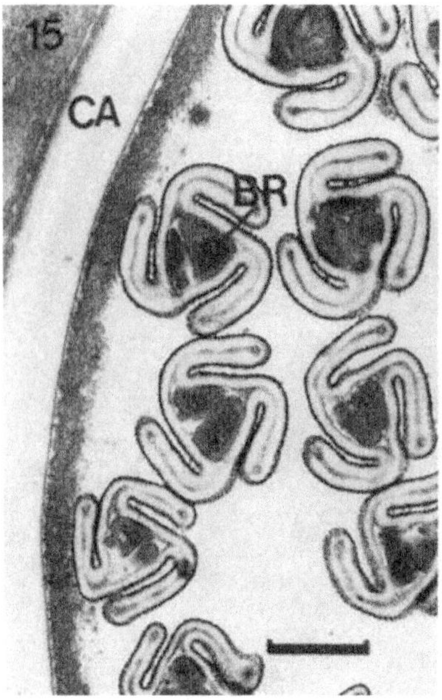


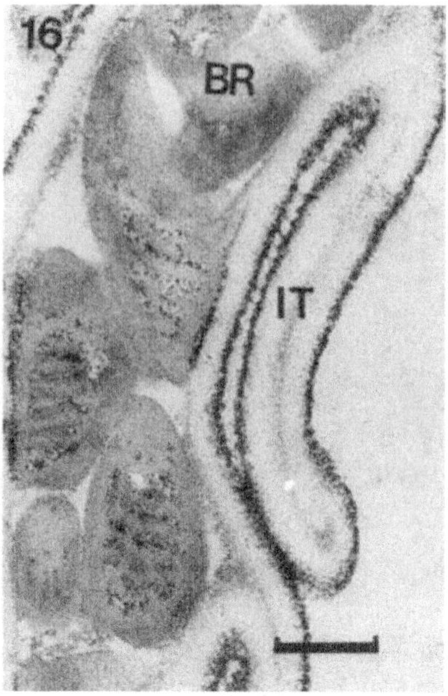


Fig. 15. Cross section of inverted tubules (IT) containing electron-dense barbs (BR). Scale bar = 0.5 μm.

Fig. 16. Longitudinal section of inverted tubule (IT) revealing honeycombed interior of barbs (BR). Bar = 0.2 μm.

structures rather than contractile. In this capacity, we believe the striated rods function to support the cnidocil apparatus of the cnidocyte.

Below the circular band of the fibrillar collar, each striated rod is encircled by a parallel array of five to ten microtubules (Figs. 5, 14) that follow the rod until the rod terminates on the fibrillar basket. The microtubules, however, continue basally beyond this point and separate laterally to encircle the nematocyst at the periphery of the cell (Fig. 6).

Also occurring in the vicinity of the microtubles are numerous small vesicles measuring approximately 40 nm in diameter (Fig. 5, 14). The small vesicles continue alongside the microtubules into the region of the fibrillar basket (Fig. 6). The site of origin of these tiny vesicles is not known. If the location of these vesicles along the microtubules represents a pathway for translocation of vesicles, then it is attractive to suggest that they originate from the deep invaginations that occur in the ledge between the bases of each pair of adjacent stereocilia (Figs.

4, 11, 12). These invaginations of approximately 0.17 μm in diameter penetrate to a depth of at least 0.5 μm. Whether the tiny vesicles are endosomal derivatives of endocytosed chemoreceptor-ligand complexes as is found for chemosensitized sea anemone tentacles chemosensitized by mucin (Watson and Hessinger, 1987; 1988) or pinocytotic vesicles transporting water to the elongate vesicles surrounding the basal half of the nematocyst will have to await further experimentation.

### D. Annular ring and associated structures

Beginning at the level of the ledge, a thick, circular bundle of horizontally-oriented fine filaments forms a ring peripheral to much of the fibrillar collar (Fig. 12). This bundle, which we call the annular ring, extends from the basal region of the circular ledge to the shoulder of the nematocyst (Fig. 5). The ring thickens basally and encircles the opercular chamber (Figs. 1a-b). The striated rods along with associated microtubules and tiny vesicles penetrate the medial aspect of the annular ring at the level of the opercular cap and exit its base (Figs. 5, 14).

Broad, electron-lucent vesicles which appear crescent-shaped in cross sections of the cnidocyte apex occur between the rim of the outer chamber and the ledge (Figs.10, 11). These vesicles are located in the cytoplasm between the vertically-oriented fibrillar bundles from the base of each stereocilium and the plasma membrane adjacent to the neighbor cell.

## IV. FUNCTIONAL SIGNIFICANCE AND CONCLUSIONS

As already noted in part, several morphological features distinguish the apical region of the small cnidocyte from non-sensitized tentacles (Cormier and Hessinger,1980b) from those of mucin-sensitized tentacles. In chemosensitized Man-of-war tentacles the gap between the stereocilia and cnidocil is decreased and replaced by extracellular bridges that connect the cnidocil to the stereocilia (Fig. 3). In addition, the opercular cap appears thicker and the bumps are more pronounced (Fig. 13) and the opercular chamber is narrower and taller (Fig. 4).

Bigger (1982) reported that the apical region of the cnidocyte thinned in the acrorhagi of certain anemones in response to stimuli from non-clonal tissues. Watson and Hessinger (in preparation) find that the stereocilia of cnidocyte sensory cones in the sea anemone, Haliplanella luciae, elongate upon mucin-sensitization in correlation with a change in the sensitivity of the mechanoreceptor (submitted). Wood and Novak (1982) using NBD-phallicidin localized actin filaments in the apical region of hydra cnidocytes. If the annular ring in the Man-of-war cnidocyte contains actin it may be contractile and thus capable of inducing morphodynamic changes observed in the present study.

We propose that the morphodynamic changes observed in the Man-of-War cnidocyte upon chemosensitization are due to shortening of the presumed contractile elements of the annular ring. Such a contraction of the annular ring against the supportive fibrillar collar would biomechanically produce the above-mentioned changes including bringing the stereocilia close enough to the cnidocil so that extracellular bridging can occur between them.

We also describe a new type of cnidocyte accessory cell, the neighbor cell, which surrounds only the apical region of the cnidocyte while the cnidocyte has its own direct attachment to the mesoglea and provides all elements of the purported cnidocyte mechanoreceptor, the cnidocil apparatus. On the basis of ultrastructural evidence, it seems that the association between the neighbor cell and the Man-of-War cnidocyte is the least intimate of any reported cnidocyte/accessory cell association. Whether the neighbor cell modulates cnidocyte responsiveness in a manner similar to the supporting cells in sea anemone tentacles (Watson and Hessinger, 1987; 1988) is not yet known. We would predict, however, from the ultrastructural evidence presented here that the Man-of-War cnidocytes have much more autonomy from the accessory cells than tentacle cnidocytes of either hydra or sea anemones.

It is our opinion that as more is learned about the sites and mechanisms of sensory functions controlling cnida discharge in different species some cnidocytes will be found possessing neither chemo- nor mechano-receptor systems of their own, such as we suspect exist on the sea anemone fishing tentacle, and will serve merely as "dependent" or "passive effectors" to adjacent sensory accessory cells. On the other extreme, however, there will also be found cnidocytes, which will function as classical "independent effectors" with no sensory and little or no modulatory influence from accessory cells or the nervous system, such as we suspect exists on the Man-of-War fishing tentacle,

ACKNOWLEDGMENTS

Wish to thank Drs. Glen Watson, Howard M. Lenhoff and Ian M. Fraser for their helpful suggestions on this manuscript. Supported in part by funds from NSF grant DCB-8609859 to DAH.

REFERENCES

Baur, P. S. and T. R. Stacey. 1977. The use of PIPES buffer in the fixation of mammalian and marine tissues for electron microscopy. J. Microscopy 109: 315-327.

Bigger, C. H. 1982. The cellular basis of aggressive acrorhagial response of sea anemones. J. Morphol. 173: 259-278.

Comis, S. D., J. O. Pickles, and M. P. Osborne. 1985. Osmium tetroxide postfixation in relation to the crosslinkage and spatial organization of stereocilia in the guinea-pig cochlea. J. Neurocytol. 14: 113-130.

Cormier, S. M. and D. A. Hessinger. 1980a. Cellular basis for tentacle adherence in the Portuguese Man-of-War (Physalia physalis). Tissue Cell 12: 713-721.

Cormier, S. M. and D. A. Hessinger. 1980b. Cnidocil apparatus: Sensory receptor of Physalia nematocytes. J. Ultrastruct. Res. 72: 13-19.

Hessinger, D. A. 1988. Nematocyst venoms and toxins. In D.A. Hessinger and H.M. Lenhoff (eds.), The Biology of Nematocysts. Academic Press, San Diego.

Hufnagel, L.A. and G. Kass-Simon. 1988. Functional anatomy of nematocyte innervation in battery cell complexes of the Hydar tentacle. In D.A. Hessinger and H.M. Lenhoff (eds.), The Biology of Nematocysts. Academic Press, San Diego.

Hyman, L. H. 1940. The Invertebrates: Protozoa through Ctenophora. Vol. 1. New York, McGraw-Hill.

Kinnamon, J. C. and J. A. Westfall. 1984. High voltage electron stereomicroscopy of the cilium-stereociliary complex of perioral sensory cells in Hydra. Tissue Cell 16: 345-353.

Lane, C. E. 1958. The toxicity of Physalia nematocysts. Biol. Bull. 115: 219-226.

Lane, C. E. 1960. The Portuguese Man-of-War. Sci. Amer. 13: 371-393.

Lenhoff, H. M. and H. A. Schneiderman. 1959. The chemical control of feeding in the Portuguese man-of-war, Physalia physalis L. and its bearing on the evolution of the Cnidaria. Biol. Bull. 116:452-460.

Mackie, G. O. (1960) Studies in Physalia physalis (L.). Part 2. Behavior and histology. Discovery Rep. 30: 371-407.

Mariscal, R. N. 1974. Nematocysts. In L. Muscatine and H.M. Lenhoff (eds.), Coelenterate Biology, pp. 129-178. Academic Press, New York.

Neugebauer, D. Ch. and U. Thurm. 1985. Interconnections between the stereocilia of the fish inner ear. Cell Tissue Res. 240: 449-453.

Pantin, C. F. A. 1942. The excitation of nematocysts. J. Exp. Biol. 19: 294-310.

Parker, G. H. 1932. Neuromuscular activities of the fishing filaments of Physalia. J. Cell. comp. Physiol. 1: 53-57.

Parker, G. H. and M. A. van Alstyne. 1932. The control and discharge of nematocysts especially in Metridium and Physalia. J. Exp. Zool. 63: 329-344.

Purcell, J. E. 1984. Predation on fish larvae by Physalia physalis, the Portuguese Man of War. Mar. Ecol. Prog. Sen. 19: 189-191.

Schultze, P. 1922. Der Bau und Entladung der Penetranten van Hydra attenuata Pallas. Arch. Zellforsch., 16: 383-438.

Tamkun, M. M. and D. A. Hessinger. 1981. Isolation and partial characterization of a hemolytic and toxic protein from nematocyst venom of the Portuguese Man-of-war, Physalia physalis. Biochim. Biophys. Acta 667: 87-98.

Tardent, P. and V. Schmid. 1972. Ultrastructure of mechanoreceptors of the polyp Coryne pintneri (Hydrozoa, Athecata). Exp. Cell Res. 72: 265-275.

Watson, G. M. and D. A. Hessinger. 1987. Receptor-mediated endocytosis of a chemoreceptor involved in triggering the discharge of cnidae in a sea anemone tentacle. Tissue Cell 19:747-755.

Watson, G. M. and D. A. Hessinger. 1988. Localization of a purported chemoreceptor involved in triggering cnida discharge in sea anemones. In D.A. Hessinger and H.M. Lenhoff (eds.), The Biology of Nematocysts. Academic Press, San Diego.

Weill, R. 1930. Essai d' une classification des nematocytes, des Cnidaires. Bull. Biol. fr. Belg. 64: 141-153.

Westfall, J. A. 1965. Nematocysts of the sea anemone Metridium. Am. Zool. 5: 377-393.

Westfall, J. A. 1970. The nematocyte complex in a hydromedusan, Gonionemus vertens. Z. Zellforch. Mikrosk. Anat. 110: 457-470.

Wood, R. L. and P. L. Novak. 1982. The anchoring of nematocysts and nematocytes in the tentacles of Hydra. J. Ultrastruct. Res. 81:104-116.

# 6

# X-RAY MICROANALYSIS AND PERSPECTIVES ON THE ROLE OF CALCIUM AND OTHER ELEMENTS IN CNIDAE

Richard N. Mariscal

Department of Biological Science
Florida State University
Tallahassee, Florida 32306-2043

## SUMMARY

At present, we know a good deal about the biochemistry of many cnidarian venoms and toxins, but very little about the possible involvement of calcium, phosphorus, sulfur, and other inorganic chemical elements in nematocyst function. A recent study has suggested a general mechanism of nematocyst discharge in which the rapid loss of large amounts of intracapsular calcium may somehow initiate discharge. Other studies, however, have shown that not all cnidae contain calcium. In addition to sulfur (found in all those so far examined), cnidae may contain either (1) primarily calcium, (2) calcium and magnesium, (3) phosphorus, or (4) sulfur only. Although sulfur is thought to be present mainly as disulfide linkages in the capsule wall and tubule collagen, there is no satisfactory hypothesis to explain the role of the other elements in cnidae. I present a new hypothesis suggesting these elements may be bound to, or associated with, the nematocyst venom. I suggest that many nematocyst venoms, like snake and other venoms, are probably metalloenzymes requiring such elements as calcium as necessary cofactors. The possible functions of other elements, including non-metals such as phosphorus, are also discussed.

## I. INTRODUCTION

Although X-ray microanalysis has been used for some time in the materials sciences such as geology and metallurgy, only in the last decade or so has the process been combined with electron microscopy and applied to the study of biological materials. The electron beam of the microscope is focused on a small point on the specimen to excite the production of X-rays from the various atoms present. The X-rays are generated when an impinging electron knocks another electron out of its characteristic orbit or shell around the nucleus of an atom. Because electrons tend to seek the lowest possible energy level, a vacancy in a lower-energy, inner shell will be filled by another electron from a higher-energy, outer shell. The difference in energy between the electron from the outer shell and that from the inner shell will then be emitted as an X-ray photon. In most electron microscopes today, this X-ray energy can be measured by an energy-dispersive spectrometer (EDS) and appears as peaks of intensity on a display screen. Because these bits of X-ray energy are characteristic of the atoms in a specimen, they provide information on what chemical elements are present in the specimen.

The primary limiting factors in the X-ray microanalysis of biological materials are in the area of specimen preparation. To eliminate spurious or overlapping peaks that might obscure the presence of naturally occurring elements in the sample, it is important, if possible, to avoid the use of fixatives, buffers, stains, adhesive compounds, heavy-metal coatings, or even metal grids or stubs, because all of these are capable of introducing artifacts into the analysis. Because the usual EDS system cannot identify the elements of low atomic number in a sample (i.e., those below sodium or atomic number less than 11), including carbon, hydrogen, oxygen, beryllium, etc., carbon is a good choice of material for mounting stubs and for coating a specimen to make it conductive in the electron microscope. Cnidae are extremely well suited for X-ray microanalysis because of their relative indestructibility. One can readily analyze undischarged, partially discharged, or completely discharged cnidae with the introduction of only carbon into the preparation procedure (Chandler, 1977; Hayat, 1980; Mariscal, 1980; Postek *et al.*, 1980; Goldstein *et al.*, 1981).

## II. X-RAY MICROANALYSIS AMONG THE CNIDARIA

The first X-ray microanalytical study of nematocysts was that of Mariscal (1980) on the discharged stenoteles of the gymnoblastic hydroid *Pennaria tiarella*. This study dealt only with the completely discharged stenoteles of *Pennaria* and not the intracapsular fluid, which had already left the capsule during discharge. It was directed at understanding the elemental composition of the discharged capsule, thread, and spines. The dominant chemical element found in the capsule wall and operculum was sulfur, thus tending to confirm prior studies by the author and others that suggested the nematocyst capsule was composed of a collagen-like protein cross-linked by disulfide bonds (e.g., Blanquet and Lenhoff, 1966; Mariscal and Lenhoff, 1969; Watson and Mariscal, 1984b).

Four samples of cnidae have been collected by the author from each species investigated. The first sample was allowed to air dry only to serve as a control. The other three samples were immersed in triple-distilled water for one, five, and fifteen minutes respectively to eliminate artifactual peaks of sodium, potassium, and especially chlorine introduced by sea-water contamination. Because the samples immersed in distilled water for one and five minutes still showed some sea-water contamination, whereas the samples immersed for 15 minutes did not, the latter have been used primarily for the comparison of elements among the various cnidae examined. It should be pointed out that comparisons with the non-immersed controls showed no evidence that the 15-minute immersion in distilled water affected the presence or concentration of the various elements within the undischarged capsule, nor did it appear to initiate the discharge of the unfired nematocysts in the sample. This finding is of interest because many workers have assumed that distilled water would routinely discharge nematocysts, based on the osmotic hypothesis of discharge. However, Lubbock and Amos (1981) have reported that the isolated, undischarged anthozoan nematocyst capsules in their preparations are "continuously permeable to water." Furthermore, some types of undischarged nematocysts can be routinely isolated for later physiological studies with distilled water as the isolating medium, as has been shown by Salleo *et al.* (1983) for the scyphozoan *Pelagia*. Hidaka and Mariscal (1985) have also found that undischarged acontial nematocysts of the sea anemone *Calliactis tricolor* can be isolated in distilled water, yet these same nematocysts still retain their capability for discharge when exposed later to various stimuli.

Lubbock et al. (1981), using X-ray microanalysis, found high concentrations of calcium within the undischarged capsules of Rhodactis and Anthopleura nematocysts, as well as in the fluid surrounding discharging nematocysts. These observations suggested that the loss of calcium from the intracapsular fluid somehow causes an increase in the intracapsular osmotic pressure to initiate discharge.

The next X-ray microanalytical study of cnidae was that of Mariscal (1984), who examined the feeding-tentacle cnidae of the sea anemone Haliplanella luciae. This study showed that some of the undischarged and partially discharged feeding-tentacle nematocysts (basitrichs and microbasic p-mastigophores) contained large amounts of sulfur and calcium (Figures 1, 2). Calcium was also detected within the lumen of the everting thread (Figures 3, 4), as well as in the partially discharged capsules of both basitrichs and microbasic p-mastigophores.

Not all of the feeding-tentacle cnidae of H. luciae contained calcium, however. For example, the undischarged holotrichous isorhizas contained relatively large amounts of phosphorus and sulfur (Figures 5, 6), whereas the undischarged spirocysts contained primarily sulfur and some aluminum (Mariscal, 1984).

The undischarged holotrichous isorhizas from the catch tentacles of H. luciae, however, contain calcium (Watson and Mariscal, 1984a). Thus, the same type of nematocyst (holotrichous isorhiza) in the same species of sea anemone may contain calcium in one developmental mode (i.e., the catch tentacle) and phosphorus in another mode (i.e., the feeding tentacle) (Watson and Mariscal, 1983a, b, 1984a). Also in H. luciae, Watson and Mariscal (1984a, 1985), using the potassium pyroantimonate calcium localization method and EGTA destaining along with transmission electron microscopy (TEM), have reported the involvement of calcium both in tubule inversion during development and in thread eversion during the discharge process.

Gupta and Hall (1984) have revised some of their previous data (Lubbock et al., 1981) and report detectable levels of zinc and copper in the uneverted and everted tubule of Rhodactis rhodostoma holotrichous isorhizas.

I have also examined undischarged holotrichous isorhizas and heterotrichous microbasic euryteles from two different scyphomedusae, Cyanea capillata and Stomolophus meleagris (unpublished). Interestingly enough, these nematocysts have a combination of elements different from those

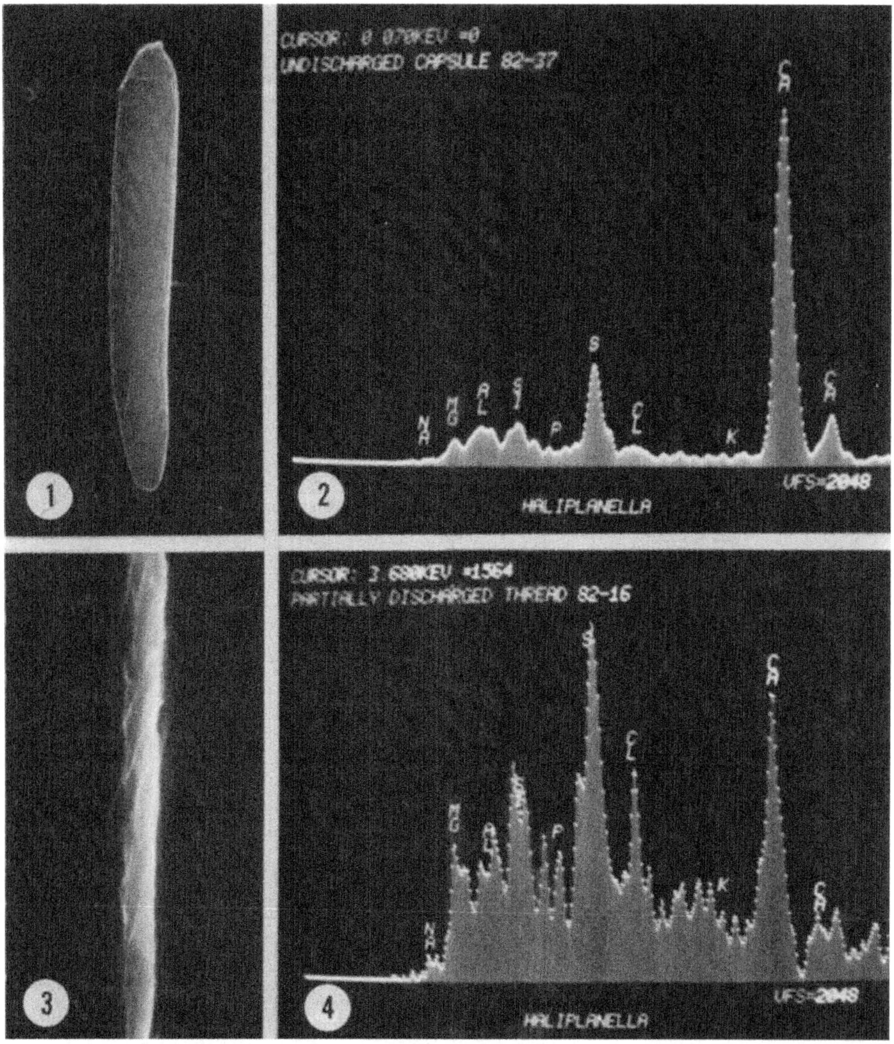


Fig. 1. SEM of an undischarged microbasic p-mastigophore capsule (length = 16.4 μm) from the feeding tentacles of the sea anemone Haliplanella luciae.

Fig. 2. X-ray spectrum showing the chemical elements present in the above nematocyst. Note the sulfur and especially the high calcium peaks.

Fig. 3. SEM of the partially everted tubule (width = .52 μm) of a microbasic p-mastigophore from the feeding tentacles of the sea anemone H. luciae.

Fig. 4. X-ray spectrum showing the chemical elements present in the above tubule. Note the high calcium level.

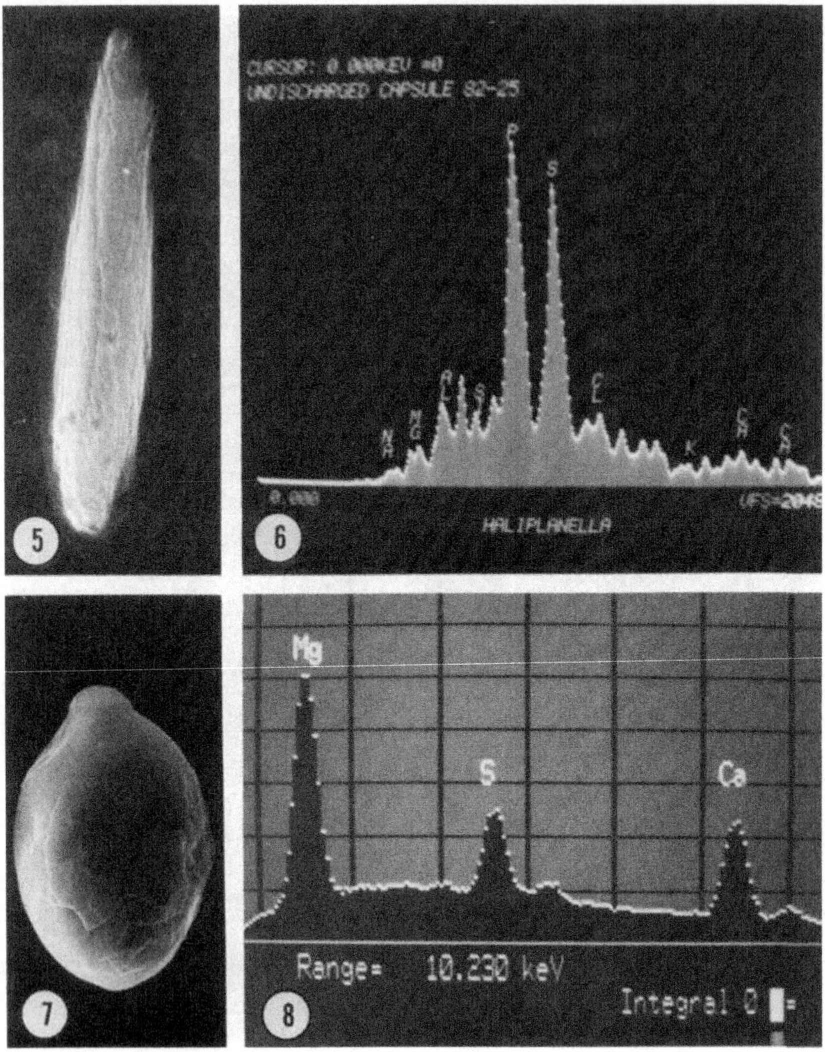


Fig. 5. SEM of undischarged holotrichous isorhiza capsule (length = 11.9 μm) from the feeding tentacles of the sea anemone <u>H. luciae</u>.

Fig. 6. X-ray spectrum showing the chemical elements in the above nematocyst. Note the high phosphorus peak.

Fig. 7. SEM of undischarged heterotrichous microbasic eurytele (length = 9.4 μm) from the scyphomedusan <u>Stomolophus meleagris</u>.

Fig. 8. X-ray spectrum showing the magnesium and calcium present in the above nematocyst.

of the anthozoans investigated to date. The scyphozoan holotrichs and euryteles contain calcium and magnesium (Figures 7, 8) as the two dominant elements (in addition to sulfur).

## III. THE POSSIBLE ROLE OF CHEMICAL ELEMENTS IN CNIDAE

X-ray microanalysis has revealed several different categories of dominant elements among the cnidae examined to date. In addition to the presence of sulfur, which shows up in all the undischarged and discharged cnidae examined, the cnidae investigated to date contain predominantly either (1) calcium, (2) calcium and magnesium, (3) phosphorus, or (4) no additional elements. A question of great interest remains as to whether or not these elements play a role in the physiology of discharge or are associated with the venom or both.

Lubbock and Amos (1981), Lubbock et al. (1981), and Gupta and Hall (1984) have developed a general hypothesis of nematocyst discharge suggesting that a rapid loss of calcium at the beginning of discharge from the capsular contents into a region surrounding the capsule somehow increases the osmotic pressure within the capsule, thereby initiating discharge. Although their hypothesis may apply to those cnidae containing calcium, X-ray microanalytical observations of additional cnidae by the author suggest that calcium and the other elements may play an additional role in the function of nematocysts. These observations include the following: (1) Partially discharged nematocyst capsules still retain relatively large amounts of calcium that theoretically should have been rapidly lost from the capsule at the beginning of discharge according to the hypothesis of Lubbock et al. (1981). (2) The everting threads of nematocysts contain relatively large amounts of calcium, suggesting that the calcium is passing up the everting thread along with the toxin, while at the same time the intracapsular calcium content appears to be proportionately reduced. (3) Large amounts of calcium have not been found surrounding the capsules of other discharging nematocysts (Watson and Mariscal, 1985). (4) Not all undischarged cnidae contain calcium.

In order to provide a possible explanation for the occurrence of calcium and the other elements found in cnidae, we have put forward a "venom hypothesis" suggesting that many of the chemical elements in nematocysts may in some way be associated with the venom and that these elements probably pass into the target tissue during envenomation. There are many studies in the literature, especially those dealing with snake and cnidarian venoms, that have reported a key role for certain chemical elements, especially calcium and zinc, in the maintenance of the toxic properties of various venom components.

## IV. THE OCCURRENCE OF CHEMICAL ELEMENTS IN VENOMS AND TOXINS

Although the snake venom literature provides a number of clear examples of the association of chemical elements with venom components, the problems associated with obtaining large amounts of pure nematocyst venom have made it difficult to do similar studies with cnidarians. Thus, much of what we know about cnidarian venoms (or perhaps, more correctly, cnidarian toxins) comes from studies in which the whole animal or its tentacles are homogenized and then analyzed for toxic properties. Although it is often assumed, probably correctly in many cases, that nematocyst venoms are responsible for the observed toxic effects, it is very difficult to prove that they are. The reason is that at least some cnidarians are known to possess an inherent tissue toxicity in addition to whatever toxic effects are provided by nematocyst venoms (e.g., see Mariscal, 1974; Blanquet, 1977; Burnett and Calton, 1977, 1987a; Beress, 1982a, b; and Russell, 1984).

### A. Calcium

As is now well known, calcium ions are essential in regulating a host of cellular processes and membrane phenomena (e.g., Scarpa and Carafoli, 1978; Berne and Hoffman, 1983). In the case of snake venoms, phospholipase A activity is dependent on the presence of calcium as a catalyst, and all the proteases from rattlesnake and pit viper venoms have been found to contain calcium as "an integrating component of the molecule" (Tu, 1977). Among *Agkistrodon* (copperheads, moccasins, etc.) venoms, workers from several different laboratories using atomic absorption spectrophotometry have found that two gram-atoms of calcium and one g-atom of zinc are present per mole of the enzymes, these being the only metals present in large amounts (e.g. Prescott and Wagner, 1976; Tu, 1977; Zhang et al., 1984). An Asiatic pit viper has one mole of calcium and one mole of zinc per mole of protein (Tu, 1977), whereas a Japanese pit viper has two g-atoms of calcium, but lacks zinc in its venom (Iwanaga et al., 1976; Tu, 1977). Prescott and Wagner (1976) find that the zinc is the catalytically important co-factor of the protease in their samples and that the calcium probably acts to stabilize the enzyme. In those venoms lacking zinc, removal of the calcium resulted in conformational changes of the enzyme and irreversible loss of the proteolytic activity, suggesting again that calcium was required to stabilize the toxic proteins. Similarly, Zhang et al. (1984) find that calcium is tightly bound to the toxic protein and that this element is essential for the hemorrhagic and proteolytic activities of the venom.

Interestingly, among non-venom systems, calcium is known to be required for the action of certain proteases (e.g., de Martino, 1981). Calcium is also known to be required for the entry into cells of a variety of toxic proteins (e.g., Sandvig and Olsnes, 1982), this being an area of relevance to the possible role of calcium in venoms.

Turning now to cnidarian systems, there is a great deal of evidence demonstrating the role of calcium in the action of nematocyst venoms and cnidarian toxins in general. For example, Hessinger and Lenhoff (1973) have demonstrated that calcium is required for the hemolytic activity of nematocyst venom phospholipase A. Other studies such as those of Lal et al. (1981a, b) and Calton and Burnett (1982, 1983) have shown that both calcium and sodium are involved to varying degrees in stimulating the enzymatic activity of *Chrysaora* and *Physalia* venoms. Endean and Henderson (1969, 1974) have shown that calcium ions in the external medium are necessary for the toxic effects exhibited by *Chironex* venom on barnacle muscles. Furthermore, these same workers found that *Chironex* toxin extracts caused a marked change in the mean calcium content of barnacle and mammalian muscle fibers. They suggest that these toxin-induced changes in the calcium content of muscle fibers are responsible for the paralytic properties commonly associated with *Chironex* venom. They further suggest that the toxin-induced interference with calcium ion transport through cell membranes may also be responsible for the dermonecrotic effects associated wtih such stingings.

It is now well known that calcium pumps, calcium channels, and calcium-activated potassium channels are associated with a variety of cell membranes (e.g., see Scarpa and Carafoli, 1978; Berne and Hoffman, 1983). In addition, many toxicological studies have provided insights into the functioning of calcium and other ionic channels in a variety of cell and membrane preparations (e.g., Berne and Hoffman, 1983; Hamilton *et al*., 1985; Schweitz *et al*., 1985). Considering the central role that calcium plays in many membrane phenomena and the fact that many venoms contain large amounts of calcium, it is possible that at least some of the deleterious effects of venoms on membranes may be due simply to the "flooding" of the target tissue with calcium ions. Thus, calcium and the various enzymes or toxins might act synergistically to produce the overall toxic effects observed. Such an idea is supported by the findings of Endean and Henderson (1969, 1974), who reported that calcium ions are necessary for *Chironex* venom to exert its typical toxic effect on muscles.

There has been a great deal of interest in recent years in the drugs that block calcium channels (i.e., verapamil, nifedipine, and diltiazem) for the treatment of hypertension and coronary artery disease as a result of their vasodilator and antiarrhythmic properties (Schwartz and Triggle, 1984). It is of interest here that calcium antagonists, such as verapamil, have therapeutic value in alleviating some of the effects of jellyfish stingings. For example, Burnett and his colleagues (Burnett and Calton, 1983, 1987a; Burnett et al., 1985) find that verapamil can delay death by reversing or inhibiting to varying degrees the cardiotoxic effects of *Chironex*, *Chrysaora*, and *Physalia* nematocyst venoms in test animals. In these experiments, verapamil did not prolong survival indefinitely, however, possibly because of other venom-induced complications. However, verapamil injections are now being recommended in the treatment of serious human stingings (Burnett and Calton, 1987a, b).

The fact that calcium ions in the vicinity have been found to be necessary for *Chironex* venom extracts to exert their toxic action on muscles (e.g., Endean and Henderson, 1969, 1974) and the fact that verapamil has been found to have therapeutic value in counteracting some of the cardiotoxic effects of *Chironex*, *Chrysaora*, and *Physalia* venoms (e.g., Burnett and Calton, 1983; Burnett *et al*., 1985) have further emphasized the possibly important role of calcium in at least some coelenterate envenomations. It should be pointed out, however, that in some other studies the cardiotoxic effect of *Chrysaora* venom was thought to be associated with low calcium (e.g., see Burnett and Calton, 1977). It should also be noted here that the exact chemical nature of the association of calcium with nematocysts or their venom remains unknown at present.

## B. Zinc and Copper

It is of interest that Gupta and Hall (1984) have reported zinc and copper to be present along with calcium in certain sea anemone nematocysts, in view of the fact that copper and especially calcium and zinc have frequently been reported to be associated with various snake venoms (e.g., Jiménez-Porras, 1970; Prescott and Wagner, 1976; Tu, 1977; Zhang *et al.*, 1984). Perhaps the most widespread occurrence of zinc in venoms is in the hemorrhagic toxins of various snakes. Nine metalloenzymes from five species of snakes have been reported to contain zinc at about unimolar concentrations (Nikai *et al.*, 1982). The zinc in these snake venoms was found to be associated with both the hemorrhagic and the proteolytic activities of the enzymes present. Lal *et al.* (1981a) believe that *Chrysaora* collagenase contains zinc, and we have occasionally found zinc in undischarged microbasic p-mastigophore nematocysts from *H. luciae*.

## C. Sulfur

Mariscal (1980, 1984) has reviewed the occurrence of sulfur in discharged and undischarged nematocysts as detected by X-ray microanalysis and has suggested that much of it probably occurs in the form of disulfide linkages in the nematocyst capsules. The reason is that a high sulfur peak remains even after the capsule has completely discharged. However, disulfide bridges are also known to occur in both snake and cnidarian toxins. For example, Jiménez-Porras (1970), Iwanaga *et al.* (1976), and Tu (1977) report the presence of disulfide bridges (usually four to five) in a variety of terrestrial snake venom enzymes and believe that they are essential for their enzymatic, cardiotoxic, and neurotoxic properties. Tu (1974) has also found four disulfide bridges in sea snake venoms, even though these venoms differ chemically in some respects from those of terrestrial snakes. Markland (1976) finds seven disulfide bonds present in crotalase from the eastern diamondback rattlesnake. Cleavage of these bonds resulted in an almost complete loss of esterase activity. Disulfide bridges also occur in *Anemonia sulcata* toxins (Beress, 1982a, b), and Lal *et al.* (1981a) find evidence of disulfide bridges in *Chrysaora* nematocyst venom collagenase.

### D. Phosphorus

Little is known about the possible role of phosphorus in nematocysts or their venoms, but phosphorus (especially in the form of various phosphates and phosphate esters) is ubiquitous among living things. At least one nematocyst, the H. luciae feeding tentacle holotrich, has been found to contain phosphorus as the dominant element (Figs. 3, 4). Three suggestions are presented here to explain the presence of phosphorus: (1) Certain nematocysts contain ATP that is injected into cells along with toxins, (2) phosphorus may be injected, probably in the form of a phosphate, to function in a biochemical degradation of some sort, and (3) phosphorus may be present in the form of toxic organophosphorus compounds.

The presence of ATP in nematocysts could conceivably affect calcium channels or other membrane ionic transport in target tissues (see Wei and Salkoff, 1986). Another possibility is that the presence of ATP or other related phosphorus-containing compounds in nematocysts could stimulate the activity and lethality of nematocyst venoms. In fact, certain spider venoms have been reported to contain not only ATP, but also ADP and AMP, with the ATP being present as the major nucleotide component of the venoms. The ATP has been reported to act synergistically to enhance the toxicity of these spider venoms (Tu, 1977).

In a recent review, Westheimer (1987) reports on the central role of phosphorus (primarily in the form of phosphate esters and anhydrides) in living things. As is well known, phosphorus occurs in both RNA and DNA, and it is also ubiquitous in the various energy sources of living things (e.g., ATP, creatine phosphate, and phosphoenolpyruvate). However, phosphorus, in the form of various phosphates, is also an essential intermediary metabolite in many biochemical syntheses and decompositions. Thus, it would not be surprising to find that nematocyst phosphorus (probably in the form of a phosphate) might be involved in a toxic degradation of some sort.

Although organophosphorus compounds have not yet been reported in nematocysts, their discovery would also be of great interest. These compounds are widely used as pesticides and are thought to be cholinesterase inhibitors in insects and mammals, as well as inhibitors of other enzyme systems (e.g., Fukuto and Metcalf, 1956; Licastro et al., 1986). The inhibition of these enzyme systems is thought to be due to an irreversible phosphorylation of the enzyme at an active site. However, organophosphorus compounds such as cyolane can also affect vertebrate and invertebrate nervous systems by increasing the level of the neurotransmitter gamma aminobutyric acid (GABA), thus affecting membrane conductance and causing an inhibitory effect on the neurones of the central nervous system (e.g. Curtis and Watkins, 1965; Hussein *et al*., 1986). In this respect, it is of interest that the venom of the South American bird spider has been reported to contain GABA, and it has been suggested that the role of this neurotransmitter may be to enhance the neurotoxic effect of the venom on the spider's prey (Tu, 1977).

It may be relevant to point out here that high levels of aspartic acid and especially glutamic acid, a precursor of GABA, are known to occur in nematocyst venoms from several cnidarian species (Lane, 1961, 1974; Blanquet, 1977; Phelan and Blanquet, 1985). Although glutamic acid is one of the most abundant amino acids in sea snake venoms, aspartic acid is present in the highest concentrations in terrestrial snake venoms (Tu, 1974, 1977).

## E. Magnesium

All of the undischarged scyphozoan holotrichs and euryteles from both *Cyanea* and *Stomolophus* examined to date in my laboratory have shown a very consistent presence of calcium and magnesium, these being the two dominant elements (Figures 7, 8). Magnesium has also been detected in small amounts in the capsular contents of undischarged and partially discharged microbasic p-mastigophores from *H. luciae* feeding tentacles. Lane (1974) has reported relatively large amounts of magnesium and calcium in *Physalia* toxin as well, with the magnesium being present at about ten times the level of calcium. The possible function of magnesium in either scyphozoan or anthozoan nematocysts remains unknown. Tu (1977), however, reports magnesium as being an important co-factor in some snake venoms.

## F. Aluminum

We have detected small amounts of aluminum in undischarged *H. luciae* spirocysts. As far as is known, aluminum has not heretofore been reported in association with either snake or cnidarian toxins; however, trace amounts of aluminum have been reported in Gila monster venoms (Tu, 1977).

## V. CONCLUSIONS

Regardless of the source, whether it be a marine invertebrate at the bottom of the phylogenetic ladder or a terrestrial vertebrate, many venoms have remarkably similar accumulations of enzymes, toxic proteins, and toxic properties. A review of the voluminous snake venom literature is almost like reading a "preview of coming attractions" in the cnidarian venom literature. The reason is that a good many of the recent discoveries concerning the toxic properties of cnidarian venoms have been preceded by similar findings about snake venoms, in some cases by many years. As but one example, collagenase was first detected in snake venoms in 1949 (Tu, 1977), but it was not reported in cnidarian venoms until 1980 (Neeman et al., 1980).

An analysis of the possible roles of the various chemical elements observed to date in association with nematocysts is made difficult by the fact that only a relatively few elements have even been described in association with cnidarian venoms. Although the situation is clearer in the case of snake venoms, even here the function of these elements is not always well understood. However, calcium and zinc are present in relatively large amounts in nearly all snake venoms, and their removal generally results in the loss of toxic activity (Tu, 1977).

In summary, the bulk of the evidence from the published literature on both invertebrate and vertebrate toxicology, as well as from the X-ray microanalytical studies presented here, suggests that the various inorganic chemical elements found in undischarged nematocysts are probably associated with the venom or its action in some way. This idea is supported by the fact that calcium occurs neither in the most common anthozoan cnida, the non-toxic spirocyst, nor in all nematocysts. Furthermore, the fact that calcium and other elements are known to be required for the toxic activity of both cnidarian and vertebrate venoms indicates that, in addition to whatever role these elements might play in the physiology of nematocyst discharge, they are most certainly involved also in the expression of at least some of the toxic properties of cnidarian venoms. This view may be further supported by the studies showing that verapamil, a calcium antagonist, can be used as a therapeutic agent in the treatment of some non-lethal nematocyst envenomations.

## VI. ACKNOWLEDGMENTS

Thanks are due Bill Miller, Sandra Silvers, and Kim Riddle for expert technical assistance, as well as the Psychobiology Program at Florida State University for the use of electron microscope facilities. Paul Greenwood read the manuscript and offered many helpful suggestions.

## REFERENCES

Beress, L. 1982a. Sea anemone toxins: a minireview. In W. Voelter, E. Wunsch, Y. Ovchinnikov, and V. Ivanov (eds.), Chemistry of peptides and proteins, pp. 121-126. W. De Gruyter, Berlin.

Beress, L. 1982b. Biologically active compounds from coelenterates. Pure Appl. Chem. 54:1981-1994.

Berne, R. M., and J. F. Hoffman. 1983. Cell and membrane physiology. Annu. Rev. Physiol. 45:301-390.

Blanquet, R. 1977. Cnidarian venoms. In A. W. Bernheimer (ed.), Perspectives in toxinology, pp. 149-167. John Wiley & Sons, New York.

Blanquet, R., and H. M. Lenhoff. 1966. A disulfide-linked collagenous protein of nematocyst capsules. Science 154:152-153.

Burnett, J. W., and G. J. Calton. 1977. The chemistry and toxicology of some venomous pelagic coelenterates. Toxicon 15:177-196.

Burnett, J. W., and G. J. Calton. 1983. Response of the box-jellyfish (Chironex fleckeri) cardiotoxin to intravenous administration of verapamil. Med. J. Aust. 2:192-194.

Burnett, J. W., and G. J. Calton. 1987a. Venomous pelagic coelenterates: chemistry, toxicology, immunology and treatment of their stings. Toxicon 25:581-602.

Burnett, J. W., and G. J. Calton. 1987b. Jellyfish envenomation syndromes updated. Ann. Emerg. Med. 16:1000-1005.

Burnett, J. W., C. J. Gean, G. J. Calton, and J. E. Warnick. 1985. The effect of verapamil on the cardiotoxic activity of Portuguese man-o'-war (Physalia physalis) and sea nettle (Chrysaora quinquecirrha) venoms. Toxicon 23:681-689.

Calton, G. J., and J. W. Burnett. 1982. Partial purification and characterization of the acid protease of sea nettle (Chrysaora quinquecirrha) nematocyst venom. Comp. Biochem. Physiol. 72B:93-97.

Calton, G. J., and J. W. Burnett. 1983. Partial purification and characterization of the alkaline protease of sea nettle (Chrysaora quinquecirrha) nematocyst venom. Comp. Biochem. Physiol. 74C:361-364.

Chandler, J. A. 1977. X-ray microanalysis in the electron microscope. North-Holland, Amsterdam.

Curtis, D. R., and J. C. Watkins. 1965. The pharmacology of amino acids related to gamma-aminobutyric acid. Pharmac. Rev. 17:347-391.

Endean, R., and L. Henderson. 1969. Further studies of toxic material from nematocysts of the cubomedusan Chironex fleckeri Southcott. Toxicon 7:303-314.

Endean, R., and L. Henderson. 1974. Some aspects of the biological activity of crude nematocyst toxin in Chironex fleckeri. In H. J. Humm and C. E. Lane (eds.), Bioactive compounds from the sea, pp. 1-13. Marcel Dekker, Inc., New York.

Fukuto, T. R., and R. L. Metcalf. 1956. Structure and insecticidal activity of some diethyl substituted phenyl phosphates. J. Agr. Food Chem. 4:930-935.

Goldstein, J. I., D. E. Newbury, P. Echlin, D. C. Joy, C. Fiori, and E. Lifshin. 1981. Scanning electron microscopy and X-ray microanalysis. Plenum Pres, New York.

Gupta, B. L., and T. A. Hall. 1984. Role of high concentrations of Ca, Cu, and Zn in the maturation and discharge in situ of sea anemone nematocysts as shown by X-ray microanalysis of cryosections. In L. Bolis, J. Zadunaisky, and R. Gilles (eds.), Toxins, drugs, and pollutants in marine animals, pp. 77-95. Springer-Verlag, Berlin.

Hamilton, S. L., A. Yatani, M. J. Hawkes, K. Redding, and A. M. Brown. 1985. Atrotoxin: a specific agonist for calcium currents in heart. Science 229:182-184.

Hayat, M. A. 1980. X-ray microanalysis in biology. University Park Press, Baltimore.

Hessinger, D. A., and H. M. Lenhoff. 1973. Assay and properties of the hemolysis activity of pure venom from the nematocysts of the acontia of the sea anemone Aiptasia pallida. Arch. Biochem. Biophys. 159:629-638.

Hidaka, M., and R. N. Mariscal. 1985. Effects of ions on nematocysts isolated from acontia of a sea anemone by different methods. Amer. Zool. 25:111A.

Hussein, M. F., N. A. Ahmed, and S. M. Rawi. 1986. The alterations of rat brain GABA and glutamine by the organophosphorus compound cyolane. Comp. Biochem. Physiol. 84C:165-170.

Iwanaga, S., G. Oshima, and T. Suzuki. 1976. Proteinases from the venom of Agkistrodon halys blomhoffi. In L. Lorand (ed.), Methods in enzymology, Vol. 45, Part B, pp. 459-468. Academic Press, New York.

Jiménez-Porras, J. M. 1970. Biochemistry of snake venoms. Clinical Toxicol. 3:389-431.

Lal, D. M., G. J. Calton, I. Neeman, and J. W. Burnett. 1981a. Characterization of Chrysaora quinquecirrha (sea nettle) nematocyst venom collagenase. Comp. Biochem. Physiol. 69B:529-533.

Lal, D. M., G. J. Calton, I. Neeman, and J. W. Burnett. 1981b. Characterization of Physalia physalis (Portugese man-o'war) nematocyst venom collagenase. Comp. Biochem. Physiol. 70B:635-638.

Lane, C. E. 1961. Physalia nematocysts and their toxin. In H. Lenhoff and W. Loomis (eds.), The biology of hydra and of some other coelenterates, pp. 169-178. Univ. of Miami Press, Coral Gables.

Lane, C. E. 1974. Nematocyst toxins of coelenterates. In H. J. Humm and C. E. Lane (eds.), Bioactive compounds from the sea, pp. 123-137. Marcel Dekker, Inc., New York.

Licastro, S. A. de, M. I. P. de Villar, E. J. Wood, F. J. Melgar, and E. N. Zerba. 1986. Toxicity and anticholinesterase activity of organophosphorus compounds in Triatoma infestans. Comp. Biochem. Physiol. 84C:131-137.

Lubbock, R., and W. B. Amos. 1981. Removal of bound calcium from nematocyst contents causes discharge. Nature 290:500-501.

Lubbock, R., B. L. Gupta, and T. A. Hall. 1981. Novel role of calcium in exocytosis: mechanism of nematocyst discharge as shown by x-ray microanalysis. Proc. Natl. Acad. Sci. USA 78:3624-3628.

Mariscal, R. N. 1974. Nematocysts. In L. Muscatine and H. M. Lenhoff (eds.), Coelenterate biology: reviews and new perspectives, pp. 129-178. Academic Press, New York.

Mariscal, R. N. 1980. The elemental composition of nematocysts as determined by x-ray microanalysis. In P. Tardent and R. Tardent (eds.), Developmental and cellular biology of coelenterates, pp. 337-342. Elsevier/North-Holland, Amsterdam.

Mariscal, R. N. 1984. Cnidaria: cnidae. In J. Bereiter-Hahn, A. G. Matoltsy, and K. S. Richards (eds.), Biology of the integument: 1 invertebrates, pp. 57-68. Springer-Verlag, Berlin.

Mariscal, R. N., and H. M. Lenhoff. 1969. Effect of a disulfide reducing agent on coelenterate nematocyst capsules. Experientia 25:330-331.

Markland, F. S. 1976. Crotalase. In L. Lorand (ed.), Methods in enzymology, Vol. 45, Part B, pp. 223-236. Academic Press, New York.

Martino, G. N. de. 1981. Calcium-dependent proteolytic activity in rat liver: identification of two proteases with different calcium requirements. Arch. Biochem. Biophys. 211:253-257.

Neeman, I, G. J. Calton, and J. W. Burnett. 1980. An ultrastructural study of the cytotoxic effect of the venoms from the sea nettle (Chrysaora quinquecirrha) and Portugese man-of-war (Physalia physalis) on cultured Chinese hamster ovary K-1 cells. Toxicon 18:495-501.

Nikai, T., H. Ishizaki, A. T. Tu, and H. Sugihara. 1982. Presence of zinc in proteolytic hemorrhagic toxin isolated from Agkistrodon acutus venom. Comp. Biochem. Physiol. 72C:103-106.

Phelan, M. A., and R. S. Blanquet. 1985. Characterization of nematocyst proteins from the sea anemones Aiptasia pallida and Pachycerianthus torreyi (Cnidaria: Anthozoa). Comp. Biochem. Physiol. 81B:661-666.

Postek, M. T., K. S. Howard, A. H. Johnson, and K. L. McMichael. 1980. Scanning electron microscopy. Ladd Research Industries, Inc., Burlington.

Prescott, J. M., and F. W. Wagner. 1976. Leucostoma peptidase A. In L. Lorand (ed.), Methods in enzymology, Vol. 45, Part B, pp. 397-404. Academic Press, New York.

Russell, F. E. 1984. Marine toxins and venomous and poisonous marine plants and animals (invertebrates). Adv. Mar. Biol. 21:59-217.

Salleo, A., G. la Spada, and M. Alfa. 1983. Blockage of trypsin-induced discharge of nematocysts of Pelagia noctiluca by $Ca^{2+}$. Molec. Physiol. 3:89-97.

Sandvig, K., and S. Olsnes. 1982. Entry of the toxic proteins abrin, modeccin, ricin and diphtheria toxin into cells. 1. Requirement for calcium. J. Biol. Chem. 257:7495-7503.

Scarpa, A., and E. Carafoli. 1978. Calcium transport and cell function. Ann. N. Y. Acad. Sci. 307:1-655.

Schwartz, A., and D. J. Triggle. 1984. Cellular action of calcium channel blocking drugs. Annu. Rev. Medicine 35:325-339.

Schweitz, H., J.-N. Bidard, C. Frelin, D. Pauron, H. P. M. Vijverberg, D. M. Mahasneh, M. Lazdunski, F. Vilbois, and A. Tsugita. 1985. Purification, sequence, and pharmacological properties of sea anemone toxins from *Radianthus paumotensis*: a new class of sea anemone toxins acting on the sodium channel. Biochemistry 24:3554-3561.

Tu, A. T. 1974. Distribution of sea snakes in southeast Asia and the Far East and chemistry of venoms of three species. *In* H. J. Humm and C. E. Lane (eds.), *Bioactive compounds from the sea*, pp. 207-230. Marcel Dekker, Inc., New York.

Tu, A. T. 1977. *Venoms: chemistry and molecular biology*. John Wiley & Sons, New York.

Watson, G. M., and R. N. Mariscal. 1983a. The development of a sea anemone tentacle specialized for aggression: morphogenesis and regression of the catch tentacle of *Haliplanella luciae* (Cnidaria, Anthozoa). Biol. Bull. 164:506-517.

Watson, G. M., and R. N. Mariscal. 1983b. Comparative ultrastructure of catch tentacles and feeding tentacles in the sea anemone *Haliplanella*. Tissue and Cell 15:939-953.

Watson, G. M., and R. N. Mariscal. 1984a. Calcium cytochemistry of nematocyst development in catch tentacles of the sea anemone *Haliplanella luciae* (Cnidaria, Anthozoa) and the molecular basis for tube inversion into the capsule. J. Ultra. Res. 86:202-214.

Watson, G. M., and R. N. Mariscal. 1984b. Ultrastructure and sulfur cytochemistry of nematocyst development in catch tentacles of the sea anemone *Haliplanella luciae* (Cnidaria, Anthozoa). J. Ultra. Res. 87:159-171.

Watson, G. M., and R. N. Mariscal. 1985. Ultrastructure of nematocyst discharge in catch tentacles of the sea anemone *Haliplanella luciae* (Cnidaria, Anthozoa). Tissue and Cell 17:199-213.

Wei, A., and L. Salkoff. 1986. Occult *Drosophila* calcium channels and twinning of calcium and voltage-activated potassium channels. Science 233:780-782.

Westheimer, F. H. 1987. Why nature chose phosphates. Science 235:1173-1178.

Zhang, J., Z. Chen, Y. He, and X. Xu. 1984. Effect of calcium on proteolytic activity and conformation of hemorrhagic toxin I from five pace snake (*Agkistrodon acutus*) venom. Toxicon 22:931-935.

7

# THE NEMATOCYTE: AN ENCAPSULATION OF DEVELOPMENTAL PROCESSES

Richard D. Campbell

Department of Developmental and Cell Biology, and
Developmental Biology Center,
University of California, Irvine, CA. 92717 USA

## SUMMARY

The formation of nematocytes encapsulates nearly all of the great questions facing developmental biologists today. Questions concerning the development of nematocytes can be explored from the molecular to the organismic levels. The history of research in this subject has been, and continues to be, dotted with brilliant observations and embroiled by controversy. This session of the symposium will deal with some of the more interesting and well-formulated areas of current research.

## I. INTRODUCTION

Questions on development were among the first to be asked by the investigators who studied nematocytes during the nineteenth century. The major steps in the development of nematocytes were established to be determination, differentiation, migration and deployment (Fig. 1). Current investigations are still pursuing these subjects. The following group of papers summarizes our current understanding of the development of nematocysts.

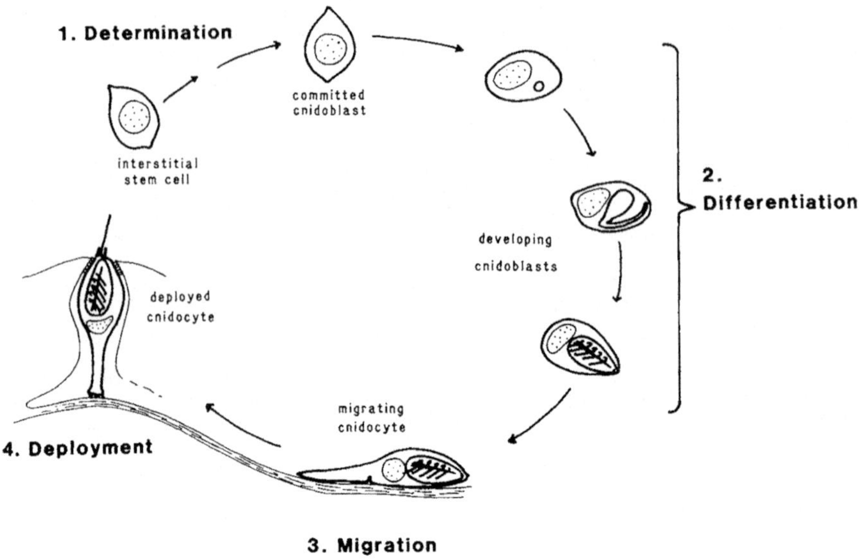


Fig. 1. Four major stages in the development of nematocytes.

## II. MAJOR EVENTS IN THE DEVELOPMENT OF NEMATOCYTES.

### A. Determination of Stem Cells

Nematocytes are products of a lineage of cells called interstitial cells, one of the classic examples of a line of multipotent stem cells. Until recently, interstitial cells came to be viewed as a single, homogeneous population of cells that serve as precursors for many of the specialized somatic and germ cells (Brien and Reniers-Decoen, 1955; Bode and David, 1978). They provide several kinds of differentiated cells during embryogenesis, during the entire normal life of the animal, and perhaps during wound repair and regeneration.

Challenges have recently eroded the concept that interstitial cells constitute a homogeneous population of cells. Littlefield (1985) presents evidence that some interstitial cells are restricted in the determination for germ line differentiation and that there are even stem cell lines for male and female gametes. Bode (1987) argues that interstitial cells comprise one or several complex lineages with numerous intermediate proliferative stages. In growth and regeneration, whereas the interstitial cells were once considered as the source of all new tissue it is now thought that interstitial cells have only minor roles other than supplying the various specialized cells such as nematocytes and nerve cells (see Kanajew, 1930; Diehl and Bouillon, 1966; Tardent, 1978; Campbell, 1979).

The most important question regarding the determination of nematocyte precursors concerns its genetic and epigenetic causes. The use of monoclonal antibodies, as Bode discusses in this volume (Bode, 1987), and the application of genetic methodology to hydra (Fujisawa and Sugiyama, 1986), are rapidly bringing the molecular biological aspects of determination into reach.

B. Nematocyst assembly and differentiation

Formation of the the nematocyst provides one of the most interesting and challenging problems in animal morphogenesis. The nematocyst is constructed largely of crystalline components. Barbs (Skaer and Picken, 1965), capsule wall (Chapman, 1961), and tubule (Cormier and Hessinger, 1981; Hausmann and Holstein, 1985) all show periodic substructure. Nematocysts organize out of materials previously sequestered into a vesicle that is isolated from the cytoplasm. Thus the morphogenesis of nematocysts has the earmarks of a process of self-assembly. Yet two aspects of nematocysts make them stand out from other structures that arise by self-assembly. First, assembly and shaping of nematocysts clearly involves mechanical distortions. For example, the early stages of capsules are round or oval but later stages are more asymmetrical. The specific pattern of coiling, shape of the spines, and shape of the capsule will probably find their origins in forces generated by the growing spines acting on the pliable capsule and tubule. Second, nematocysts are variable in form and size: slight variation from one capsule to the next; greater variation between similar types of capsules in different species; and still greater variation between the different classes of nematocysts (Mariscal, 1987). These experiments of nature will doubtless be important in our unraveling developmental mechanisms. Thus nematocysts display richer patterns of morphogenesis than do other systems of self-assembly, and the experimental material available makes investigation of them more attractive.

The aspect of nematocyst differentiation that has always attracted the most attention concerns whether or not the tubule develops outside or inside the capsule. This question has been argued since the developmental stages of nematocyst were just discovered in the 19th century (see Weill, 1934). The answer is now being sketched out. A temporary, long extension of the capsule does form externally, as seen by some early investigators. But the external tubule does not invert to become the internal tubule as these investigators had maintained, but withdraws in some intimate association with the formation of the internal tubule (Skaer, 1973; Carré and Carré, 1973; Carré, 1974; Westfall, 1966; Schmidt, 1981; Holstein, 1981). Dr. Watson reveals in his paper in this volume (Watson, 1987) that we are now in a position to carry the problem forward at a histochemical, and perhaps chemical, level.

C. Migration of Nematocytes

Cell locomotion receives so much attention in biology and biomedicine that it is surprising how little is known about the migration of nematocytes. These cells would seem to present a perfect set of characters for studying the problem. More so than many other cells they have well defined sites of origin and deployment. Their paths of migrations may be very long. They migrate quickly. Many cnidarians have continuous streams of migrating cells. Finally, the cell has a nice optical marker in the form of the refractile capsule that announces both the position and polarity of the cell. The work so far done on the migration of nematocytes seems to pose enigmas. How can the cells travel so rapidly with such a simple cell form? How do they manage to so efficiently get into the tentacles when, at least in some cases, their actual movement does not seem to be polarized? Only in hydrozoans have these migrations been studied carefully, as discussed in this volume (Campbell, 1987). Robson (Robson, 1987) defines and organizes the problem relative to the Anthozoans. There is essentially no information at hand about the situation in scyphozoans.

D. Maturation and deployment of nematocysts

It is likely that the most complex parts of nematocyst development take place during the positioning, maturation, and deployment of the nematocyte at its destination. We know some of these events: final maturation of the capsule (Günzl, 1968, known to electron microscopists through the sudden change in appearance of the matrix inside the capsule); penetration of the nematocyte to the surface of the ectoderm;

formation of specialized junctions with adjacent cells (Wood and Novak, 1982); innervation of the nematocytes and surrounding battery cells (Westfall, 1973); reorganization of the nematocyte as its capsule is positioned at the surface of the tentacle; growth of the cnidocil apparatus; and physiological alterations that prepare the cell for discharging its nematocyst. These events are rapid and involve simultaneous and coordinated maneuvers by the epithelial cells that invest the nematocytes and perhaps also by neighboring nematocytes.

The processes leading to deployment extend well into the range of tissue organization. They probably vary between species more than do other aspects of nematocyst development. The deployment of nematocysts, touched upon by many papers in the present volume, has never been formulated and investigated as a single scientific problem, however. The paper in this volume by Skaer (1987) represents one of the most detailed glimpses at this process in what is surely the most complex example of deployment thus far examined in the cnidarians. The remarkable dancing of the nematocytes of siphonophores will shed light and enthusiasm on the formulation of a general understanding of the process of deployment.

## III. CONCLUSION

During the last 15 years each of these four phases of the development of nematocytes has received increasing attention by experimental biologists. Each can now be formulated to various degrees in terms of molecular, cellular, and tissue phenomena, even though in none do we have a deep level of understanding. We seem poised on the edge of exploration in each area, and the papers in this volume furnish a good base from which to proceed.

## REFERENCES


Bode, H. R. 1987. (This volume).

Bode, H. R. and C. N. David. 1978. Regulation of a multipotent stem cell, the interstitial cell of hydra. Progr. Biophys. Mol. Biol. 33:189-206.

Brien, P. and M. Reniers-Decoen. 1955. La signification des cellules interstitielles des hydres d'eau douce et le problème de la réserve embryonnaire. Bull. Biol. France Belgique 89: 258-325.

Campbell, R. D. 1979. Development of hydra lacking interstitial and nerve cells ("epithelial hydra"). Symp. Soc. Dev. Biol. 37:267-293.

Campbell, R. D. 1987. Migration of nematocytes in hydrozoans. (This volume).

Carré, C. and D. Carré. 1973. Etude du cnidome et de la cnidogénèse chez Apolemia uvaria (Lesueur, 1811)(Siphonophore Physonecte). Exp. Cell Res. 81:237-249.

Carré, D. 1974. Formation, migration et maturation des nématoblastes et des nématocystes chez les siphonophores I - Mise en evidence et formation des clones de nématocystes. Ann. Embryol. Morphogen. 7:205-218.

Chapman, G. B. 1961. The fine structure of the stenoteles of hydra. In H. M. Lenhoff and W. F. Loomis (eds.), The Biology of Hydra and of some other coelenterates:1961, pp. 131-151. Univ. of Miami Press, Coral Gables, Florida.

Diehl, F. and J. Bouillon. 1966. Observations sur les potentialités morphogénétiques des tissus irradiés de Cordylophora caspia (Pallas). Bull. Acad. Roy. Belge Classe Sci. 52:138-146.

Fujisawa, T., C. Nishimiya, and T. Sugiyama. 1986. Nematocyte differentiation in Hydra. Curr. Top. Develop. Bio. 20:281-290.

Günzl, H. 1968. Ueber die Reifung der Nesselkapseln bei Dipurena reesi Vannucci (Hydrozoa). Z. Zellforsch. 89:509-518.

Hausmann, K. and T. Holstein. 1985. Bilateral symmetry in the cnidocil-nematocyst complex of the freshwater medusa Craspedacusta sowerbii Lankester (Hydrozoa, Limnomedusae). J. Ultrastruct. Res. 90:89-104.

Holstein, T. 1981. The morphogenesis of nematocytes in Hydra and Forskålia: An ultrastructural study. J. Ultrastruct. Res. 75:276-290.

Kanajew, J. 1930. Zur Frage der Bedeutung der Interstitiellen Zellen bei Hydra. Wilhelm Roux' Archiv 122:736-759.

Littlefield, C. L. 1985. Germ cells in Hydra oligactis males I. Isolation of a subpopulation of interstitial cells that is developmentally restricted to sperm production. Develop. Biol. 112:185-193.
Mariscal, R. 1987. (This volume).
Robson, E. 1987. (This volume).
Schmidt, H. 1981. Die Cnidogenese der Octocorallia (Anthozoa, Cnidaria): I. Sekretion und Differenzierung von Kapsul und Schlauch. Helgol. Wiss. Meeresunters. 34:463-484.
Skaer, R. J. and L. E. R. Picken. 1965. The structure of the nematocyst thread and the geometry of discharge in Corynactis viridis Allman. Proc. Roy. Soc. London Ser B 250:131-164.
Skaer, R. J. 1973. The secretion and development of nematocysts in a siphonophore. J. Cell Sci. 13:371-393.
Skaer, R. J. 1987. (This volume).
Tardent, P. 1978. Coelenterate, Cnidaria. In F. Seidel (ed.), Morphogenese der Tiere. Lief I:A-I, pp. 69-415. Gustav Fischer, Stuttgart.
Watson, G. 1987. (This volume).
Weill, R. 1934. Contribution à l'étude des Cnidaires et de leurs nématocystes. Trav. Stat. Zool. Wimereux 10-11:1-700.
Westfall, J. A. 1966. The differentiation of nematocysts and associated structures in the Cnidaria. Z. Zellforsch. 75:381-403.
Westfall, J. A. 1973. Ultrastructural evidence for neuromuscular systems in Coelenterates. Amer. Zool. 13:237-246.
Wood, R. L. and P. L. Novak. 1982. The anchoring of nematocysts and nematocytes in the tentacles of hydra. J. Ultrastruct. Res. 81:104-16.

# 8

# MIGRATION OF NEMATOCYTES IN HYDROZOANS

Richard D. Campbell

Department of Developmental and Cell Biology, and
Developmental Biology Center,
University of California, Irvine, CA. 92717 USA

## SUMMARY

Nematocytes generally differentiate far from the tentacles or other sites where they are deployed. Then they migrate, translocate or somehow get moved to the tentacles. This phenomenon represents a superb situation for the cell biologist to study directed cell movements. The mechanics of nematocyte movement and the cues that guide the cells have been investigated for a century. Nematocytes almost always move as individual amoeboid cells. It is possible that a variety of exceptional pathways exist for migration in individual groups or species. There is evidence both for and against the possibility that the direction of migration is controlled by chemoattractants or other environmental cues. The subcellular basis of the amoeboid movements and factors that control the direction of nematocyte migration will be the main subjects for future research in this area.

## I. INTRODUCTION

Early microscopists recognized the disparity between sites of production and utilization of nematocytes in hydrozoans and deduced the existence of intermediate mechanisms of transport (Hadži, 1907). By the end of the 19th century, three such mechanisms had been postulated and supported by observations:

1) independent migration of mature nematocytes as amoeboid cells, usually along the base of the ectoderm (Hadži, 1907);
2) translocation through the gastric cavity (Hadži, 1907);
3) passive displacement outwards along a tentacle by growth movements of the tentacular epithelium (Perkins, 1902).

These three alternatives continue to be studied with evidence supporting and opposing each. The predominant mode of transport is by amoeboid migration of individual nematocytes. The two other modes, with additional unknown ones, may apply in particular instances.

## II. INDEPENDENT MIGRATION OF NEMATOCYTES AS AMOEBOID CELLS

### A. Pathways

The most detailed analyses of how nematocytes travel show that the cells actively move in an amoeboid fashion in between the ectodermal cells (Fig. 1). This old idea is based on superb observations by many workers using live specimens of hydroids (both polyps and medusae), siphonophores, and hydra (Murbach, 1894, 1895; Schneider, 1894, 1900; Hadži, 1907, 1911; Lipin, 1911; Gelei, 1927; others cited by Weill, 1934; Hauenschild, 1957; Skaer, 1987). More recently, these movements have been filmed (Günzl, 1971; Rahat and Campbell, 1974; Campbell and Marcum, 1980).

In colonial hydroids the nematocytes develop, at least in part, in the stolons and migrate into the hydranths. In *Polypodium* (Lipin, 1911) nematocytes form near the bases of the tentacles and migrate distally within the tentacles. In siphonophores (Schneider, 1894; Carré, 1974*a*; Skaer, 1987) nematocytes arise in a thickening of the polyp near the base of the tentacle, and migrate out along the tentacle to the complex batteries. In hydra (Campbell, 1974) nematocytes arise in the column and move to the tentacles as well as elsewhere on the body. In hydromedusae (Hadži, 1911; Weiler-Stolt, 1960) nematocytes form near the bases of the tentacles and migrate outwards along the tentacles. In all of these instances the nematocytes travel at the base of the ectoderm.

These reports on the nature of moving nematocytes describe many common features. The cells are amoeboid;

Murbach (1895) wrote that one such cell he observed "might well have been taken for an Amoeba which had swallowed a nematocyst." A single, varying pseudopod leads.

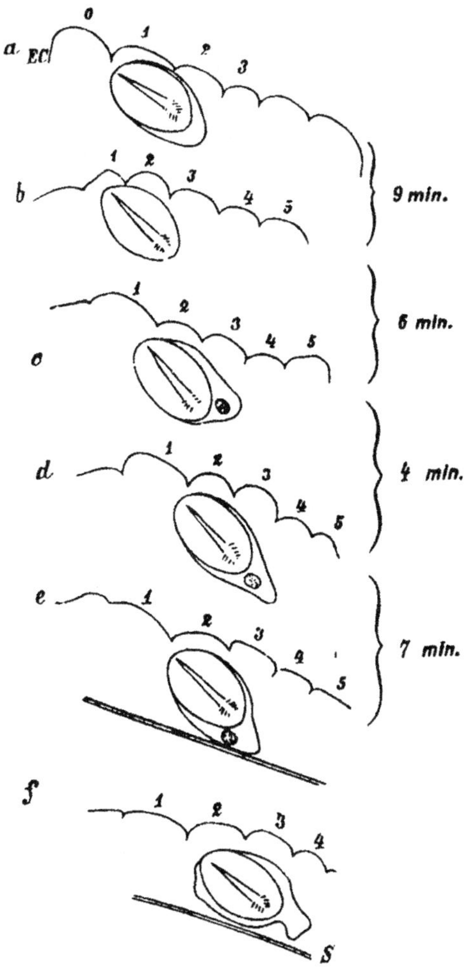


Fig. 1. Stenotele nematocyte migrating along the tentacle of Pennaria Cavolinii, as observed and drawn by Murbach (1894). Frames c-e show relative positions of the single anterior pseudopod, nucleus, and backwards-directed capsule. Frames e and f show the mesoglea.

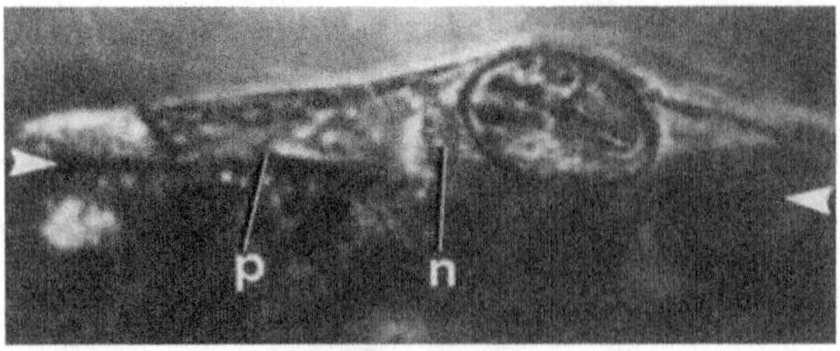


Fig. 2. Stenotele nematocyte migrating along the stolon of Dipurena reesi, from a movie made by Günzl (1971). This nematocyte is much thicker than the ectoderm, which is therefore distorted as the nematocyte moves. The mesoglea is indicated by arrowheads, with ectoderm above and endoderm below. p: tip of pseudopod. n:nucleus. The nematocyte is migrating towards the left.

The nucleus is in the middle of the cell. The capsule trails with its apical pole always pointing backwards (Figs. 1, 2). The cell is situated at the base of the ectoderm (Fig. 3) and it moves between the epithelial cells. This movement distorts the nematocyte and the surrounding epithelial cells, for sometimes nematocytes squeeze along an ectoderm that is thinner than the diameter of the nematocyst capsule (as in Fig. 2). Movement is rapid, often tens of micrometers per minute, and is oriented along the axis of the stolon, column or tentacle.

An exception to this normal situation concerns the nematocytes that are deployed along the rim of the mouth in athecate hydromedusae. Evidently these nematocytes develop in the endoderm of the manubrium and migrate along the base of the endoderm, not of the ectoderm, to their destination (Boulenger, 1910; Hadži, 1911). A similar situation has been described for migrating nematocytes of the siphonophore Vellela (Murbach, 1894).

Experimental analysis of this migration started in earnest only in the last several decades. Lenhoff (1959) first used a radiolabel and autoradiography to demonstrate the rapid migration of nematocytes to the tentacles of hydra. Burnett (1959-60a,b) and Burnett and Lentz (1959-60) verified the speed of movement by using chimeric hydra, constructed either by staining or grafting, in which nematocytes arising from the lower body region could be identified in the tentacles. Tardent and Morgenthaler (1966), Campbell, 1967b), Vögeli (1972) and Herlands and Bode (1974b) grafted lower portions of hydra whose cells had been labeled with $^{3}$H-thymidine to upper portions of unlabeled hydra. Within

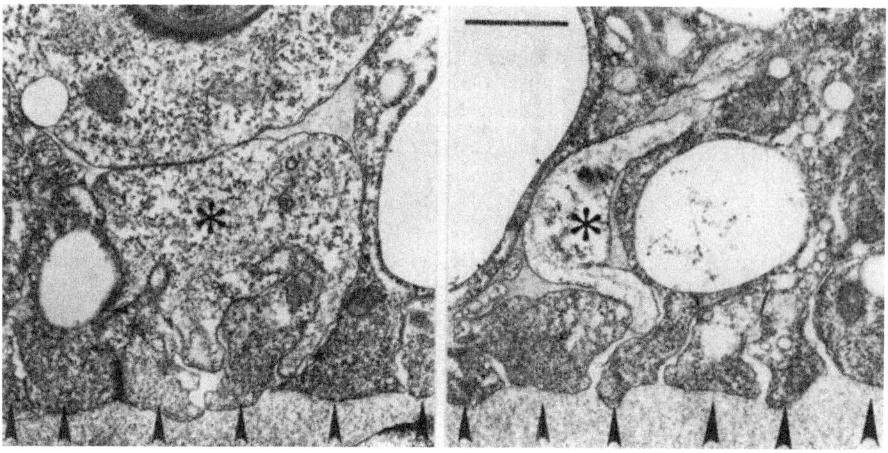

Fig. 3. Nematocytes migrating along the muscle processes of ectodermal epithelial cells in hydra, as seen in transverse electron micrographs through the body column. These sections are cut through the middle of one pseudopod (asterisk, left) and through the tip of another pseudopod (asterisk, right) of separate nematocytes. Arrowheads point to muscle processes. Mesoglea is at the bottom of the photomicrographs. The pseudopods of the nematocytes are in intimate contact with the muscle processes. Bar = 0.9 um.

hours, labeled nematocytes appeared in the previously unlabeled upper halves of the polyps, always at the base of the ectoderm. Subsequently the labeled nematocytes appeared mounted in the tentacles.

## B. Directionality

Migrating nematocytes seem to be an excellent system to study the control over directionality of cell migration because nematocytes reach specific destinations. The first question to address, and one which is not yet satisfactorily answered, is whether the actual locomotion of individual cells is directed. Prevailing statements of investigators who have watched nematocytes migrating in situ support the idea that the movements of individual cells is directed, but generally these statements are not supported by precise data. A few observers, however, are very clear about the unidirectional nature of cell migration in circumstances where it would be easy to tell if cells did move in more than one direction (Hauenschild, 1957; Skaer, 1987).

A number of workers, most of them using hydra, have attempted to study the directionality of migration by methods other than simple observation. Typically, radioactively labeled donor tissue is grafted to unlabeled tissue and after a time the distribution of labeled nematocytes is determined in histological sections or in macerated pieces of the host tissue. Campbell (1967a) and Herlands and Bode (1974b) found much larger numbers of labeled nematocytes in host tissue when the donor tissue was grafted in a basal position than when the donor tissue was implanted apically. These results have been widely interpreted as indicating that nematocytes migrate unidirectionally and apically.

On the other hand, other similar studies uncovered concomitant downward migrations. Herlands and Bode (1974a) ascribed the unexpected, basal migrations to cells moving towards developing buds. In this way a migration down the polyp could still be viewed as towards a new apex and hence fit into the scheme of unidirectional migration. But Campbell and Marcum (1980) implanted pieces of donor tissue into various positions of hydra that lacked their own

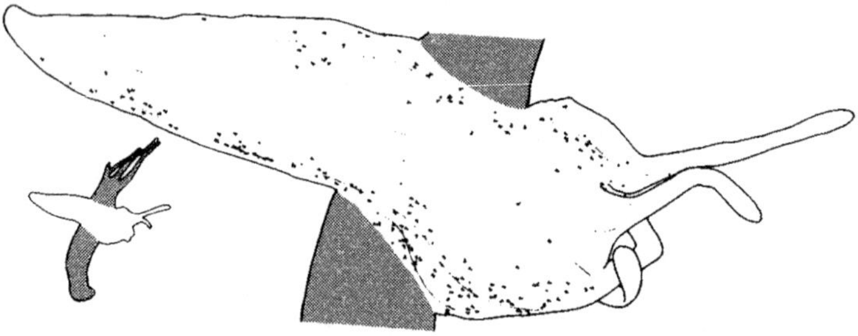

Fig. 4. Nematocytes (points) that have migrated into the sides of a nematocyte-free hydra (clear design in middle) from apical and basal halves of a normal polyp (stippled) grafted to it 3 hours earlier. Fine lines represent the general directions of the ectodermal muscle processes in the host. Many nematocytes have traveled both apically and basally for the length of the host polyp, but only slightly around its circumference. The direction of nematocyte migration usually follows the muscle processes, even in places where the processes have been deflected from their normal axial direction (near the lower right side of the figure). From Campbell and Marcum (1980).

nematocytes and found no bias in migration towards the tentacles; massive numbers of nematocytes migrated basally as well as apically (Fig. 4). These experimental situations, although they may have induced abnormalities in migration by the pretreatments and grafting procedures, show that nematocytes can travel in both directions along the body, and suggest that evidence for unidirectional locomotion be carefully reviewed.

Three time-lapse movies that show migrating nematocytes leave no doubt that nematocytes migrate bidirectionally along the axis of a polyp and tentacles. Günzl's (1971) film shows massive freeway-like traffic of nematocytes in both directions along stolons of _Dipurena reesi_. Campbell and Marcum's (1980) film taken of _Hydra attenuata_ shows nematocytes shuttling back and forth, sometimes moving short distances in each direction, sometimes moving the length of the tentacle in each direction. A film of the minute polyp _Calpasoma dactyloptera_ (Rahat and Campbell, 1974) similarly shows nematocytes shuttling along the polyp, towards and away from the tentacles. Thus at the outset of the discussion of directionality, there is considerable uncertainty as to whether cells do migrate unidirectionally. Many workers believe that migration of nematocytes is oriented and much evidence has been accumulated for such movement. Yet this simply does not fit the pattern of migrations made visible in published films.

If the migration of nematocytes is directed, one might imagine three possible causes of it: contact guidance, chemotaxis, and tissue polarity.

## 1. Contact Guidance

The migration of nematocytes is strongly oriented along the axes of both stolons and hydranths. Campbell and Marcum (1980) concluded that the migration of nematocytes is constrained along the longitudinal axis of the hydra by means of the muscle processes (and other geometry) of the epithelial cells with which they are in intimate contact (Fig. 3). In polyps whose muscle processes were experimentally deflected, the polarity and migration of nematocytes were similarly altered locally (Fig. 4). As viewed _in situ_, nematocytes only rarely move circumferentially and then only for short distances. Contact guidance along muscle processes seems to be a definite and strong influence of surrounding tissue on the migration of nematocytes.

2. Chemotaxis

One view put forth to explain the directioality of nematocyte movement is that tentacles emit a chemical that attracts nematocytes. In support of this view, Weber et al. (1978) found that supernumerary tentacles grafted into the side or base of a hydra rapidly accumulated nematocytes and Herlands and Bode (1974a) found that tentacles and hypostomes grafted to the base of a hydra accumulated nematocytes there, a site which otherwise rarely acquired nematocytes. Nematocytes are said to migrate preferentially towards tentacles, buds and hydranths in a number of hydrozoans (Hadži, 1907; Hauenschild, 1957; Herlands and Bode, 1974b).

What could be the source of a chemoattractant? Amputation of the crown of tentacles, leading to regeneration, has long been supposed to provide an attraction for interstitial cells (see review by Kanajew, 1930; Tardent, 1963; Webster, 1971). In some cases regeneration seems to stimulate the migrations of nematocytes (Jones et al., 1941; Tardent and Eymann, 1959) while in other cases it has no effect or even decreases immigration into tentacles (Tardent and Morgenthaler, 1966; Herlands and Bode, 1974b). Herlands and Bode (1974b) found that removal of the hydranth reduced the directed migration of stenotele nematocytes without affecting the movements of desmoneme nematocytes. This finding suggested to the authors that tentacles are the source of an attractant, an idea shared by others. The arrangement of nematocytes within a battery suggested to Gelei (1927) that the central stenotele nematocyte is responsible for attracting the other nematocytes around itself.

Weber et al. (1978) found that nematocytes labeled with $^{3}$H-proline accumulated in supernumerary tentacles grafted into the columns of labeled hydra. Three times as many stenoteles migrated into supernumerary tentacles from which stenoteles had just been depleted. Removal of stenoteles from normal tentacles similarly provoked increased nematocyte migration (Zumstein, 1973). Thus the ability of a tentacles to attract nematocytes is dependent on its physiological and cytological condition. This fits in with Günzl's (1971) well documented observation that feeding a hydranth stimulates nematocyte migration into the hydranth and also increases the percentage of migrating nematocytes whose movement is directed rather than undirected. But nematocytes do not appear to be attracted to tentacles of hydra that have been depleted of all non-epithelial cells (Campbell and Marcum, 1980). Also, in determinate nematocyst batteries of the siphonophores, experimental elimination of nematocytes does not lead to additional immigration (Carré, 1974b) indicating that there is a phase of maturation of the tentacle tissue after which nematocytes do not enter.

One of the most interesting studies on the stimulus for nematocyte migration is that of Hauenschild (1957). In the medusa Eleutheria dichotoma, large penetrant nematocytes develop in cushions at the bases of the tentacles whence they migrate to the tips of the capitate tentacles. Migration occurs normally only after the medusa captures prey during feeding. Even a single Artemia being caught by a tentacle, from which it is removed shortly afterwards, stimulates nematocyte migration into that same tentacle. Typically 10-30 nematocytes would migrate into a stimulated tentacle within 2 hours following capture of prey whereas fewer than one nematocyte would migrate into an unstimulated tentacle. Treatment of tentacles with an extract of Artemia sufficed to stimulate migration, suggesting that tentacular receptor cells rather than the discharge of nematocysts were the source of the stimulus. Achermann and Tardent (1973), on the other hand, found that a purely mechanical induction of stenotele discharge in hydra inhibited nematocyte immigration. Hauenschild's study included other interesting data as well. For example, each tentacle is forked and the nematocytes migrate only into the branch of the tentacle that bears the battery of nematocytes.

Thus, there is scattered albeit not entirely consistent evidence that nematocyte migration can be attracted by the tentacles (or hypostome) and that this attraction is dependent upon the physiological condition and perhaps cellular composition of the tissue.

3. Tissue polarity

The proximo-distal developmental polarity as manifested by polarized regeneration (Tardent, 1963; Webster, 1971) is intrinsic to hydrozoan tissue and this polarity would offer another possible way for guiding nematocytes. If nematocytes could sense the local, intrinsic polarity of the tissue rather than some diffusible attractant, they could migrate according to it and generally arrive at the tentacles. Vögeli (1972), Herlands and Bode (1974a) and Campbell and Marcum (1980) tested this possibility in hydra by comparing the rates at which nematocytes move through grafted tissue that was implanted into the hydra in either normal or reversed polarity. Herlands and Bode (1974a) found that desmoneme nematocytes moved more readily along the intrinsic tissue polarity than against it (Fig. 5) even though stenotele nematocytes showed no such preference. The

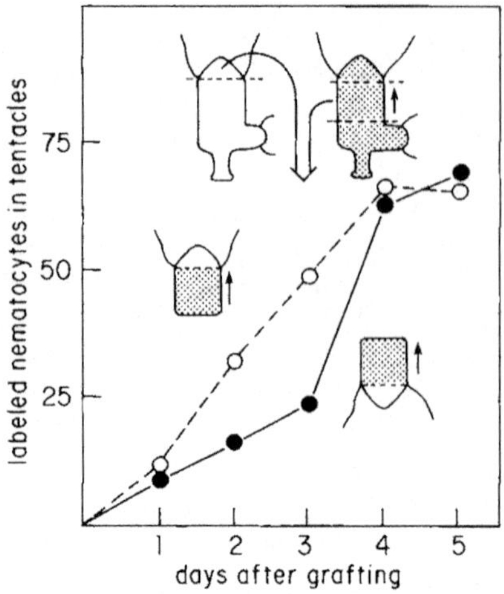


Fig. 5. Accumulation of $^{3}$H-proline-labeled nematocytes in tentacles grafted onto the top or bottom of a segment of labelled column of hydra. --o-- nematocytes in apical head; --●-- nematocytes in basal head. Somewhat fewer nematocytes have accumulated in the basal head on days 2 and 3, suggesting that they are sensitive to the direction of tissue polarity, but many nematocytes migrate in both directions. This data is only for desmoneme nematocytes. Redrawn from Herlands and Bode (1974c).

differences between the number of nematocytes going in the two directions were not large, however, and they disappeared soon after the tissues were grafted, possibly as the polarity of the reversed tissue altered to conform to the polarity of the rest of the animal. These investigators carried out many experiments of this general type and always found considerable migration in both directions. Campbell and Marcum (1980) found that nematocytes migrated both against and with the polarity of the tissue, in both implanted segments and in the host hydra (Fig. 4). Thus experimentation reveals that many nematocytes migrate without regard for the intrinsic developmental polarity of the surrounding tissue, but yields some quantitative disagreement about the degree of directional bias. Quantitative reconciliation of these various observations is difficult because all of the experimental studies involved some degree of abnormality and because it is always difficult to know in an experimental situation exactly what the tissue polarity will be at any time (Herlands and Bode, 1974a). One can conclude that there is some evidence for nematocytes being directed by intrinsic tissue polarity, but this seems to be at most a minor effect in the direction of nematocyte migration.

4. Conclusions Regarding Directionality

In reviewing all the evidence on control of directionality of nematocyte migration, one is impressed by the meager amount of clear evidence that the amoeboid migration itself is directed towards the tentacles. Most experimentation has measured rates at which nematocytes accumulate rather than the rates at which the cells migrate. In hydra, at least, there is extensive traffic in both directions. It is clear that migration is strongly directed axially, apparently by a phenomenon of contact guidance by the epithelio-muscular processes. Within this constraint nematocytes shuttle back and forth. Perhaps they arrive at their destination by chance and are then trapped there. Or perhaps there is a slight bias in their movement, not obvious in most observation, that statistically brings them to the tentacles more often than elsewhere. Or perhaps, as Günzl (1971) suggests, only a small fraction of migrating nematocytes are responsive to chemoattraction. Other observations support the idea that only some of the many nematocytes present are in the process of directed immigration (Hauenschild, 1957; Tardent and Eymann, 1959; Zumstein, 1973). Within an individual, different types of nematocytes may be guided by different cues. Some observations indicate strictly unidirectional migration. The beautiful system of the nematocyte migration in the hydrozoa currently presents a mixed picture and has not yet been successfully developed as a model for studying the guidance of cell locomotion.

C. Translocation through the gastric cavity.

Hadži (1907) observed that nematocytes in *Tubularia* migrate actively through the mesoglea into the endoderm and then into the gastric cavity. From there they are carried by fluid movements to the tentacles, at which point they crawl back through the endoderm and mesoglea and into the ectoderm. Hadži (1907) made it clear that this was an exceptional route, for he observed it only in one species out of the 20 that he had studied. And even in *Tubularia*, most nematocytes accumulate in the hydranth during the period of regeneration using the typical amoeboid form of movement (Hădzi, 1907; Tardent and Eymann, 1959).

While Hadži's work on Tubularia has not been reassessed by subsequent observation, the gastrocoelic route of transport has spawned considerable contention with regard to how widespread it is among the hydrozoans. Many workers cite Hadzi as proposing this route to be a general one. The idea has led several workers to suggest that it is the major route in hydra (Jacobsohn, 1912; Kepner et al., 1937; Jones, 1941; Burnett, 1959-60a,b; Burnett and Lentz, 1959-60). Evidence in hydra is based largely on the fact that nematocytes can be observed in the gastric fluid of living animals and in endodermal epithelial cells in histological sections; however no one has actually watched cells passing from the body to the tentacles along this route. In hydra, endodermal nematocytes are located inside the endodermal epithelial cells, rather than between the cells as is reported in other instances of endodermal migrations (Hadži, 1907, 1911). While Jacobsohn (1912) described the nematocytes in the gastric cavity as being mature (even with partially formed cnidocils), the other workers have described a chaotic variety of developmental and degenerating stages of nematocysts. Kepner et al. (1937) and Jones (1941) attempted to exclude other possible explanations for endodermal and gastrocoelic nematocysts, such as by removing tentacles and starving the hydra to avoid nematocytes being ingested, but they were unable to marshall any direct evidence that nematocytes actually took the gastrocoelic route.

That nematocytes take such a route has been vigorously contested (see esp. Schlottke, 1930, pp. 503-507). One type of evidence against it is the frequent observation by histologists (Schneider, 1890; Gelei, 1927; Schlottke, 1930; Weill, 1934; Bouillon, 19456-57; Burnett, 1959-60b; Campbell, 1967b) that the abundant nematocysts observable in the endoderm and gastric cavity are degenerative in appearance. Hence these nematocysts might be ones that are not destined for the tentacles. Campbell (1967) and Vögeli (1972) showed that nematocytes labeled in the gastric region move to the tentacles before they appear in the endoderm. They probably get into the endoderm by subsequent ingestion (Campbell, 1967b). The tips of tentacles are constantly sloughing (Tripp, 1928; Brien, 1950) and the broken tips get eaten with the food to which they adhere (Hartog, 1880). Other investigators report that excess unused nematocytes may be drawn into the endoderm within the tentacles (Schlottke, 1930). Another explanation for the presence of nematocytes in the endoderm is that they are engulfed by endodermal cells that reach through the mesoglea into the ectoderm; transfer of materials and cells between tissue layers may be a process in the regulation of normal tissue composition (eg. Campbell, 1976; Muscatine and McAuley, 1982). Fujisawa and David (1984) have shown that some nematoblasts are taken up by the endoderm and digested in response to injury or decapitation of the hydra. Thus Kepner *et al*. (1938) and Jones (1941), who had cut off tentacles to remove extraneous nematocytes, could have stimulated withdrawal of nematocytes into the endoderm. Evidently the presence of nematocytes in the endoderm does mean that the cells are in transit to the tentacles.

For nematocytes to take a gastrocoelic route, they must cross the mesoglea twice. The extent to which cells do this in hydrozoans has never been well understood. Many microscopists have described histological profiles that presumably represent such crossings (e.g. Brien, 1943; Bouillon, 1956-57; Nagao, 1965; Noda, 1968; Summers and Haynes, 1969; and Glätzer, 1971). In most hydrozoans, interstitial cells and nematoblasts generally arise in the endoderm during embryogenesis but end up in the ectoderm. Therefore centrifugal migrations of both cell types across the mesoglea are a part of normal development (e.g. van der Vyver, 1964, 1967; Tardent, 1978). Hydrozoan germ cells have long been considered to cross the mesoglea (Weismann, 1883; Harm, 1903; Brien, 1943; Tardent, 1985). But good histologcal profiles of cells crossing the mesoglea are rare. It remains for a quantitative study of the rates and frequencies of mesoglea crossings to be made before we can determine whether the number of nematocytes migrating across the mesoglea represents a significant proportion of those migrating to the tentacles.

Hence the gastrocoelic route of nematocyst transport may be restricted to *Tubularia*, as originally described, and even there as a minor route. It could conceivably also represent a minor pathway in other organisms, but the data in the literature is not compelling. It would be worthhile even to repeat Hadži's original observations on living *Tubularia*.

## III. PASSIVE TRANSPORT BY DISPLACEMENT OF EPITHELIA

Semal-van Gansen (1951) elaborated on a mechanism by which nematocytes reach the tentacles without migrating. Tripp (1928) and Brien (1950) had demonstrated that the tentacles of hydra are derived by outward displacement of the body tissue. Would not, then, the nematocytes also be carried out, and already enmeshed with epithelial cells just the way the early nematoblasts are? This idea goes back to Murbach (1895) who deduced that growing tentacles of the medusa *Gonionemus Murbachii* acquired their nematocytes by a proximal aggregation of precursors being swept outward during growth of tentacles.

A quantitative assessment of transport of nematocytes by epithelial displacement in hydra indicates that it is much too feeble to account for the numbers of nematocytes in the tentacles (see data in Bode *et al.*, 1973; Zumstein, 1973). Lenhoff (1959) first showed the rapidity of the movement of nematocytes into the tentacles; massive translocation from the body column to the tentacles occurs in less than 48 hours, a process much too fast to be explained by the slower displacement of tissue. Others have found that nematocytes traverse the length of the polyp in just hours (Campbell and Marcum, 1980; Tardent and Morgenthaler, 1966; Novak and Wood, 1983). Brien (1950) showed, as has been abundantly documented by others (Tripp, 1928; Campbell, 1967*a*; Otto and Campbell, 1977) that only the uppermost portion of the body column displaces into the tentacles; most of the column moves downwards away from the tentacles. Epithelial displacement would carry most nematocytes *away* from the tentacles. Furthermore, the uppermost portion of the column is devoid of developing nematoblasts (Schneider, 1890; David and Challoner, 1974). Hence the epithelium displacing into the tentacles of hydra simply does not have enough nematocytes to account for the numbers of stinging cells farther out along the tentacles. Such quantitative analyses have not yet been carried out with other hydrozoans.

Although passive displacement cannot account for the movements of nematocytes in hydra, wherever nematocytes are fixed in position they must be carried with any tissue displacements that occur, and tissue displacements are probably common in all hydrozoans. These passive transports could be viewed as supplementary mechanisms of transport that coexist with other modes (Burnett and Lentz, 1959-60). Leuckart (1853) first realized this in connection with renewal of tentacles in siphonophores. Murbach (1894) and Perkins (1902) asserted that during early outgrowing of tentacles in *Gonionemus*, nematocytes were translocated by displacement while later in life additional nematocytes entered by active migration. Lipin (1911) described how a new cluster of positioned nematocytes that formed at the midpoint of a tentacle suddenly became a terminal cluster when the tentacle pinched in half. There are many cases of elongation of tissue during which nematocytes are carried along with the tissue; it remains to be discovered to what extent passive displacement operates in more drastic morphogenetic situations.

## IV. FUTURE PROSPECTS

We can look forward to two major types of research in the area of nematocyte movements:

1. Obtaining further information on the amoeboid migratory mode of translocation. Very little ultrastructural work has yet been carried out on this subject, and we have little evidence of the mode of movement. Does the pseudopod of the nematocyte pull the cell forward, as in filopodial or fibroblast-like types of migration? Or are the cells more like mammalian leukocytes (Haston et al., 1982) or deep cells of the fish blastoderm (Trinkaus, 1984) that move by pouring cytoplasm forward into the pseudopod? Electron micrographs of migrating nematocytes fail to show points of attachment to surrounding cells, suggesting that movement occurs by the latter mechanism (Fig. 3). To get answers to this question, more careful in situ observations are needed. Excellent matierial for such observations may be found in the tentacles of siphonophores (Carre, 1974a; Skaer, 1987), nerve-free polyps (Campbell, 1979), the exumbrella of hydromedusae, and the stolon mat or network of colonial hydrozoans (Günzl, 1971). Once good model systems have been identified, experimental analyses of force application and the cytoplasmic localization of cytoskeletal proteins should lead to an understanding of the cellular basis of migration.

2. Analysis of the means by which the migration of nematocytes leads to the sites of deployment. This will probably be the most important area of research on nematocyte movement in the coming decades and the one that will tie in most closely to related studies in vertebrates. By what means do the muscle processes of the epithelial cells influence the nematocyte migrations? Are there gradients of chemoattractants that guide nematocytes? Are nematocytes sensitive to the axial tissue gradients of hydrozoan tissues? The many provocative experiments cited in this review lead to differing interpretations because they employ different methodologies and organisms. Focus on one or several good model systems will yield a sudden wealth of information on cellular movement.

## REFERENCES

Achermann, J. and P. Tardent. 1973. Untersuchungen zum Problem des Nematocyten-Nachschubs bei Cladonema radiatum (Hydrozoa). Rev. Suisse Zool. 80:704-712.

Bode, H., S. Berking, C. N. David, A. Gierer, H. Schaller and E. Trenkner. 1973. Quantitative analysis of cell types during growth and morphogenesis in hydra. Wilhelm Roux' Archiv, 171: 269-285.

Bouillon, J. 1957. Etude monographique du genre Limnocnida (Limnomeduse). Ann. Soc. Roy. Zool. Belgique, 87:251-500.

Boulenger, C. L. 1910. On the origin and migration of the stinging-cells in craspedote medusae. Quart. J. Microscop. Sci. 55:763-782.

Brien, P. 1943. Etudes de la régénération et de la rénovations de l'apareil sexuel chez les hydroïdes (Clava squamata O.F. Muller). Arch. Biol. 54:409-475.

Brien, P. and R. Reniers-Decoen. 1950. Etude de l'hydre verte. Ann. Soc. Roy. Zool. Belgique 81: 33-110.

Burnett, A. L. 1959-60a. The nematocyst of hydra (Part IV). The inhibition of the nematocyst formation process in hydra. Ann. Soc. Roy. Zool. Belgique 90:295-305.

Burnett, A. L. 1959-60b. The nematocyst of hydra (Part V). The exchange of cnidoblast cells between two different species of hydra. Ann. Soc. Roy. Zool. Belgique 90:307-315.

Burnett, A. L. and T. Lentz 1959-60. The nematocyst of hydra (Part III). The migration pathways of nematocysts in hydra. Ann. Soc. Roy. Zool. Belgique, 90:281-293.

Campbell, R. D. 1967a. Tissue dynamics of steady state growth in Hydra littoralis II. Patterns of tissue movement. J. Morphol. 121:19-28.

Campbell, R. D. 1967b. Tissue dynamics of steady state growth in Hydra littoralis III. Behavior of specific cell types during tissue movements. J. Exp. Zool. 164:379-391.

Campbell, R. D. 1974. Cell movements in Hydra. Amer. Zool. 14:523-535.

Campbell, R. D. 1976. Elimination of Hydra interstitial and nerve cells by means of colchicine. J. Cell Sci. 21:1-13.

Campbell, R. D. 1979. Development of hydra lacking interstitial and nerve cells ("epithelial hydra"). Symp. Soc. Dev. Biol. 37:267-293.

Campbell, R. D. and B. A. Marcum. 1980. Nematocyte migration in hydra: evidence for contact guidance in vivo. J. Cell Sci. 41:33-51.

Carré, D. 1974a. Formation, migration et maturation des nématoblastes et des nématocystes chez les siphonophores. II. -- Migration. Ann. Embryol. Morphogen. 7:221-232.

Carré, D. 1974b. Formation, maturation et migration des nématoblastes et des nématocystes chez les siphonophores. III.-- Maturation des nématoblastes et des nématocystes. Ann. Embryol. Morphogen. 7:233-242.

David, C. N. and D. Challoner. 1974. Distribution of interstitial cells and differentiating nematocytes in nests in Hydra attenuata. Amer. Zool. 14:537-542.

Fujisawa, T. and C. N. David. 1984. Loss of differentiating nematocytes induced by regeneration and wound healing in Hydra. J. Cell Sci. 68:243-255.

Gelei, J. von 1927. Das Rätsel der Nesselzellen. Sitzber. Ges. Naturf. Freunde Berlin. (1927):12-16.

Glätzer, K. H. 1971. Die Ei- und Embryonalentwicklung von Corydendrium parasiticum mit besonderer Berücksichtigung der Oocyten-Feinstruktur während der Vitellogenese. Helgol. Wiss. Meeresunters. 22:213-280.

Günzl, H. 1971. Dipurena reesi (Hydrozoa). Wanderung der Cnidoblasten in den Rhizostolonen. Encyclopaedia Cinematographica (G. Wolf, ed.), Institut fur den Wissenschaftlichen Film, Göttingen, 15 pp. (text accompanying Film # 1106/1966).

Hadži, J. 1907. Ueber die Nesselzellwanderung bei den Hydroidpolypen. Arb. Zool. Inst. Univ. Wien 17:65-94.

Hadži, J. 1911. Ueber die Nesselzellverhältnisse bei den Hydromedusen. Zool. Anz. 37:471-478.

Harm, K. 1903. Die Entwicklungsgeschichte von Clava squamata. Z. Wiss. Zool. 73:115-165.

Hartog, M. M. 1880. On the mode in which Hydra swallows its prey. Quart. J. Microscop. Sci. 20:243-244.

Haston, W. S., J. M. Shields, and P. L. Wilkinson. 1982. Lymphocyte locomotion and attachment of two-dimensional surfaces and in three-dimensional matrices. J. Cell Biol. 92:747-752.

Hauenschild, C. 1957. Versuche ueber die Wanderung der Nesselzellen bei der Meduse von Eleutheria dichotoma. Z. Naturforsch. 12b:472-477.

Herlands, R. L. and H. R. Bode. 1974a. The influence of tissue polarity on nematocyte migration in Hydra attenuata. Develop. Biol. 40:323-339.

Herlands, R. L. and H. R. Bode. 1974b. Oriented migration of interstitial cells and nematocytes in Hydra attenuata. Wilhelm Roux' Arch. 176:67-88.

Herlands, R. L. and H. R. Bode. 1974c. Nematocyte migration in Hydra influenced by tissue polarity. Nature 248:387-390.

Jacobsohn, A. 1912. Die Nesselzellen. Archiv Naturg. Berlin 78(8)A:111-144.

Jones, C. S. (1941) The place of origin and the transportation of cnidoblasts in Pelmatohydra oligactis (Pallas). J. Exp. Zool. 87:457-475.

Kanajew, J. 1930. Zur Frage der Bedeutung der Interstitiellen Zellen bei Hydra. Wilhelm Roux' Arch. 122:736-759.

Kass-Simon, G. 1987. This volume.

Kepner, W. A., R. A. Goodwin, and A. L. Ingles. 1938. The place of origin and the manipulation of cnidoblasts in Chlorohydra viridissima. Zool. Anz. 121:85-94.

Lenhoff, H. M. 1959. Migration of $^{14}$C-labeled cnidoblasts. Exp. Cell Res. 17:570-573.

Leuckart, R. 1853. Zoologische Untersuchungen. I. Die Siphonophoren. Giessen, 95 pp.

Lipin, A. (1911) Die Morphologie und Biologie von Polypodium hydriforme Uss. Zool. Jahrb. Abt. Anat. 31:317-426.

Murbach, L. 1894. Beiträge zur Kenntnis der Anatomie und Entwickelung der Nesselorgane der Hydroiden. Arch. Naturg. Berlin. 60:217-254.

Murbach, L. 1895. Observations on the development and migration of the urticating organs of sea nettles, Cnidaria. Proc. U. S. Nat. Mus. 18:733-740.

Muscatine, L. and P. J. McAuley. 1982. Transmission of symbiotic algae to eggs of green hydra. Cytobios 33:111-123.

Nagao, Z. 1965. Studies on the development of Tubularia radiata and Tubularia venusta (Hydrozoa). Publ. Akkeshi Mar. Biol. Stat. 15:9-35.

Noda, K. 1968. Migratory cells traversing the mesoglea in hydra. J. Fac. Sci. Hokkaido Univ. Ser VI. Zool. 16:353-358.

Novak, P. L. and R. L. Wood. 1983. Development of the nematocyte junctional complex in hydra tentacles in relation to cellular recognition and positioning. J. Ultrastruct. Res. 83:111-121.

Otto, J. J. and R. D. Campbell. 1977. Tissue economics of hydra: regulation of cell cycle, animal size and development by controlled feeding rates. J. Cell Sci. 28:117-132.

Perkins, H. F. 1902. The development of Gonionema Murbachii. Proc. Acad. Nat. Sci. Phila. 54:750-790.

Rahat, M. and R. D. Campbell. 1974. Nematocyte migration in the polyp and one-celled tentacles of the minute fresh-water Coelenterate Calpasoma dactyloptera. Trans. Amer. Microscop. Soc. 93:379-385.

Schlottke, E. 1930. Zellstudien an Hydra. I. Altern und Abbau von Zellen und Kernen. Z. Mikroskop. Anat. Forsch. 22: 493-532.

Schneider, K. C. 1890. Histologie von Hydra fusca mit besonderer Berücksichtigung des Nervensystems der Hydropolypen. Arch. Mikroskop. Anat., 35:321-379.
Schneider, K. C. 1894. Mittheilungen über Siphonophoren. I. Nesselzellen. Zool. Anz. 17:461-471.
Schneider, K. C. 1900. Mitteilungen über Siphonophoren. V. Nesselzellen. Arb. Zool. Inst. Wien, 12:133-242.
Semal-van Gansen, P. 1951. Le cnidosome de l'hydre et le bouton urticant. Bull. Classe Sci. Acad. Roy. Belgique Ser. 5. 37:650-664.
Skaer, R. J. 1987. (This volume).
Summers, R. G. and J. F. Haynes. 1969. The ontogeny of interstitial cells in Pennaria tiarella. J. Morphol. 129:81-88.
Tardent, P. 1954. Axiale Verteilungs-Gradienten der Interstitiellen Zellen bei Hydra und Tubularia und ihre Bedeutung fur die Regeneration. Wilhelm Roux' Archiv. 146:593-649.
Tardent, P. 1963. Regeneration in the Hydrozoa. Biol. Rev. 38:293-333.
Tardent, P. 1978. Coelenterata, Cnidaria. In F. Seidel (ed.), Morphogenese der Tiere. Lief 1:A-I, pp. 69-415. Gustav Fischer, Stuttgart.
Tardent, P. 1985. The differentiation of germ cells in Cnidaria. In H. O. Halvorson and A. Monroy (eds.), The Origin and Evolution of Sex, pp. 163-197. Alan R. Liss, New York.
Tardent, P. and H. Eymann. 1959. Experimentelle Untersuchungen über Regenerationshemmende Faktor von Tubularia. Wilhelm Roux' Arch. 151:1-37.
Tardent, P. and U. Morgenthaler. 1966. Autoradiagraphische Untersuchungen zum Problem der Zellwanderungen bei Hydra attenuata Pall. Rev. Suisse Zool. 73:468-480.
Trinkaus, J. P. 1984. Cells Into Organs; the Forces that Shape the Embryo, 543 pp. Prentice-Hall, Englewood Cliffs, N.J.
Tripp, K. 1928. Die Regenerationsfähigkeit von Hydren in den vershiedenen Körperregionen. Nach Regnerations- und Transplantationsversuchen. Z. Ziss. Zool. 132:476-525.
Van de Vyver, G. 1964. Etude histologique du developpement d'Hydractinia echinata (Flem.). Cah. Biol. Mar., 5: 295-310.
Van de Vyver, G. 1967. Etude du developpement embryonnaire des hydraires athecates (gymnoblastiques) à gonophores. Arch. Biol. 78:451-518.

Vögeli, G. 1972. Autoradiographische Untersuchungen zur kurzzeitigen Zellmarkierung sowie Polarität und Ausmass der individuellen Zellwanderung bei Hydra attenuata (Pall.). Rev. Suisse Zool. 79:649-674.

Weber, G., T. Honegger, and P. Tardent. 1978. Neuorientierung der Nesselzellwanderung bei Hydra attenuatta Pall. durch transplantierte Tentakel. Rev. Suisse Zool. 85:768-774.

Webster, G. 1971. Morphogenesis and pattern formation in hydroids. Biol. Rev. 46:1-46.

Weiler-Stolt, B. 1960. Ueber die Bedeutung der Interstitiellen Zellen fur die Entwicklung und Fortpflanzung mariner Hydroiden. Wilhelm Roux' Arch. 152:398-454.

Weill, R. 1934. Contributions a l'étude des Cnidaires et de leurs nematocystes. Trav. Stat. Zool. Wimereux. 10,11:1-700.

Weismann, A. 1883. Die Entstehung der Sexualzellen bei den Hydromedusen. Jena.

Zumstein, A. 1973. Regulation der Nematocyten-Produktion bei Hydra attenuata Pall. Wilhelm Roux' Arch., 173:294-318.

# 9

# ULTRASTRUCTURE AND CYTOCHEMISTRY OF DEVELOPING NEMATOCYSTS

Glen M. Watson

Department of Physiology and Pharmacology
Loma Linda University
Loma Linda, California 92350

## SUMMARY

Nematocysts develop from large vesicles called "nematocyst primordia" which form by the fusion of smaller vesicles derived from the Golgi. A cage-like array of microtubules surrounds the nematocyst primordium, but the exact function of these microtubules remains unclear. The nematocyst primordium enlarges first to form a capsule and then to form a tubule continuous with the capsule. After the tubule forms, it inverts into the capsule. The possibility that the tubule assembles spontaneously inside the capsule is considered to be unlikely. A wall assembles onto the limiting membrane of the nematocyst primordium soon after Golgi-derived vesicles fuse with it. This wall thickens throughout nematocyst formation and then thins after the tubule inverts into the capsule. Certain nematocysts accumulate large amounts of calcium throughout their formation via the Golgi, while other nematocysts accumulate calcium without involving the Golgi.

## I. INTRODUCTION

Nematocysts, as well as other types of cnida, are complex intracellular structures, each consisting of a capsule containing a folded, inverted tubule. Capsules vary from spherical to ovoid, depending on the type of nematocyst. Tubules can differ in diameter at various points along their length. Sizes, shapes and abundances of spines on the surface of everted tubules vary according to nematocyst type. Matrix contents within nematocysts can consist of extremely potent toxins (reviewed by Hessinger, this volume) as well as high concentrations of certain ions such as calcium and magnesium (Lubbock et al., 1981; Mariscal, 1984).

In view of such a high degree of structural and chemical complexity, it is difficult to comprehend that each nematocyst is produced within a single cell. Numerous studies of developing nematocysts have provided insight into how nematocysts are assembled. Despite these efforts and perhaps attesting to the complexity of the processes involved, many aspects of nematocyst development remain unclear.

In this review, I discuss aspects of nematocyst development that now seem to be resolved and identify aspects of nematocyst development that still remain poorly understood.

## A. Early Changes in Interstitial Cells

Interstitial cells, pluripotent stem cells, possess small Golgi apparatus and sparse rough endoplasmic reticulum (RER). Interstitial cells that differentiate into cnidoblasts form well-developed Golgi apparatus (Fig. 1) and proliferate RER in the cytoplasm (Figs. 1,2) (Chapman and Tilney, 1959; Slautterback and Fawcett, 1959).

## B. The Golgi-Microtubule Complex

The nematocyst primordium forms just beyond trans cisternae of the Golgi apparatus. This vesicle becomes surrounded by 50 to 150 microtubules arranged in a cage-like array (Fig. 3) (Slautterback, 1961, 1963). Microtubules originate from a centriole pair located near the Golgi (Westfall, 1966) and measure from four to ten μm in length (Westfall, 1966; Skaer, 1973; Holstein, 1980, 1981; Watson and Mariscal, 1984b). The array changes width such that it is widest as it surrounds the capsule and narrowest as it surrounds the preinverted tubule. More microtubules are present in wide arrays than in narrow arrays, indicating that the number of microtubules within the array changes as the array changes width (Holstein, 1981). Such changes probably comprise complex processes because microtubules are interconnected by an amorphous material (Figs. 3,4) (Holstein, 1981) and separated from the membrane forming the outer boundary for the nematocyst primordium by an electron-dense material (Fig. 3) (Holstein, 1981). Microtubules added to the array would have to form proper associations with other microtubules and with the nematocyst primordium, presumably, via materials described above. Following a removal of microtubules from the array, remaining microtubules would have to form anew proper associations with any "new neighbors." Holstein (1981) raised the interesting possibility that the microtubule array changes size to determine the shape of the forming nematocyst.

No study has yet focused on this important question, "What determines the shape of the nematocyst as it develops from a single vesicle?"

### C. Nematocyst Formation by the Golgi

The region of the nematocyst primordium that becomes the capsule forms first, followed by that which becomes the tubule (Slautterback and Fawcett, 1959; Slautterback, 1961, 1963). Small vesicles having bristle coats (Fig. 5) and larger, uncoated (smooth) vesicles are present between the Golgi and nematocyst primordium (Watson and Mariscal, 1984a,b). Coated vesicles also are visible in micrographs in Germer and Hundgen, 1980; Holstein, 1981; and Schmidt, 1981. It is not clear in all cases which type of vesicle, coated or uncoated, fuses with the nematocyst primordium to cause it to become larger.

It appears that Golgi-derived vesicles fuse with the nematocyst primordium before the wall forms. In some nematocysts, the forming wall is first observed at some distance from the end of the nematocyst primordium closest to the Golgi (Fig. 5) (Germer and Hundgen, 1980; Watson and Mariscal, 1984b). This distance has been measured to be approximately one µm (Watson and Mariscal, 1984b). Materials tentatively identified as precursors of the nematocyst wall are present in the matrix contents (Fig. 5) (Schmidt, 1981; Watson and Mariscal, 1984b). Thus, Golgi-derived vesicles fuse with the nematocyst primordium and release their contents into it. Some of these contents are wall precursors which assemble onto the limiting membrane of the nematocyst primordium. Invariably, the wall forms along regions of the nematocyst primordium surrounded by the microtubule array (Fig. 5). Unfortunately, the function of this array is not known.

Several hypotheses stating possible functions for the microtubule array have been published. The first proposed that microtubules transport materials to the site of nematocyst assembly (Slautterback, 1963). Another suggested that microtubules prevent Golgi-derived vesicles from fusing at an incorrect locus (Westfall, 1966). Two hypotheses have postulated that microtubules support the developing nematocyst: (i) until the wall matures (i.e.,hardens) (Westfall, 1966), or (ii) so that wall precursors assemble onto the limiting membrane to form a template of material in the proper shape of the mature nematocyst (Watson and Mariscal, 1984b). Areas of the capsule wall that extend beyond the microtubule array can appear to be wrinkled, while areas of the capsule wall surrounded by microtubules remain straight (Fig. 6). Presumably, additional wall material later binds this template to strengthen it (illustrated in Fig. 7). Unfortunately, data are unavailable at present to support or to reject any of these hypotheses.

Once the capsule forms, the Golgi-microtubule complex "moves off" the capsule and forms a tubule continuous with the capsule. The majority of reports have suggested that the tubule is synthesized in a fashion that, at least superficially, is the same as for the capsule. The most obvious differences between the synthesis of capsules and tubules are that (i) tubules have a smaller diameter than capsules, (ii) tubule walls are thinner than capsule walls, and (iii) tubules are longer than capsules. A few papers have indicated that contents of tubules have a different appearance in the electron microscope than capsule contents (Ivester, 1977; Schmidt, 1981). The significance of such differences is not known at the present time.

FIG. 1. Longitudinal section of a cnidoblast showing a forming capsule (c) at the Golgi apparatus (g). The cytoplasm contains extensive rough endoplasmic reticulum (rer). Scale bar = 0.5 μm.

FIG. 2. Longitudinal section of a cnidoblast showing rough endoplasmic reticulum (rer) in the cytoplasm. Scale bar = 0.5 μm.

FIG. 3. Cross section of a forming nematocyst capsule (c) surrounded by numerous microtubules (mt). An amorphous material (am) is present between adjacent microtubules. Microtubules lie just outside of the nematocyst membrane (nm). This tissue was fixed in glutaraldehyde, then poststained with Alcian blue. Scale bar = 100 nm.

FIG. 4. Tangentially sectioned capsule (c) of a developing nematocyst. Microtubules (mt) that surround the nematocyst during its formation are parallel to one another. Scale bar = 0.5 μm.

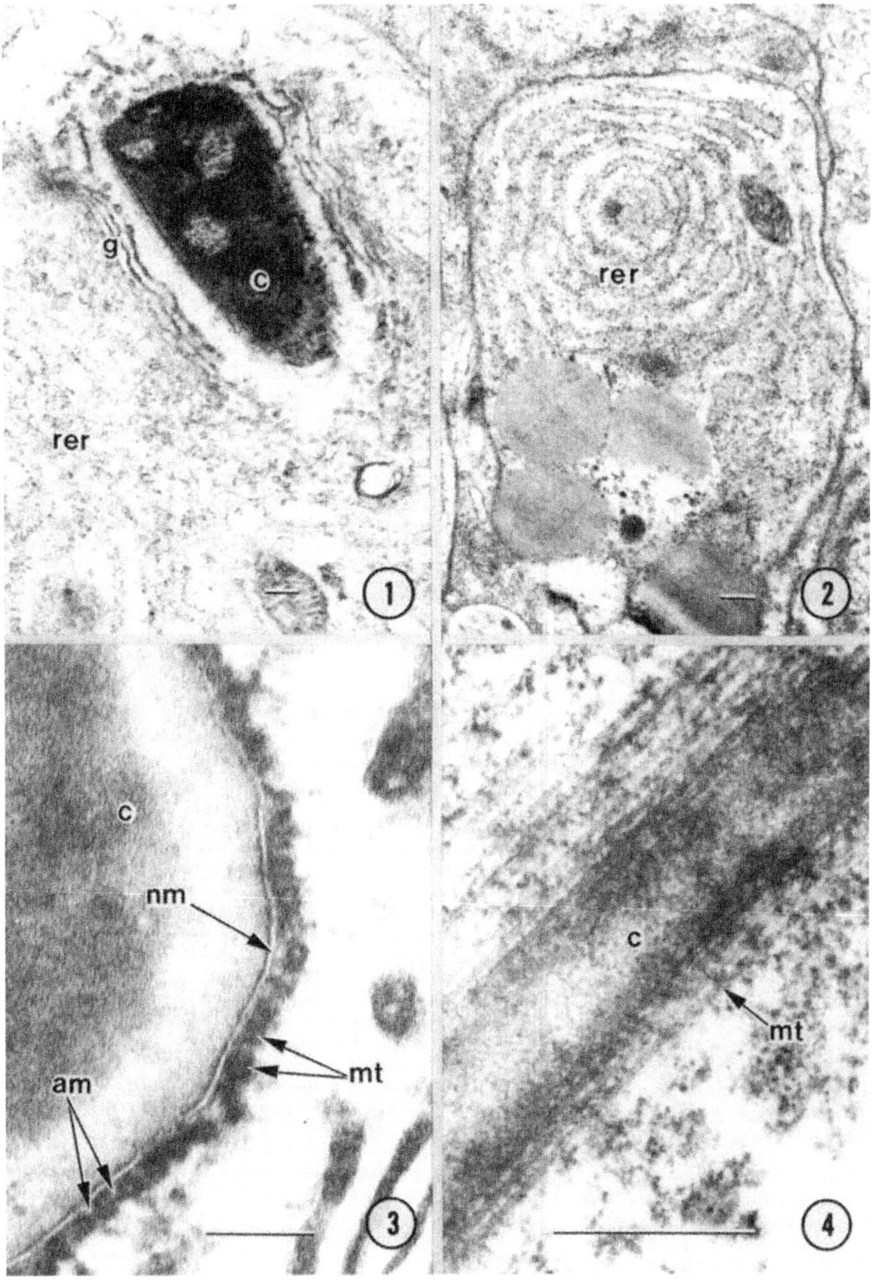
g
c
rer
1
rer
2
c
nm
mt
am
3
c
mt
4

FIG. 5. Longitudinal section of a cnidoblast fixed during capsule formation. Coated vesicles (cv) are present in the cytoplasm between the Golgi (g) and the capsule (c). A centriole (ce) is visible near the edge of the forming capsule. The nematocyst wall (nw) is apparent at some distance from the end of the capsule closest to the centriole. Small, electron-lucent particles, similar in appearance to the electron-lucent layer of the nematocyst wall are present (circles) within the capsule. The electron-lucent layer has a mottled appearance. Scale bar = 0.5 μm.

FIG. 6. Longitudinal section of a cnidoblast fixed during formation of the capsule. Portions of the capsule (c) that extend beyond the microtubule array (arrowhead) appear to be wrinkled (i.e., undulate), whereas portions of the capsule surrounded by microtubules (between the arrowhead and Golgi (g)) remain straight. Scale bar = 0.5 μm.

FIG. 7. Schematic drawing for a hypothetical function of microtubules in nematocyst formation. (A) Microtubules (mt) surround the nematocyst primordium during its formation via fusion of vesicles (v) derived from the Golgi. A template of wall material (ow, outer wall; iw, inner wall) assembles onto the nematocyst membrane (nm) after vesicles fuse with the nematocyst primordium. This event takes place in the presence of microtubule support. (B) In the absence of microtubule support, the template of wall material collapses (appears to undulate). Additional precursors bind the inner layer to strengthen it. (C) Ultimately, sufficient material binds the wall template so that it regains its original shape.

## D. Tubule Inversion

Once the tubule forms, the Golgi-microtubule complex detaches (Holstein, 1980), and the tubule inverts into the capsule (Fig. 8) (Skaer, 1973; Carré, 1972, 1980; Carré and Carré, 1973; Holstein, 1980, 1981; Schmidt, 1981; Watson and Mariscal, 1984a,b). Prior to inverting into the capsule, the tubule is designated, "preinverted." Several workers independently have denied that the tubule inverts into the capsule (Lentz, 1965; Westfall, 1966; Ivester, 1977; Germer and Hundgen, 1980). These workers suggest that the inverted tubule arises from precursors in contents of the capsule and "preinverted" tubule. According to this interpretation, the preinverted tubule is incorporated into the capsule just before the inverted tubule appears within the capsule. Thus, the capsule enlarges at about the same time as the inverted tubule appears within the capsule.

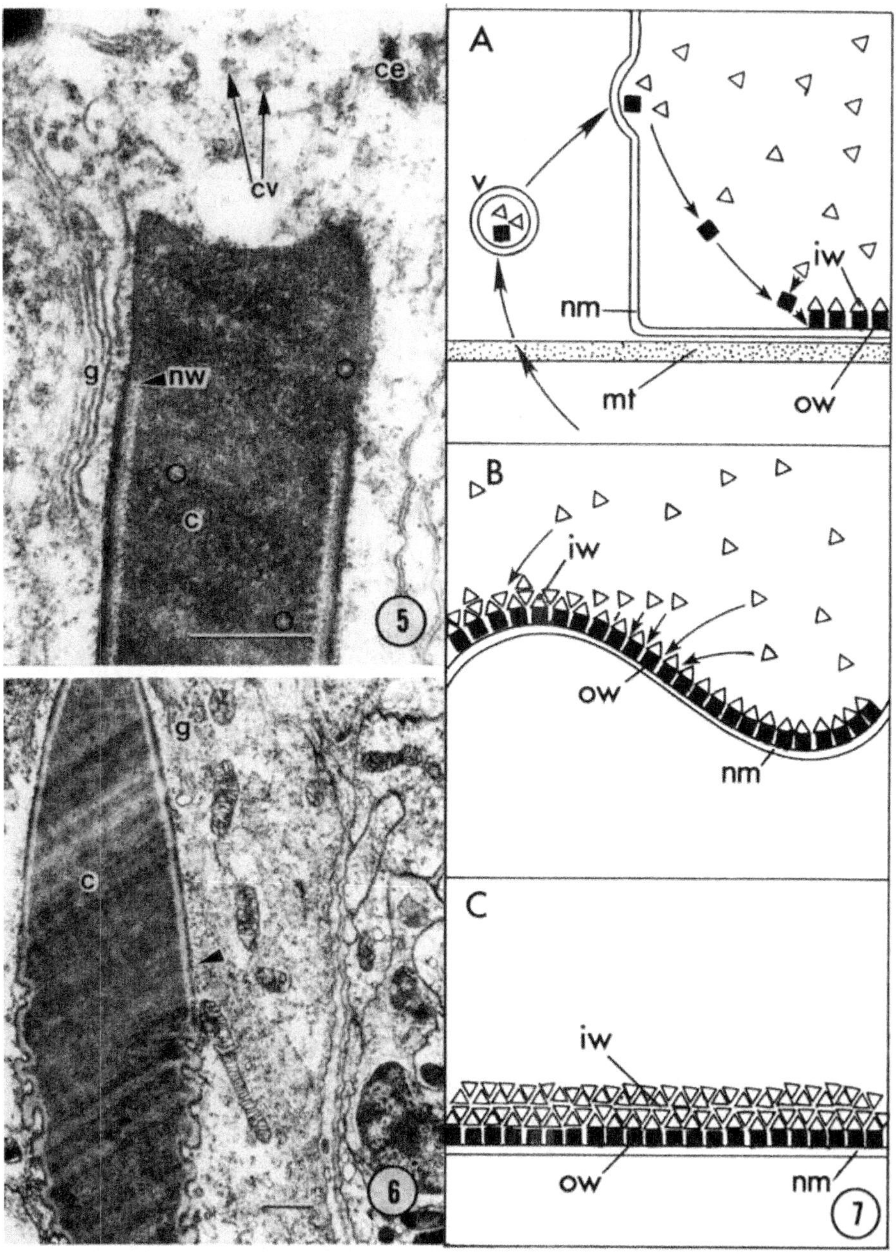
ce
cv
g
nw
c
5
g
c
6
A
v
nm
iw
mt
ow
B
iw
ow
nm
C
iw
ow
nm
7

I propose that tubule inversion occurs in all developing nematocysts and that observations of the above workers can be explained by the following:

Observations of living cnidoblasts using light microscopes have shown that tubules rapidly invert into capsules (Skaer, 1973; Carré, 1972, 1980; Carré and Carré, 1973; Holstein, 1980). In one type of nematocyst, the entire process is completed within two hours (Carré, 1972). Considering the difficulty encountered in attempting to survey a large number of fixed-cells using thick- and thin-sectioning methods, combined with the short time period during which the tubule may invert, it is not surprising that some workers may not have obtained sections of inverting tubules.

The observation that the capsule enlarges in some nematocysts at approximately the same time as the tubule appears within the capsule can be explained by a decrease in volume of the nematocyst caused by tubule inversion. Prior to tubule inversion, the total volume of the nematocyst consists of the volume of the capsule plus the volume of the preinverted tubule. After the tubule has inverted, the total volume of the nematocyst consists only of a fraction of the original volume of the capsule (i.e., the capsule volume minus that displaced by the inverted tubule). Thus, the total volume of the nematocyst decreases as a result of tubule inversion. Because fluids are incompressible, tubule inversion must be accompanied by an enlargement of the capsule to accommodate the fluid previously contained in the preinverted tubule, by a removal of the fluid from the nematocyst, or by both processes. An enlargement of the capsule coincidental to the appearance of the tubule within the capsule has been measured for one type of nematocyst (Skaer, 1973). Another report has provided micrographs suggesting osmotic stress occurs in the cnidoblast at the time of tubule inversion. Measurements from such micrographs indicated that no change in capsule dimensions accompanies tubule inversion (Watson and Mariscal, 1984a).

In certain types of nematocyst, matrix contents undergo a dramatic change in appearance at about the same time as the tubule appears within the capsule. This change can be described as an accumulation of matrix contents along the capsule wall and also along the wall of the inverted tubule, if present. Matrix contents of certain immature nematocysts are electron-dense, whereas matrix contents of these nematocysts at maturity are electron-transparent (Skaer, 1973). Somewhat less dramatic events have been described in other nematocysts. Laminar granules within anemone nematocysts solubilize just prior to tubule inversion (Watson and Mariscal, 1984a). The occurrence of such processes at approximately the same time as tubule inversion suggests that this stage of nematocyst development is complex and hence, difficult to interpret. The significance of an accumulation of matrix contents along the nematocyst wall or of granule solubilization for certain developing nematocysts is not known.

## E. The Mechanism of Tubule Inversion

Little is known about the mechanism of tubule inversion. Virtually all of the published electron micrographs of inverting tubules have shown that they are pleated and folded in essentially the same manner as are mature inverted tubules. The basis for such folding is unknown. Small fibrils are present along the inside and outside of inverting tubules in some nematocysts (Fig. 9, and illustrated in Fig. 10) (Watson and Mariscal, 1984a). Some of these structures interconnect pleats of the tubule (Fig. 9). Such structures persist in the mature nematocyst albeit in a more spherical configuration (Watson and Mariscal, 1984a, 1985). Moreover, these structures appear to bind calcium (Watson and Mariscal, 1985). Morphologically similar structures are visible in published electron micrographs of nematocyst tubules from various cnidarians (Skaer and Picken, 1965; Carré, 1980; Cormier and Hessinger, 1980). Recently, the presence of calcium-binding fibrils on tubules of immature nematocysts in Physalia was demonstrated using potassium antimonate (Fig. 11). Many fibrils in Figure 11 interconnect pleats of the tubule. Thus, a variety of nematocysts have structures on the surfaces of the tubule, some of which interconnect pleats. The function of these structures is unknown.

## F. Changes in the Cnidoblast after Tubule Inversion

The RER and Golgi regress in cnidoblasts after the tubule inverts into the capsule (Slautterback and Fawcett, 1959; Slautterback, 1961). In most cnidocytes these organelles no longer are present and presumably have been degraded.

## G. Tubule Maturation

The inverted tubule apparently becomes more tightly folded after it arrives in the capsule (Skaer, 1973; Carré, 1972, 1980; Carré and Carré, 1973). In addition, spines form along the inner layer of the wall of the inverted tubule (Skaer,

1973; Holstein, 1980). Almost nothing is known about how spines are formed except that they can begin to develop as the tubule inverts into the capsule and can completely form within several hours after the tubule has inverted into the capsule (Holstein, 1980; Skaer, 1973; Carré, 1972; Schmidt, 1981).

### H. Changes in the Nematocyst Wall Throughout Nematocyst Development

Mature nematocysts have a wall consisting of an outer electron-dense layer and an inner electron-lucent layer. In octocoral nematocysts, the electron-lucent layer thickens at the same time as honeycomb-shaped arrays of electron-lucent material within the capsule decrease in size and ultimately disappear (Schmidt, 1981). A dramatic reorganization of electron-dense and electron-lucent layers takes place in these nematocysts after the tubule inverts into the capsule (Schmidt, 1981). In anemones, the electron-lucent layer in walls of immature nematocysts has a mottled appearance with electron-dense areas within the electron-lucent layer (Watson and Mariscal, 1984b). The electron-lucent layer thickens as an amorphous mass of electron-lucent material within the capsule (Fig. 12) decreases in size and ultimately disappears.

FIG. 8. Longitudinal section of a cnidoblast fixed during tubule inversion. Segments of inverting tubule (it) are present within segments of preinverted tubule (= external tubule, et). Scale bar = 0.5 μm.

FIG. 9. Longitudinal section of a cnidoblast fixed during tubule inversion. This cell was fixed in the presence of potassium antimonate to localize calcium. The inverting tubule (it) and preinverted tubule (= external tubule, et) are covered by small, electron-lucent fibrils, some of which are indicated by arrowheads. Scale bar = 0.5 μm.

FIG. 10. A tracing of Figure 9 in which the fibrils covering the tubule have been highlighted by showing them against a solid-black background.

FIG. 11. Cross section of a fully inverted tubule in a nematocyst from *Physalia*. This cell was fixed in the presence of potassium antimonate to localize calcium. Small fibrils are present (arrowheads) that interconnect pleats of the tubule. Other fibrils (f, arrowhead) are visible on the surface of the tubule. Scale bar = 0.5 μm.

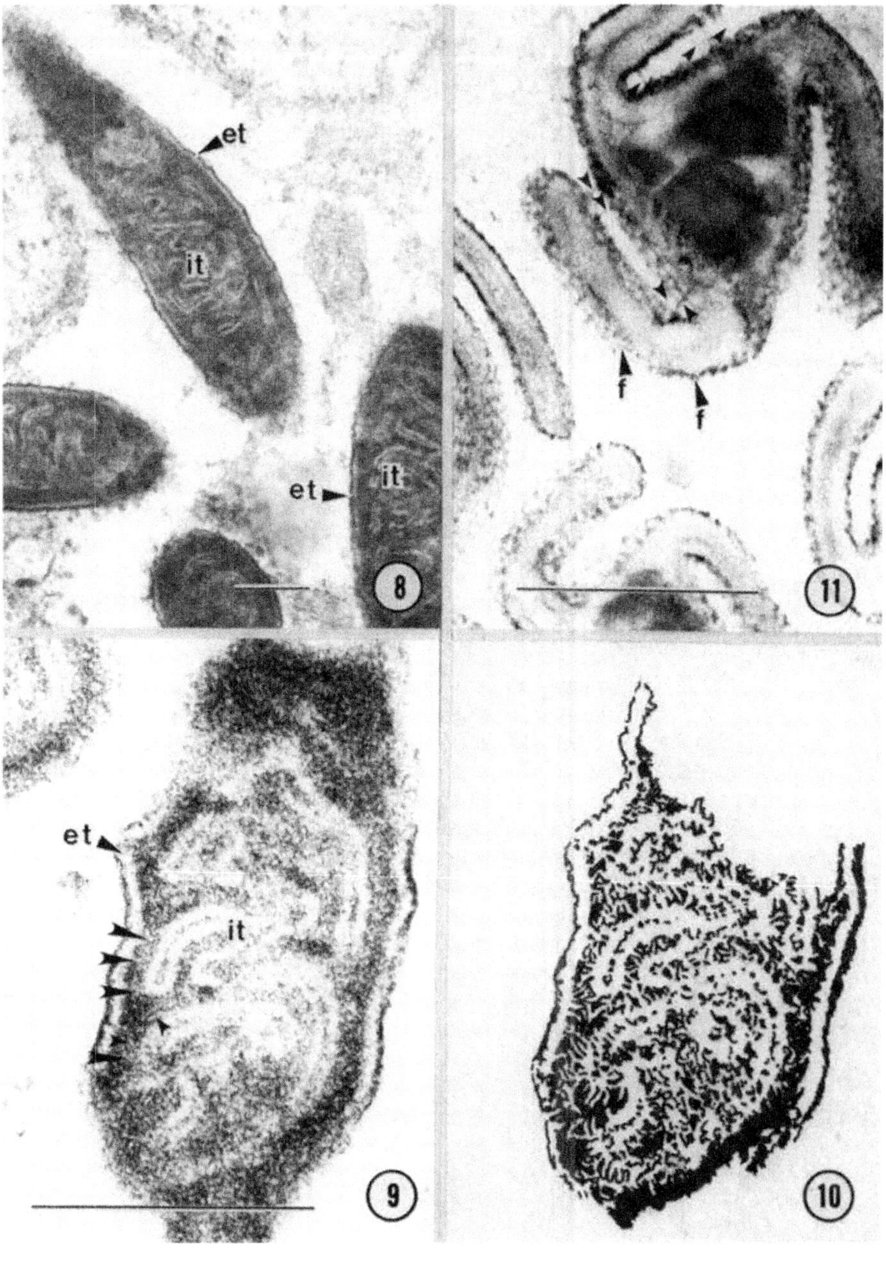
et
it
et
it
8
f
f
11
et
it
9
10

FIG. 12. Longitudinal section of a cnidoblast showing a mass of electron-lucent particles (pm) within the capsule (c). The electron-lucent layer of the nematocyst wall (nw) no longer is wrinkled. At this stage of development, this layer of the nematocyst wall still has a mottled appearance with electron-dense areas within the predominately electron-lucent layer. This tissue was fixed in glutaraldehyde and then poststained with Alcian blue. Scale bar = 0.5 μm.

FIG. 13. Longitudinal section of a capsule (c) from a mature nematocyst. The electron-lucent layer of the nematocyst wall (nw) is thinner than in previous stages of development. Furthermore, this layer now is uniformly electron-lucent (i.e., is no longer mottled). Apical flaps (a) also are visible. Scale bar = 0.5 μm.

FIG. 14. Longitudinal section of a mature nematocyst fixed during discharge. The tissue was exposed to performic acid, then stained with Alcian blue. The electron-lucent layer of the nematocyst wall (nw) is thicker than in control nematocysts (e.g., Figs. 13,16). The electron-lucent layer of the wall has a mottled appearance. Electron-dense material in the capsule (c) represents matrix contents within the capsule. Apical flaps (a) are visible. Scale bar = 0.5 μm.

FIG. 15. Longitudinal section of a capsule (c) from a fully discharged nematocyst. The tissue was exposed to performic acid, then stained with Alcian blue. The electron-lucent layer of the nematocyst wall (nw) is thick and lined with electron-dense precipitate. Apical flaps (a) are visible. Scale bar = 0.5 μm.

FIG. 16. Longitudinal section of a mature nematocyst fixed after discharge with glutaraldehyde and then poststained with Alcian blue. The capsule wall (cnw) lacks electron-dense precipitate present along the tubule wall (tnw). Apical flaps (a) are visible. Scale bar = 0.5 μm.

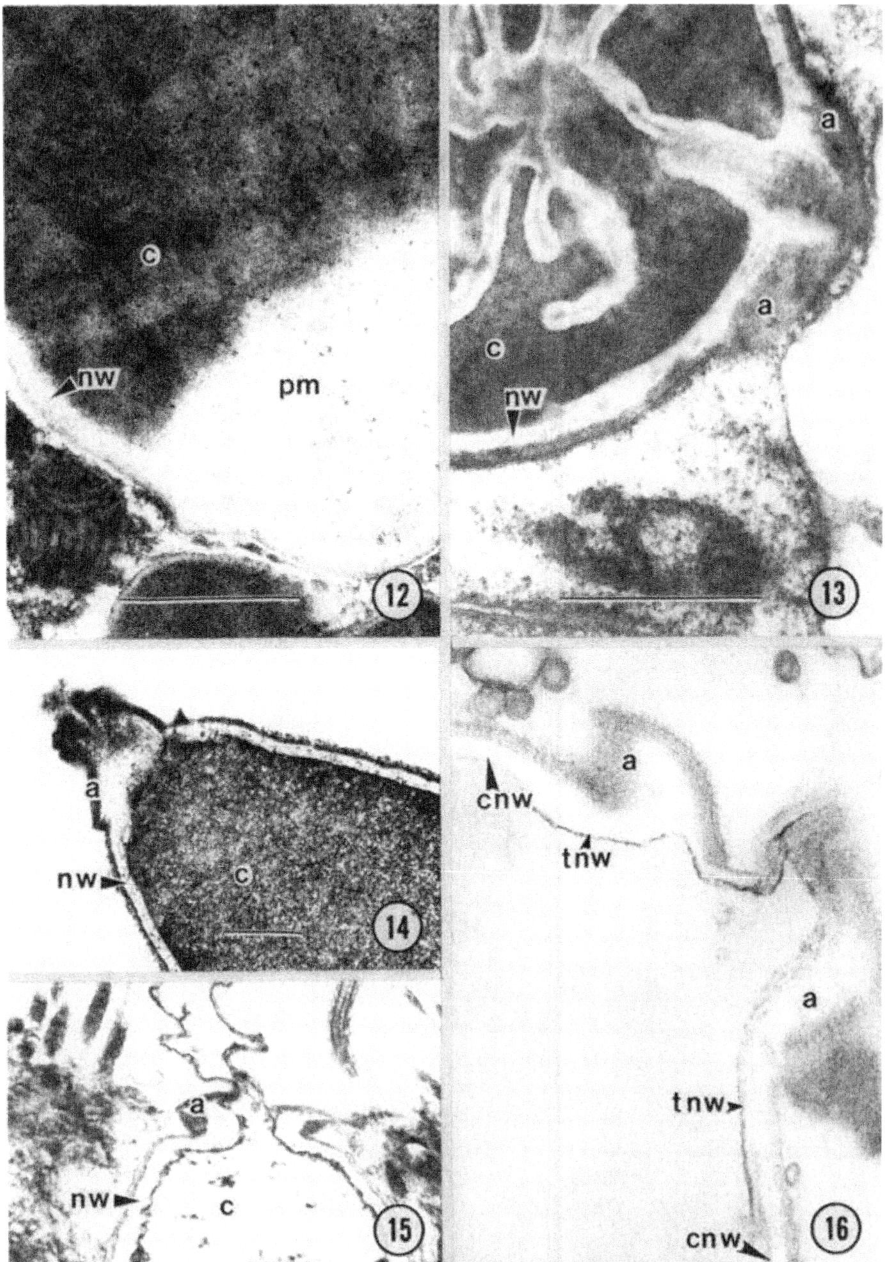
c
nw
pm
12
a
a
c
nw
13
a
nw
c
14
a
cnw
tnw
a
tnw
cnw
16
a
nw
c
15

Wall thickening is accompanied by a straightening of the wall (Watson and Mariscal, 1984b); recall that the capsule wall can wrinkle in certain nematocysts once the Golgi-microtubule complex "moves off" the capsule. After the tubule inverts into the capsule, the electron-lucent layer thins to approximately one-half the maximum width observed during nematocyst development (Watson and Mariscal, 1984b). This thinning of the electron-lucent layer is accompanied by a change in its appearance such that it no longer is mottled, and instead is uniformly electron-lucent (Fig. 13). In siphonophores, the capsule wall thins in maturing nematocysts (Skaer, 1973).

Nematocyst walls are composed of a collagen-like protein (Lenhoff _et al_., 1957; Fishman and Levy, 1967) crosslinked by disulfide bonds (Blanquet and Lenhoff, 1966; Mariscal and Lenhoff, 1969). Mature nematocysts treated with performic acid undergo substantial thickening of the electron-lucent layer (Figs. 14,15) to dimensions comparable to the maximum

FIGS. 17,18. Schematic drawings of several possible patterns for formation of disulfide bonds among molecules of the nematocyst wall. FIG. 17. Black circles represent nucleating centers for disulfide bond formation. (A) Random nucleation of disulfide bond formation (arrows). The wall remains mottled with "spaces" (arrowheads) among molecules comprising the wall. Such spaces would be occupied by matrix substances within the nematocyst (stipple). (B) A random nucleation of disulfide bond formation follows an establishment of a template of wall material along the membrane forming the outer boundary for the wall (arrowhead). The wall remains mottled. (C) An organized pattern of disulfide bond formation is shown (arrows) resulting in a uniform appearance of the nematocyst wall. FIG. 18. (A) Lined circles represent a template of wall material formed while the nematocyst was surrounded by a cage of microtubules. Formation of disulfide bonds begins at the template at many different sites and moves inward (arrows). The nematocyst wall has an uniform appearance. (B) Elongate blade-shaped lines represent polymers of wall material crosslinked by disulfide bonds (arrows) in an orderly fashion (_i.e._, outside to inside). Short arrows indicate the pattern of disulfide bond formation in progress between adjacent polymers (still curved). Once disulfide bonds are present between polymers (long arrows), polymers are straight. Ultimately, these processes would result in an uniform wall.

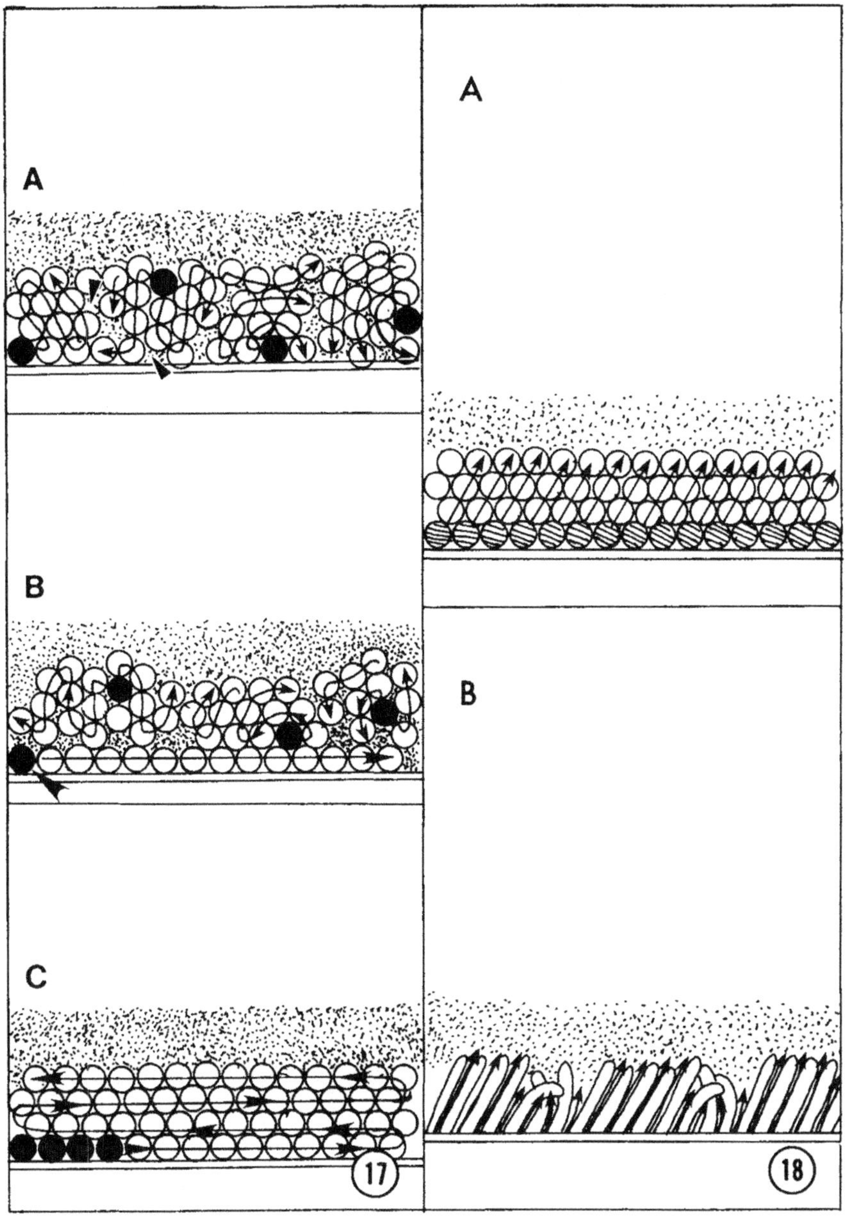
A
B
C
17
A
B
18

thickness observed in developing nematocysts (Watson and Mariscal, 1984b). Performic acid oxidizes disulfide bonds to sulfonates (*e.g.*, Stryer, 1981). Capsule walls of mature nematocysts treated with performic acid can appear to be mottled (Fig. 14) and are heavily stained by Alcian blue (Fig. 15), a stain specific for "free" sulfur groups (Pearse, 1968; Hayat, 1975). In contrast, capsule walls of control nematocysts are unstained by Alcian blue (Fig. 16).

It has been suggested that thinning of the nematocyst wall during its maturation is caused by formation of disulfide bonds among molecules in the nematocyst wall (Watson and Mariscal, 1984b). If this interpretation is correct, then one can predict that the thinning of the wall must occur in an orderly fashion. It seems likely that the appearance of the wall could change from mottled to homogeneous only if disulfide bonds formed in an orderly fashion (*e.g.*, from the outside to the inside of the wall, Figs. 17c, 18). Should disulfide bonds form more randomly (*e.g.*, Fig. 17a,b) the wall would be expected to become more rigid as a result of numerous disulfide bond cross-links, but not to change from a mottled appearance to a homogeneous appearance. While almost nothing is known about the interactions which occur between molecules as the nematocyst wall forms and then matures, this topic promises to be among the most fascinating for study in the future.

FIG. 19. Longitudinal section of a cnidoblast fixed during formation of the capsule. Potassium antimonate was included with the fixative to localize calcium. Electron-dense precipitate in Golgi cisternae (g) and within the capsule (c) is caused by the antimony salt. Scale bar = 0.5 um.

FIG. 20. Longitudinal section of a cnidoblast fixed during formation of the tubule. Antimonate deposits are present in Golgi cisternae (g) and within the preinverted tubule (= external tubule, et). Scale bar = 0.5 um.

FIG. 21. Longitudinal section of a cnidoblast from *Physalia* fixed after the tubule (t) has inverted into the capsule. Potassium antimonate was included with the fixative to localize calcium. Capsule (c) contents are virtually devoid of antimonate deposits. Scale bar = 0.5 um

FIG. 22. Longitudinal section of a cnidoblast from *Physalia* fixed after the tubule (t) has inverted into the capsule. Antimonate precipitate is present within the capsule (c) and also in the cytoplasm (arrowhead) of the cell. Scale bar = 0.5 um.

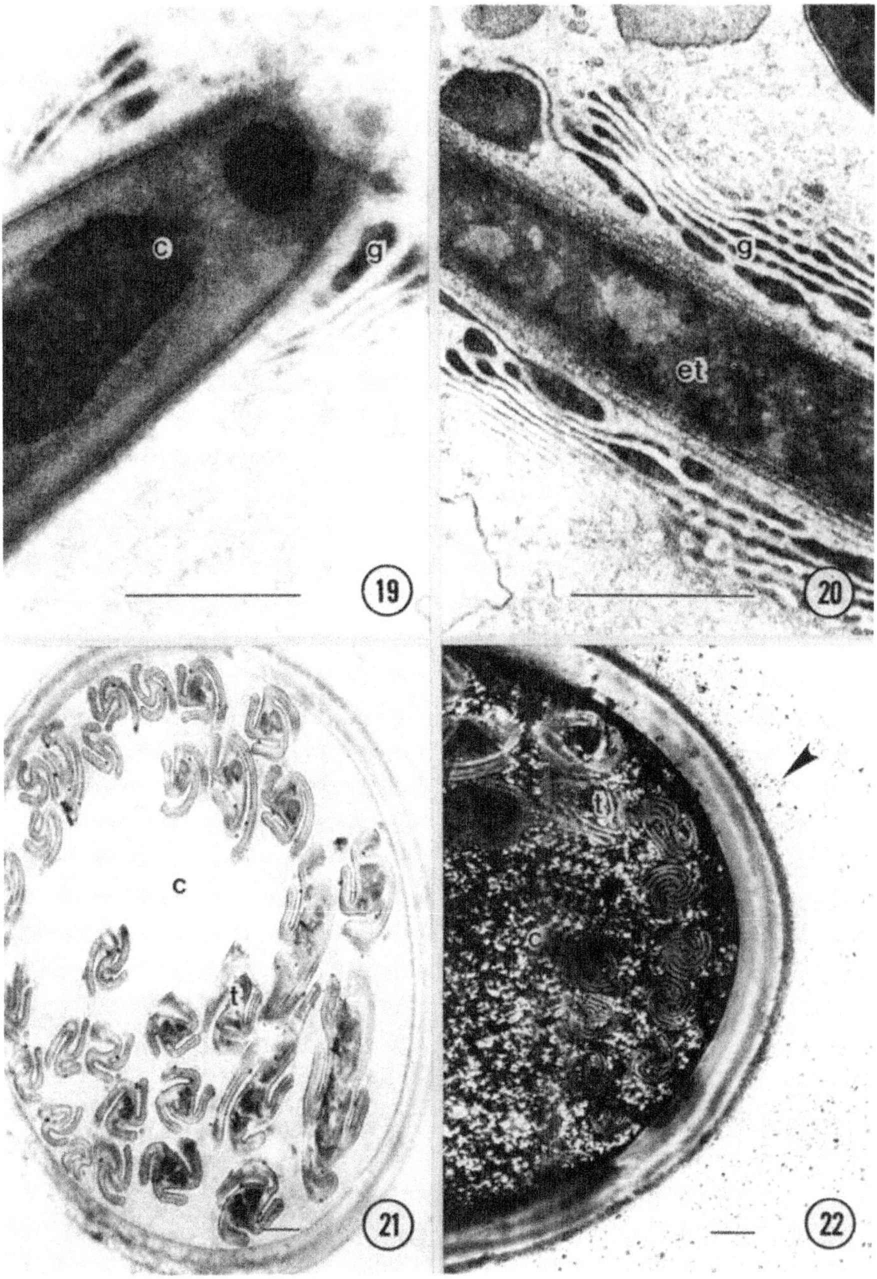
c
g
19
g
et
20
c
t
21
t
c
22

## I. The Appearance of Calcium in Immature Nematocysts

Large amounts of calcium occur in matrix contents of certain mature nematocysts (Lubbock *et al*., 1981; Mariscal, 1984). In certain nematocysts, potassium antimonate has been used to show cytochemically that calcium can be translocated from the Golgi to the nematocyst primordium (Figs. 19,20) (Watson and Mariscal, 1984a). In contrast, X-ray microanalysis on sections of frozen material has revealed a low concentration of calcium in certain immature nematocysts and a high concentration of calcium in fully mature nematocysts of the same type (Lubbock *et al*., 1981). Based on these observations as well as on some previously unpublished data, it was speculated that the endoplasmic reticulum transports calcium into immature nematocysts during final stages of maturation (Gupta and Hall, 1984). Potassium antimonate was used on immature nematocysts from *Physalia* to show cytochemically that calcium accumulates in these nematocysts after the tubule inverts into the capsule (Figs. 21,22). Because the Golgi apparatus no longer is associated with the nematocyst at this stage of development (Holstein, 1980), it appears that two different pathways may exist for calcium entry into immature nematocysts: (i) via the Golgi apparatus throughout the formation of the nematocyst and (ii) via an unknown source, possibly the ER, after the tubule inverts into the capsule.

In *Physalia* nematocysts, the accumulation of matrix contents along the capsule and tubule walls occurs soon after the appearance of calcium within the capsule (Fig. 23). The relationship, if any, between the appearance of calcium and this accumulation of matrix contents along the wall of the nematocyst is not yet known. At maturity, nematocysts from *Physalia* demonstrate calcium-binding material along the capsule wall and tubule wall (Fig. 24).

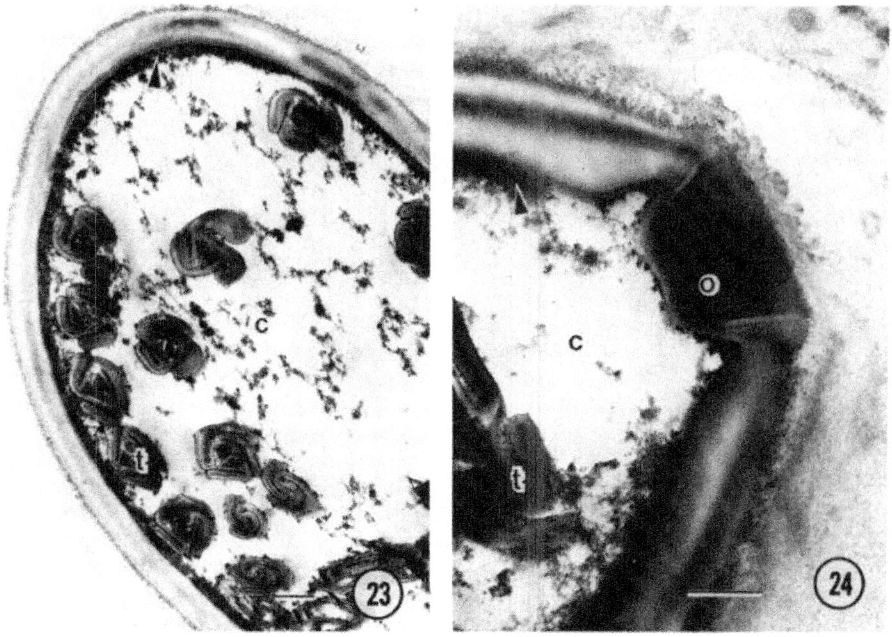


FIG. 23. Longitudinal section of a cnidoblast from *Physalia* fixed after some matrix contents have accumulated along the wall of the capsule. Antimonate deposits now are distributed along the capsule wall (arrowhead) instead of throughout the capsule lumen (c) (as was seen in Fig. 22). Almost no antimonate is apparent in the cytoplasm of the cnidoblast. (t = tubule). Scale bar = 0.5 um.

FIG. 24. Longitudinal section of a mature cnidocyte from *Physalia* with a nematocyst at the apical cell surface. Capsule contents (c) are similar in appearance to those of the immature nematocyst shown in the previous figure. Antimonate precipitate is distributed along the capsule wall (arrowhead) and along the wall of the tubule (t). The operculum (o) is visible. Scale bar = 0.5 um.

## II. CONCLUSIONS

In this paper, I have attempted both to present ultrastructural data and the interpretations, often speculative, of investigators. At this point, the sequence of events involved in the development of nematocysts appears to be fairly well resolved. However, the molecular basis of nematocyst development is almost completely unknown. In the future, the most important papers on nematocyst development will likely be those that attempt to reveal the interactions of molecules during the assembly and maturation of nematocysts. Undoubtedly, such interactions will be complex and fascinating to study.

## ACKNOWLEDGEMENTS

I am grateful to Drs. Richard Mariscal, David Hessinger and Leo Minasian for providing ample food for thought, and to Drs. Howard Lenhoff and David Hessinger for constructive criticism on the manuscript. Some micrographs have appeared in the Journal of Ultrastructure Research and are reproduced here with permission of the publisher.

## REFERENCES

Blanquet, R., and H. M. Lenhoff. 1966. Disulfide-liked collagenous protein of nematocyst capsules. Science 154: 152-153.

Carré, C., and D. Carré. 1973. Etude du cnidome et de la cnidogenese chez *Apolemia uvaria* (Lesueur, 1811) (Siphonophore Physonecte). Exp. Cell Res. 81:237-249.

Carré, D. 1972. Etude du developpement des cnidocystes dans le gastrozoide de *Muggiaea kochi* (Will, 1844) (Siphonophore calycophore). C. R. Acad. Sci. Paris 275: 1263-1266.

Carré, D. 1980. Hypothese sur le mechanisme de l'evagination du filament urticant des cnidocystes. Europ. J. Cell Biol. 20: 265-271.

Chapman, G. B., and L. G. Tilney. 1959. Cytological studies of the nematocysts of *Hydra* II. The stenoteles. Biophysic. Biochem. Cytol. 5: 79-95.

Cormier, S. M., and D. A. Hessinger. 1980. Cellular basis

for tentacle adherence in the Portuguese man-of-war (Physalia physalis). Tiss. Cell 12: 713-721.

Fishman, L., and M. Levy. 1967. Studies on nematocyst capsule protein from the sea anemone Metridium marginatum. Biol. Bull. 133: 464-465.

Germer, T., and M. Hundgen. 1980. The fine structure of developing and mature cnidoblasts of Eirene viridula. In P. Tardent and R. Tardent (eds.) Developmental and Cellular Biology of Coelenterates, pp. 325-330. Elsevier/ North Holland, Amsterdam.

Gupta, B. L., and T. A. Hall. 1984. Role of Ca, Cu, and Zn in the maturation and discharge in situ of sea anemone nematocysts as shown by X-ray microanalysis of cryosections. In Bolis et al. (eds.) Toxins, Drugs, and Pollutants in Marine Animals, pp. 77-95. Springer-Verlag, Berlin, Heidelberg.

Hayat, M. A. 1975. Positive Staining for Electron Microscopy. Van Nostrand Reinhold Company, New York.

Holstein, T. 1980. An ultrastructural and kinematographic study of the nematocyst morphogenesis in selected coelenterate species. In P. Tardent and R. Tardent (eds.) Developmental and Cellular Biology of Coelenterates, pp. 313-318. Elsevier/ North Holland, Amsterdam.

Holstein, T. 1981. The morphogenesis of nematocytes in Hydra and Forskalia: An ultrastructural study. J. Ultrastruct. Res. 75: 276-290.

Ivester, M. S. 1977. Nematocyst differentiation in the anthozoon Renilla reniformis (Pallas). Trans. Amer. Micros. Soc. 96: 238-247.

Lenhoff, H. M., E. S. Kline, and R. Hurley. 1957. A hydroxyproline-rich, intracellular, collagen-like protein of Hydra nematocysts. Biochem. Biophys. Acta 26: 204-205.

Lentz, T. L. 1965. The fine structure of differentiating interstitial cells in Hydra. Z. Zellforsch. 67: 547-560.

Lubbock, R., B. L. Gupta, and T. A. Hall. 1981. Novel role of calcium in exocytosis: Mechanism of nematocyst discharge as shown by X-ray microanalysis. Proc. Natl. Acad. Sci. USA 78: 3624-3628.

Mariscal, R. N. 1984. Cnidaria: Cnidae. In J. Bereiter-Hahn, A. G. Maltoltsy, and K. S. Richards (eds.) Biology of the Integument, vol. I, Invertebrates, pp. 57-68. Springer-Verlag, Berlin.

Mariscal, R. N., and H. M. Lenhoff. 1969. Effect of a disulfide reducing agent on coelenterate nematocyst capsules. Experientia 25: 330-331.

Pearse, A. G. E. 1968. Histochemistry: Theoretical and Applied. Little, Brown and Company, Boston.

Schmidt, H. 1981. Die Cnidogenese der Octocorallia (Anthozoa, Cnidaria): I. Sekretion und Differenzierung von

Kapsel und Schlauch. Helgol. Wiss. Meeresunters 34: 463-484.

Skaer, R. J. 1973. The secretion and development of nematocysts in a siphonophore. J. Cell Sci. 13: 371-393.

Skaer, R. J., and L. E. R. Picken. 1965. The structure of the nematocyst thread and the geometry of discharge in Corynactis viridis (Allman). Phil. Trans. R. Soc. Lond. 250: 131-164.

Slautterback, D. B. 1961. Nematocyst development. In H. M. Lenhoff, and W. F. Loomis (eds.) The Biology of Hydra, pp. 77-129. Univ. of Miami Press, Coral Gables, Florida.

Slautterback, D. B. 1963. Cytoplasmic microtubules. I. Hydra. J. Cell Biol. 13: 367-388.

Slautterback, D. B., and D. W. Fawcett. 1959. The development of the cnidoblasts of Hydra: An electron microscope study of cell differentiation. J. Biophysic. Biochem. Cytol. 5: 441-451.

Stryer, L. 1981. Biochemistry. W. H. Freeman and Company, San Francisco.

Watson, G. M., and R. N. Mariscal. 1984a. Calcium cytochemistry of nematocyst development in catch tentacles of the sea anemone Haliplanella luciae (Cnidaria: Anthozoa) and the molecular basis for tube inversion into the capsule. J. Ultrastruct. Res. 86: 202-214.

Watson, G. M., and R. N. Mariscal. 1984b. Ultrastructure and sulfur cytochemistry of nematocyst development in catch tentacles of the sea anemone Haliplanella luciae (Cnidaria: Anthozoa). J. Ultrastruct. Res. 87: 159-171.

Watson, G. M., and R. N. Mariscal. 1985. Ultrastructure of nematocyst discharge in catch tentacles of the sea anemone Haliplanella luciae (Cnidaria: Anthozoa). Tiss. Cell 17: 199-213.

Westfall, J. A. 1966. The differentiation of nematocysts and associated structures in the Cnidaria. Z. Zellforsch. 75: 381-403.

# 10
# THE FORMATION OF CNIDOCYTE PATTERNS IN SIPHONOPHORES

R. J. SKAER

Peterhouse
Cambridge
England

## SUMMARY

Groupings of cnidoblasts form simple, precise patterns in many cnidarians but particularly in siphonophores. The advantages of using cnidoblasts for the study of pattern formation are described. The information gained allows one to eliminate certain models of pattern formation and to concentrate on behavioral maneuvers of cnidoblasts that sort these cells, align them and merge simpler patterns to make more complex ones. Thus the fundamental constructional units are the clump, the row, and the ability to sort or merge rows. Parameters that are important in the rapid and accurate construction of patterns are discussed in the light of the behavior of migrating cells and of cells undergoing 'contact reassortment.'

## INTRODUCTION

One way by which we hope to learn how tissues are built up is by studying pattern formation. Patterns, particularly if they are simple yet precise, allow certain predictions to be made about where particular cells originate, where they go, how, and possibly why. Cnidoblasts are particularly useful in the study of certain types of pattern formation for they often form very striking yet simple arrays. In siphonophores the arrays are often geometrically exact, are usually made up of fixed numbers of cells, and are set up quite late in differentiation by which time the nematocyst is clearly visible and its type recognizable. The nematocyst thus acts as a label in the cell and in siphonophores one can distinguish at least 7 different categories of cnidoblasts based on nematocyst type, size, etc. Moreover, the nematocyst, by its orientation, can tell us whether the cnidoblast is migrating and the direction of migration (Günzl, 1971; Skaer, 1973). In migrating cnidoblasts the capsule trails behind and the operculum is oriented towards the rear of the cell. At the end of migration, however, the capsule rotates through 90° so it lies apically in the cell and orients with its axis normal to the surface (Wood & Novak, 1982; Novak & Wood, 1983). Thus one can distinguish between cells that are migrating and cells that are positioning themselves after migrating. Moreover, the orientation of the nematocyst about its long axis, as shown by the packing of the thread in the capsule, allows us to see the left/right asymmetry of the cells i.e. the direction in which the cells are facing when they have their nematocysts pointing outwards. Because of the conspicuous nature of the nematocysts one can observe them in the living animal. This advantage is enhanced by the fact that patterns of cnidoblasts in siphonophores form very rapidly – with a microscope one can watch it happen.

Geometrically striking arrangements of nematocysts in siphonophores were depicted by Thomas Henry Huxley in his Ray Society Monograph of 1858 and by Keferstein & Ehlers (1861), but neither they, nor, so far as I can discover, anyone else has studied how these patterns are set up. Semal-van Gansen (1954) speculated that, in hydra, the arrangement of nematocyst types around a central stenotele was based on cell size. She implied that as the battery cell flattened down on the cnidocytes during growth of the tentacle, mechanical factors distributed the nematocysts appropriately.

On the other hand, Wood & Novak (1982) and Novak & Wood (1983) suggested that the developing cnidoblasts in hydra migrate to preordained places associated with the battery cell. They showed that a unique junctional complex forms basally between the cnidoblast, battery cell and mesogloea; even in the absence of cnidoblasts, a structure that may be equivalent to the junctional complex can be found. They suggested that these hemidesmosome-like sites are the predetermined places for cnidoblast attachment. In this article I describe a completely different mechanism that appears to operate in siphonophores. More details of this process will be published in Skaer (1987a & b).

Here I will describe some of the cell patterns that occur in siphonophores and I will summarize some of the contributions to setting up the patterns made by cell migration, by the cell sorting that occurs during and as a result of migration, and by the repositioning movements of adherent cells. The latter result in alignments or realignments of the adherent cells which then may be followed by synchronised but rapid changes in orientation and relative positions of the cells involved. These relative movements I call 'contact reassortment' (Skaer 1987a,b). They may involve the relocation of the whole cell, or just changes in the relative orientation, alignment or rotation of the cells about their axes. They represent a novel form of cell behavior that may perhaps be regarded as a series of stereotyped cell 'dances', distinct from, say, contact guidance or contact inhibition of movement (Abercrombie, 1970). These movements result in further cell sorting and the formation of cell rows (and in some cases the merging of rows to form longer rows that may be made up of different types of cell). From the shapes of the cell clumps built up by migration one can speculate about the precision of the cnidoblast navigation, the information required by different mechanisms of pattern formation, and the stages that might require positional information.

## RESULTS

### *a. Cnidosac patterns*

Mature nematocysts of most siphonophores occur only in knobbed structures on the tentacles called cnidosacs (Fig. 1a). Within each cnidosac they form five major patterns (Fig. 1b):

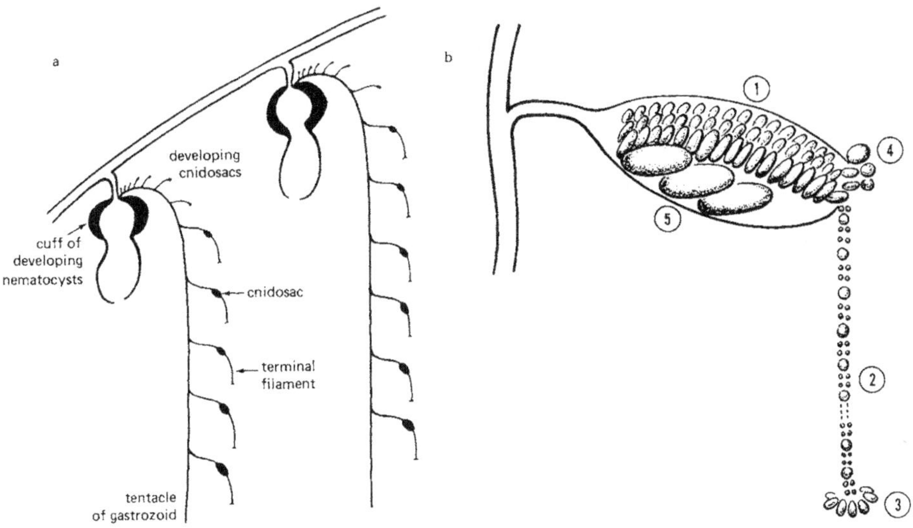


*Fig. 1a. Diagram of a typical arrangement of gastrozoids in a siphonophore, with each gastrozoid having one tentacle. The tentacle grows near its site of insertion on the gastrozoid and the cnidosacs form from outpushings of the tentacle. The cnidosacs are filled with cnidoblasts that have originated in the thick cuff around the top of the gastrozoid and then migrated into those cnidosacs that are at an appropriate stage of differentiation. Thus, the single tentacle represents a 'frozen' time sequence in the development of cnidosacs and their nematocyst patterns (youngest cnidosac proximally; oldest distally i.e. at the free end of the tentacle). Formation of the patterns is normally complete in cnidosacs a few millimetres down the tentacle.*

*Fig. 1b. Diagram of a typical Calycophoran cnidosac showing dorsal cnidoband (1); terminal filament (2); sinker (3); dorsal array of desmonemes (4); large lateral microbasic mastigophores (5).*

1. Cnidoband. This is a large array of anisorhizas that only differ from each other in size, if at all. The nematocysts form a wide band of highly aligned nematocysts that occupies most of the dorsal surface of the cnidosac. In *Calycophorae* such as *Chelophyes*, *Lensia* and *Abylopsis* the cnidoband nematocysts are arranged in seven straight rows, in *Forskalia*, 14 rows and in *Physophora* multiple rows – at least 19.
2. Terminal filament. The terminal filament is a thin, highly contractile filament attached to the distal end of the cnidosac. It contains two types of nematocysts only: rhopalonemes and desmonemes. In the genera considered here in detail (*Chelophyes* & *Abylopsis*) the arrangement of these nematocysts is a linear pattern whose repeat unit is 2 rhopalonemes, 1 desmoneme, 2 rhopalonemes. This pattern may be repeated 60–70 times along the terminal filament (Fig. 1b, 2, 3). In other genera (e.g. *Hippopodius*, *Forskalia*) the repeat unit is 2 rhopalonemes, 2 small desmonemes, 2 rhopalonemes (Fig. 4).

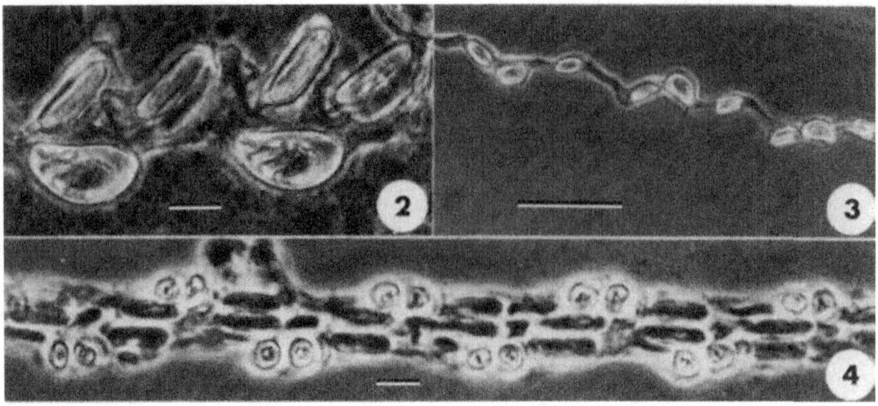


*Fig. 2. High power photomicrograph of a slightly squashed terminal filament of* Abylopsis *showing the paired rhopalonemes on each side of the (larger) desmoneme. Two repeat units are shown of the pattern 2 rhopalonemes, 1 desmoneme, 2 rhopalonemes. Phase contrast. The scale bar represents 10 μm.*

*Fig. 3. The same extended in* vivo. *The scale bar represents 100 μm.*

*Fig. 4. Terminal filament of* Forskalia *showing the repeat unit 2 rhopalonemes, 2 desmonemes, 2 rhopalonemes. Note that the relative size of the desmonemes in this genus is much smaller than in* Abylopsis. *Phase contrast. The scale bar represents 10 μm.*

3. Sinker. This is a circular disk containing large desmonemes only, situated at the distal end of the terminal filament.
4. Dorsal array of desmonemes. This pattern occurs at the distal end of the cnidoband near the origin of the terminal filament. It consists of large desmonemes similar to those on the sinker.
5. Lateral large nematocysts (usually microbasic mastigophores or stenoteles). These occur in fixed numbers (three on each side of the mature cnidosac in *Chelophyes*; eight on each side in *Abylopsis*). This pattern is ideal for studying the supply of cnidoblasts to the cnidosacs; individual nematocysts are large and their numbers are small. The numbers of nematocysts can be compared on the two sides of the cnidosac as well as in the cnidosacs immediately above and below.

*b. Accuracy of cnidosac patterns*

Cnidosac patterns are composed of cnidocytes that adhere to each other and in which the nematocysts are very highly oriented, not only with respect to the surface but also with respect to each other. This allows nematocysts of the same type to be compared in identical orientation. From this it is clear that the pattern of packing the tubule in the capsule follows the same course in nematocysts of the same type. In, for example, the large mastigophores of *Physophora* (Fig. 5) the thread emerging from the shaft immediately reverses its direction and moves forward towards the operculum on the right hand side of the shaft, it crosses above the shaft and when approximately 3/4 of the way along the shaft it forms two large loops. These loops are on the upper side of the shaft and the anterior part of the second loop on the righthand side is the part of the tubule nearest the operculum. The remaining tubule forms loops beneath the shaft moving progressively to the rear of the capsule.

The invariance of cnidoblast pattern and orientation and the precision of construction of the nematocysts within the pattern is well shown in the cnidoband of *Physophora* (Fig. 6). The loops of tubule in optical section occupy identical places in all the nematocysts shown.

The numerical accuracy of the large nematocysts in pattern 5 has been mentioned and the precise repeat unit in the terminal filament and the manner of its formation is described in detail in Skaer (1987b) and will be summarized later in this article.

*c. The source of the cnidoblasts that form the patterns*

In siphonophores, all cnidoblasts develop in a conspicuous cuff, many cells thick, that is situated at the top of the gastrozoid (Carré, 1972; Skaer, 1973). Unlike the situation in hydra where developing cnidoblasts are linked by cytoplasmic bridges so the progeny of a stem cell form a nest (Fawcett, Ito and Slautterback, 1959; Fujisawa and David, 1981); daughter cnidoblasts in siphonophores do not remain in contact during development; they merely remain moderately close (Skaer, 1973). Moreover it is very difficult to see any topographical ordering of type or developmental stage in the cuff. Cnidoblasts of all types and states occur cheek-by-jowl. From this source, a primary sorting out occurs by the migration of individual cnidoblasts out of the gastrozoid and into the developing cnidosacs; a distance of 1–2 mm.

*d. Migration*

Enormous numbers of cnidoblasts migrate, initially along a broad common pathway, then into separate pathways leading to the 3–7 cnidosacs that are being filled at any one time. There is no obvious separation into types in these pathways nor in the stalks that lead to the cnidosacs, but in the cnidosac the anisorhizas of the cnidoband move along the *dorsal* surface, whereas the rhopalonemes together with all the desmonemes whether destined for terminal filament, sinker or dorsal array of desmonemes, move *ventrally* along the cnidosac. The large desmonemes of the dorsal array finally move dorsally at the end of the cnidosac. There is clearly some sorting into types in those cnidobands where there is a variation in size of nematocysts. (e.g. *Chelophyes* and *Abylopsis*) since each row contains nematocysts of a particular size only, but it is not yet clear whether this happens during migration. Nor is it clear at what stage the sorting of the large nematocysts of pattern 5 into right-&-left hand clumps occurs. It is clear that very few cnidoblasts migrate back down the stalk of the cnidosac. Saturation of the pattern with excess cells as shown by the withdrawal of the superfluous cnidoblasts down the stalk does not occur to a significant extent.

*e. The end of migration*

Migration ends with the cells sorted into identical types that form adherent clumps (Skaer, 1987a,b). Unpublished work with Dr. A. Spencer has shown that the

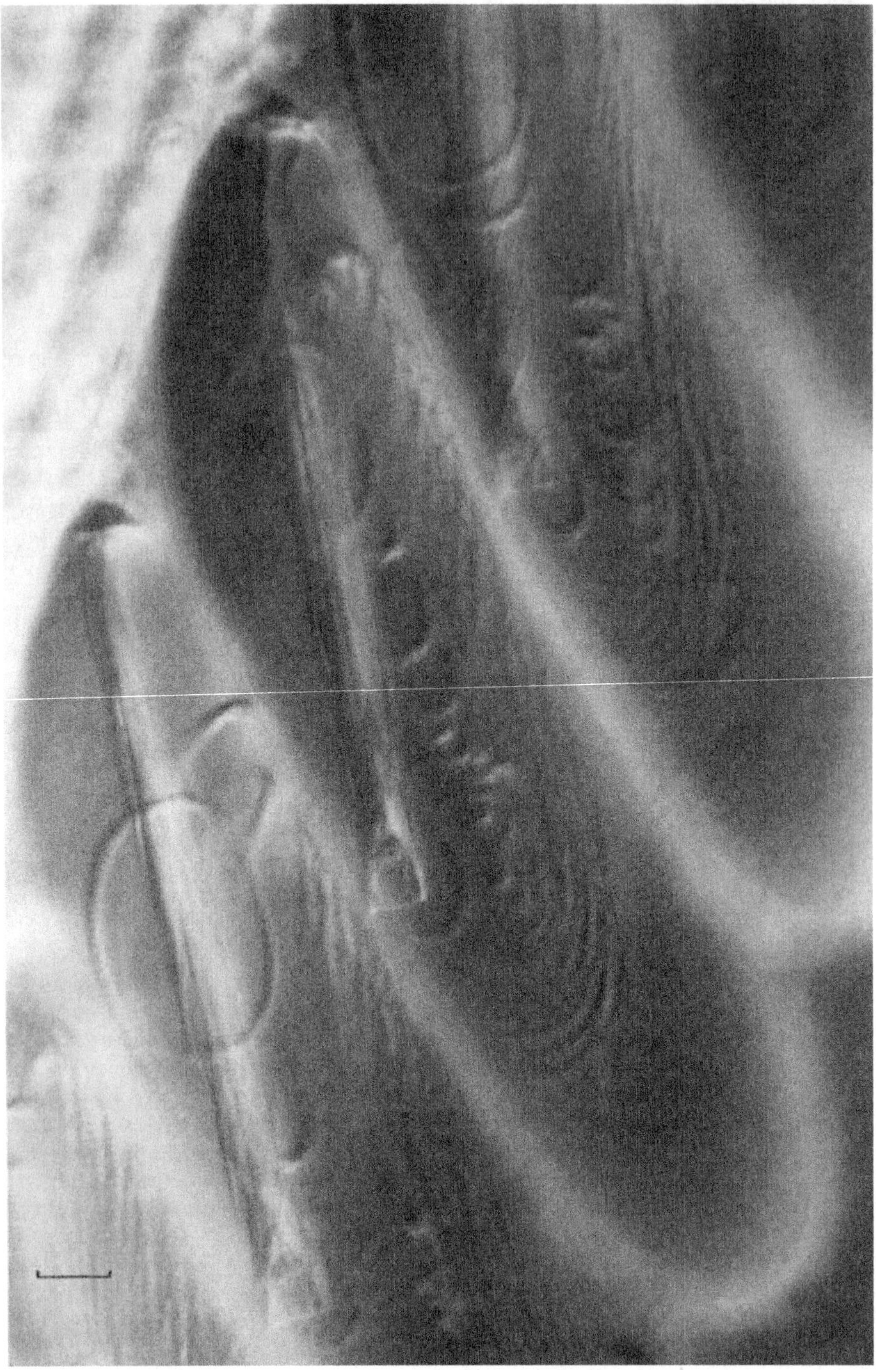

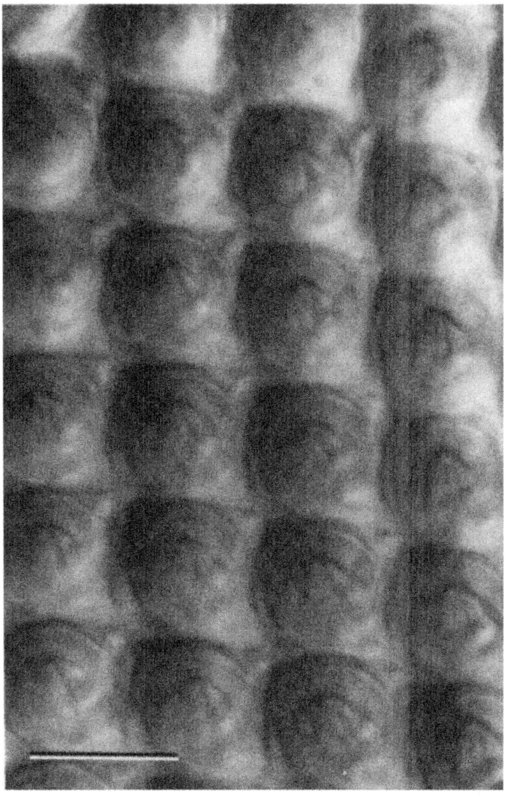

*Fig. 6. Cnidoband of* Physophora *seen from above. Four rows of cnidoblasts run from top to bottom of the picture in what is virtually a square lattice pattern. The geometrical precision of the pattern, the invariant orientation of the nematocysts within the pattern and the identical packing of thread within the capsules are shown. Nomarski optics. The scale bar represents 10 μm.*

cnidoblasts in the clumps are dye coupled for Lucifer yellow throughout all the subsequent movements of pattern formation. The end of migration is signalled by the rotation of the nematocyst through 90° so it comes to lie normal to the surface. The clumps form in the appropriate areas for pattern formation but are often of rather indefinite shape; for example a large, approximately helical clump of rhopalonemes

*Fig. 5. Large mastigophores of* Physophora *showing the invariance of orientation of the nematocysts and the similarity in packing of the shaft and tubule in the capsule in these identically oriented nematocysts. These nematocysts are approximately 180 μm long. Nomarski optics. The scale bar is 10 μm.*

and a rather smaller, approximately helical clump of small desmonemes form in the anlage of the terminal filament (Fig. 7, 8, 9). Further building of the patterns in the cnidoband and the terminal filament is described in more detail in Skaer (1987a,b) and will be summarized here.

In the cnidoband of *Physophora*, where all cnidoblasts are identical, an initially hexagonal packing of the adherent cnidoblasts in a clump is transformed into rows having a square lattice arrangement. Row formation is associated with rotation of the nematocysts (possibly of the cnidoblasts?) about their axes normal to the surface so

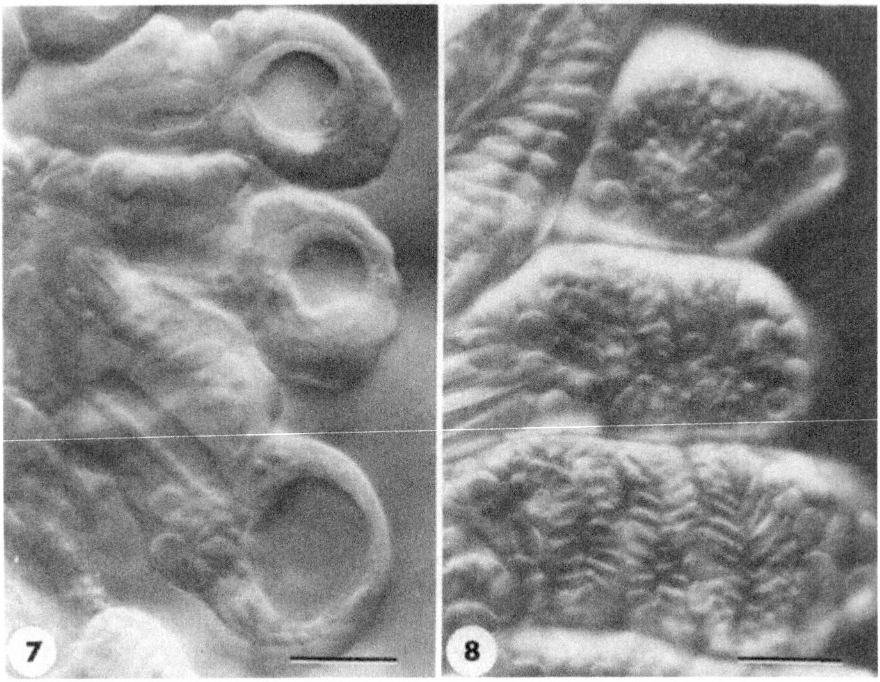


*Fig. 7. Three developing cnidosacs of* Chelophyes. *The lowermost one has just started to fill with cnidoblasts. The spherical anlage on the right-hand side are the precursors of the terminal filament. Nomarski optics. The scale bar is 100 μm. Reproduced by permission of The Company of Biologists Ltd.*

*Fig. 8. Three developing terminal filaments of* Chelophyes *at different stages of filling with cnidoblasts. The earliest stage (at the top) shows an irregular clump of small, cigar-shaped rhopalonemes flanked by a small number of spherical desmonemes. In the middle filament the rhopalonemes have formed a clump of two helical gyres in which the cells are arranging themselves into rows even though filling is not complete. In the bottom terminal filament the clump of rhopalonemes has been transformed into three helical gyres interspersed with desmonemes. The herring-bone pattern within each gyre is due to row formation and the merging of rows. The clump of large desmonemes at the far right-hand side will form the sinker. Nomarski optics. The scale bar is 100 μm. Reproduced by permission of The Company of Biologists Ltd.*

they coincide exactly with the orientation of neighbors. Row formation is fundamental to pattern-building in siphonophores. Row formation in the cnidoband of *Chelophyes* and *Abylopsis*, where the cnidoband nematocysts are all the same general type but differ in size, is more complex. Each row contains nematocysts of one particular size class. Sorting within the clump by a side-to-side pendulum movement of the cnidoblasts is aided by a certain precision of migration. Thus the cnidoblasts of the small anisorhizas that lie in the outermost row on each side of the cnidoband migrate near to the edge of the band of migrating cnidoblasts in the stalk of the cnidosac. Similarly the larger anisorhizas, that form a row on each side next to the smallest, do indeed tend to end their migration towards the edge of the cnidoband. It is not clear how the left-right symmetry on either side of the three rows of medium-sized anisorhizas is set up. Despite this relative accuracy of migration in building the cnidoband, cell sorting of adherent cells is a major factor in the precision of the procedure. The pendulum-like movements in directions at right angles to the forming row take place after migration, i.e. within the adherent clump. The movements cannot be given a periodicity since some nematocysts remain obliquely oriented until row formation is complete (Skaer, 1987a). The nematocysts may depart up to 45° from their post-migration position (normal to the surface). It is assumed that some form of differential adhesion in this position achieves this sorting. After the pendulum movements have stopped and the nematocysts again lie normal to the surface, the row is a zig-zag arrangement whose members progressively align themselves as row formation proceeds from the distal end of the cnidoband (Skaer 1987a)

Row formation and the pendulum-swinging at right angles to the row are particularly apparent in the terminal filament. Formation into the four rows of rhopalonemes and one row of desmonemes precedes the one-way pendulum maneuver that merges row with row (a type of reverse sorting) (Fig. 10, 11, 12, 13). The simultaneous merging of two distal rows of rhopalonemes and of two proximal rows of rhopalonemes effectively doubles the length of the filament which then undergoes a further elongation as the resulting two rows of rhopalonemes merge with the row of desmonemes to give the final repeat unit 2 rhopalonemes; 1 desmoneme; 2 rhopalonemes that is found in *Chelophyes*, *Abylopsis* and *Lensia* (Fig. 13). In the mature filament each repeat unit is separated from its neighbor by a length of contractile filament approximately equal to half a repeat unit. As this final elongation and merging of unlike rows occurs, the terminal filament which hitherto had been helically coiled, collapses into a long and highly contractile filament. In genera such as *Hippopodius* and *Forskalia* where the repeat unit is (2 rhopalonemes; 2 desmonemes; 2 rhopalonemes)$_n$ (Fig. 4) the paired desmonemes are smaller than in *Chelophyes* and are derived from two rows per gyre rather than the one as described above. The regularity of the repeat unit would appear to depend in part on the relative spacing of the cells in the rows.

## DISCUSSION

The results show that in siphonophores the patterns are not built by direct migration to particular sites, rather, migration builds clumps of identical cell types in

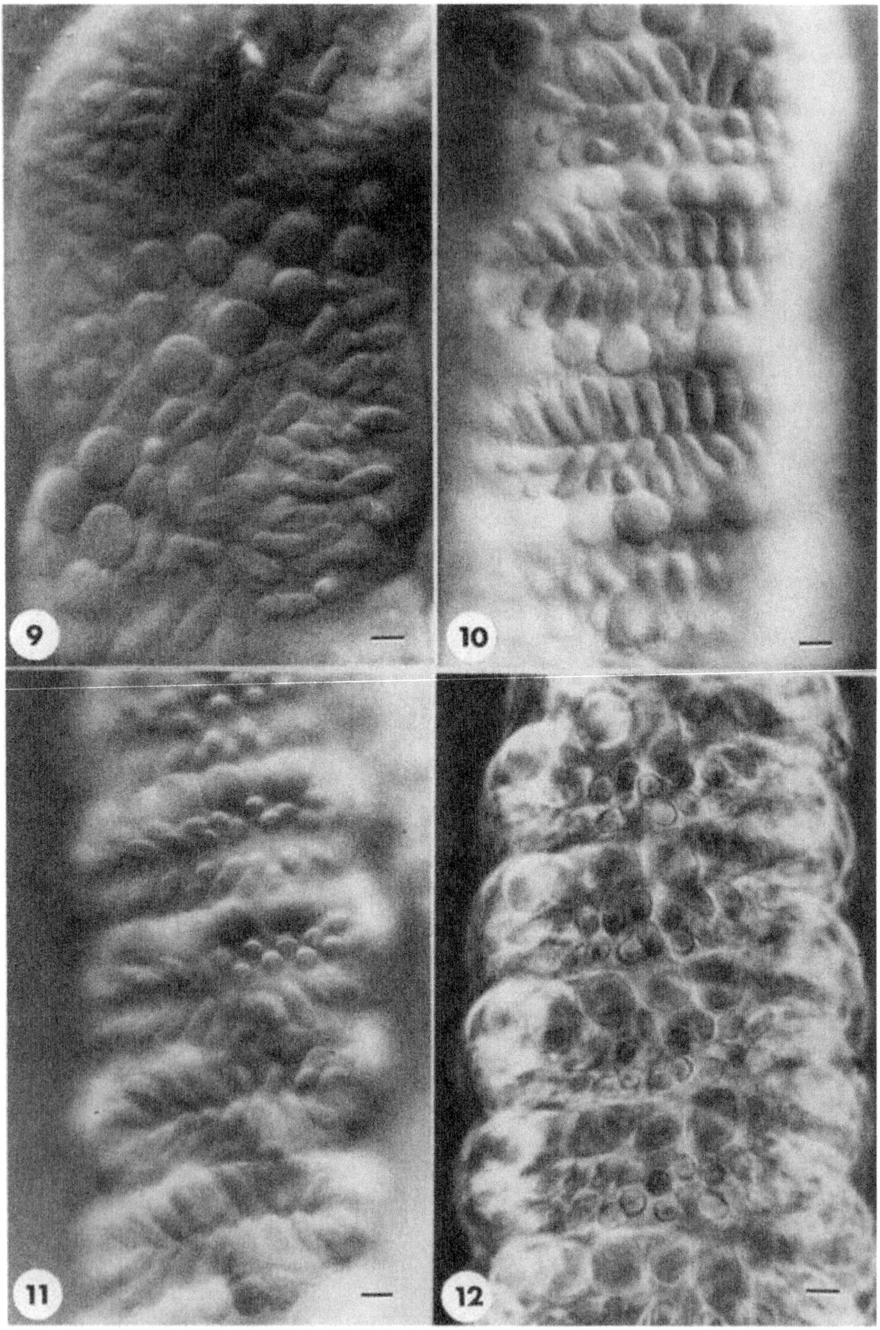
9
10
11
12

particular areas. In other words migration sorts cells into specific categories of cnidae (or nematocysts). Migration also appears to provide the clump with the correct number of cells. After migration, when the cells are adherent and dye coupled, they then perform the remarkable sorting, aligning and interacting behavior patterns that control the detailed placing of the cells. Elsewhere (Skaer, 1987b) I have called these maneuvers 'contact reassortment' to distinguish the movements of cells in contact (and dye-coupled) from the locomotion of migrating cells. The placement of cells by 'contact reassortment' is both geometrically and numerically invariant. This exactitude depends on precise alignment of the cells into rows, precise control of cell size (or, effectively, cell spacing) and the precise orientation of the cells within the rows with respect to their intrinsic asymmetry. This asymmetry is expressed, for example, in the exact orientation of the nematocysts about their longitudinal axis – an asymmetry that is revealed only because of the precision of construction of the nematocysts themselves.

We do not know whether the migrating cnidoblasts need or have positional information continuously available; it could be that all cells of a particular type arrive in a clump simply because all have identical navigational instructions. In the case of the adherent cells, however, alignment into rows, the intermeshing of the rows and the subsequent rotation through 180° of cells from one row to face in the same direction as the row they have entered, would seem to require the cells to know where they are in the pattern and, therefore, the direction they should be facing. Although one could envisage that identical cell types could develop identical cell surface patterns of adhesion, with cells adhering in a straight line by self-assembly, rather like LEGO bricks snapping together to build a straight row, the fact remains that this would still require one cell – the 'founder' cell – to be correctly oriented so it would

*Fig. 9. Higher power view of a developing terminal filament of* Ablyopsis *at a stage corresponding to that in the top anlage of Fig. 8. Two gyres of the helical clump of rhopalonemes are traversed by the oblique clump of spherical desmonemes. These are small desmonemes but are nevertheless substantially larger than the rhopalonemes. At this stage cnidoblasts of both types are still entering the terminal filament. Nomarski optics. The scale bar represents 10 μm.*

*Fig. 10. Terminal filament of* Abylopsis *showing later stage with four gyres of the helically coiled filament. Within each gyre are four rows of rhopalonemes and one row of desmonemes. The two proximal rows are merging with each other as are the two distal rows giving rise to the herring-bone pattern. This corresponds to the bottom filament in Fig. 8. Nomarski optics. The scale bar represents 10 μm. Reproduced by permission of The Company of Biologists Ltd.*

*Fig. 11. Terminal filament of* Abylopsis *at a slightly later stage showing the zig-zag positioning rhopalonemes as the two pairs of rows merge to produce two rows per gyre. The depth of each gyre is thus reduced and the length of the filament increased. Approximately five gyres are shown. Nomarski optics. The scale bar represents 10 μm.*

*Fig. 12. Terminal filament of* Abylopsis *at an even later stage. There are now only two rows of rhopalonemes in each gyre and the gyres have diminished in depth still more. The filament has increased correspondingly in length. Phase contrast. The scale bar represents 10 μm.*

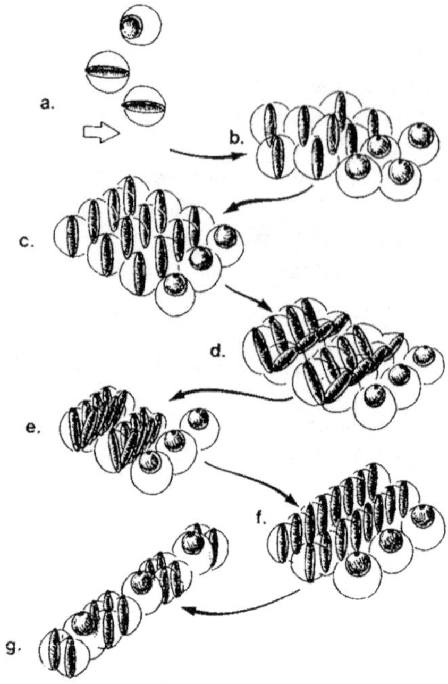


*Fig. 13. Diagram showing the formation of the pattern of nematocyst types in the terminal filament of* Chelophyes *or* Abylopsis.
*a. Cnidoblasts migrate to the site of pattern formation, nematocyst and operculum trailing. The open arrow indicates the direction of migration.*
*b. Formation of clumps of identical types of cell at the site of pattern formation. This marks the end of migration – nematocyst normal to surface, operculum at the surface. Compare with Fig. 9.*
*c. Row formation within the clump of adherent cells. Four rows of rhopalonemes (3 shown here) and one row of desmonemes formed.*
*d. Merging of rows of rhopalonemes – compare with Fig. 10 and the herring-bone pattern in Fig. 8.*
*e. Later stage of merging, with rhopaloneme orientation returning nearly normal to the surface and forming two long zig-zag rows from four shorter straight rows. Compare with Fig. 11.*
*f. Later stage still – only two long rows of rhopalonemes alongside the single row of desmonemes. Compare with Fig. 12.*
*g. Mature terminal filament – merging of unlike rows after a process similar to that in d. to form the repeat unit 2 rhopalonemes, 1 desmoneme, 2 rhopalonemes. Compare with Fig. 2.*

initiate correct orientation of the row. It is perhaps simpler to assume that all cells have this positional information.

Building a pattern cell-by-cell by the mechanism of migration (Wood & Novak, 1982; Novak & Wood, 1983) is an essentially *serial* activity requiring large amounts of information to be fed back to the migrating cells. The type of pattern building that occurs in the terminal filament proceeds simultaneously and in parallel throughout the pattern. It is thus a very rapid way to build a pattern. In some genera, the construction of the cnidoband proceeds sequentially but is not dependent upon particular cell types arriving at particular times, or on cells finding and binding to preordained sites. Migration to the cnidoband appears to carry out an approximate sorting, so the smallest anisorhizas arrive towards the edge of the pattern, and the medium sized ones in the centre. It is not yet clear whether migration, or a cell sorting procedure preceding migration, or a possible right/left distribution of daughter cells (Albrecht-Buehler 1977a,b; Levinstone, Eden & Bell, 1983) allocates identical cell types to the two sides of the cnidoband.[1] Investigations into this are proceeding with the large microbasic mastigophores on each side of the cnidosac of *Chelophyes*.

It is tempting to regard the mechanisms described here as general mechanisms in animal development, that have been observed in siphonophores because of the exceptionally favourable circumstances. Nevertheless, one must not forget that siphonophores are creatures that grow very rapidly (Biggs, 1976) and some of the mechanisms may be related to this. However, rapid and accurate formation of patterns is often a general feature of developing systems, whether in embryos or adults.

## ACKNOWLEDGMENTS

This work was financed by grants from the Royal Society of London (Marshall & Orr Bequest & Browne Research Fund). I thank the Director and Staff of the Station Zoologique, Villefranche-sur-Mer where this work was performed. The Drs. Carre and Cosson were particularly kind and helpful. Dr. A. Spencer carried out the Lucifer yellow dye coupling experiments while we were at Villefranche.

Figures 7, 10 and 8 are reproduced with permission of The Company of Biologists Limited.

---

[1] It does not seem that the latter is a likely explanation for all cases of handedness described here – it cannot account, for example, for the odd number of rows (7) in the cnidoband of *Calyphorae*.

## REFERENCES

Abercrombie, M. 1970. Contact inhibition in tissue culture. *In vitro* 6:128–142.

Albrecht-Buehler, G. 1977a. Daughter 3T3 cells: are they mirror images of each other? J. Cell Biol. 72:595–603.

Albrecht-Buehler, G. 1977b. The phagokinetic tracks of 3T3 cells. Cell 11:395–404.

Biggs, D. C. 1976. Nutritional Econology of *Agalma okeni* (Siphonophora: Physonectae) In G.O. Mackie (Ed) Coelenterate Ecology and Behaviour pp.201–210. Plenum Press, New York & London.

Carré, D. 1972. Etude du developpement des cnidocystes dans le gastrozoide de *Muggiaea kochi* (Will, 1844) (Siphonophore calycophore). C.r. hebd. Seanc. Acad. Sci., Paris. 275:1263–1266.

Fawcett, D. W., Ito, S. & Slautterback, D. 1959. The occurrence of intercellular bridges in groups of cells exhibiting synchronous differentiation. J. biophys. biochem. Cytol. 5:453–460.

Fujisawa, T. & David, C. N. 1981. Commitment during nematocyte differentiation in *Hydra*. J.Cell Sci. 48:207–222.

Günzl, H. 1979. *Dipurena reesi* (Hydrozoa) Wanderung der Cnidoblasten in den Rhizostolonen. In G. Wolff (ed.) *Encyclopaedia Cinematographia* pp.1–15 Film E 1106/1966. Institut für den Wissenschaftlichen Film, Göttingen.

Huxley, T. H. 1858. The Oceanic Hydrozoa; a description of the Calycophoridae and Physophoridae observed during the voyage of H.M.S. 'Rattlesnake' in the years 1846–1850. Plate V. The Ray Society, London.

Keferstein, W. & Ehlers, E. (1861). *Beobachtungen über die Siphonophoren von Neapel und Messina. Zoologische Beiträge gesammelt im Winter, 1859–1860 in Neapel und Messina* pp.1–34. *Wilhelm Engelmann, Leipzig.*

Levinstone, D., Eden, M. & Bell, E. 1983. Similarity of sister-cell trajectories in fibroblast clones. J.Cell Sci. 59:105–119.

Novak, P. L. & Wood, R. L. 1983. Development of the nematocyte junctional complex in hydra tentacles in relation to cellular recognition and positioning. J. Ultrastruct. Res. 83:111–121.

Semal-van Gansen, P. 1951. Le cnidosome de l'hydre et le bouton urticant. Bull. Acad. roy. Belgique Sci. 53 ser. 37: 650–664.

Skaer, R. J. 1973. The secretion and development of nematocysts in a siphonophore. J. Cell. Sci. 13:371–393.

Skaer, R. J. 1987a. The formation of cell rows. A study of the cnidoband of siphonophores. Submitted to 'Development'.

Skaer, R. J. 1987b. A cell sorting procedure in pattern formation. The terminal filament of siphonophores. Submitted to 'Development'.

Wood, R. L. & Novak, P. L. 1982. The anchoring of nematocysts and nematocytes in the tentacles of hydra. J. Ultrastruct. Res. 81:104–116.

11

# PROBLEMS OF SUPPLY AND DEMAND FOR CNIDAE IN ANTHOZOA

*Elaine A. Robson*

Department of Pure and Applied Zoology
The University
Reading RG6 2AJ, England

## SUMMARY

The Anthozoa are a large and diverse class of coelenterates. Their general organization, which includes the formation of local, histologically specialized structures, suggests that patterns of peripheral (i.e. centrifugal) growth in both solitary and colonial forms should be taken into account in trying to resolve questions raised by the development of cnidae. No developmental explanation has yet been put forward in respect of the differential distribution of cnidae within a sea anemone, although it is of both functional and taxonomic importance.

Considerable numbers of cnidae are expended in feeding, aggression, defence, in locomotion or attachment (e.g. larvae) and their discharge is an important aspect of behaviour. But there are few quantitative studies of the replacement of cnidae.

Evidence about precursor cells is reviewed. It is concluded that in anemone tentacles and acontia, and in the oral disk of octocoral polyps, cnidoblasts develop locally at the base of the ectoderm. Interstitial cells have been identified among these cnidoblasts, but they have not yet been found in connection with other tissues of Anthozoa. The only example of extensive migration described so far is in octocorals, where tentacle pinnules are supplied with nematocysts which migrate as cnidoblasts from ectodermal depots in the scapus. Elsewhere, cnidae simply move up to the epithelial surface at their site of development.

The differentiation of a special cnidome in catch tentacles (anemones) and sweeper tentacles (corals) as they develop has been examined but not yet explained. Current work on the growth of anemone tentacles shows that the concept of tissue renewal is important in relation to cnidogenesis.

## INTRODUCTION

Cnidae are expendable secreted devices continually liable to depletion and replacement. The numbers produced by a large anthozoan polyp during its lifetime are considerable: in the case of the famous 70-year old specimens of the sea anemone *Cereus pedunculatus* (Stephenson, 1935), for example, even the simplest estimate runs into millions (such as 100 new cnidae per day for 25,550 days). The cost of cnidae to the overall energy budget of a large cnidarian has not yet been measured but it may perhaps be quite small, possibly comparable to the production of mucus or of digestive enzymes (Hadzi, 1963; Tidball, 1984; Coffroth, 1984). In hydra, on the other hand, whose polyps are small and can be maintained in rapid growth, the production of nematocysts is regarded as a major synthetic activity (Bode and David, 1978). The utilization of cnidae fluctuates in natural conditions according to demand and is governed essentially by the animal's behaviour towards environmental stimuli (Picken and Skaer, 1966; Mariscal, 1974; Shelton, 1982). Restoration of the supply appears to follow any depletion almost as a matter of course.

How the utilization of cnidae and their replacement are linked, however, is specified as yet for very few Anthozoa (Schmidt, 1982; Schmidt and Moraw, 1982; Watson and Mariscal, 1983*a*,*b*) and the problems are considered here. Relevant information about the Scyphozoa and Cubozoa is also sparse (Chapman, 1963 (cited in Picken and Skaer, 1966), 1974, 1978; Rifkin and Endean, 1983). Fortunately many aspects of this dynamic balance have been illuminated by work on *Hydra* (Lenhoff, 1983; Heimfeld *et al*, 1985; Toshitaka *et al*., 1986).

## ANTHOZOA: OVERVIEW

The Anthozoa are all marine, and include a wide range of solitary and colonial forms, some of which attain great size and longevity. The term sea anemone refers primarily to the Actiniaria, but used loosely it includes also Corallimorpharia, Ptychodactiaria and Ceriantharia (Carlgren, 1949). These are essentially benthic animals which do not form a skeleton and are sessile or burrowing, almost universally distributed in the oceans and well known as members of the intertidal fauna. Most reproduce sexually, and a number of species multiply asexually as well (Chia, 1976).

The hard corals or Scleractinia are related to the first three of these groups in their structure and origins (Hand, 1966; Robson, 1985). Hermatypic or reef-forming species are of wide biogeographical significance in tropical and subtropical waters, while ahermatypes (lacking zooxanthellae) are confined to cooler regions, where they may produce modest reefs. Both types include solitary as well as colonial species. The possession of cnidae, as well as of symbiotic algae, undoubtedly contributes to the success of modern reef corals.

The octocorals, comprising alcyonarians, sea fans, sea whips and sea pens are all colonies with distinctive 8-rayed polyps whose tentacles are pinnate. This subclass contributes richly to coral reefs and to the fauna of deeper and colder waters. Many of these colonies are adapted to filter-feeding in currents (Wainwright and Koehl, 1976; Sebens and Koehl, 1984) and possess a multitude of small polyps. Ceriantharia are probably related to the octocorals, but their affinities with Antipatharia or black corals (they are usually linked as Ceriantipatharia) do not seem close (Schmidt, 1974).

Anthozoan polyps may be thought of as predators which utilize cnidae, but this is not universal since a few tropical species with symbiotic algae may take no food (e.g. Gohar, 1940) and many reef corals feed using mucus nets (Lewis and Price, 1975; Coffroth, 1984).

Spirocysts and nematocysts are present in the early larval stage (planula), and may assist settlement (Bonnenfant, 1961; Chia and Bickell, 1978).

## SIZE, STRUCTURE AND COMPLEXITY

The diameter of anthozoan polyps ranges from perhaps a millimeter to almost a meter in the largest anemones. For the record, the smallest cnidae are under 10μm long, the largest 300μm (den Hartog, 1980). In Cnidaria the large size of certain cells, however, does not reflect the size of the animal but indicates functional specialization (e.g. muscle fibres, oocytes), also occasional polyploidy (Totton and Mackie, 1960) should not be ruled out.

Partly in consequence of the regular array of radial mesenteries which support the disk and carry retractor muscles, digestive filaments and gonads, anthozoan anatomy is more complex than that of any hydrozoan polyp save perhaps the giant *Branchiocerianthus imperator*. This complexity is a consequence of regionalization of the body for different functions, which in the Cnidaria is expressed as differentiated zones rather than as distinct organs (Pantin, 1960). As is well known the complement of cnidae is in turn distributed differentially among these anatomical structures. It seems clear that the utilization of cnidae should be studied with reference to their anatomical distribution.

Although sea anemones are simple compared to many other free-living animals, 20 or 30 different structural parts are discerned by taxonomists as a matter of course (Stephenson, 1928; Manuel, 1981). The crown consists of the oral disk and tentacles, with a central mouth and pharynx. There is an adhesive pedal disk sometimes used in locomotion, or in burrowing species a rounded physa. The column has functional subdivisions, depending on the development of a marginal sphincter, and of accessory structures such as acrorhagi, suckers, verrucae and pores (cinclides). The mesenteries are associated with several functions: they carry muscles, gonads and a mesenteric filament which is involved in digestion and excretion (Van-Praët, 1985). The filament has a central cnidoglandular tract and in certain anemones this part is continuous with a thread-like, contractile acontium.

The cnidae of different functional regions such as the tentacles, pharynx, filaments, column and basal disk show a characteristic distribution which is often important taxonomically (England, 1986). The functional advantages which anthozoan cnidae of different types and dimensions must be assumed to possess, on the other hand, are only just beginning to be understood (Mariscal, 1974, 1984).

The biradial pattern of growth evident in anemones may be reflected in the orderly multiplication of centrifugal structures such as cycles of mesenteries and tentacles, and derivatives of the column such as acrorhagi, pseudotentacles, vesicles and so forth. Such structures are localised and repeated, and some of them contain characteristic nematocysts. A good example are the nemathybomes of certain burrowing anemones (Edwardsiidae: Stephenson, 1928, 1935; England, 1986), ectodermal sacs with nematocysts which form endocoelic rows within the mesogloea of the lower column. One may ask how and where these nematocysts originate, and whether the nemathybomes, in quite orderly linear array between the mesenteries, perhaps represent foci of migratory precursor cells. Questions of this kind need to be addressed more broadly, however, in respect not only of structures containing special cnidae such as acrorhagi, but of all secondary anatomical elements including column derivatives. What governs the development of small ectodermal structures specialized for secretion, for example, suckers or tenaculi (Stephenson, 1928; Carlgren, 1942; Bigger, 1982)?

From a simple anatomical point of view, a salient feature of large polyps such as sea anemones is the extent to which localised functional differentiation of structures occurs. The mesenteric filament, for example, has a central cnidoglandular tract bordered by phagocytic digestive cells. Towards the pharynx the filament is trilobed, with lateral ciliated tracts, and there are reticular tracts which deal with indigestible materials (Van-Praët, 1985). But there is no reason to suppose that the Anthozoa are more advanced in their histology than the Hydrozoa or Scyphozoa. The number of different types of cell is probably about the same in each class, and throughout the Cnidaria, as pointed out by Bonner (1965), it is significantly fewer than in triplobastic animals. In Anthozoa such as anemones, well marked radial symmetry is expressed as a propensity for developing localised peripheral structures, and associated cellular differentiation, such as the appearance of special cnidae, is evidently part of a broader pattern of morphogenesis.

In the case of <u>Hydra</u>, studies of tissue and cell dynamics have illuminated many aspects of its biology (Lenhoff, 1983). Here different types of cell are attributed to one of three cell lineages. Cnidoblasts, for example, like nerve cells and germ cells usually arise from interstitial cells. Although this kind of interpretation extends to other members of the Hydrozoa, work on transdifferentiation suggests that lineages are not invariable: in some conditions isolated striated muscle from the medusa of <u>Podocoryne</u> gives rise to many other types of cell including cnidoblasts (Schmid and Alder, 1984, 1986; see Bouillon and Massin, 1974). With regard to Anthozoa, however, few comparable conclusions have been drawn and relatively few studies have yet sought to elucidate the dynamics of cell populations (Singer, 1971; Doumenc, 1976, 1977; Minasian, 1980). The same holds for Scyphozoa.

## BEHAVIOUR AND THE UTILIZATION OF CNIDAE

The functions of cnidae are closely integrated with the animal's behaviour, which is governed by the neuromuscular system. In many Anthozoa, the musculature is endodermal except for the crown (Ceriantharia excluded). As in other Cnidaria musculoepithelium is usually but not always present (e.g. sphincter muscle fibres). The nervous system, a synaptic nerve net above the muscle, receives the neurites of numerous epithelial sensory cells. Besides the through-conducting nervous system whose properties are well known (Pantin, 1952; Shelton, 1982) there are two slow conduction systems, in the ectoderm and endoderm, respectively, whose action is integrated with that of the nerve net (McFarland, 1982). The SSI excites or coordinates muscle activity and may be associated with the discharge of cnidae (Lubbock and Shelton, 1981), but it has not yet been identified histologically.

Behavioural activities which involve the deployment of cnidae include feeding (tentacles, oral disk, pharynx, mesenteric filaments, acontia), locomotion (using tentacles), adhesion (e.g. larval settlement), aggression (acrorhagi and catch tentacles in anemones; sweeper tentacles in corals, and also mesenteric filaments but the role of their cnidae is less clear), and perhaps to a lesser extent defence (e.g. mutualism of anemones and hermit crabs: Ross, 1983).

Rates of utilization of cnidae will depend on whether deployment is infrequent, as is probably the case for nemathybomes and acrorhagi, or recurrent, causing more regular depletion, as in the case of tentacles and acontia which are used in feeding, and the ptychocysts used in tube-building in Ceriantharia (Mariscal, Conklin and Bigger, 1977). Elements concerned with feeding need to have cnidae constantly available. On the other hand, in those protecting an individual or a colony by aggression cnidae will be expended at longer intervals. The importance of mechanisms controlling the discharge of cnidae is especially evident where tentacles are able to carry out several functions, e.g. feeding, locomotion or aggression in *Gonactinia* and *Protanthea* (Robson, 1971 and unpublished observations); feeding, shell-climbing or defence in *Calliactis* (Shelton, 1982; Mariscal, 1974; Ross, 1983).

The differential use of cnidae may well be matched by selective replacement, but almost nothing is known about this in Anthozoa.

On the other hand, the specialization of tentacles for different functions, in particular for aggression instead of feeding, leads to changes in their behaviour, and in their structure and cnidome. The catch tentacles of anemones and the sweeper tentacles of corals (Williams, 1975; den Hartog, 1977; Purcell, 1977; Watson and Mariscal, 1983a,b; Hidaka and Yamazato, 1984) thus provide developmental material of very great interest.

## PRECURSOR CELLS: ANTHOZOA

Cnidoblasts can be recognized by the developing cnidae which they contain, from the earliest time at which the vacuole destined later to form the capsule can be seen. In the Anthozoa it has not yet been established how cnidoblasts or their precursors may be recognized before this stage. In sea anemones a population of cells equivalent to the interstitial cells of *Hydra* has been referred to quite often (Westfall, 1966; Singer, 1971; Van-Praët and Doumenc, 1975; Quero, 1980; Watson and Mariscal, 1983a) but little related experimental work has yet been carried out. The first two authors illustrated interstitial cells found among cnidoblasts at the base of the ectoderm in tentacles. Ivester (1977) showed comparable cells in tentacle ectoderm of the octocoral Renilla. In anemones, autoradiography following the use of tritiated thymidine suggested a widespread distribution of labelled nuclei in both ectoderm and endoderm (Minasian, 1980), which probably included amoebocytes ("granulocytes"). Singer (1971, 1974) identified as interstitial cells some of the cells which became labelled with triated thymidine or proline. According to Lentz (1966), the interstitial cells of Hydra are characterized by abundant ribosomes and relative scarcity of other organelles, and the nucleolus is small or absent.

It is perhaps worth recalling what Hyman wrote in 1940: "Characteristic of coelenterates are free undifferentiated interstitial cells, which lie among the epithelial cells, give rise to sex cells and nematocysts and participate in regenerative and reproductive processes". Where the embryological origin of interstitial cells is known (i.e. among Hydrozoa), it is in many cases endodermal (Tardent, 1978; Niewkoop and Satasura, 1981).

To ask what are the precursor cells of cnidoblasts in Anthozoa and where they are located is evidently to address a broader subject than that of cnidae alone, for the antecedents of most of the histologically different types of cell have not yet been ascertained in these animals. A brief list of them follows. In anemones the ectoderm includes supporting cells, musculo-epithelial cells and often subepithelial muscle cells, sensory cells, nerve cells, gland cells such as those secreting mucus or cuticular material, and cnidocytes. Among colonial Anthozoa, skeleton-forming cells adopt specialised features (e.g. Tidball, 1985; Kingsley and Watabe, 1985). But no nests of interstitial cells have been reported, and Van-Praët and Doumenc (1975) could find no population of basophilic cells which might correspond to them. In the endoderm, similar categories occur, and as already noted, certain areas are specialized for the secretion of digestive enzymes, and phagocytosis. In addition to general ciliation (with exceptions) of ectoderm and endoderm, specialized tracts of ciliated cells occur (e.g. siphonoglyphs, filaments). Within the animal, cnidae develop only in the cnidoglandular tracts of filaments and acontia when present. These structures are probably endodermal derivatives but the question has not been altogether resolved (Dawydoff, 1928; Doumenc, 1976, 1977). In sections, unused cnidae may be seen in the enteron, or engulfed into endodermal cells, perhaps in a state of degeneration (e.g. Schmidt and Moraw, 1982; Watson and Mariscal, 1984b). As in other Anthozoa gonads develop into the mesogloea behind the mesenteric filaments. In addition, granular amoeboid cells are found in the mesogloea and epithelia (Larkman 1984b, as amoebocytes) but their functions are not fully understood.

Several categories of cells appear to be involved in repair and regeneration of tissues, though studies of regeneration in anemones have not attempted to establish relationships between lines of cells (e.g. Singer 1971, Young 1974). In regenerating tentacles of Actinia, cnidoblasts have been observed still containing lysosomes, which suggests that here they were derived from de-differentiated epithelial cells (van-Praët and Doumenc, 1975). Regenerating tissues show amoeboid cells distinct from ordinary amoebocytes (ibid; Young 1974).

Whether the interstitial cells associated with cnidoblasts also give rise to nerve cells and to germ cells, as in Hydra (Campbell, 1979; Tardent, 1978) cannot yet be said. After a careful study of gametogenesis in Actinia fragacea Larkman (1983, 1984a) considered that in this species at least, morphological evidence for the precursor cells of gametocytes was not easily to be found. The origins of neurons are equally obscure.

In sum, interstitial cells have been found among cnidoblasts, particularly in anemone tentacles, but their affiliations remain to be discovered.

## CNIDOMIGRATION: WEILL 1934

The following represents Weill's admirable synopsis, as a starting point.

A nematocyst is never born at the site of its ultimate location: so a nematoblast always carries out a more or less long and complicated migration. The origin of nematoblasts, their migration, settlement, and fate after discharge all require study (p.304).

With prescience, Weill pointed out that since nematoblasts could not be recognized morphologically before the development of a nematocyst rudiment, in the absence of cell-lineage studies (sic) only fairly advanced nematoblasts could be examined, without it being possible to trace their first origins and any migration(s) taking place before the nematocyst began to form. At that time there was some evidence for migration from live observation, and a good deal from anatomical work: the migration routes of nematoblasts are inferred from trails leading from sites of cnidogenesis ("foyers cnidogènes" or nurseries) to the organs bearing mature nematocysts, within which nematocysts, orientated with the apical pole hindmost, are sometimes least developed when nearest to their point of departure.

According to Weill, such migration routes vary from one group to another. In the simplest case, when the nematocyst rudiment appears in the interstitial cell of a tentacle, the nematoblast has only to insinuate itself between (or through) a few ectodermal cells to reach its final site in contact with the outside. When the nematoblast is of endodermal origin, thanks to its amoeboid pseudopodia it can cross the mesogloea and may undergo quite long journeys in ectodermal organs before reaching its destination. Finally, in the most complex instances, the nematoblast, born of a subectodermal interstitial cell located in a deep-lying organ, passes in turn through the mesogloea and the endoderm, and thus falls into the gastrovascular cavity through which it is driven by the flagella of the endodermal cells, re-enters the endoderm of the upper part of the body, crosses the mesogloea and finally penetrates the ectoderm, where it takes its final seat. Weill points out that in species which possess several types of nematocyst, their migration may be determined with startling precision. He later comments that to express a disturbing phenomenon in known terms does not in itelf constitute a satisfactory explanation, since it can be tested only in the light of direct observations and experiments (p.308).

## MIGRATION IN ANTHOZOA

Schmidt (1981; Schmidt and Moraw, 1982) has reviewed the literature concerning development of cnidae and migration in Anthozoa, which remains sparse compared to the substantial number of contributions focussed on *Hydra* and other species of Hydrozoa. His research on 4 species of octocorals, selected because of the simple cnidome (rhabdoid heteronemes only) has revealed hitherto-unknown accumulations of nematoblasts in the distal scapus ectoderm, below the bases of tentacles. Consisting solely of developing stages, they are the main source of supply to the tentacles and their pinnules. In the pinnules, only mature nematocysts are found. Migrations of the order of several millimeters are implied (the polyps being fairly small) during which the nematocysts continue to develop. Cnidoblasts migrate individually from such depots into the tentacles and finally reach the pinnules, where they appear as nematocytes ready for discharge. These octocorals probably feed on small zooplankton captured with the help of nematocysts (e.g. Sebens and Koehl, 1984).

This example of quite extensive migration within the ectoderm is the only one so far reported from Anthozoa. In the same specimens, Schmidt and Moraw (1982) also showed that in the oral disk of the polyps, where fewer nematocysts are found, cnidoblasts develop on site as is thought to be the case in the tentacles of *Renilla* (Ivester, 1977) and in anemones. In these Anthozoa nematocysts are not produced by the mesenteric filaments and the only ones which were found in the endoderm had been phagocytosed and were degenerating.

In Weill's terms, cnidoblasts developing within columnar epithelium need to undergo a short vertical migration to the surface before taking their place and functioning as mature cnidocytes (Van-Praët and Doumenc, 1975). As mentioned above, studies of tentacles in anemones show cnidoblasts developing between the bases of ectodermal cells (Westfall, 1966; Singer, 1971; Watson and Marsical 1983b). Yanagita and Wada (1959) and Wada (1972, Fig. 10) described comparably localised development of cnidae in the acontia of *Diadumene* (= *Haliplanella*) *luciae*, where developing stages are invariably found at the base of the epithelium.

It is worth mentioning that according to Bonnenfant (1960), developing acrorhagi in *Actinia equina* acquire atrichs only after these nematocysts appear in mesenteric filaments, and she raised the possibility that atrichs could migrate from the filaments to acrorhagi. *Actinia* shows aggressive behaviour and on contact with an antagonist, the ectoderm of inflated acrorhagi dehisces, and is later replaced. It would be very interesting to study cnidogenesis at this stage and earlier (Bigger, 1982).

Yanagita and Wada (1959) carried out experiments with acontia which did not indicate immigration of cnidoblasts. Acontia were depleted of cnidae and left in the anemones (D. luciae). In laboratory experiments they were examined after one and two weeks and in a field experiment after 4 weeks. Developmental stages of cnidae were produced throughout the acontia, in the depths of the epithelium, while maturing and normal cnidae were present nearer the surface. In 3 anemones acontia were removed altogether, and after 2 weeks their regenerating acontia contained both immature and normal cnidae.

Experiments on cnidogenesis in anemone tentacles have so far revealed little evidence for migration. Schmidt (1982) found that tentacles of Anemonia sulcata (= viridis) when depleted of cnidae regained their normal complement after 5 or 6 days. The normal ratio of spirocysts to nematocysts (basitrichs, b-rhabdoids) is approximately 2:1. The tip has more cnidae than the base, but the rates of replacement were approximately the same, suggesting localised cnidogenesis.

Watson and Mariscal (1983a,b) have studied the distribution of cnidae in feeding tentacles, and in catch tentacles at different stages of development in Haliplanella luciae. In ordinary tentacles spirocysts are up to twice as common as nematocysts; of the latter, microbasic p-mastigophores predominate and are more numerous near the tip and basitrichs are the least abundant. Cnidoblasts are common throughout the tentacle, and are found at the base of the ectoderm. Catch tentacles, which are longer and wider, offer a complete histological contrast and their composition varies markedly from base to tip. Near the tip the only cnidae are holotrichs (large and small, in almost equal proportions), and cnidoblasts are absent. Towards mid-tentacle level microbasic p-mastigophores and basitrichs are found as well, and fewer holotrichs but many late cnidoblasts. At the base of the tentacle there is a high proportion of cnidoblasts, and the only cnidae are a few spirocysts, mastigophores and basitrichs. The tips of catch tentacles sampled at different stages of development show that as the cnidae typical of feeding tentacles are lost, cnidoblasts appear in large numbers, and are succeeded by the holotrichs typical of mature catch tentacles. These findings agree with Purcell's (1977) study of the development of the special cnidome in catch tentacles of Metridium.

As aggressive organs catch tentacles show distal autotomy (Fukui, 1986) and the lost tissue is replaced. Watson and Marsical (1983b) discuss the possible role of tissue migration in maintaining both ordinary and catch tentacles. They suggest that if structural tissues were to migrate from the oral disk to the bases of both kinds of tentacle, some ripe nematocysts might then be supplied as well. They note phagocytosis and cellular degradation, which would be consistent with tissue reorganization, at the base of feeding tentacles and more distally in catch tentacles. Independent migration of nematocytes, however, is not implied.

The question of cellular migration in Anthozoa may be considered more generally. In anemones, granular amoebocytes aggregate rapidly towards wounds (Young, 1974; Patterson and Landolt, 1979), a reaction discovered in other coelenterates by Metschnikoff (e.g. 1892). In colonial Anthozoa the formation or placement of spicules may possibly involve the migration of scleroblasts (see Kingsley and Watabe, 1984). Cell movements in the hydroplasm (i.e. fluid in the coelenteron), well documented for hydroids (Hale, 1964; Braverman, 1974), could possibly occur also in colonies of Anthozoa. And cell migrations always form part of larval development (Doumenc, 1976, 1977).

## THE SUPPLY OF CNIDAE: ANEMONE TENTACLES

In ordinary anemone tentacles mature cnidae, cnidoblasts and sometimes interstitial cells are found together, and this suggests that cnidogenesis is localised. Axial migrations have not been discerned and it seems that they would not be extensive. On the other hand, cnidae may be transported partly owing to ectodermal displacements during growth, as suggested by Watson and Mariscal (1983b). In what follows it will be seen that the tentacles of certain anemones undergo steady renewal and cnidae will be part of tissues in slow centrifugal movement (see Campbell, 1974). Two

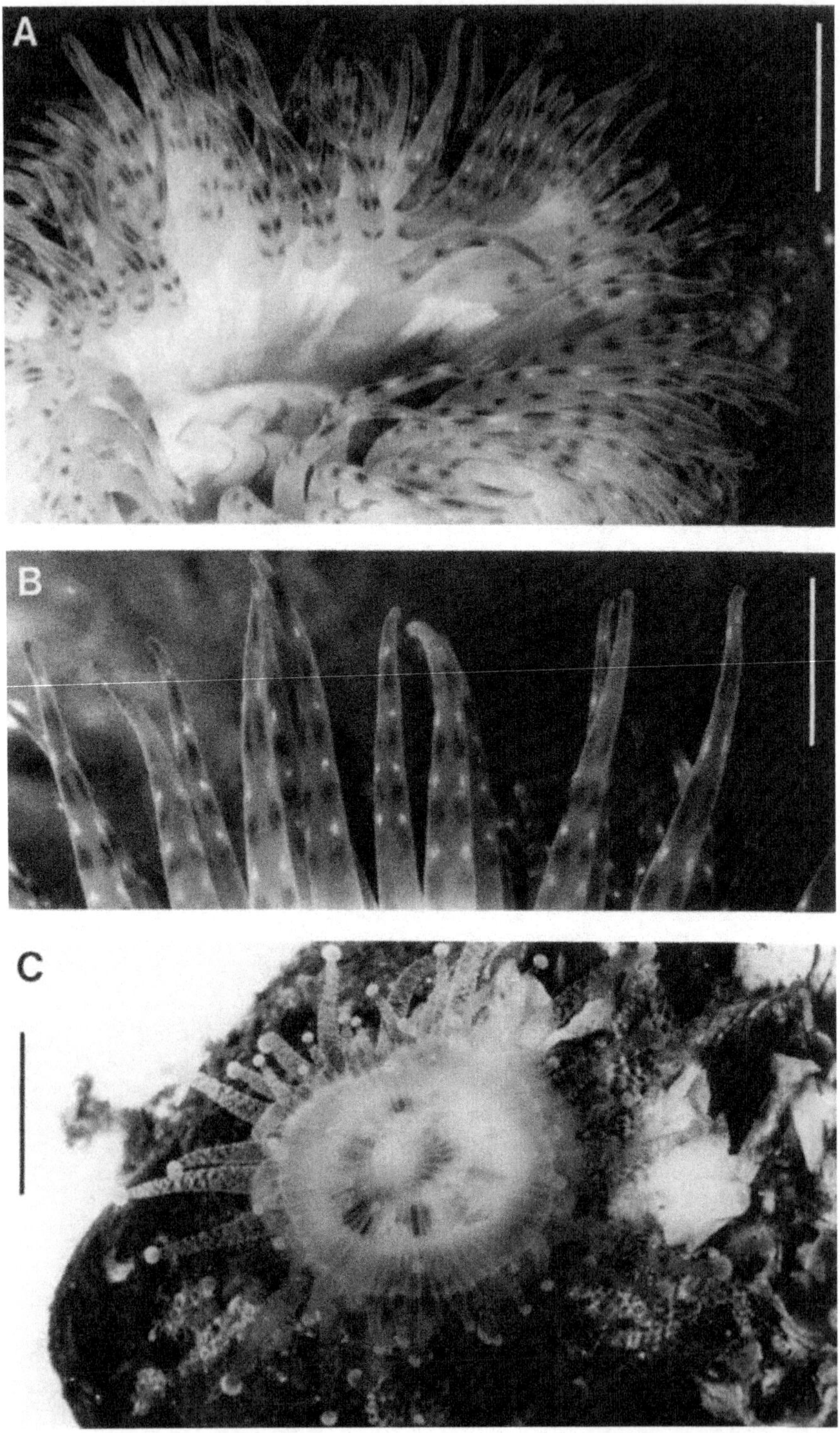
A
B
C

examples are considered briefly, the tentacles of Calliactis parasitica (Hormathiidae), an acontiarian anemone commensal with the hermit crab Eupagurus bernhardus, and those of the corallimorpharian Corynactis viridis, which forms clones on hard substrates (Figs. 1A,C).

Recent work (Robson, 1987) shows that all the tentacles of Calliactis are in a state of perpetual growth, comparable in some degree to those of Hydra (Campbell, 1974). In most specimens there is a simple banding pattern of the tentacles, due to endodermal pigments (Figs. 1A,B; 3C), and these bands can be used to monitor growth. C. parasitica is particularly sensitive to poor aquarium conditions and it is quite well known that it may lose its tentacles when kept for any length of time (Stephenson, 1935). The condition is reversible, and tentacles regenerate as soon as the anemones are placed in a favourable environment (see Fig. 2A,B,C,D). Figure 2D shows tentacles regenerated in the course of 10 weeks, with a normal banding pattern.

Regeneration experiments show that bands are added proximally, near the tentacle base, and disappear by attrition of tissue at the tentacle tip (Figs. 1B, 3C). Evidence that the unpigmented ectoderm grows at about the same rate as the endoderm has been obtained by recording the progressive terminalization of occasional branched tentacles, which occur as in Figure 3D (see also Shostak, 1981), and by marking the ectoderm with injected spots of ink (Otto and Campbell, 1983). Furthermore, since the tentacles taper, ectodermal movements must accommodate not only tissue growth, which is most evident proximally, but distal realignment to achieve a narrowing profile (Fig. 1B). The apical pore at

Fig. 1A. Calliactis parasitica. Oblique view of oral disk and tentacles from above (mouth lower left) to show banding pattern of tentacles. Bar is 1 cm. B. Calliactis. Banding of distal parts of main tentacles in another specimen. Bar is 0.5 cm. C. Corynactis viridis, expanded specimen on rock. Tentacles show acrospheres, and stems have opaque spirals of adhesive pads (mucus cells and cnidae). Bar is 0.5 cm.

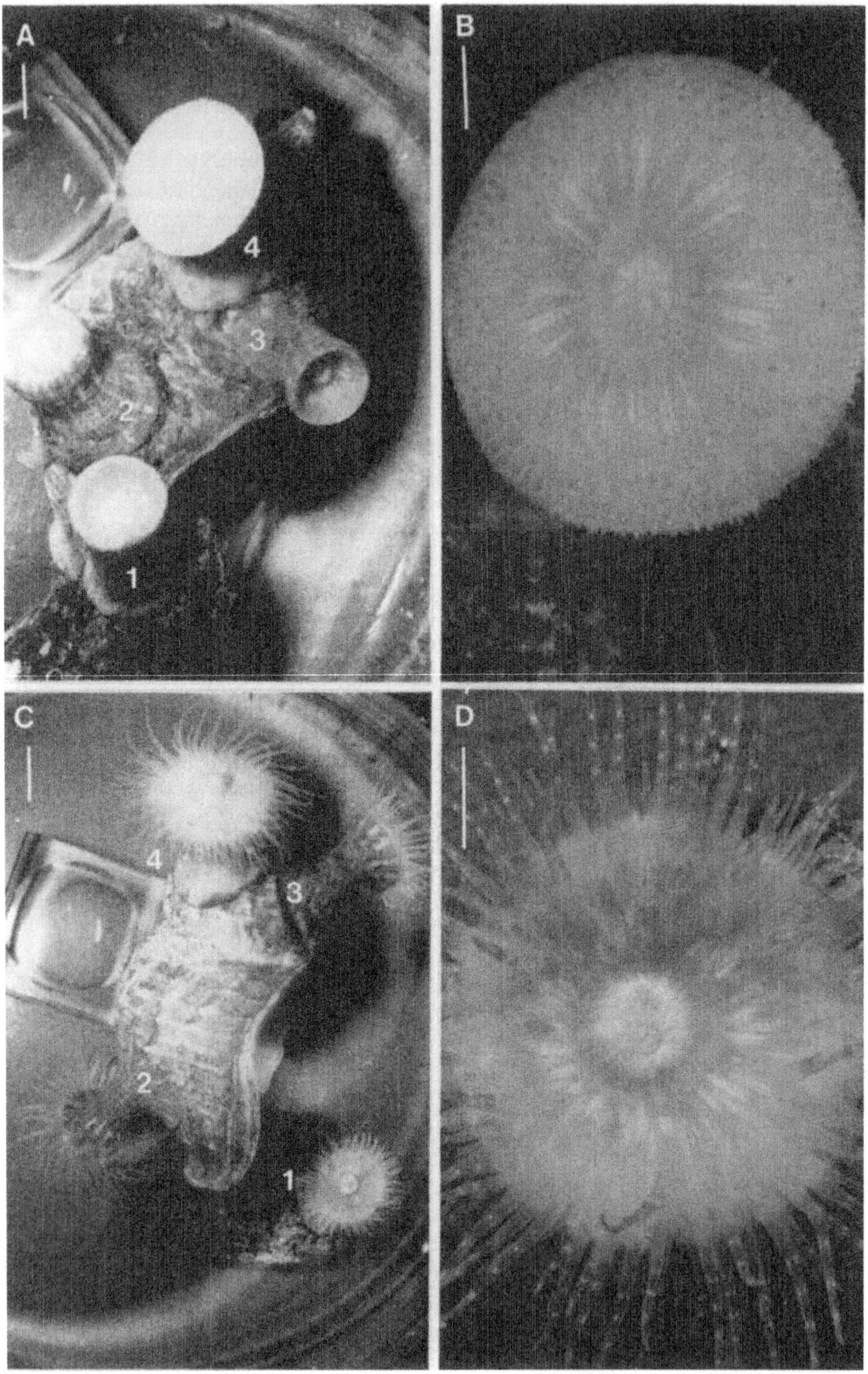
A
4
3
2
1
B
C
4
3
2
1
D

the tip of the tentacle is a zone from which endoderm, ectoderm and mesogloea are lost continually (Figs. 3A,C).

The steady renewal of tissues proceeds at modest rates, of perhaps several millimeters per month. Longitudinal sections of growing tentacles (Fig. 3B) show that the ectoderm is filled with numerous cnidae, superficial to the underlying zone of dense cnidoblasts, to within a short distance of the terminal pore. It is difficult to see what advantages might be conferred here by any extensive migrations of individual cnidoblasts within the tentacle ectoderm, although they would perhaps be more probable in the basal regions and in the vicinity of the oral disk.

*Corynactis*, in contrast, possesses knobbed tentacles like those of corals and the acrospheres lack any suggestion of an apical pore (Fig. 4A,B,C). In acrospheres the mesogloea has a firm elastic texture and is traversed by scattered cells. The endoderm within an acrosphere has many irregular inclusions and often shows phagocytosed cnidae (Fig. 4A). The hemispherical cushion of ectoderm contains spirocysts and nematocysts (holotrichs, and microbasic mastigophores or p- and b-rhabdoids) whose developing stages are near the mesogloea (Fig. 4C). Expended cnidae may therefore be replaced by cnidoblasts differentiating on site. In the corallimorpharian *Discosoma sanctithomae* cnidoblasts were also found in the acrospheres of marginal tentacles (den Hartog, 1980).

The acrospheres of *Corynactis* act as adhesive batteries which are extremely effective in catching zooplankton, especially Crustacea. They are borne on hollow, contractile stems, covered with spirals of

Fig. 2. *Calliactis parasitica*, specimens 1, 2, 3, 4. A. Start of experiment. All specimens without tentacles (see text). The whelk shell is 11 cm long. Bar is 1 cm. B. Specimen 4 to show oral disk at start of experiment. Tentacles are minimal. Bar is 0.5 cm. C. End of experiment: the same specimens after 10 weeks, all with regenerated tentacles. Bar is 1 cm. D. Specimen 2 which showed well-marked tentacle banding after regeneration. Time as Figure 1C. Bar is 0.5 cm.

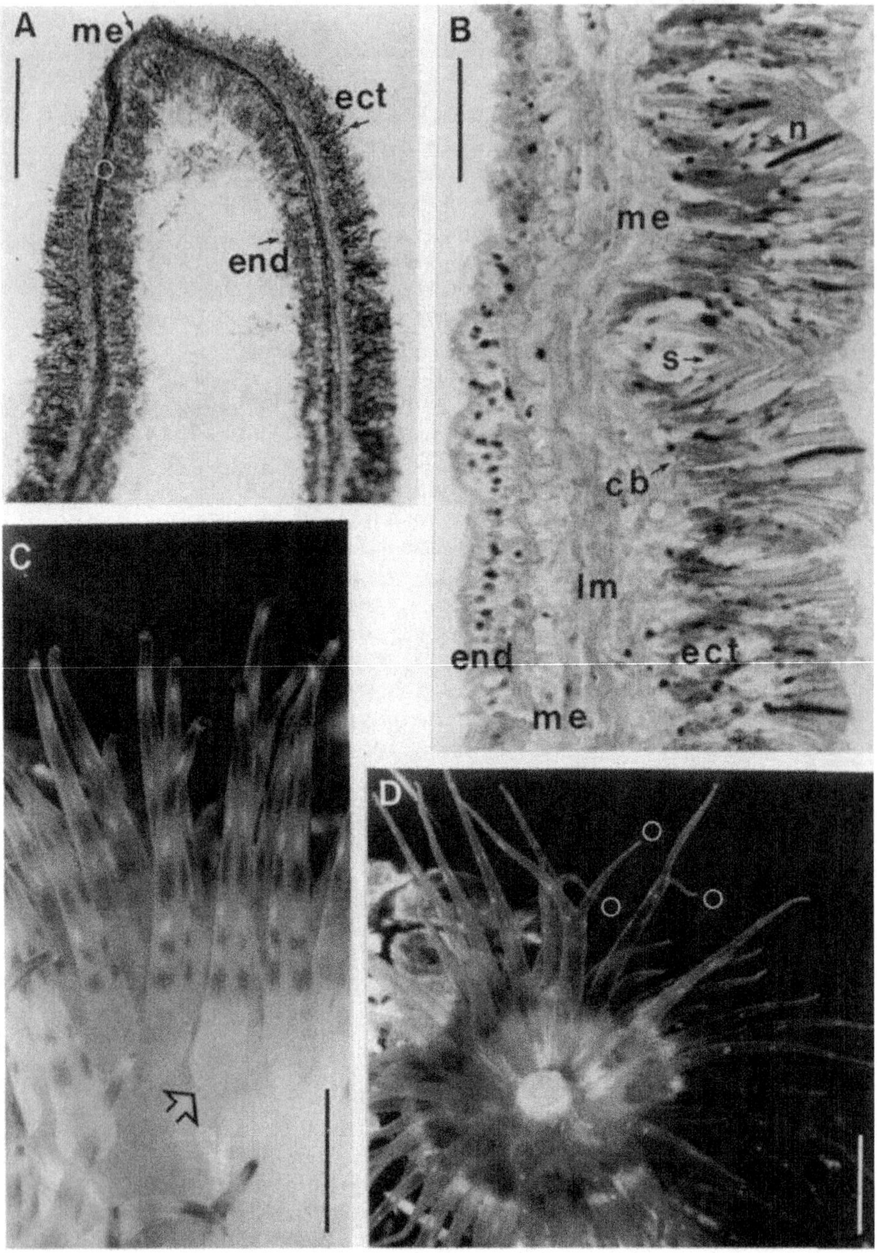
A
me
ect
end
B
n
me
s
cb
lm
end
ect
me
C
D

large adhesive pads containing gland cells (mucus) and cnidae, which are spirocysts and small nematocysts only. The tentacles of Corynactis may increase in length and are able to regenerate acrospheres, but they do not appear to be in a state of steady centrifugal growth as in Calliactis. While it is possible that cells are being continually replaced in both ectoderm and endoderm, the pattern of growth seems more determinate than in Calliactis. As far as cnidae are concerned, it is always possible that interstitial cells or early cnidoblasts may migrate to the acrosphere from more proximal sources.

Unlike Calliactis, Corynactis polyps often undergo longitudinal fission and so multiply asexually. The organization of cnida-bearing structures should be examined in more of the colonial Anthozoa, and especially perhaps in Scleractinia. In the octocorals, where recent studies show that extrinsically-developed cnidoblasts migrate from depots in the scapus to supply the numerous pinnules of tentacles (Schmidt and Moraw, 1982), colonies grow by the addition of new polyps and established polyps

Fig. 3. Calliactis parasitica. Abbreviations: cb, cnidoblasts; ect, ectoderm; end, endoderm; lm, longitudinal muscle; me, mesogloea; n, nematocyst; s, spirocyst. A. Longitudinal radial section of a tentacle just to one side of the apical pore. Near the apex the ectoderm thins out, the mesogloea (white ring) is in contact with the exterior (arrow) and is disintegrating, and the endoderm has numerous mucus cells (not shown here). Note cellular debris and mucus in the coelenteron. Celloidin/Masson. Courtesy of Mr G. Best. Bar is 100µ. B. Longitudinal section, oriented as right hand of Figure 3A, of tentacle from a young specimen. Longitudinal muscle fibres are sunk into mesogloea. Ectodermal epithelium with numerous cnidae at the surface (chiefly spirocysts in this section), and an underlying denser zone representing cnidoblasts. Bar is 25µ. C. Tentacles of a large specimen, recently collected, to show junctions with oral disk (arrow) and tentacle tips, some of them perhaps losing material at the apical pore. Bar is 0.5 cm. D. Specimen with 3 forked tentacles (rings show the shorter branch in each case). Terminalisation of branches took several weeks. Bar is 0.5 cm.

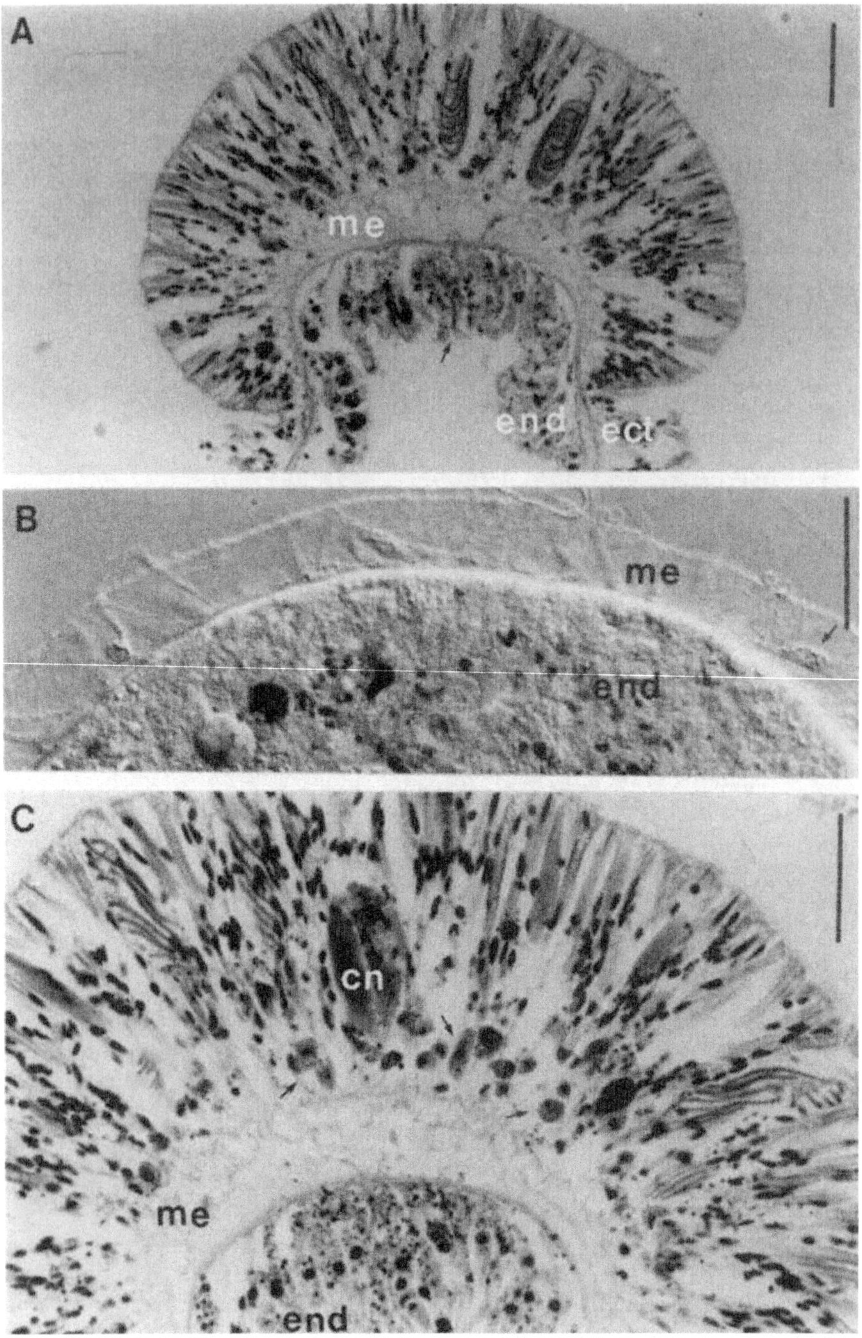
A
me
end
ect
B
me
end
C
cn
me
end

do not appear to show continuous renewal of tentacle tissues comparable to Calliactis. In one sense cnidomigration may be an alternative expression of perpetual growth.

## CONCLUSION

Interstitial cells and cnidoblasts have been reported in the base of the ectoderm in anemone tentacles, suggesting localised cnidogenesis, and cnidae also develop locally in acontia. In octocorals, however, cnidae supplying the tentacles migrate from accumulations in the distal scapsu. In Anthozoa the origins and affinities of interstitial cells are not yet known. In anemones, the differential distribution of the cnidome, often a taxonomic characteristic, reflects an aspect of cnidogenesis about which almost nothing is known: what determines the development of different types and sizes of cnidae in certain tissues? It is not known at what stage, for example, spirocysts and nematocysts are committed to develop. Cnidogenesis is known to take several days but there have been no reports of earlier mitoses and the sequence of cellular and tissue events preceding the appearance of mature cnidae at the surface of epithelia needs to be elucidated further. The growth and maintenance or renewal of many cnida-bearing structures also needs to be taken into account. The question how, or

Fig. 4. Corynactis viridis, acrospheres of tentacles. Bars are 25μ. Abbreviations: cn, cnidoblasts; ect, ectoderm; end, endoderm; me, mesogloea. A. Median longitudinal section, showing thickened ectoderm of acrosphere with numerous cnidae (3 holotrichs visible), and apical endoderm with inclusions and a phagocytosed nematocyst (arrow). Masson stain. B. Macerated preparation (osmic-acetic) showing dome of an acrosphere from which ectoderm has been removed. This shows its continuous thick mesogloea with some ramifying cells visible, and inclusions in the endoderm beneath. C. Median longitudinal section showing cnidoblasts of all sizes (arrows) in the proximal (basal) zone of the ectoderm. Masson stain.

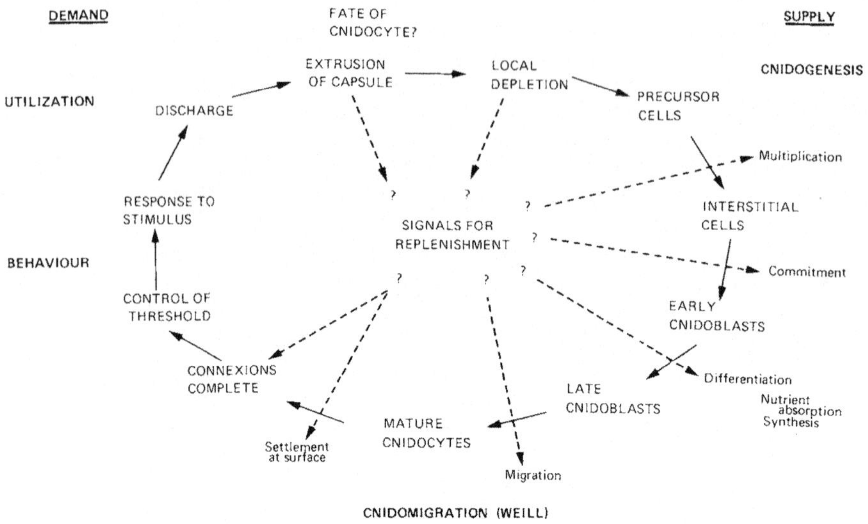


Fig. 5. Explanatory scheme to illustrate a possible sequence of supply and demand for cnidae in Anthozoa.

whether, upon discharge, signals reach the factory floor to engender the manufacture and delivery of further cnidae (as specified) has yet to be answered.

A provisional scheme incorporating some of the questions discussed here is given in Fig. 5. It will be clear that itemization and sequence in respect of Anthozoa are tentative, and that the location of developing cnidae is known in relatively few instances. Precursor cells remain elusive in Anthozoa, and "signals for replenishment" are an unknown quantity. No attempt is made to indicate the differentiation of diverse cnidae, or to depict energy flow.

ACKNOWLEDGEMENTS

Some of this work was carried out at the laboratory of the Marine Biological Association, Plymouth, and it is a pleasure to thank the Director and Staff for their help and for the use of facilities. Financial support from the Research Board of the University of Reading and the Marshall and Orr Bequest of the Royal Society is gratefully acknowledged. Mrs E. Wheeler is thanked for help with histological sections, and Mrs G. Bogue for preparing the typescript.

## REFERENCES

Bigger, C.H. 1982. The cellular basis of the aggressive acrorhagial response of sea anemones. J. Morph. 173: 259-278.

Bode, H. and C.N. David. 1978. Regulation of a multipotent stem cell, the interstitial cell of Hydra. Progr. Biophys. Molec. Biol. 33: 189-206.

Bonnenfant, J. 1961. Recherches sur l'evolution larvaire et la variation physiologique du cnidome chez Actinia equina. Bull. Soc. Zool. Fr. 85: 157-165.

Bonner, J.T. 1965. Size and Cycle. Princeton University Press, New Jersey.

Bouillon, J. and Massin, C. 1974. La cnidogenese. Ann. Soc. R. zool. Belg. 104: 7-38.

Braverman, M. 1974. The cellular basis of morphogenesis and morphostasis in hydroids. Oceanogr. mar. Biol. Ann. Rev. 12: 129-221.

Campbell, R.D. 1974. Cell movements in Hydra. Am. Zool. 14: 523-535.

Campbell, R.D. 1979. Development of Hydra lacking interstitial and nerve cells. Symp. Soc. Dev. Biol. 37: 267-293.

Carlgren, O. 1942. Actiniaria. Part II. Danish Ingolf-Exped. Vol. 5(12): 1-92.

Carlgren, O. 1949. A survey of the Ptychodactiaria, Corallimorpharia and Actiniaria. Kungl. svensk. Vetensk. Akad. Handl. (4) 1: 1-121.

Chapman, D.M. 1974. Cnidarian histology. In L. Muscatine and H.M. Lenhoff (eds.), Coelenterate biology, pp. 1-92. Academic Press, New York.

Chapman, D.M. 1978. Microanatomy of the cubopolyp, Tripedalia cystophora (Class Cubozoa). Helgo. Wiss. Meeresunters. 31: 128-168.

Chia, F.S. 1976. Sea anemone reproduction: Patterns and adaptive radiations. In G.O. Mackie (ed.) Coelenterate ecology and behaviour, pp.261-270. Plenum Press, New York.

Chia, F.S. and L.R. Bickell. 1978. Mechanisms of larval attachment and the induction of settlement and metamorphosis in coelenterates: A review. In F.S. Chia and M. Rice (eds.), Settlement and metamorphosis of marine invertebrate larvae, pp.1-12. Elsevier North Holland, New York.

Coffroth, M.A. 1984. Ingestion and incorporation of coral mucus aggregates by a gorgonian soft coral. Mar. Ecol. Progr. Ser. 17: 193-199.

Dawydoff, C. 1928. Traité d'embryologie des invertebrés. Masson, Paris.

Doumenc, D.A. 1976. Etude ultrastructurale des stades parenchymula et actinella de l'actinie Cereus pedunculatus Pennant. Arch. Zool. exp. gén. 117: 295-324.

Doumenc, D.A. 1977. Etude dynamique de la morphogenèse au cours des phases actinella et edwardsia de l'actinie Cereus pedunculatus Pennant. Arch. Zool. exp. gén. 118: 79-102.

England, K.W. 1986. Certain Actinaria (Coelenterata, Anthozoa) from the Red Sea and tropical Indo-Pacific Ocean. Bull. Brit. Mus. Nat. Hist. (Zool.). In press.

Fujisawa, T., C. Nishimiya and T. Sugiyama. 1986. Nematocyte differentiation in Hydra. Curr. Topics Dev. Biol. 20: 281-290.

Fukui, Y. 1986. Catch tentacles in the sea anemone Haliplanella luciae. Mar. Biol. 91: 245-251.

Gohar, H.A.F. 1940. Studies on the Xeniidae of the Red Sea. Publ. mar. biol. Sta. Ghardaqa 2: 23-118.

Hadzi, J. 1963. The evolution of the Metazoa. Pergamon Press, Oxford.

Hale, L.J. 1964. Cell movements, cell division and growth in the hydroid Clytia johnstoni. J. Embr. exp. Morph. 12: 517-538.

Hand, C. 1966. On the evolution of the Actiniaria. Symp. Zool. Soc. London, 16: 135-146.

den Hartog, J.C. 1977. The marginal tentacles of Rhodactis sanctithomae (Corallimorpharia) and the sweeper tentacles of Montastrea cavernosa (Scleractinia); their cnidom and possible function. Proc. Int. Symp. Coral Reefs 3: 463-469.

den Hartog, J.C. 1980. Carribean shallow water Corallimorpharia. Zool. Verhand., Leiden. 176: 1-183.

Heimfeld, S., L.C. Javois, J.F. Dunne, C.L. Littlefield, L. Huang and H.R. Bode. 1985. Monoclonal antibodies: a new approach to the study of Hydra development. Arch. Sci. Geneve, 38: 429-438.

Hidaka, M. and K. Yamazato. 1984. Intraspecific interactions in a scleractinian coral, Galaxea fascicularis: induced formation of sweeper tentacles. Coral Reefs 3: 77-85.

Hyman, L.H. 1940. The Invertebrates: Protozoa through Ctenophoza. McGraw-Hill, New York.

Ivester, M.S. 1977. Nematocyst differentiation in the anthozoon Renilla reniformis (Pallas). Trans. Am. micr. Soc. 96: 238-247.

Kingsley, R.J. and N. Watabe. 1985. An autoradiographic study of calcium formation in the gorgonian Leptogorgia virgulata (Lamarck) (Coelenterata: Gorgonacea). Cell Tiss. Res. 239: 305-310.

Larkman, A.U. 1983. An ultrastructural study of oocyte growth within the endoderm and entry into the mesogloea in Actinia fragacea (Cnidaria, Anthozoa). J. Morph. 178: 155-177.

Larkman, A.U. 1984a. An ultrastructural study of the establishment of the testicular cysts during spermatogenesis in the sea anemone Actinia fragacea (Cnidaria: Anthosoa). Gamete Res. 9: 303-327.

Larkman, A.U. 1984b. The fine structure of granular amoebocytes from the gonads of the sea anemone Actinia fragacea (Cnidaria: Anthozoa). Protoplasma 122: 203-221.

Lenhoff, H.M. 1983. Hydra: Research Methods. Plenum Press, New York.

Lentz, T.L. 1966. The cell biology of Hydra. North-Holland, Amsterdam.

Lewis, J.B. and W.S. Price. 1975. Feeding mechanisms and feeding strategies of Atlantic reef corals. J. Zool. Lond. 176: 527-544.

Lubbock, R. and G.A.B. Shelton. 1981. Electrical activity following cellular recognition of self and non-self in a sea anemone. Nature 289: 59-60.

Manuel, R.L. 1981. British Anthozoa. Academic Press, London.

Mariscal, R.N.. 1973. The control of nematocyst discharge during feeding by sea anemones. Publ. Set. mar. biol. Lab. 20, 695-702.
Mariscal, R.N. 1974. Nematocysts. In L. Muscatine and H.M. Lenhoff (eds.), Coelenterate biology, pp.129-178. Academic Press, New York.
Mariscal, R.N. 1984. Cnidaria: Cnidae. In J. Bereiter-Hahn, A.G. Matoltsy and K.S. Richards (eds.). Biology of the Integument Vol. I, p.57-68. Springer-Verlag, Berlin.
Mariscal, R.N., E.J. Conklin and C.H. Bigger. 1977. The ptychocyst, a major new category of cnida used in tube construction by a cerianthid anemone. Biol. Bull. 152: 392-405.
McFarlane, I.D. 1982. Calliactis parasitica. In G.A.B. Shelton (ed.), Electrical conduction and behaviour in 'simple' invertebrates, pp.243-265. Clarendon Press, Oxford.
Metchnikoff, E. 1892. Leçons sur la pathologie comparée de l'inflammation. G. Masson (Paris).
Minasian, L.L. 1980. The distribution of proliferating cells in an anthozoan polyp, Haliplanella luciae (Actiniaria: Acontiaria), as indicated by $^3$H-thymidine incorporation. In P. Tardent and R. Tardent (eds.), Developmental and cellular biology of coelenterates, pp.409-420. Elsevier/North-Holland Biomedical Press, Amsterdam.
Nieuwkoop, P.D. and L.A. Sutasurya. 1981. Primordial germ cells in the invertebrates. Cambridge University Press.
Otto, J.J. and R.D. Campbell. 1983. Marking epithelial cells in living Hydra with India ink. In H.M. Lenhoff (ed.), Hydra: Research Methods, pp.183-186. Plenum Press, New York.
Pantin, C.F.A. 1952. The elementary nervous system. Proc. Roy. Soc. B, 140: 147-168.
Pantin, C.F.A. 1960. Diploblastic animals. Proc. Linn. Soc. 171: 1-34.
Patterson, M.J. and M.L. Landolt. 1979. Cellular reaction to injury in the anthozoan Anthopleura elegantissima. J. inv. Path. 33: 189-196.
Picken, L.E.R. and R.J. Skaer. 1966. A review of researches on nematocysts. Symp. Zool. Soc. London, 16: 19-50.
Purcell, J.E. 1977. Aggressive function and induced development of catch tentacles in the sea anemone Metridium senile (Coelenterata, Actiniaria). Biol. Bull. 153: 355-368.

Quero, A. 1980. Formacion de los espirocistos de Actinia equina L. Bol. R. Soc. Espan. Hist. nat. (Biol.). 78: 43-52.

Rifkin, J. and R. Endean. 1983. The structure and function of the nematocysts of Chironex fleckeri Southcott, 1956. Cell Tiss. Res. 233: 563-577.

Robson, E.A. 1971. The behaviour and neuromuscular system of Gonactinia prolifera, a swimming sea-anemone. J. exp. Biol. 55: 611-640.

Robson, E.A. 1985. Speculations on coelenterates. In S. Conway Morris, J.D. George, R. Gibson and H.M. Platt (eds.). The origins and relationships of lower invertebrates, pp.60-77. Clarendon Press, Oxford.

Robson, E.A. 1987. The dynamics of tentacle growth in sea anemones. In preparation.

Ross, D.M. 1983. Symbiotic relations. In D. Bliss (ed.). The biology of Crustacea, Vol. 7, pp.163-212. Academic Press, New York.

Schmid, V. and H. Alder. 1984. Isolated, mononucleated, striated muscle can undergo pluripotent transdifferentiation and can form a complex regenerate. Cell 38: 801-809.

Schmid, V. and H. Alder, 1986. The potential for transdifferentiation of differentiated medusa tissues in vitro. Curr. Topics dev. Biol. 20: 117-135.

Schmidt, G.H. 1982. Replacement of discharged cnidae in the tentacles of Anemonia sulcata. J. mar. biol. Ass. U.K. 62: 685-691.

Schmidt, H. 1974. On evolution in the Anthozoa. Proc. Int. Symp. Coral Reefs 2(1): 533-560.

Schmidt, H. 1981. Die Cnidogenese der Octocorallia (Anthozoa, Cnidaria): I. Sekretion und Differenzierung von Kapsel und Schlaunch. Helgol. Meeresunters. 34: 463-484.

Schmidt, H. and B. Moraw. 1982. Die Cnidogenese der Octocorallia (Anthozoa, Cnidaria): II. Reifung, Wanderung und Zerfall von Cnidoblast und Nesselkapsel. Helgol. Meeresunters. 35: 97-118.

Sebens, K.P. and M.A.R. Koehl. 1984. Predation on zooplankton by the benthic anthozoans Alcyonium siderium (Alcyonacea) and Metridium senile (Actiniaria) in the New England subtidal. Mar. Biol. 81: 255-271.

Shelton, G.A.B. 1982. Electrical conduction and behaviour in 'simple' invertebrates. Clarendon Press, Oxford.
Shostak, S. 1981. Regeneration in young Bunodactis verrucosa Pennant (Actiniaria, Anthozoa). Roux' Archiv. Entwicklungsmech. 190: 274-282.
Singer, I.I. 1971. Tentacular and oral-disc regeneration in the sea anemone, Aiptasia diaphana III. Autoradiographic analysis of patterns of tritiated thymiding uptake. J. Embryol. exp. Morph. 26: 253-270.
Singer, I.I. 1974. An electron microscopic and autoradiographic study of mesogloeal organization and collagen synthesis in the sea anemone Aiptasia diaphana. Cell Tiss. Res. 149: 537-554.
Stephenson, T.A. 1928. The British Sea Anemones, Vol. I. Ray Society, London.
Stephenson, T.A. 1935. The British Sea Anemones, Vol. II. Ray Society, London.
Tardent, P. 1978. Coelenterata, Cnidaria. In F. Seidel (ed.), Morphogenese der Tiere, Reihe I, 1: 69-415. Gustav Fischer Verlag, Stuttgart.
Tidball, J.G. 1984. Cnidaria: Secreted surface. In J. Bereiter-Hahn, A.G. Matoltsy and K.S. Richards (eds.), Biology of the Integument Vol. I, p.69-78. Springer-Verlag, Berlin.
Totton, A.K. and G.O. Mackie. 1960. Studies on Physalia physalis (L.) Discovery Reports 30: 301-408.
Van-Praët, M. 1985. Nutrition of sea anemones. Adv. mar. Biol. 22: 65-99.
Van-Praët, M. and D. Doumenc, 1975. Morphologie et morphogenèse expérimentale du tentacule chez Actinia equina L. J. microsc. Biol. cell. 23: 29-38.
Wada, T. 1972. Contractile activities of the acontial filaments of Metridium senile var. fimbriatum Verrill. J. Fac. Sci. Hokkaido Univ. Ser. VI, Zool. 18: 387-399.
Wainwright, S.A. and M.A.R. Koehl. 1976. The nature of flow and the reaction of benthic Cnidaria to it. In G.O. Mackie (ed.) Coelenterate ecology and behaviour, pp.3-21. Plenum Press, New York.

Watson, G.M. and R.N. Mariscal. 1983a. The development of a sea anemone tentacle specialized for aggression: Morphogenesis and regression of the catch tentacle of Haliplanella luciae (Cnidaria, Anthozoa). Biol. Bull. 164: 506-517.
Watson, G.M. and R.N. Mariscal. 1983b. Comparative ultrastructure of catch tentcles and feeding tentacles in the sea anemone Haliplanella. Tiss. Cell. 15: 939-953.
Weill, R. 1934. Contribution à l'etude des cnidaires et de leurs nématocystes. Trav. Stat. Zool. Wimereux, 10, 11: 1-702.
Westfall, J.A. 1966. The differentiation of nematocysts and associated structures in Cnidaria. Z. Zellforsch. 75: 381-403.
Williams, R.B. 1975. Catch tentacles in sea anemones: occurrence in Haliplanella luciae (Verrill) and a review of current knowledge. J. nat. Hist. 9: 241-248.
Yanagita, T.M. and T. Wada. 1959. Physiological mechanism of nematocyst responses in sea anemone. VI. A note on the microscopical structure of acontium, with special reference to the situation of cnidae within its surface. Cytologia 24: 81-97.
Young, J.A.C. 1974. The nature of tissue regeneration after wounding in the sea anemone Calliactis parasitica (Couch), J. mar. biol. Ass. U.K. 54: 599-617.

12

# CONTROL OF NEMATOCYTE DIFFERENTIATION IN HYDRA

Hans Bode

Developmental Biology Center
University of California
Irvine, California

## SUMMARY

In hydra the four nematocyte populations are in a steady state balancing high rates of production and loss. Their distribution in the animal, population dynamics, and differentiation pathways are fairly well understood. A review of the fragmentary knowledge of the control of nematocyte differentiation suggests the following. Stem cells among the interstitial cells become committed to nematocyte differentiation based on decisions internal to the cell. The environment apparently plays no role in determining either the type of nematocyte to be formed, or the number of stem cells to be committed to nematocyte differentiation. Once committed, the cell is subject to environmental influences to maintain the homeostasis of the nematocyte population. At stages further down the pathway, differentiation intermediates are subject to environmental influences governing pathway traverse, and the observed position-dependent pattern of differentiation.

## I. INTRODUCTION

Nematocyte differentiation is a major and continuing activity in hydra. It is a major activity because these cells and their differentiation intermediates make up more than half of all the cells in the adult animal. It is a continuing activity because the nematocyte populations are in a steady-state in which continual loss of mature nematocytes is balanced by continual production of new ones. This overall process

The Biology of Nematocysts

has been extensively described. The size, distribution and dynamics of each of the nematocyte populations has been measured. The differentiation pathways have been characterized in terms of the intermediate cells and the kinetics of traverse for each nematocyte type.

Of current interest are the elements of control of nematocyte differentiation that maintain the steady-state population sizes of each nematocyte type. Not a great deal is known, but what is known excludes certain types of control mechanisms and leads to a first approximation of how nematocyte differentiation is regulated. Before presenting that body of information, a short summary of the relevant facts about nematocyte differentiation will help provide a context in which to deal with plausible elements of control.

## A. Nematocytes and Their Differentiation Pathways

Four types of nematocytes have been found in the many species of hydra examined: desmonemes, stenoteles, atrichous isorhizas, and holotrichous isorhizas (Weill, 1934; Werner, 1965). Their relative population sizes are unequal, and their distributions in the animal non-uniform. Typically, as in H. attenuata, desmonemes are the most abundant accounting for 82% of all nematocytes. The remainder are divided among the atrichous isorhizas, 11%, stenoteles, 5%, and holotrichous isorhizas, 2% (Zumstein and Tardent, 1971; Bode and Flick, 1976). 95% of the nematocytes are mounted in the battery cells of the tentacles. This includes all of the desmonemes and atrichous isorhizas, and about half of the stenoteles and holotrichous isorhizas (Bode and Flick, 1976). The remaining stenoteles and holotrichous isorhizas are scattered throughout the body column mounted within the epithelio-muscular cells of the ectoderm. The nematocyst capsule is fairly uniform in size throughout the population for three of the four types, but not for the stenoteles. The capsules of this latter type range from 7.5 - 13.0 microns in length (Lehn, 1951; Rich and Tardent, 1969), with the length dependent upon the axial location where the stenotele developed. The largest develop near the apical end of the body column, and the size decreases monotonically down the length of the animal (Bode and Smith, 1977).

As diagrammed in Figure 1, nematocytes arise by differentiation from multipotent stem cells (Im, Fig. 1) among the interstitial cell population (Lehn, 1951; Brien, 1961; David and Murphy, 1977). An interstitial cell committed to nematocyte differentiation (Inc) undergoes 2-5 divisions, the progeny forming a syncytial cluster, or nest, of 4 - 32 cells (Lehn, 1951; Rich and Tardent, 1969; David and Challoner,

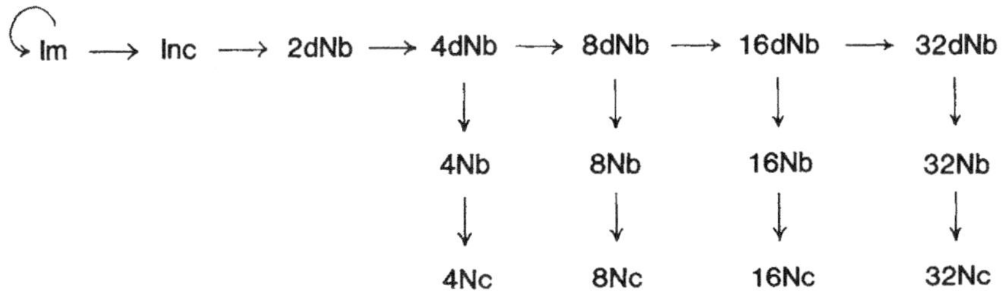


Fig. 1. Schematic of the nematocyte differentiation pathways. Im: multipotent stem cell; Inc: interstitial cell committed to nematocyte differentiation; dNb: dividing nematoblast; Nb: non-dividing nematoblast differentiating nematocyst; Nc: nematocyte. The intermediates in this schematic have been designated in functional terms as described in part by Campbell and Bode (1983). In the morphological terms defined by David (1973), Im, Inc, and the nests of 2 and 4 dNb correspond to large (big) interstitial cells, while dNb in nests of 8, 16, and 32 correspond to small (little) interstitial cells. Nb correspond to nests of nematoblasts of any size with a capsule in some stage of development, and Nc refers to single fully differentiated nematocytes.

1974). Thereafter, all cells of a nest differentiate synchronously, each forming the same type of nematocyte (Slautterback and Fawcett, 1959). Upon completion the cells separate from one another and migrate primarily into the tentacles where they are mounted and become functional (Tardent and Morgenthaler, 1966; Campbell, 1967b; Herlands and Bode, 1974). The time required to traverse the pathway from interstitial cell to completion of differentiation, depending on nest size, is about 5-7 days for desmonemes, 5 1/2 - 7 1/2 days for isorhizas, and 7-8 days for stenoteles (Campbell and David, 1974; David and Gierer, 1974).

An interesting aspect of nematocyte differentiation is its dependence on axial location. Although interstitial cells are found throughout the animal except in the tentacles, nematocyte differentiation is confined to the body column (Bode et al, 1973; David and Challoner, 1974). It does not occur in either the head or the foot. Further, different types are made preferentially in different parts of the body column. For example, stenoteles differentiate primarily in the lower half of the body column, while desmonemes are formed mostly in the upper half (Bode and Smith, 1977; Nishimiya et al, 1986).

Finally, the tissue dynamics in the adult are such that cells of a tentacle are completely replaced in 7-10 days

(Campbell, 1967a; Bode and Flick, 1976). Since the vast majority of the nematocytes are mounted in the tentacles, they also turn over within this time. The rate of loss is actually higher as the nematocyte density is much higher near the tip of the tentacle resulting in a loss of as much as 25% of the nematocytes per day (Bode and Flick, 1976). The stenoteles and holotrichous isorhizas mounted on the body column turn over more slowly. In addition to replacing nematocytes in the adult, nematocytes are also continuously produced for developing buds.

The relatively high rates of production and loss indicate a very active system that would seem to require considerable regulation to keep it functioning properly. Focusing on the control of differentiation in this overall process, a number of questions have been raised. How does a multipotent stem cell become committed to a particular nematocyte pathway? Related to that, are there quantitative controls on commitment to maintain the differences in the relative population sizes of the four nematocyte types? Is commitment to a particular nematocyte pathway a single-step or multi-step process? What is the basis of position-dependent differentiation? Following commitment of the stem cell is traverse of a differentiation pathway also regulated, and if so, how? In the following the evidence bearing on each of these questions will be reviewed.

## II. COMMITMENT OF STEM CELLS TO NEMATOCYTE DIFFERENTIATION

### A. Is Control of Commitment to Type of Nematocyte Internal or External to The Stem Cell?

In principle, the decision of a multipotent stem cell to enter a specific nematocyte pathway could be based on influences external to, or on processes wholly internal to, the stem cell. An example of an external influence on commitment is the induction of the chick ectoderm to form either feathers or scales depending on the source of the mesoderm (e.g. Rawles, 1963). Cell autonomous differentiation in many cell lineages of the nematode is an example of internal control (e.g. Sternberg and Horvitz, 1984). A very prominent characteristic of the development of the four nematocyte pathways is their axial-dependent distribution of nests of differentiating nematoblasts. Thus, the question of internal or external control of commitment to type of nematocyte has focused on understanding the basis of this position-dependence.

The position-dependent differentiation of nematocyte type has suggested for some time (e.g. Bode and David, 1978) that the stem cells were "reading" their axial location, that is, they were reacting to an external cue and responding accordingly by entering a specific nematocyte pathway. For example, interstitial cells in the lower half of the animal might react to that environment and preferentially form stenoteles. The few indirect experiments carried out to test this correlation are open to multiple interpretations, while the one direct test is negative.

The direct test was carried out by Sugiyama and his colleagues (Nishimiya et al, 1986). They examined the possibility that the stem cells among the interstitial cells were "reading" one of the developmental gradients implicated in head formation. Both the head activation and head inhibition gradients are maximal at the apical end of the body column and decrease down the body column (e.g. MacWilliams, 1983a,b). If the stem cells were responding to one or the other, then altered distributions of nematoblast nests should occur in mutants with altered gradients. They examined these distributions in two mutants. One was reg-16, which has elevated levels of inhibition and reduced levels of activation (Achermann and Sugiyama, 1985) while the other, mh-1, has reduced levels of activation and elevated levels of inhibition (Sugiyama, 1982). Although the developmental consequences of the altered gradients is profound in both mutants, the axial distributions of the nests of nematocytes by type was normal. Hence, the stem cells were not responding to either of these gradients to make decisions about nematocyte type.

The position-dependent pattern of nematocyte differentiation may have nothing to do with the commitment decision of the stem cells at all. Instead, these distributions could be based on the known migration behavior of interstitial cells (Tardent and Morgenthaler, 1966; Campbell, 1967b; Herlands and Bode, 1974). If so, any external controls would affect cells further down the pathway, but not the stem cells.

Single large interstitial cells (and possibly pairs) emigrate from all regions of the animal, and on average 15% of the interstitial cells leave a particular region within 24 hours (Heimfeld and Bode, 1984a). Migration is biased in an apical direction with more interstitial cells moving towards the head as towards the foot. There is an important difference between migrating and non-migrating interstitial cells. The proportion of migrating interstitial cells that are stem cells is only half that of the non-migrating interstitial cells (Heimfeld and Bode, 1984b). In other words, a higher fraction of the migrating interstitial cells are committed to differentiation.

Thus, the position-dependent differentiation of nematocytes may be due to selective migration of committed interstitial cells. For example, stem cells may become committed to stenotele formation anywhere along the body column. These committed cells would migrate preferentially in a basal direction, settle somewhere in the lower half of the body column and undergo differentiation. Since the numbers of interstitial cells that migrate are sufficient to account for the number of nests initiated daily (Heimfeld and Bode, unpublished results), the non-uniform distributions of nematoblast nests can be explained entirely in this manner. However, there is no direct evidence for this possibility.

In contrast, there is evidence for the migration of differentiation intermediates in other pathways which supports the idea that this is a general mechanism. One example concerns nerve cell differentiation.

The distribution of nerve cells along the body axis is highly non-uniform. The density of nerve cells in the head and foot are respectively 6-fold and 2-fold higher than in the body column (Bode et al, 1973). Correspondingly, the rates of nerve cell differentiation are also higher at the extremities than in the body column (David and Gierer, 1974; Yaross and Bode, 1978a). Recently single small interstitial cells, which have a morphology similar to that of small interstitial cells of the nematocyte pathway and are distinct from the large interstitial cells (David, 1973), have been shown to be intermediates in the nerve cell differentiation pathway (Bode and Chow, in prep.). These single small interstitial cells migrate in numbers comparable to the large interstitial cells (Heimfeld and Bode, 1984a). Their numbers, final regional distribution, and labelling indices correlate strongly with the subsequent appearance of nerve cells (Heimfeld and Bode, 1984b).

These results suggest the following explanation for the position-dependent pattern of nerve cell differentiation. Stem cells become committed to nerve cell differentiation anywhere along the body axis. These committed divide to form separate small interstitial cells, which are capable of migration, and subsequently migrate preferentially into head and foot and complete nerve cell differentiation.

A second example concerns spermatogenesis. In H. oligactis, as well as in other species of hydra, asexual males can be induced to form sperm by lowering the culture temperature (Brien, 1961; 1963) After 2-3 weeks in the cold, the animals develop testes that are filled with sperm and differentiation intermediates. This oft observed phenomenon suggested that the lowered temperature induced interstitial cells in specific locations to undergo spermatogenesis, namely at those sites where testes eventually appeared.

Recent evidence making use of a monoclonal antibody that recognizes an antigen on all cells of the spermatogenic pathway and no other cell types again suggests migration plays a role. Littlefield et al (1985) have shown that in asexual animals, grown at the usual culture temperature of 18C, early spermatogenic intermediates are found all over the animal. After the temperature is lowered, the numbers of intermediates increase throughout the body column, and only later become localized in the testes. Logically, the accumulation of intermediates in the testes would be due to migration.

Thus, the weight of evidence currently suggests that position-dependent pattern of interstitial cell differentiation is not due to external positional influences on stem cells. Instead, the initial mechanism by which the stem cells become committed to a particular nematocyte pathway may well be internal. Thereafter, the committed cells migrate, most likely following external cues to specific regions to form the observed distributions of nests of developing nematocytes.

## B. Are There Controls On The Numbers of Stem Cells Committed to Specific Nematocyte Pathways?

The four nematocyte populations are in a steady-state that is the result of high rates of production balanced by high rates of loss. To maintain the population size of each type in this dynamic situation one would expect quantitative controls affecting the production of nematocytes. These could be, for example, feedback loops affecting either the numbers of stem cells committed to each nematocyte pathway, or the traverse of a particular differentiation pathway by already committed cells. Here also the evidence indicates that external controls affecting the stem cells they do not exist, or are weak.

A good indication of this is the observation that the total number of nematocytes can vary between animals of similar size by a factor of four (Bode et al, 1977). This suggests that controls on the population size are not very tight. Fluctuations in cell number of this magnitude in, for example, the blood cells of a mammal would result in serious disease.

The possibility of a negative feedback loop controlling stem cell commitment to stenoteles was examined directly. The stenotele population of the tentacles was reduced by 80-90% either by mechanical (Zumstein and Tardent, 1971; Zumstein, 1973) or electrical stimulation (Smith et al, unpublished results), and the commitment behavior of stem cells assayed. An increase in the numbers of nests of stenotele nematoblasts, but not any other nematocyte type, was found 7-9 days after stimulation. For two reasons this indicated that either the stem cells or committed interstitial cells had responded to the reduction in stenoteles. The time of increase in nest number corresponds to the time required for a committed cell to traverse the pathway to reach a nest of recognizable stenotele nematoblasts (David and Gierer, 1974). And second, the distribution of nest sizes was unchanged indicating the perturbation must have affected only cells early in the pathway (Zumstein, 1973; Smith et al, unpub. results).

However, the size of the effect did not correspond with the extent of the loss. Both Zumstein (1973) and Smith et al (unpub. results) found a 40-50% increase in stenotele nest number. The effect was also transient. By periodically stimulating animals the stenotele population was reduced to 10% of normal for a month. This resulted in the initial transient increase, but thereafter, the numbers of nests were no different than in the controls (Fig. 2). Had there been a strong feedback loop from the stenotele population in the tentacles

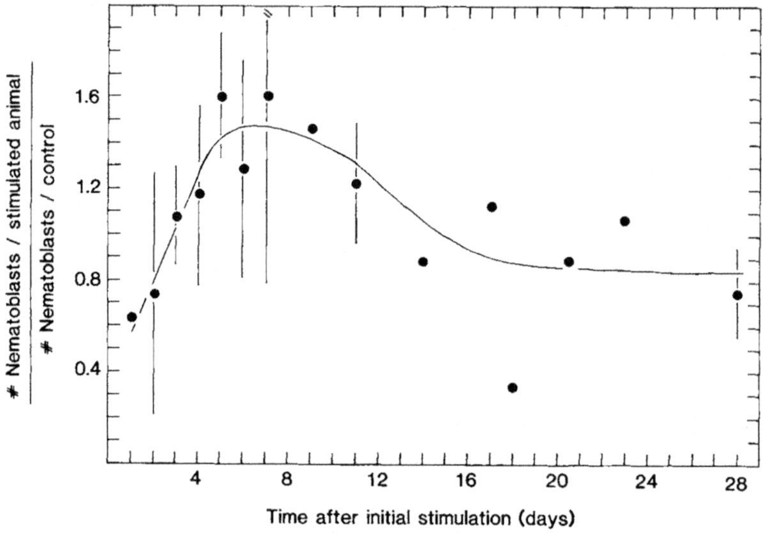


Fig. 2. Transient rise in number of stenotele nematoblasts in animals in which the stenotele nematocyte population has been maintained at 10% of normal for a month. Animals were electrically stimulated briefly every other day for 28 days. The animals remained healthy during the entire period of the experiment. The number of nests of differentiating stenotele nematoblasts was determined in fixed whole mounts using the lead nitrate-thioacetic acid (David and Challoner, 1974). 10 stimulated and 10 control animals were analyzed for each time point. (G.S. Smith, K. Nadeau and H.R. Bode; unpublished results)

affecting stem cell commitment to this pathway one would have expected a larger initial response and a continuously raised level of stenotele production. Hence, a strong negative feedback loop acting on the stem cell does not exist.

Feedback control on stem cell commitment might also be exerted earlier in the pathway either by the differentiation intermediates such as the nematoblasts in nests, or at the level of the interstitial cells. The evidence indicates that neither is the case. Treatment with hydroxyurea can eliminate all nematoblasts and reduce the interstitial cell population in the animal (Bode et al, 1976). The elimination of the nematoblasts had no effect on the fraction of stem cells entering the nematocyte pathways (Heimfeld and Bode, 1986a). Were there a simple and strong negative feedback from the nematoblasts one would have expected an increase in the numbers of stem cells committed to nematocytes.

Hydroxyurea treatment can reduce the interstitial cell population to a few percent of normal (Bode et al, 1976). Earlier evidence suggested that such a reduction led to an increase in stem cell proliferation (Im -> Im, Fig. 1), and a corresponding reduction in the fraction of stem cells forming nematocytes (Im -> Inc, Fig. 1; Bode et al, 1976; David and MacWilliams, 1978). This was interpreted in terms of a negative feedback loop from the stem cells on themselves affecting the probability that they would undergo self-renewal, or cell division. However, recent more detailed studies indicate that there is no change in the overall rate of proliferation of the interstitial cells (including the stem cells) after reducing the size of the interstitial cell population (Heimfeld, 1985; Heimfeld and Bode, 1986a). Furthermore, the cells affected by the reduction are not stem cells (Im, Fig. 1), but the committed interstitial cells (e.g. Inc, Fig. 1; Heimfeld and Bode, 1986a,b).

Hence, there do not seem to be any strong external controls exerted by the mature nematocytes (although this has been demonstrated only for the stenoteles), nematoblasts, or the interstitial cell population which affect stem cell commitment to nematocyte differentiation.

In striking contrast, one aspect of nematocyte differentiation does appear to be under some form of control. The ratios of the population sizes of the four nematocyte populations are much more constant than the total number of nematocytes per animal. The total number may vary 4-fold, but the desmonemes always make up 80%, and the other three types their typical fractions (Bode et al, 1977). Placing animals on different feeding regimes resulted in a 7-fold variation in the total number of nematocytes per animal among the regimes. In contrast, there was only a <20% difference in the fraction that formed desmonemes, and a 2-3 fold difference for the atrichous isorhizas. Hence, the ratios of the population sizes are much more constant than the any of the individual nematocyte population sizes.

Such a control could be exerted on the stem cells by determining the fraction allocated to each of the four nematocyte pathways. As there is no evidence for external control, this could also be due to some form of internal control.

## C. Does Commitment to a Specific Nematocyte Pathway Involve a Single Decision or Multiple Decisions?

The process of commitment of a multipotent stem cell to a particular nematocyte differentiation pathway can be imagined in several ways. It could be a single decision in which the stem cell is directly committed to a particular type of nematocyte (Fig. 3A). It could involve two decisions: an initial decision that commits the stem cell to form nematocytes, and a later decision as to type (Fig. 3B). The latter decision could also involve more steps. For example, the initial decision might be the general commitment to any of four types, a second decision restricting the choice to either of two types, and a third narrowing the choice to a particular pathway.

Efforts to examine the question have focused on whether one or two decisions are made. In particular the question has been posed in the following form: does the choice as to type occur in the committed cell at the beginning of the pathway, or someplace along the pathway in one of the differentiation intermediates (Fig. 3A vs. Fig. 3B)? As described above (Fig. 1), the early part of the pathway involves several cell divisions that lead to the formation of a syncytial nest of 4, 8, 16, or 32 dividing nematoblasts (Lehn, 1951; Rich and Tardent, 1969; David and Challoner, 1974). If the decision as to type occurs later than the 4-cell stage, it is likely that commitment involves two decisions: commitment to nematocyte differentiation in general before the 4-cell stage, and commitment to type after the 4-cell stage. The 4-cell stage is a good landmark because in _H. attenuata_, where this question was examined, nests of 4 interstitial cells are considered to be part of a nematocyte pathways (Fig.1). There is evidence for both points of view.

Fujisawa and David (1981) have evidence for commitment to nematocyte type at a late stage in the pathway. They isolated the upper quarter of animals where few stenotele nests are formed. Such animals regenerated a peduncle and foot at the basal end, as well as exhibiting an increase in the number of stenotele nests as would be expected for the lower end of the animal. The increase began sharply 4 days after bisection. Since 7-8 days are required for a committed stem cell to

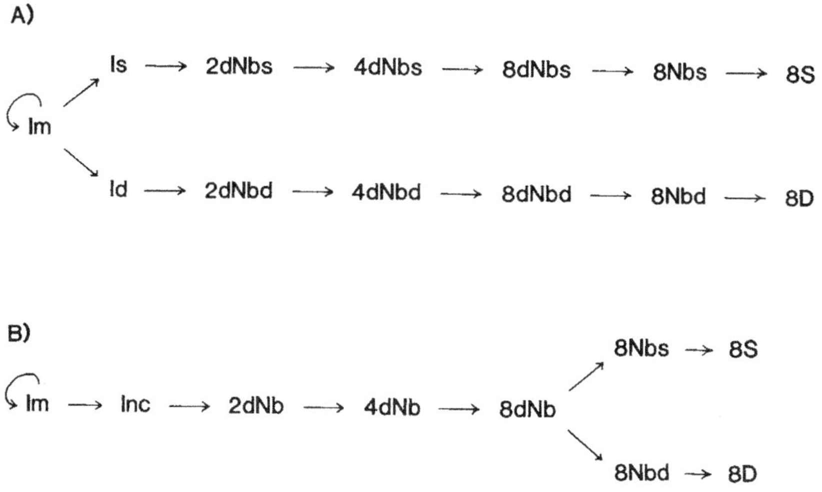


Fig. 3. Two plausible schemes describing when during the nematocyte differentiation pathway a cell becomes committed to type of nematocyte. In this example, S = stenotele and D = desmoneme. A. Commitment to type is part of the initial commitment of the stem cell. B. Commitment to type occurs to nematocyte in general early, and to type later at the dNb stage. Is: interstitial cell committed to stenotele differentiation; dNbs and Nbs: intermediates in the stenotele pathway; S: mature stenotele; Id: interstitial cell committed to desmoneme differentiation; dNbd and Nb: intermediates in the desmoneme pathway; D: mature desmoneme. Other cell type designations as in Figure 1. The branch point indicates that a cell (or nest) enters one or the other pathway.

traverse this differentiation pathway, the sharp increase could not be due to an increase in stem cells committed to stenotele formation. Instead, the commitment to stenoteles apparently occurred at the stage when cell division is completed or during the last cell division. Since the great majority of stenoteles occur in numbers of nests of 8 or 16 (Rich and Tardent, 1969; David and Challoner, 1974), commitment would have occurred at the 8-and 16-cell stage, or at the 4- and 8-cell stage.

Other evidence is difficult to reconcile with a late commitment. One aspect concerns the relationship between the number of cell divisions undergone by a committed cell and the type of nematocyte it will form. The distribution of nest sizes varies among the types (Rich and Tardent, 1969; David and Challoner, 1974). The majority of stenoteles occur in nests of 8 with less in nests of 16, and only a few percent found in nests of 4. More than 80% of the desmoneme nests contain 16 cells with the remainder equally divided among nests of 8 and 32. Most of both types of isorhiza are evenly split in nests of 8 or 16 cells. The small range of cell divisions, 2-5, suggests a mechanism controlling the number of cell divisions that begins to function at the beginning of the pathway. An example would be the dilution of a specific metabolite required for cell division that is no longer produced once the committed cell enters the pathway. This is plausible since the total volume of the cells in a nest of nematoblasts is greater than that of the committed interstitial cell (Lehn, 1951).

A priori, the mechanisms determining the number of cell divisions and commitment to type could be either independent, or causally related. If commitment to type occurred during or after the last cell division, one would expect the two mechanisms to be independent since they would be operating at different times. Then, it is not easy to understand why the distribution of nest sizes should vary between the types. The control on cell division initiated early in the pathway would result in a distribution of nest sizes of 4-32 cells. Thereafter, commitment to type would occur, and one would expect a similar distribution of nest sizes for each type. However, the distributions of nest sizes are dissimilar among the types. Because the distribution of size is peculiar to type, it is more likely that the cell division and commitment mechanisms are either closely related, or part of the same mechanism. This is most simply understood if both operate early in the pathway.

A second argument against early commitment is based on the differences in morphology of the intermediates at different stages in the nematocyte pathways. The nests of little interstitial cells, the dividing intermediates of the early part of the pathway, are not uniform. Many nests of 8 or 16 cells contain small cells with highly condensed nuclei. Other nests of 8 or 16 cells contain much larger cells with nuclei in which the nucleolus is visible.

These differences are correlated with differences at a slightly later stage when the developing capsules are not yet recognizable by type. Developing stenoteles can easily be distinguished from the others because the developing capsule is distinctly larger than those of the other nematocytes. The morphology of these cells also corresponds with those in nests of little interstitial cells that are large and have a clear nucleolus. Similarly, nematoblasts with small capsules destined to be desmonemes or atrichous isorhizas occur in nests in which the cells are much smaller and have condensed nuclei. These correlations strengthen the view that the differences in morphology of the cells in nests of dividing little interstitial cells are important and reflect earlier restrictions in commitment to type.

Though there is evidence for both points of view, my bias is that commitment to type occurs early in the pathway, probably by the stage of the committed interstitial cell. This does not resolve the issue of the number of steps in the commitment decision, but suggests that whatever number exist they occur between the uncommitted stem cell and the cell about to enter the differentiation pathway.

### D. Summary

For a long time the common assumption has been that commitment of the multipotent stem cell to one of the nematocyte pathways was controlled by the environment. The number committed was indirectly controlled by a feedback loop affecting the fraction of stem cells that underwent proliferation, and the type was dependent on the axial location of the stem cell (Bode and David, 1978; Bode, 1980). Recent data has raised considerable doubt about this view. The numbers and nature of migrating interstitial cells has provided a more likely explanation for position-dependent differentiation based on the selective migration of committed interstitial cells. Further it is now apparent that stem cell commitment to nematocyte differentiation is unaffected by alterations in the sizes of the nematocyte, nematoblast or interstitial populations. This is true for both the fraction of stem cells committed as well as for the ratios within this fraction allocated to each of the four types. Hence, there is no evidence that stem cell differentiation decisions are based on external influences such as position, or cell-cell interactions. The current simplest view is that the nematocyte commitment decisions are internal to the stem cell.

## III. TRAVERSE OF THE NEMATOCYTE PATHWAYS IS CONTROLLED BY THE ENVIRONMENT

While the multipotent stem cells may not be subject to external influences, there is considerable evidence that the committed cell and differentiation intermediates traversing the pathway are subject to environmental control. Such controls are exerted at several different points along the pathway.

## A. Control of Cell Division of the Committed Cell

An early external influence is exerted on the committed interstitial cell about to enter the pathway. Normally a committed cell (Inc, Fig. 4A) undergoes 2-5 cell divisions to form a single syncytial nest. If, however, the nematoblast and interstitial cell populations are reduced to a small fraction of normal by hydroxyurea or nitrogen mustard treatment, a committed cell undergoes extra divisions before carrying out the differentiative divisions (Heimfeld and Bode, 1986a,b). As shown in Figure 4B, each cell arising from these extra divisions will then undergo the usual 2-5 cell divisions yielding several syncytial nests of cells. The result is an increase in the number of nematocytes produced from a single stem cell, and represents an amplification mechanism.

Such a mechanism could also have been operating after the reduction of the stenotele populations described earlier. The observed increased number of nests of stenotele nematoblasts

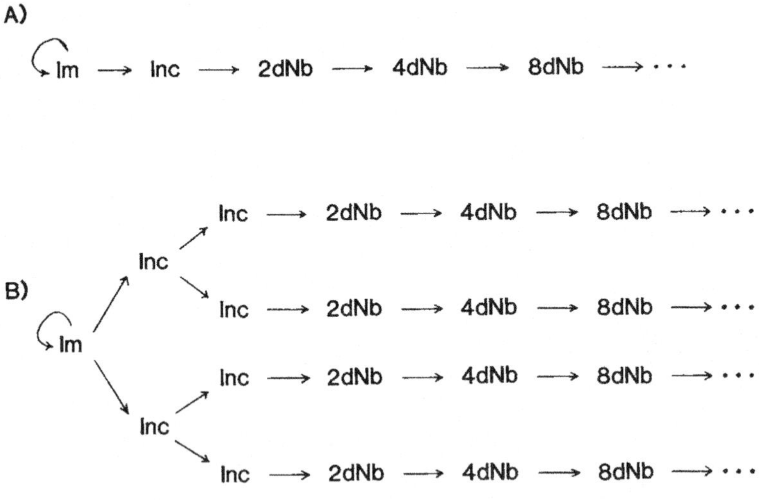


Fig. 4. Alteration in the differentiation pathway due to amplification divisions of a committed interstitial cell. A. Normal pathway. B. Altered pathway with amplification divisions. The cell types are defined as in Figure 1.

would have been due to an amplification of the interstitial cells already committed to form stenoteles by undergoing additional divisions. One piece of data suggests that flexibility in terms of cell division is restricted to the committed cells, and is not characteristic of the later dividing intermediates traversing the pathways. Reduction of the stenotele population had no effect on the distribution of nest sizes of stenotele nematoblasts (Zumstein, 1973; Smith et al, unpub. results). Were the intermediates responsive to such alterations, changes in the distribution of nest sizes would have occurred.

Hence, the stem cells may not respond to environmental perturbation, but the committed cells clearly do. This kind of mechanism provides a quantitative, or homeostatic, control that would be expected to help maintain the steady state population sizes in the context of the high rates of production and loss of these cell types.

## B. Control of Pathway Traverse and the Position-Dependent Pattern of Differentiation

Once having begun the cell divisions that will form the nest, the cells are known to be subject to other types of external control. Formally, one kind includes factors which influence the traverse of the pathway. Two examples are known for this kind.

In the first, a factor(s) exerts a local control whose absence leads to cell death of nematocyte intermediates. It was demonstrated as follows. Since nematocyte differentiation does not take place in the hypostome (Bode et al, 1973) even though stem cells capable of forming nematocytes occur there, it was plausible that the environment did not support nematocyte differentiation. In principle this possibility could be tested by moving nematocyte differentiation intermediates into the head and following their fate. To do this hydra were bisected which commonly results in the regeneration of a head at the apical end of the lower half. This occurs in a morphallactic manner by remodeling the body column tissue at the apical end of the stump into tissue of a head (Park et al, 1971; Hicklin and Wolpert, 1973; Cummings and Bode, 1984). As regeneration proceeded, the nematocyte intermediates in this region were essentially moved from the body column into the head.

During this process nests of large and small interstitial cells, the dividing intermediates, and nematoblasts that were at the apical end declined rapidly in number within 48 hours (Yaross and Bode, 1978b). As the decline was not due to a decrease in stem cells entering these pathways, emigration of intermediates in a basal direction, or an alteration of the cell cycle of these intermediates, the simplest explanation is that the intermediates died. The effect is specific since only nematocyte differentiation, but not nerve cell differentiation declined, and it is local as early nematocyte intermediates in the region directly basal to the regenerating tip were unaffected (Yaross and Bode, 1978b).

The second control operates on traverse of late stages of the pathway when the nematocyst capsules of the cells in nests are well developed. By removing both head and foot as well as the regenerating tips 12 hours later, Fujisawa and David (1984) demonstrated the numbers of late desmoneme and stenotele nematoblasts, but not the early dividing intermediates, declined 3 and 4 days later respectively. Thereafter, they recovered rapidly. As the decline occurred throughout the body column the factor must be a global in nature. Further, the results suggest the nature of the factor(s) is complex. The decline in desmonemes is consistent with a maintenance factor emanating from one or both extremities. However, interference with the traverse of the stenotele, but not the desmoneme, differentiation pathway also occurred when animals were simply injured, or intact animals were exposed to regenerating animals. These results indicate the release of a substance that inhibits traverse of the stenotele pathway.

The other type of external control is an influence on the position-dependent pattern of nematocyte differentiation. In two instances the factors also affect pathway traverse and have been described. The absence of a local nematocyte-maintenance in the head contributes to the absence of nematocyte differentiation occurring there (Yaross and Bode, 1978b), and thus, to the position-dependent pattern. In this instance, position-dependence refers to the formation of nematocytes in general in the body column as opposed to the head.

In very recent work, Fujisawa (pers. comm.) has shown that a factor inhibiting traverse of the stenotele pathway is graded along the body column in concentration with a maximum in the head. Hence, this factor may be involved in the distribution of the stenotele nematoblasts along the column. Large numbers of stenotele nematoblasts at the lower end may correspond to low concentrations of this stenotele-inhibiting factor.

Finally, the head has an influence on the stenotele capsule size. As mentioned earlier, stenotele capsules are 13 microns in length near the head and decrease in size in a graded manner reaching a value of 7.5 microns at the lower end (Rich and Tardent, 1969; Smith and Bode, 1977). Decapitation and subsequent removal of the regenerating tip on three successive days resulted in the formation of small capsules, 7-8 microns in length, throughout the body column several days later (Tardent et al, 1971). Removal of the foot and the regenerating tip on three successive days had no effect on stenotele size. This result indicates that the head exerts an influence on the size of the capsule formed. The graded distribution of capsule sizes suggests the factor is global in nature, and that its influence (concentration?) decreases with distance from the head. A transplantation experiment indicated that the alteration in capsule size can occur as soon as 5 days after the perturbation (Tardent et al, 1971). This suggests the influence exerted by the head acts on intermediates shortly before the time of capsule formation.

## IV. CONCLUSIONS

Although the data are fragmentary, there is enough to piece together a first approximation of the nature of the controls governing nematocyte differentiation. The elements that lead to the view are several. [1] There is no evidence for the external control of the multipotent stem cell commitment to nematocyte differentiation. [2] The position-dependent pattern of nematocyte differentiation is most likely due to the selective migration of committed interstitial cells. [3] There are no controls in the form of feedback loops from cells anywhere along the nematocyte pathways affecting the numbers of stem cells committed to forming nematocytes. Consequently, the total number of nematocytes per animal can vary over a four-fold range. [4] In contrast, the ratios of the population sizes of the four types of nematocytes are quite constant suggesting some form of control. [5] The cell division behavior of the committed interstitial cell before entering the pathway is correlated with alterations in the size of committed cell, nematoblast and/or nematocyte populations. [6] Traverse of a differentiation pathway by a committed cell is influenced by the environment at several stages along the pathway. Some of these influences are global in nature, and others local.

Since there is no evidence for any external controls on the commitment of a stem cell to nematocyte differentiation, it is logical to consider that such decisions are based on mechanisms wholly internal to the cell. Such a mechanism could be a fixed developmental program. For example, a stem cell could continually undergo asymmetric cell divisions giving rise to a stem cell and to a specific sequence of differentiation products. Once the sequence is complete, it starts over. The fixed lineages observed during nematode development provide an example (Sternberg and Horvitz, 1984). Another mechanism, completely opposite in nature, is one in which the decision is stochastic and the choice is based on probabilities. The probability for commitment to each nematocyte type would be fixed and would differ between the types. The *in vitro* behavior of hemopoietic stem cells and their derivatives is consistent with such a mechanism (Ogawa *et al*, 1983).

The major consequence of an internal commitment mechanism is that the pattern of commitment behavior of the multipotent stem cells would be constant. The pattern would be the same regardless of axial location of the stem cell, and independent of any sort of environmental perturbation. Two of the findings are consistent with such a mechanism. Internal control would explain why stem cell commitment behavior is independent of alterations in the population sizes of nematocytes, nematoblasts or committed interstitial cells. It would also account for the constant ratios of stem cells committed to the four types of nematocytes, and the independence of these ratios to environmental perturbation.

If control of stem cell commitment is internal, then many aspects of nematocyte differentiation such as the position-dependent pattern, the recovery of reduced nematocyte populations, and the steady state population sizes must all be regulated at stages after commitment of the stem cell. There is ample evidence for this view.

Position-dependent nematocyte differentiation would be explained as follows. Commitment of a stem cell to a particular nematocyte type would be internal, and could occur anywhere along the body column. Then, the committed cell would selectively migrate, be it in an apical or basal direction, to the appropriate part of the body column, and carry out the differentiation. For example, cells committed to stenoteles would migrate to the lower end, settle, and enter the pathway. In this scheme the selective migration is clearly under some form of external regulation.

Additional environmental controls such as maintenance factors or inhibitory substances acting on differentiation intermediates would play refining roles in the position-dependent pattern. The absence of nematocyte differentiation in the head could be due to the absence of maintenance factors. Such a control would provide an additional means for preventing an errant interstitial cell committed to nematocyte differentiation from undergoing differentiation in the head. The graded distribution of the stenotele inhibitor could contribute to the non-uniform axial pattern of stenotele differentiation by affecting the rate of traverse of that pathway. Finally, the gradient in stenotele sizes is under the control of a global factor emanating from the head.

The other prominent feature of nematocyte differentiation is the steady state of the nematocyte population sizes in the context of the high rates of production and loss. To maintain this steady state as part of the tissue dynamics of the animal, at least one stage along the pathway must be sensitive to environmental perturbations of population size so that some form of homeostatic mechanism can operate. That stage is most likely the committed interstitial cell, and only this cell, as neither the uncommitted stem cell nor dividing intermediates respond to environmental perturbation by dividing more or less. One can envision negative feedback loops from one or more of the nematocyte-related populations affecting the cell division behavior of the committed cell. If the populations are reduced, the committed cell would undergo additional, or amplification, divisions which would rapidly allow the population size to recover. Additional controls on population size would be in the form of maintenance factors or inhibitory factors, such as the demonstrated global and local ones. They would act on intermediates at various stages along the pathway.

In sum, a current view of the controls exerted on nematocyte differentiation can be described in two parts. The initial commitment of a multipotent stem cell to a particular nematocyte pathway would be due to a mechanism internal to the cell. Thereafter, the committed cell would be subject to variety of environmental influences which would establish the position-dependent character of nematocyte differentiation and maintain the steady state populations of these four cell types.

Many questions remain. For example, if the commitment behavior of the stem cell is wholly internal, does the population of stem cells recover after reduction by drug treatment, and if so is this due to an external or internal influence? Are there progenitor cells which are intermediates between the uncommitted stem cell and the committed interstitial cell about to enter the differentiation pathway? What is the nature of the control of the selective migration of the committed interstitial cells? The fact that this view leaves many questions unanswered and raises others emphasizes that it is a first approximation of the control of nematocyte differentiation.

## V. ACKNOWLEDGEMENTS

I would like to thank Patricia Bode and Shelly Heimfeld for stimulating discussions and their thoughtful comments about the manuscript, and acknowledge the support provided by the National Institutes of Health (HD 08086).

## VI. REFERENCES

Achermann, J., and Sugiyama, T. 1985. Genetic analysis of developing mechanisms in hydra. X. Morphogenetic potentials of a regeneration-deficient strain (reg-16). Dev. Biol. 107: 13-27.

Bode, H.R. 1980. Control of interstitial cell and differentiation in **H. attenuata.** In P. Tardent and R. Tardent (eds.) **Development and Cellular Biology of Coelenterates.** pp. 307-312. Elsevier/North Holland Biomedical Press.

Bode, H.R., and K.M. Flick. 1976. Distribution and dynamics of the nematocyte populations in **Hydra attenuata.** J. Cell Sci. 21: 15-34.

Bode, H.R., and G.S. Smith. 1977. Regulation of interstitial cell differentiation in **Hydra attenuata.** II. Correlation of the axial position of the interstitial cell with nematocyte differentiation. Wilhelm Roux Arch. EntwMech. Org. 181: 203-213.

Bode, H.R., and C.N. David. 1978. Regulation of a multipotent stem cell, the interstitial cell of hydra. Prog. Biophys. Molec. Biol. 33: 189-206.

Bode, H.R., K.M. Flick, and G.S. Smith. 1976. Regulation of interstitial cell differentiation in **Hydra attenuata.** I. Homeostatic control of interstitial cell population size. J. Cell Sci. 20: 29-46.

Bode, H.R., K.M. Flick, and P.M. Bode. 1977. Constraints on the relative sizes of the cell populations in **Hydra attenuata.** J. Cell Sci. 24: 31-50.

Bode, H.R., S. Berking, C.N. David, A. Gierer, H. Schaller, and E. Trenkner. 1973. Quantitative analysis of cell types during growth and morphogenesis in hydra. Wilhelm Roux Arch. EntwMech. Org. 171: 269-285.

Brien, P. 1961. Etude d'**Hydra pirardi** (Nov. spec). Origine et repartition des nematocystes. Gametogenese. Involution postgametique. Evolution reversible des cellules interstitielles. Bull. Biol. France Belg. 95: 301-363.

Brien, P. 1963. Contribution a l'etude de la biologie sexuelle chez les Hydres d'eau douce. Induction gametique et sexuelle par la methode des greffes en parabiose. Bull.

Biol. France Belg. 97: 213-283.

Campbell, R.D. 1967a. Tissue dynamics of steady state growth in **Hydra littoralis**. II. Patterns of tissue movement. J. Morph. 121: 19-28.

Campbell, R.D. 1967b. Tissue dynamics of steady state growth in **Hydra littoralis**. III. Behavior of specific cell types during tissue movements. J. exp. Zool. 164: 379-391.

Campbell, R.D., and C.N. David. 1974. Cell cycle kinetics and development of **Hydra attenuata**. II. Interstitial cells. J. Cell Sci. 16: 349-358.

Campbell, R.D., and H.R. Bode. 1983. Terminology for morphology and cell types. In H. Lenhoff (ed.) **Hydra: Research Methods**. pp. 5-14. Plenum Press, New York.

Cummings, S.G., and H.R. Bode. 1984. Head regeneration and polarity reversal in **Hydra attenuata** can occur in the absence of DNA synthesis. Roux's Arch. Dev. Biol. 194: 79-86.

David, C.N. 1973. A quantitative method for maceration of hydra tissue. Wilhelm Roux Arch. EntwMech. Org. 171: 259-268.

David, C.N., and D. Challoner. 1974. Distribution of interstitial cells and differentiating nematocytes in **Hydra attenuata**. Am. Zool. 14: 537-542.

David, C.N., and A. Gierer. 1974. Cell cycle kinetics and development of **Hydra attenuata**. III. Nerve and nematocyte differentiation. J. Cell Sci. 16: 359-375.

David, C.N., and S. Murphy. 1977. Characterization of interstitial stem cells in hydra by cloning. Dev. Biol. 58: 372-383.

David, C.N., and H. MacWilliams. 1978. Regulation of the self-renewal probability in hydra stem cell clones. PNAS (USA) 75: 886-890.

Fujisawa, T., and C.N. David. 1981. Commitment during nematocyte differentiation in hydra. J. Cell Sci. 48: 207-222.

Fujisawa, T., and C.N. David. 1984. Loss of differentiating nematocytes induced by regeneration and wound healing in hydra. J. Cell Sci. 68: 243-255.

Heimfeld, S., and H.R. Bode. 1984a. Interstitial cell migration in **Hydra attenuata.** I. Quantitative description of cell movements. Dev. Bio. 105: 1-9.

Heim feld, S., and H.R. Bode. 1984b. Interstitial cell migration in **Hydra attenuata.** II. Selective migration of nerve cell precursors as the basis for position-dependent nerve cell differentiation. Dev. Bio. 105: 10-17.

Heimfeld, S. 1985. Growth regulation of the interstitial cell population in hydra. II. A new mechanism for the homeostatic recovery of reduced interstitial cell populations. Dev. Bio. 111: 499-509.

Heimfeld, S., and H.R. Bode. 1986a. Growth regulation of the interstitial cell population in hydra. III. Interstitial cell density does not control stem cell proliferation. Dev. Bio. 115: 51-58.

Heimfeld, S., and H.R. Bode. 1986b. Growth regulation of the interstitial cell population in hydra. IV. Control of nerve cell and nematocyte differentiation by amplification of non-stem interstitial cells. Dev. Bio. 116: 59-68.

Herlands, R.L., and H.R. Bode. 1974. Oriented migration of interstitial cells and nematocytes in **Hydra attenuata.** Wilhelm Roux Arch. EntwMech. Org. 176: 67-88.

Hicklin, J. and L. Wolpert. 1973. Positional information and pattern regulation in hydra: the effect of radiation. J. Embryol. exp. Morph. 30: 741-752.

Lehn, H. 1951. Teilungsfolgen und Determination von I-zellen fuer die Cnidenbildung bei Hydra. Z. Naturforsch. B 6: 388-391.

Littlefield, C.L., J.F. Dunne, and H.R. Bode. 1985. Spermatogenesis in **Hydra oligactis.** I. Morphological description and characterization using a monoclonal antibody specific for cells of the spermatogenic pathway. Dev. Bio. 110: 308-320.

MacWilliams, H.K. 1983a. Hydra transplantation phenomenon and the mechanism of hydra head regeneration. I. Properties of the head inhibition. Dev. Biol. 96: 217-238.

MacWilliams, H.K. 1983b. Hydra transplantation phenomenon and the mechanism of hydra head regeneration. II. Properties of the head activation. Dev. Biol. 96: 239-257.

Nishimiya, C., T. Fujisawa, N. Wanek, and T. Sugiyama. 1986. Genetic analysis of developmental mechanisms in hydra. XV. Nematocyte differentiation in strains with altered developmental gradients. Develop. Growth and Differ. 28(1): 17-30.

Ogawa, M., P.N. Porter, and T. Nakahata. 1983. Renewal and commitment to differentiation of hemopoietic stem cells: an interpretive review. Blood 61: 823-829.

Park, H.D., A.B. Ortmeyer, and D.P. Blankenbaker. 1970. Cell division during regeneration in hydra. Nature 177: 617-619.

Rawles, M.E. 1963. Tissue interactions in scale and feather development as studied in dermal-epidermal recombinations. J. Embryol. exp. Morph. 11: 765-789.

Rich, F., and P. Tardent. 1969. Untersuchungen zur Nematocyten-Differenzierung bei **Hydra attenuata** Pall. Rev. suisse Zool. 76: 779-787.

Slautterback, D.B., and D.W. Fawcett. 1959. The development of the cnidoblasts of hydra. An electron microscopic study of cell differentiation. J. biophys. biochem. Cytol. 5: 441-452.

Sternberg, P.W., and H.R. Horvitz. 1984. The genetic control of cell lineage during nematode development. Ann. Rev. Genetics 18: 489-524.

Sugiyama, T. 1982. Roles of head-activation and head-inhibition potentials in pattern formation of hydra: analysis of a multi-headed mutant strain. Amer. Zool. 22: 27-34.

Tardent, P., and V. Morgenthaler. 1966. Autoradiographische Untersuchungen zum Problem der Zell Wanderung bei **H. attenuata**. Rev. suisse Zool. 73: 468-480.

Tardent, P., F. Rich, and V. Schneider. 1971. The polarity of stenotele in **Hydra attenuata** Pall. Dev. Bio. 24: 596-608.

Weill, R. 1934. Contribution a l'etude des cnidaires et de leurs nemtocytes. Trav. Stat. Zool. Wimereux 10,11: 1-700.

Werner, B. 1965. Die Nesselkapseln der Cnidaria mit besonderer Berucksichtigung der Hydroidea. I. Klassifikation und

Bedeutung fur die Systematik und Evolution. Helgolander wiss. Meeresunters. 12: 1-39.

Yaross, M.S., and H.R. Bode. 1978a. Regulation of interstitial cell differentiation in **Hydra attenuata**. III. Effects of I-cell and nerve cell densities. J. Cell Sci. 34: 1-26.

Yaross, M.S., and H.R. Bode. 1978b. Regulation of interstitial cell differentiation in **Hydra attenuata**. V. Inability of regenerating head to support nematocyte differentiation. J. Cell Sci. 34: 39-52.

Zumstein, A. 1973. Regulation der Nematocyten Produktion bei **Hydra attenuata** Pall. Wilhelm Roux Arch. EntwMech. Org. 173: 468-480.

Zumstein, A., and P. Tardent. 1971. Beitrag zum Problem der Regulation der Nematocyten Produktion bei **Hydra attenuata** Pall. Rev. suisse Zool. 78: 705-714.

13

# CONTROL OF DISCHARGE: FACTORS AFFECTING DISCHARGE OF CNIDAE

Glyne U. Thorington and David A. Hessinger

Department of Physiology and Pharmacology
School of Medicine
Loma Linda University
Loma Linda, California 92350

## SUMMARY

Concomitant mechanical and chemical stimulation of cnidocytes is required to initiate or "trigger" the discharge of cnidae. The triggering of *in situ* cnidae to discharge is under the direct control of the cnidocyte, a specialized secretory cell that is also assumed to be a sensory cell. Cnidocytes are chemosensitized by certain substances which cause them to become more sensitive to triggering by mechanical stimuli. Using a novel and sensitive method to quantify the discharge of cnidae in tentacles of the sea anemone *Aiptasia pallida*, two groups of low molecular weight substances have been identified which chemosensitize *in situ* cnidocytes; these are a wide variety of amino compounds and N-acetylated sugars. These substances are without effect on isolated cnidae which only discharge in response to harsh chemicals and to conditions not normally encountered in the natural environment. It has been shown that these sensitizing substances act through at least two distinct classes of chemoreceptors. These chemoreceptors presumably detect substances originating on or from prey and participate with appropriate prey-derived tactile stimuli to initiate the discharge of cnidae. In addition, the *in situ* discharge of cnidae is influenced by ions in the ambient medium, ambient light intensity, time elapsed between feeding, and by other factors influencing the physiological state of the animal. Classically, cnidocytes have been considered to be independent effectors, acting both as receptors and effectors without intervention of the nervous system. This view, however, is being challenged in two respects by observations which indicate (i) that the nervous system may modulate the responsiveness of cnidocytes; and (ii) that some of the sensory receptors which control discharge reside on adjacent accessory cells.

## I. INTRODUCTION

The process of cnida discharge may be divided into four sequential and hierarchal components: (i) The mechanics of discharge entails changes in the structural relationships among the various parts of the cnida during eversion of the tubule. (ii) The cause of discharge includes changes within or to the cnida that generate or release the force causing the tubule to evert. (iii) The trigger of discharge refers to the action of the cnidocyte on the cnida to initiate tubule eversion. (vi) The control of discharge comprises those cellular functions which determine whether or not the cnidocyte "triggers" discharge. Of these four components the control of discharge is the least studied and least understood, and undoubtedly the most complex. At the present we consider the control of cnida discharge to involve several constituent processes: the sensitizing chemical stimulus and the triggering mechanical stimulus; the respective receptors for these stimuli; the transduction and conduction mechanisms associated with these receptors; the morphodynamic and biochemical changes produced by the stimulated receptors; and the cellular mechanisms that integrate incoming stimuli with respect to the threshold of triggering and the mechanism of triggering. In addition, the control of discharge includes modulating signals originating from remote sensory receptors and conducted to the cnidocytes by the nervous system.

There are two widely accepted views regarding the control of cnida discharge: (i) the triggering of cnida discharge requires two kinds of stimuli, mechanical and chemical, acting in concert; and (ii) cnidocytes behave as independent effectors. Parker and Van Alstyne (1932) using the Portuguese Man-of-war and Pantin (1942a,b) using sea anemones demonstrated that neither mechanical stimulation alone (i.e. touching with clean grass rods) nor chemical stimulation alone (i.e. soluble food extracts) can trigger the discharge of *in situ* cnidae. While these findings have been extended to other species (Williams, 1968; Lubbock, 1979) and are generally accepted, it should be noted that there are published observations (Williams, 1968; Conklin and Mariscal, 1976; Thorington and Hessinger, 1988) which indicate that gentle mechanical stimulation alone can elicit significant cnida discharge in anemone tentacles. However, in anemones more cnida discharge occurs upon combined chemical and mechanical stimulation than with mechanical stimulation alone. Thus, is appears that chemosensitization lowers the mechanical threshold of cnida discharge.

In multicellular organisms independent effectors are defined as individual cells in which causally related sensory and effector systems are both located and controlled without the need for an intervening nervous system. Parker and Van Alstyne (1932) and Pantin (1942a,b) observed that electrical stimulation of local areas of tentacles causes restricted and local discharge. These investigators concluded that nervous conduction to the cnidocytes does not occur. Hence, the concept of cnidocytes as independent effectors was founded. These workers, however, did not seem to consider the possibility of the nervous system exerting either a modulatory or an inhibitory influence on discharge. Their experiments demonstrated only that electrical stimulation does not cause generalized activation of discharge.

In this paper we will primarily discuss the nature of the chemosensitizing stimuli, the characterization of the cnidocyte-associated chemoreceptors, and the effects of various ions and physiological changes on the discharge of cnidae since these are essentially the only aspects of the control of discharge that have been reported. We will discuss these findings in terms of evaluating the adequacy of the independent effector hypothesis as it applies to cnidocytes. Finally, we will present a comprehensive model depicting what is known about the complex interactions that occur between the prey and the feeding tentacle in triggering cnidae to discharge.

## II. IDENTIFICATION OF CHEMORECEPTORS

Any study of the control of cnida discharge must take into account the two types of sensory receptors known to trigger discharge. Most of what is known about cnidocyte receptors deals with the chemoreceptors since chemical stimuli are fairly easy to control and define. Cnidocyte mechanoreceptors, on the other hand, have been the subject of very little research, possibly because they have been assumed to respond nonspecifically to mechanical stimuli following specific chemical stimulation, and possibly because mechanical stimuli are generally difficult to control and direct at specific cellular sites. The chemoreceptors, first proposed to exist by Parker and Van Alsytne (1932), are generally assumed to confer specificity to the recognition of appropriate prey and to sensitize cnidocytes to nonspecific mechanical stimuli.

## A. *In vivo* Tentacles

Nagel (1892) first indicated that prey-derived substances are important in the discharge of cnidae when he demonstrated that the tentacles of *Calliactis parasitica* and *Actinia equina* seize sardine meat, while inert objects, such as filter paper, were rejected unless presoaked in food extracts. Later, Parker and Van Alstyne (1932) working on the acontial filaments of *Metridium* observed that application of an extract of the skin of *Fundulus* caused limited discharge of cnidae.

Using the sea anemone *Anemonia sulcata*, Pantin (1942b) tested a variety of food extracts, amino acids, proteins, carbohydrates, and fatty acids for the ability to sensitize cnidae to discharge in response to triggering mechanical stimuli. He found that in some cases mechanical stimulation of the cnidocytes without chemical stimulation caused a limited number of cnidae to discharge. However, for optimal discharge to occur the presence of food was required. Pantin concluded that the maximum response of cnidocytes requires both chemical and mechanical stimulation. If the tentacle was touched by a clean glass rod, there was no observable discharge. If, however, it was touched by a glass rod coated with appropriate food extract or with human hair or with a piece of human skin, there was massive discharge. Furthermore, the discharge was localized, occurring just at the point of contact. As a rule, aqueous extracts of food applied alone were ineffective, but if applied concomitantly with mechanical stimulation significant local discharge occurred. Thus, food extracts appeared to lower the threshold to mechanical stimulation.

Analysis of the active food extracts indicated that the effect was not caused by amino acids, carbohydrates, or de-fatted proteins. Pantin reported that the active substance could be extracted by acetone or alcohol, but not by diethyl ether, and concluded that the active substance was a "lipoid contact chemical" strongly adsorbed to protein. He suggested that the "lipoid" substance must come in direct contact with the lipid surface layer of the cnidocyte in order to exert its effect (Pantin, 1942a,b).

## 1. Semi-quantitative Method

In 1979 Lubbock attempted to identify the substance(s) present on the surface of prey that sensitize cnidae to discharge. Using two species of anemones, Stichodactyla haddoni and Gyrostoma hertwigi, he coated glass rods with each of 32 different substances including proteins, glycoproteins, polysaccharides, and lipids, and presented them to the tentacles of the experimental animals. By noting the relative extent to which the tentacles adhered to the coated rod, he was able to qualitatively rank the various substances tested for the ability to sensitize cnidocytes to discharge.

Unlike Pantin, Lubbock found that for the most part, proteins and glycoproteins elicited a stronger tentacle-adhesion response than did other substances. Lipids and polysaccharides, with the exception of hyaluronic acid, were not stimulatory. Comparisons of the responses of S. haddoni and G. hertwigi indicated a marked similarity between these two species, although G. hertwigi showed increased sensitivity to certain compounds, such as starch and especially glycogen. Of the 15 substances eliciting a positive response, 11 were proteins. Pantin had concluded from his study with Anemonia sulcata that proteins did not elicit discharge, but that an active lipid was strongly adsorbed to protein. Whether these differences are due to species specificity or to different experimental methods is not known.

Most of the published reports on cnidocyte responsiveness either consist of observations of organismic behavior in uncontrolled settings or of qualitative results. Consequently, little is known of the cellular and molecular mechanisms involved. Two difficulties have consistently hampered such studies of cnidocyte responsiveness. They are (i) a lack of standardizable experimental animals and uncontrolled experimental conditions, and (ii) an absence of accurate and sensitive assays to quantitate the extent to which the cnidae discharged.

We began by using a rapid and semiquantitative screening method to detect the discharge of cnidae in response to a large number of test substances. This method involved touching a hand-held "probe" to the tentacles of the sea anemone, Aiptasia pallida. Test probes consisted of either gelatin (30% w/v) or agarose (1% w/v) pellets that had been presoaked in test solutions of specified biochemicals in filtered sea water, adjusted to pH 7.65, or else glass rods dipped into the test solutions. The pellets were of defined size and shape (cylindrical, 5 mm diam. and 3 mm length) with each fastened to a fine-wire wand. The relative adhesion of the tentacle to the probe mediated by discharged cnidae gave an estimate of cnida discharge. Using this method we screened a wide variety of more than 60 naturally-occurring biochemicals (Thorington and Hessinger, 1988). It was observed that two groups of low-molecular weight substances, namely a variety of amino compounds, including amino and imino acids and histamine, and a N-acetylated sugars sensitize tentacle cnidocytes to tactile triggering. Some high molecular weight compounds, such as mucins, glycoproteins, and acid mucopolysaccharides, which have oligosaccharide chains of N-acetylated sugars, were also very effective.

## 2. Quantitative Method

Recognizing that the semiquantitative method described above could not give detailed information on the activities of the sensitizing agents (agonists), we developed a more sensitive and accurate method of measuring cnida discharge. To quantitate the discharge of cnidae we accurately measured cnida-mediated adherence between the tentacle and a test probe using a strain guage (Thorington and Hessinger, 1988). The probe consisted of a small (0.80 mm) diameter, nylon bead attached to a strain gauge (Grass Model FT-03) via a narrow steel shaft. Variations in probe diameter were restricted to $\pm$0.01 mm by screening probes with a drill gauge. Resistance springs were removed from the strain gauge to increase sensitivity, and the strain gauge was connected to a chart recorder. Adhesive force was expressed in hybrid units of milligrams-force (mgf) which is the force required to separate the tentacle from the probe (Lubbock, 1979). Calibrations were made with gravimetric weight standards and a sensitivity of 1 mgf with a 5% error was achieved.

To minimize variations between animals asexually cloned sea anemones, *A. pallida*, were used (Hessinger and Hessinger, 1981). Animals were kept in a temperature-controlled room at 24$^{o}$C on a 12:12 hr light/dark cycle and fed daily on freshly hatched *Artemia nauplii* and cleaned five to six hr later. Prior to each experiment anemones were starved 48 hr and kept under constant fluorescent light (4,500 lux). Test solutions were made up in filtered seawater and adjusted to pH 7.62 with 1N HCl or 1N NaOH. Individual animals were immersed in test solutions and permitted to acclimate for 10 min before testing occurred. Contact between the probe and the tip of a tentacle was permitted to last 2 sec. Each probe was used on 4 separate tentacles, and each animal was used for a total of 20 measurements. Measurements taken in the absence of agonist were subtracted from measurements in the presence of agonist to give values for the chemosensitizing effect.

All dose-response curves are biphasic, having a sigmoidal

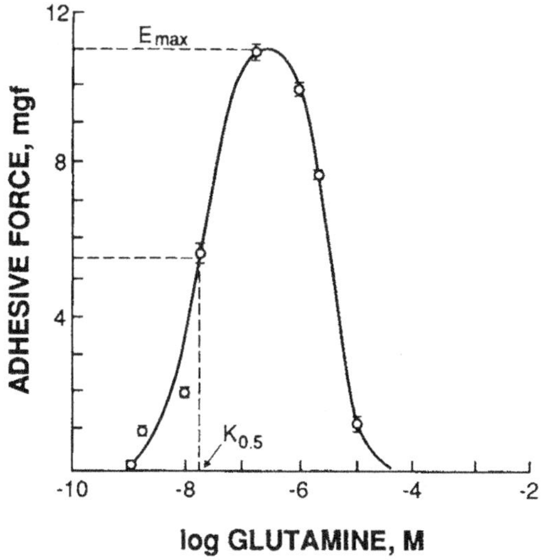


Fig. 1. Dose-response of the discharge of cnidae to glutamine (mean n = 53). Results are expressed as means of adhesive force (mgf) after correcting for adhesion with seawater alone, with vertical bars representing standard errors of means (95% confidence limits).

region of sensitization at low doses of agonist, a region of maximum response or effect ($E_{max}$) at optimal concentrations, and a region of desensitization at higher concentrations. The dose-response curve of glutamine is representative of all chemosensitizers tested (Fig. 1).

The sensitization region of dose-response curves can be linearized on double-reciprocal plots to calculate $K_{0.5}$ and $E_{max}$ values (Lineweaver-Burke, 1934). The $K_{0.5}$ value represents the molar concentration of chemosensitizing agent that half-maximally sensitizes cnidocytes. These values ranged from 0.1 to 1.5 $\mu$M for N-acetylated sugars and 1 to 3 x $10^{-2}$ $\mu$M for amino agonists. $E_{max}$ values ranged from 9.75 to 14.4 mgf for "sugars" and 7.95 to 14.2 mgf for amino agonists (Thorington and Hessinger, 1988). It is not known what causes the apparent desensitization observed at higher concentrations of agonists.

As determined from double-reciprocal and Dixon-type plots the response to amino agonists is competitively inhibited by certain antihistamines such as diphenhydramine (Thorington and Hessinger, 1988), while the response to N-acetylated sugars and mucin is unaffected. On this basis we conclude that at least two classes of chemoreceptors are involved in affecting cnidae to discharge.

The Ki value for a competitive inhibitor is the molar dissociation constant of the inhibitor for the receptor (Segel, 1976). Significantly different Ki values for a particular inhibitor indicate that it binds to different receptors. We suspect the existence of more than one amino receptor since the $K_i$ value for diphenhydramine in the presence of glycine (Ki = 0.13 $\mu$M) as compared to other amino agonists such as proline, glutamine and histamine (Ki = 2.4-3.1 $\mu$M) is more than an order of magnitude lower (Thorington and Hessinger, 1988). The existence of more than one amino receptor is also suggested by the markedly different patterns of competitive inhibition on dose-response curves for these two groups of amino agonists (Fig. 2). Unlike the other amino agonists tested in which the entire biphasic dose-response curve is shifted to the right, only the regions of sensitization for glycine are shifted to the right while the regions of desensitization are shifted to the left. Thus, because of differences in $K_i$ values and the differences in patterns of inhibition in the presence of diphenhydramine, we believe there exist two distinct "amino" receptors.

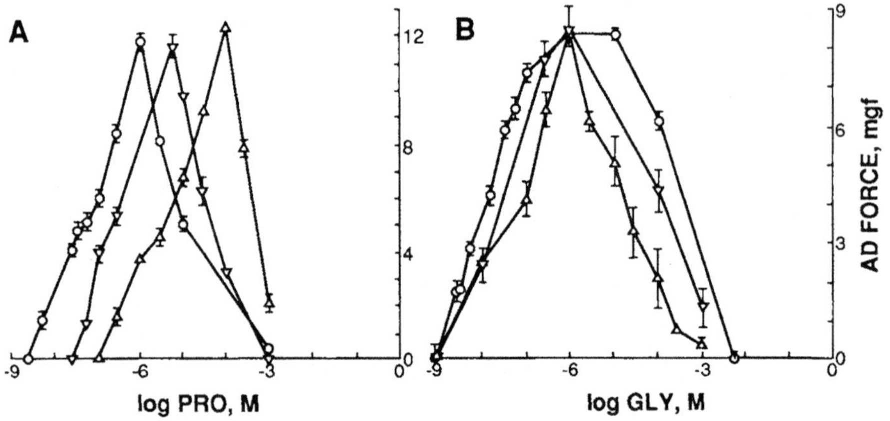


Fig. 2. Dose-responses of cnida discharge to proline (A) and to glycine (B) in the presence of different concentrations of diphenhydramine. Molar concentrations of diphenhydramine used with proline (A) were: none (n = 79; ○ ); 1.65 x $10^{-5}$ M (n = 60; ▽ ); and 3.3 x $10^{-5}$ M (n = 48; △ ). Concentrations used with glycine (B) were: none (n = 57; ○ ); 5 x $10^{-8}$ M (n = 49; ▽ ); and 1 x $10^{-6}$ M (n = 57; △ ). Data expressed as in Figure 1.

## B. Acontial Filaments

Yanagita and his co-researchers (1960a) performed experiments similar to Pantin's on the acontial filaments of the orange-striped anemone, _Haliplanella_ (= _Diadumene_) _luciae_. The acontial filaments of certain sea anemones contain predominantly one type of nematocyst, the microbasic p-mastigophores (Weill, 1934; Carlgren, 1954). For these experiments, the acontial filaments were freshly excised from the animal and placed on a glass slide with a small amount of sea water. Following application of the potential stimulating agent the response was observed under a dissecting microscope.

For the most part, results obtained with acontial filaments, confirmed Pantin's findings (Pantin, 1942a,b). For example, contact with certain solid materials such as human hair and skin or with some food objects elicited discharge. Soluble substances, such as food extracts, egg white, albumen, and ethyl stearate, did not cause discharge by themselves but did sensitize cnidae to the slightest mechanical stimulus. On the other hand, a pronounced difference occured between the acontia of _H. luciae_ and the tentacles of _A. sulcata_ with respect to responses to lipids (Yanagita, 1959a,b; 1960 b,c). While shorter chained fatty acids (acetic and butyric acids) induced no response in acontia except at high concentrations (e.g., 0.5%), longer chained fatty acids were very effective in causing discharge of acontial cnidae, especially stearic acid. In addition, while steroids were ineffective on the tentacles of _A. sulcata_ they induced a marked local discharge accompanied by extrusion from the acontia of _H. luciae_. Responses such as these from steroids and aliphatic lipids are quite different from the more localized discharge obtained from natural food contact (Yanagita, 1960a). The responses to lipids, unlike those to food substances, required a high chloride ion content in the surrounding medium. These differences coupled with those made from _in situ_ tentacles of _H. luciae_ (Williams, 1968) suggest that lipids and surface active agents do not act via the same mechanisms as do prey-derived substances.

## C. _In vitro_ Cnidae

Undischarged cnidae free of their cnidocytes, can be isolated from many cnida-bearing tissues and will still retain the ability to discharge. Acontial filaments of _H. luciae_ (Yanagita, 1943, 1951, 1959a,b) and _A. pallida_

(Blanquet, 1970) can be placed in either distilled water or 1M glycerin or in 1 M sodium citrate, respectively, to cause massive extrusion of undischarged nematocysts. Cnidae isolated in this manner can be tested for the ability of various substances to cause discharge. Food extracts, human hair and skin, cholesterol, and others are ineffective on isolated nematocysts (Yanagita, 1960). This is also true for cnidae isolated from the tentacles of other cnidarians (Glaser and Sparrow, 1909; Parker and Van Alstyne, 1932; Weill, 1934; and Pantin, 1942a). Isolated acontial nematocysts are, however, discharged by trypsin and by thioglycollate at pH 8 (Yanagita and Wada, 1954).

Individual surface-active agents, such as saponin and various bile salts, evoked no discharge even at high levels (e.g., 10% w/v). However, in 50/50 mixtures of surface-active agents in the presence of high concentrations of chloride salts, copious discharge occurred. Yanagita proposed that bile salts act in synergy with chloride ions to dislodge the apical flaps of nematocysts to cause discharge.

It thus appears that the mechanisms by which cnidocytes initiate the discharge of cnidae are quite different from the means by which isolated cnidae can be induced to undergo discharge. *In situ* cnidocytes are sensitive and responsive to mild and biologically specific stimuli, such as slight mechanical contact in the presence of appropriate chemosensitizer(s). Isolated cnidae, apparently lacking associations with sensory receptors, require comparatively harsh treatments, such as extreme pH values, surface active active agents, proteolytic enzymes and disulfide reducing agents. Yanagita (1960a) concluded that the "parent cell" (cnidocyte) plays an intermediary role in receiving and transforming the stimulus to the intact tissue.

## III. OTHER FACTORS AFFECTING THE DISCHARGE OF CNIDAE

### A. Modulation By Nervous System

The cnidocyte is classically viewed as an "independent effector," a cell which receives both appropriate mechanical and chemical stimulation in order to effect discharge of the cnidae without the intervention of the nervous system. A strict interpretation of the independent effector hypothesis, in which the nervous system is excluded from affecting discharge of cnidae, must be critically questioned as it applies to sea anemones and to hydra. The cnidae used by the swimming anemone, *Stomphia coccinea*, do not capture prey when the pedal disc is detached from the substratum (e.g., during swimming) (Ross and Sutton, 1964). Tactile stimulation of the body column of the anemone, Anemonia sargassensis, increases the discharge of tentacle cnidae to coverslips coated with food extract (Conklin and Mariscal, 1976). Cnidarians fed to repletion cannot capture prey, even if the prey are placed directly on the tentacles (Wagner, 1905; Parker, 1917; Pantin, 1942b; Pantin and Pantin, 1943; and Glumac, 1953). For instance, in the anemone, Calliactis tricolor (Sandberg et al., 1971) and in hydra (Burnett et al., 1960; Bouchet, 1961; and Smith et al., 1974) there is a marked reduction in discharge of cnidae after heavy feeding.

The conclusion from these observations is that the animal exerts some degree of control over the discharge of cnidae independently of the cnidocyte-associated sensory systems, presumably via the nervous system. The problem associated with this conclusion is two-fold: the implicated involvement of the nervous system has never been directly documented; and the effects on cnidocyte responsiveness have never been quantitatively determined. In the following sections the quantitative effects on cnidocyte responsiveness by feeding and starving, and by differences in ambient light intensity are reported.

## 1. Effects of feeding and starving

To determine whether the time since last feeding influences the responsiveness of tentacle cnidocytes in A. pallida we starved animals for durations of one to three days before testing their responsiveness to proline, a chemosensitizer (Fig. 3). There is about a two-fold increase in the adhesive force measured at optimum concentrations of proline between animals starved 24 versus 48 hr, but only about a 10% increase between 48- and 72-hr starved animals. Other sensitizers gave similar results. Interestingly, the $K_{0.5}$ values did not reflect any change in sensitivity to the sensitizer (Fig. 3). We interpret these findings to indicate that between day 1 and day 2 the number of cnidae discharged increases, but that the apparent sensitivity of the chemoreceptor to the sensitizing agents does not change. We do not know whether the increased responsiveness of the cnidocytes with duration of starvation is due to an increase in the number of chemoreceptors and/or cnidocytes on the cell surface, to an increase in the activity of some second-messenger-generating process, or to a decrease in some inhibitory signal mediated by the nervous system.

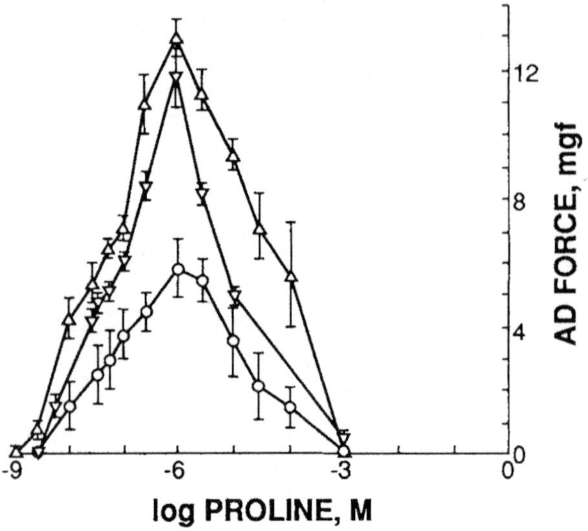


Fig. 3. Dose-responses of cnida discharge to proline in sea anemones starved for 1-day ( O ), 2-days ( ∇ ), and 3-days ( Δ ). Data expressed as in Figure 1.

2. Effects of light

Cnidocytes are not considered to be photosensitive. It is well documented, however, that sea anemones respond to light while not containing any recognizable photosensitive organ or cells. Millot (1968) viewed the reaction of these animals to light as a perception of a "widespread photic sense that is not mediated by eyes or eyespots and in which light does not act directly on an effector." He suggested that photosensitivity may be associated with nerve concentrations found in or just beneath the ectodermal layer.

In A. pallida, cnidocyte responsiveness and the ability to capture prey is biphasically dose-responsive to light intensity. Both complete darkness and bright light (e.g. 40 klux) depress the responsivenss of cnidocytes to tactile stimulation, totally inhibit chemosensitization, and prevent capture of natural prey. Only under intermediate light intensities (e.g. 4 to 6 klux) is cnidocyte responsiveness and prey capture optimal (Baldwin and Hessinger, in preparation).

Thus, it seems that the cnidocytes of certain anemones and hydra do not behave as independent effectors in the strict sense. Rather chemosensitization and cnidocyte responsiveness seem to be subject to modulatory signals emanating from sensory receptors remote from the cnidocytes.

## B. Ions

Various inorganic anions and cations are known to affect the discharge of cnidae, but the effects depend upon species and whether or not the cnidae are *in situ* or isolated. In assessing the effect of ions on the discharge of cnidae it should be kept in mind that the tested ions may be acting at any or all of several steps in the process of discharge.

### 1. Hydrogen ions (pH)

When acontial filaments and isolated cnidae (Blanquet, 1970) from *A. pallida* are placed in seawater of different pH values *in situ* nematocysts are induced to discharge at pH values below pH 4 and above pH 11. On the other hand, isolated nematocysts are stable at all pH values above pH 2.

### 2. Anions

All anions tested as sodium salts on nematocysts isolated from the acontia of *H. luciae* possessed the ability to induce discharge, with the exception of sulfate and tartarate (Yanagita, 1959a). However, of the salts tested, the chloride salts of monovalent and divalent cations were most effective. Yanagita postulated that reactive ions changed a protein component of the apical flaps to loosen the flaps and allow the inverted tubule to evert.

*In situ* nematocysts in the acontia of *H. luciae* required chloride ions for discharge. Substances that elicited normal discharge in seawater produced "misfiring extrusion" in a chloride-poor medium. The requirement for chloride ions, however, is not universal. Blanquet (1966, 1968) reported that high chloride concentrations were not required for nematocyst discharge in the acontial filaments of *A. pallida*. Salts of sulfate, nitrate and citrate also had the ability to cause discharge. The greatest discharge was induced by salts of univalent cations and divalent anions, such as sodium and potassium sulfate. Intermediate in effectiveness were salts of univalent cations and univalent anions, such as sodium and potassium chloride, whereas salts of divalent cations and univalent anions, such as magnesium chloride and calcium chloride, were least effective.

### 3. Cations

Increased potassium concentrations in the medium elicit cnida discharge from the lips of *Metridium* (Parker, 1905). This has been confirmed by others (Pantin, 1942a; Yanagita, 1959a; 1960a). Neither the excitatory effects of potassium and ammonium ions nor the inhibitory effect of magnesium ions upon *in situ* cnidae of the acontia, is observed with isolated nematocysts (Parker, 1909; Pantin, 1942b; and Yanagita, 1959a, 1960a). Magnesium prevents discharge of *in situ* acontial nematocysts by all of the previously mentioned stimulants, with the exception of the surface-active agents (Yanagita, 1959a, 1960b,c) as well as *in situ* tentacle nematocysts (Parker and Van Alstyne, 1932; Pantin, 1942b; Jones, 1947). This inhibitory effect of magnesium is fully reversible within 30 min by placing the experimental tissue in fresh seawater (Yanagita, 1959b, 1960b,c; Blanquet, 1970). Since calcium is required for the discharge of cnidae from the tentacles of *Hydra littoralis* (Lenhoff and Bovaird, 1959) it is likely that high magnesium inhibits by its ability to compete with required calcium.

The cnida is a secretory product (Slauterback, 1961; Skaer, 1973; Watson, 1988) and it has been suggested (Lubbock *et al*., 1981) that cnida discharge in anemones is preceded by an exocytotic event in which the membrane that surrounds the cnida fuses with the apical plasma membrane. Calcium influx is generally associated with stimulus-secretion coupling in many secretory cell systems (Bennett *et al*., 1979). Since membrane fusion requires free calcium (Poste *et al*., 1973); and since the discharge of *in situ* cnidae also requires extracellular calcium (Lenhoff and Bovaird, 1959), it is plausible that calcium channels are involved in cnidocyte-associated chemosensory and/or mechanosensory transduction that leads to the discharge of cnidae.

## IV. CONCLUSIONS

Many factors, both exogenous and endogenous, influence the discharge of cnidae. Some of these factors affect the mechanisms by which the cnidocyte normally controls discharge. Others affect the cnidocyte indirectly via the nervous system, and still others affect the cnidae themselves. A limited variety of low molecular weight substances acting via two specific classes of chemocreceptors sensitize cnidocyte to tactile triggering of discharge. These include several amino compounds and N-acetylated sugars. Such substances occur as common constituents of the natural prey. The amino agonists occur in the body fluids of prey which are presumed to leak into the ambient medium from wounds produced by penetrant nematocysts. The N-acetylated sugars occur as conjugates of the surface mucins and/or external chitin of prey.

Factors which influence the physiological state of the animal such as light, feeding regime, mechanical stimuli, pH, ions, and temperature also seem to affect cnidocyte responsiveness. These influences are presumed to affect cnidocyte responsiveness indirectly via the nervous system. The influences exerted by light and by feeding, as presented here, renew the question of whether the cnidocyte is an independent effector in the <u>strict sense</u>, that is, a cell which acts as both a sensory receptor and effector without the involvement of the nervous system (Wagner, 1905; Parker and Van Alstyne, 1932; Pantin, 1942; Ewer, 1947; Jones, 1947; Bullock and Horridge, 1965; Picken and Skaer, 1966). Although the critical experiments have not yet been performed in which parts of the nervous system are stimulated directly and the effects on cnidocyte responsiveness measured, the preponderance of evidence indicates that the discharge of cnidae is not under the exclusive control of the cnidocyte, but shared with remote sensory elements which communicate through the nervous system. In fact, it seems more appropriate now to ask if there are any cnidocytes which possess exclusive control of cnida discharge.

At another level, the question may be raised concerning whether all cnidocytes should be considered independent effectors in the <u>broad sense</u>. This question asks whether all cnidocytes act as both sensory receptors and effectors of discharge, regardless of modulatory influences from the nervous system. Another way to express this is to ask if the cnidocyte itself possesses the chemo- and mechano-sensory systems that trigger discharge or whether these receptors are located on different cells. With the recent identification of specific chemoreceptors which regulate chemosensitization of cnidocytes on sea anemone feeding tentacles (Thorington and Hessinger, 1988) and the evidence that such chemoreceptors are located on supporting cells adjacent to the responding cnidocytes (Watson and Hessinger, 1987; 1988a) there now exists, for the first time, direct refutation of the independent effector hypothesis in the broad sense. Whether these findings will be applcable to any or all species and be of more general import has yet to be determined. Yet it must be clearly stated that not all cnidocytes are independent effectors in either the strict or the broad senses.

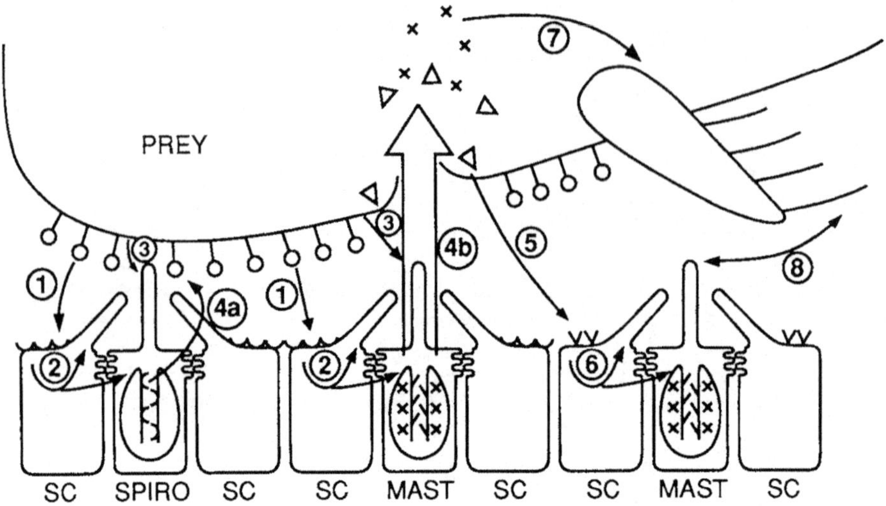


Fig. 4. Schematic diagram of the sequence and nature of interactions between prey, and the supporting cells and cnidocytes of the sea anemone feeding tentacle. Diagram explained in text.

From our present knowledge we can schematically represent the kinds of interactions that occur between the prey and the feeding tentacle of sea anemones. In feeding tentacles chemoreceptor-mediated cnida discharge involves two kinds of cells, the cnidocyte and the adjacent supporting cells. These two types of cells and the prey are involved in many chemical and mechanical interactions as described below (Fig. 4): (1) Mucins and/or chitin on the surface of prey activate chemoreceptors specific for N-acetylated sugars (Thorington and Hessinger, 1988; Geibel et al., 1988) located on adjacent supporting cells (Watson and Hessinger, 1987; 1988a). (2) The ligand-receptor complex stimulates generation of second messenger which sensitizes the associated cnidocyte to mechanical stimuli (Watson and Hessinger, in preparation). (3) Mechanical stimuli originating from the prey trigger "sensitized" cnidocytes (Pantin, 1942a). (4a) The triggered spirocyte discharges a spirocyst and the everting spirocyst tubule adheres to the surface of the prey. (4b) The triggered nematocyte discharges a penetrant nematocyst (microbasic p-mastigophore) and the everting tubule punctures the prey (Tardent, 1988) and "injects" a potent venom (Hessinger, 1988). (5) Body fluids from the prey containing low molecular weight amino compounds leak into the ambient medium and bind to "amino" receptors (Thorington and Hessinger, 1988; Geibel _et al._, 1988) presumed to be on supporting cells. (6) Second messenger which is presumed to be generated sensitizes the associated cnidocytes to mechanical stimuli. (7) Neurotoxic protein of the venom from penetrant mastigophores initially causes increased motor activity in prey (Hessinger _et al._, 1973). (8) Venom-induced motor activity in stung prey stimulates additional mechanoreceptors associated with sensitized cnidocytes. Thus, the cnidocyte with its associated supporting cells forms a multicellular receptor-effector complex (Watson and Hessinger, 1988b).

It is evident that there are still many questions to be answered concerning chemoreception, mechanoreception, sensory transduction and sensory integration as it relates to the control of cnida discharge. Elucidation of these processes at the cellular and molecular levels is within reach using the current ability to precisely stimulate and control cnidocyte-associated chemocreceptors (Thorington and Hessinger, 1988) and mechanoreceptors (Watson and Hessinger, in preparation) and with methods to rapidly and accurately count discharged nematocysts (Geibel _et al._, 1988) and spirocysts (Thorington and Hessinger, in preparation).

## ACKNOWLEDGMENTS

We thank Drs. H.M. Lenhoff and I.M. Fraser for helpful suggestions on this manuscript. Supported in part by funds from NSF grant DCB-8609859 to DAH.

## REFERENCES

Bennett, J. P. S., S. Cockcroft, and B. D. Gomperts. 1979. Ionomycin stimulates mast cell histamine secretion by forming a lipid-soluble calcium complex. Nature 282: 851-853.

Blanquet, R. S. 1966. The acontia nematocysts of the sea anemone _Aiptasia pallida_: Mechanism of discharge and toxin. Ph.D. dissertation, pp. 1-102. Duke University, Durham, North Carolina (University Microfilms, Ann Arbor, Michigan).

Blanquet, R. S. 1968. Properties and composition of the nematocyst toxin of the sea anemone, Aiptasia pallida. Comp. Biochem. Physiol. 25: 893-902.

Blanquet, R. S. 1970. Ionic effects on discharge of the isolated and in situ nematocysts of the sea anemone, Aiptasia pallida: A possible role of calcium. Comp. Biochem. Physiol. 35: 451-461.

Bouchet, C. 1961. Le Controle de la decharge nematocystique chez l'Hydre. C. R. Acad. Sc. 252: 327-328.

Bullock, T. H. and G. A. Horridge. 1965. Structure and Function in the Nervous System of Invertebrates. Vol. 1. San Francisco: W. H. Freeman and Co.

Burnett, A. L., T. Lentz, and M. Warren. 1960. The nematocyst of Hydra. (Part 1). The question of control of the nematocyst discharge reaction by fully fed hydra. Ann. Soc. Roy. Zool. Belg. 90: 247-267.

Calgren, O. 1954. A contribution to the knowledge of the structure and distribution of the cnidae in the Anthozoa. Lunds. Univ. Arsskr. N. F. 36: 1-62.

Conklin, E. J. and R. N. Mariscal. 1976. Increase in nematocyst and spirocyst discharge in a sea-anemone in response to mechanical stimulation. In G. O. Mackie (ed), Coelenterate Ecology and Behaviour, pp. 549-558. Plenum Press, New York.

Davenport. D., D. M. Ross, and L. Sutton. 1961. The remote contol of nematocyst discharge in the attachment of Calliactus parasitica to shells of hermit crabs. Vie Milieu 12: 197-209.

Ellis, V. L., D. M. Ross, and L. Sutton. 1969. The pedal disc of swimming sea anemone Stomphia coccinea during detachment and resettlement. Canad. J. Zool. 47: 332-342.

Ewer, R. F. 1947. On the functions and mode of action of the nematocysts of hydra. Proc. Zool. Soc. London 117: 365-376.

Geibel, G. M., G. U. Thorington, R. Y. Lim, and D. A. Hessinger. 1988. Control of cnida discharge: II. Microbasic p-mastigophore nematocysts are regulated by two classes of chemoreceptors. Biol. Bull. 175: 132-136.

Glaser, O. C. and C. M. Sparrow. 1909. The physiology of nematocysts. J. Exp. Biol. 6: 361-382.

Glumac, S. 1953. Contribution a la connaissance du fonctionnement des cellules a nematocystes chez l'Hydre d' eau douce. Glasn. Mus. Zrpsk. Zeml (B) 5-6: 503-511.

Hessinger, D. A. 1988. Nematocyst venoms and toxins. In D.A. Hessinger and H.M. Lenhoff (eds.), The Biology of Nematocysts. Academic Press, San Diego.

Hessinger, D. A., H. M. Lenhoff and L. B. Kahan. 1973. Haemolytic, phospholipase A and nerve-affecting activities of sea anemone venom. Nature New Biol. 241: 125-127.

Hessinger, D. A. and J. A. Hessinger. 1981. Methods for rearing sea anemones in the laboratory. In, Committee on Marine Invertebrate (ed.) Marine Invertebrates, pp. 153-179. National Academy Press, Washington, DC.

Hyman, L. H. 1963. The Invertebrates: Protozoa through Ctenophora. Vol. 1. McGraw-Hill, New York and London.

Jones, C. S. 1947. The control and discharge of namatocysts in hydra. J. Exp. Zool. 105: 25-60.

Lenhoff, H. M. and J. Bovaird. 1959. Requirement of bound calcium for the action of surface chemoreceptors. Science 130: 1474-1476.

Lineweaver, H. and D. Burke. 1934. The determination of enzyme dissocation constants. J. Am. Chem. Soc. 56: 558-666.

Lubbock, R. 1979. Chemical recognition and nematocyte excitation in a sea anemone. J. Exp. Biol. 83: 283-292.

Lubock, R. and W. B. Amos. 1981. Removal of bound calcium from namatocyst contents causes discharge. Nature 290: 500-501.

Millot, N. 1968. The dermal light sense. Symp. Zool. Soc. Lond. 23: 1-36.

Nagel, W.A. 1892. Das geschmacksinn der Actinien Zoo. Ang. 1: 334-338.

Pantin, C. F. A. 1942a. Excitation of nematocysts. Nature 149: 109.

Pantin, C. F. A. 1942b. The excitation of nematocysts. J. Exp. Biol. 19: 294-310.

Pantin, C. F. A., and A.M.P. Pantin. 1943. The stimulus to feeding in Anemonia sulcata. J. Expt. Biol. 20: 6-13.

Parker, G. H. and M. A. Van Alstyne. 1932. The control and discharge of nematocysts especially in Metridium and Physalia. J. Exp. Zool. 63: 329-344.

Picken, L. E. R. and R. J. Skaer. 1966. A review of researches on nematocysts. In W. J. Rees (ed.), The Cnidarians and their evolution. Symp. Zool. Soc. Lond. Vol. 16, pp. 15-50. Academic Press, London, New York.

Poste, G. and A. C. Allison. 1973. Membrane fusion. Biochem. Biophy. Acta 300: 421-465.

Ross, D. M. and L. Sutton. 1964. Inhibition of the swimming response by food and of nematocyst discharge during swimming in the sea-anemone, Stomphia coccinea. J. Exp. Biol. 4(4): 7451-757.

Sandberg, D. M., P. Kanciruk, and R. N. Mariscal. 1971. Inhibition of nematocyst discharge correlated with feeding in a sea anemone, Calliactus tricolor(Leseur). Nature 232: 263-264.

Segel, I. N. 1976. Biochemical Calculations, 2nd. edtn., p. 252. Wiley, New York

Skaer, R. J. 1973. The secretion and development of nematocysts in a siponophore. J. Cell Sci. 13: 371-393.

Slautterback, D. B. 1961. Nematocyst development. In H. M. Lenhoff and W. F. Loomis (eds.), The Biology of Hydra, pp. 77-130. Univ. of Miami Press, Coral Gables.

Smith, S., J. Oshida, and H. Bode. 1974. Inhibition of nematocyst discharge in hydra fed to repletion. Biol. Bull. 147: 186-202.

Thorington, G.. U. and D. A. Hessinger. 1988. Control of cnida discharge: I. Evidence for two classes of chemoreceptor. Biol. Bull. 174: 163-171.

Wagner, G. 1905. One some movements and reactions of Hydra. Quart. J. Micros. Sci. 48: 585-622.

Watson, G. M. and D. A. Hessinger. 1987. Receptor-mediated endocytosis of a chemoreceptor involved in triggering the discharge of cnidae in a sea anemone tentacle. Tissue Cell 19: 747-755.

Watson, G. M. and D. A. Hessinger. 1988a. Localization of a purported chemorecptor associated with cnida discharge in sea anemones. In D.A. Hessinger and H.M. Lenhoff (eds.), The Biology of Nematocysts. Academic Press, San Diego.

Watson, G. M. and D. A. Hessinger. 1988b. Cnidocytes and adjacent supporting cells constitute receptor-effector complexes in anemone tentacles. Tissue Cell (in press).

Weill, R. 1934. Contribution a l'letude des cnidaires et de leurs nematocystes. Trav. Sta. Zool. Wimereaux 10, 11: 1-701.

Williams, R. B. 1968. Control of the discharge of cnidae in Diadumene luciae (Verrill). Nature 219: 959.

Yanagita, T. M. 1943. Discharge of nematocysts. J. Fac. Sci., Tokyo Univ., Section IV, 6: 97-108.

Yanagita, T. M. 1951. The influence of immersion media on the "longevity" of isolated nematocysts of sea anemone. Nat. Sci. Rep., Ochanomizu Univ. 2: 117-123.

Yanagita, T. M. 1959a. Physiological mechanism of nematocyst responses in sea-anemone. II. Effects of electrolyte ions upon the isolated cnidae. J. Fac. Sci. Tokyo (4) Zool. 8: 381-400.

Yanagita, T. M. 1959b. Physiological mechanism of nematocyst responses in sea-anemone. VII. Extrusion of resting cnidae; its nature and its possible bearing on the normal nettling response. J. Exp. Biol. 36: 478-494.

Yanagita, T. M. 1960a. Physiological mechanism of nematocyst responses in sea-anemone. III. Excitation and anaesthetization of the nettling response system. Comp. Biochem. Physiol 1: 123-139.

Yanagita, T. M. 1960b. Pysiological mechanism of nematocyst responses in sea-anemone. IV. Effects of surface-active agents on the cnidae in situ and in isolation. Comp. Biochem. Physiol. 1: 140-154.

Yanagita, T. M. 1960c. The physiological mechanism of nematocyst responses in sea-anemone. V. The effects of lipoid solvents on the cnidae in situ and in isolation. Annornes. Zool. Jap. 33: 203-210.

Yanagita, T. M. and T. Wada. 1954. Effects of trypsin and thioglycollate upon the nematocysts of sea anemone. Nature 173: 171

# 14

# LOCALIZATION OF A PURPORTED CHEMORECEPTOR INVOLVED IN TRIGGERING CNIDA DISCHARGE IN SEA ANEMONES

Glen M. Watson
David A. Hessinger

Department of Physiology and Pharmacology
Loma Linda University
Loma Linda, California 92350

## SUMMARY

Contrary to the independent effector hypothesis, in which it is proposed that the mechanoreceptor and chemoreceptors involved in triggering cnida discharge are located on the cnidocyte, we have found that chemoreceptors involved in triggering cnida discharge in a sea anemone are located on supporting cells adjacent to the cnidocyte. Binding sites for mucin were visualized in the electron microscope using colloidal gold coated with bovine submaxillary mucin (BSM). We observed bound BSM-gold to be distributed randomly on the surface of supporting cells, accessory cells that surround cnidocytes in anemone tentacles. Bound BSM-gold is internalized and then translocated through several intracellular compartments including endosomes, multivesicular bodies and lysosomes.

After one hour of exposure to BSM-gold, supporting cells adjacent to two different types of cnidocyte bind comparable amounts of BSM-gold. However, supporting cells adjacent to one type of cnidocyte translocate less BSM-gold through intracellular compartments than supporting cells adjacent to the other type of cnidocyte. Thus, the extent of translocation of BSM-gold varies with the identity of the cnidocyte adjacent to a supporting cell. In addition, supporting cells adjacent to the same type of cnidocyte, but located at different points along the length of the tentacle, translocate different amounts of BSM-gold to the

The Biology of Nematocysts

multivesicular body. Thus, the extent of translocation of BSM-gold varies with the position of the supporting cell along the long axis of the tentacle.

## I. INTRODUCTION

Cnidarians use cnidae to capture prey (Ewer, 1947), to defend against predators, to attack (and to defend against attack by) unrelated, neighboring cnidarians (Francis, 1973) and to attach to suitable substrates (Mariscal, 1972). In situ discharge of cnidae requires appropriate mechanical and chemical stimuli (Parker and Van Alstyne, 1932; Pantin, 1942). However, the specific chemoreceptors and mechanoreceptors involved in triggering discharge of cnidae have not yet been identified. Studies using the light microscope have shown that clusters of "cilia" are present at the surface of the cnidocyte (e.g., Hyman, 1940; Pantin, 1942). Studies using the electron microscope have revealed that such clusters consist of conical arrays of stereocilia, microvilli and cilia. The microvilli and stereocilia converge onto a single cilium located at the center of the cluster. The cilium exhibits a 9+2 microtubule arrangement in cnidocytes from anthozoans (Westfall, 1965; Watson and Mariscal, 1985), but not in cnidocytes from hydrozoans and scyphozoans. Cnidocytes in hydrozoans and scyphozoans have a modified cilium called a cnidocil (Chapman and Tilney, 1959; Slautterback, 1967; Westfall, 1970). The entire apparatus is called a ciliary cone in anthozoans and a cnidocil apparatus in hydrozoans and scyphozoans (Mariscal et al., 1978).

Ciliary cones in anthozoans originate from several cells rather than solely from the cnidocyte. Ciliary cones are formed by stereocilia from the two to five supporting cells that surround the cnidocyte, and microvilli and a cilium that originate from the cnidocyte (Westfall, 1965; Watson and Mariscal, 1985). In contrast, the cnidocil apparatus originates solely from the cnidocyte (Westfall, 1970; Cormier and Hessinger, 1980; Carré and Carré, 1980; Hausmann and Holstein, 1985). A combined chemo- and mechano-receptor function has been attributed to the ciliary cone and also to the cnidocil apparatus (Westfall, 1970; Mariscal et al., 1978; Cormier and Hessinger, 1980). Direct evidence for such functions is lacking.

Recently, two different chemoreceptors involved in triggering discharge of cnidae in a sea anemone have been discovered and characterized (Thorington and Hessinger, 1984; 1987). One of these chemoreceptors is specific for certain amino-containing compounds including various amino acids and histamine. The other chemoreceptor is specific for N-acetylated sugars including those on certain glycoproteins, acid mucopolysaccharides and mucins.

In the present report, we describe the location, distribution and mode of intracellular translocation of the "sugar" receptor involved in triggering cnida discharge in a sea anemone.

We used colloidal gold coated with the glycoprotein, bovine submaxillary mucin (BSM) to localize the sugar receptor in feeding tentacles of the sea anemone, Haliplanella luciae. Bovine submaxillary mucin was selected because it is a potent sensitizer of cnida discharge in anemones (Thorington and Hessinger, 1987) and has a molecular weight exceeding 30,000 daltons (Bettelheim and Dey, 1965); the reported lower limit for the successful adsorption of macromolecules onto colloidal gold (Horisberger and Rosset, 1977). Gelatin-coated probes incubated in seawater containing BSM-gold and then touched to sea anemone tentacles adhere more tightly to the tentacle than do gelatin-coated probes incubated in seawater alone (Thorington, personal communication). Thus, BSM-gold sensitizes anemones to discharge cnidae. Methods used in this study are presented elsewhere in detail (Watson and Hessinger, in prep).

## A. Specificity of BSM-gold binding

There are seven types of cell in the tentacle ectoderm with surfaces in contact with the ambient seawater and thus, available to bind BSM-gold. These include a presumed mechano-sensory cell (Peteya, 1975) (Fig. 1), two types of gland cell (Figs. 2,3), a specialized ectodermal cell called the supporting cell (Figs. 4,5,6) and three types of cnidocyte: the spirocyte (Fig. 4), a nematocyte containing the microbasic p-mastigophore (MPM) (Fig. 5) and a nematocyte containing the basitrichous isorhiza (Fig. 6). In H. luciae, the spirocyte is the most abundant cnidocyte in the tentacle ectoderm, followed by the MPM nematocyte, which is more abundant than the basitrich nematocyte (Watson and Mariscal, 1983).

To determine which cells bind BSM-gold, we exposed anemones to BSM-gold and then processed the animals for electron microscopy. Over 99% of the bound BSM-gold particles appeared on or within supporting cells (Table I).

Table I. Distribution of BSM-gold among Cells in the Tentacle Ectoderm

| CELL TYPE | % TOTAL BSM-GOLD |
|---|---|
| Supporting Cell | 99.37 |
| Spirocyte | 0.25 |
| MPM Nematocyte | 0.13 |
| Basitrich Nematocyte | 0.00 |
| Mucous Gland Cell | 0.25 |
| Dense-vesicle Gland Cell | 0.00 |
| Mechano-sensory Cell | 0.00 |

Binding of BSM-gold can be blocked by including 2 mg/ml free BSM in the labeling medium. The supporting cell is virtually the only type of cell in the tentacle ectoderm that binds BSM-gold. Thus, the mucin-induced sensitization to cnida discharge involves a selective binding of mucin to supporting cells. This idea is morphologically consistent because mature cnidocytes in anemone tentacles always are surrounded by supporting cells (Watson, unpublished). In a functional context, supporting cells may be more significant to the cnidocyte than is suggested by the term, "supporting cell." We believe that BSM-gold binds to the sugar receptor rather than to some other site for the following reasons: (i) binding is specific for a single cell-type in the tissue; (ii) binding can be blocked by including free BSM in the labeling medium and, (iii) BSM-gold is biologically active.

FIG. 1. Longitudinal section of a presumed mechano-sensory cell (msc) in the tentacle ectoderm. All tissues were fixed using conventional methods (glutaraldehyde and osmium tetroxide) unless otherwise indicated. Scale bar = 0.5 μm.

FIG. 2. Longitudinal section of a mucous gland cell (mgc). Scale bar = 0.5 μm.

FIG. 3. Longitudinal section of a type of gland cell (gc) containing electron-dense vesicles (v). Scale bar = 0.5 μm.

FIG. 4. Longitudinal section of a spirocyte (sp) containing a mature spirocyst at the cell surface. The adjacent supporting cell (sc) also is visible. Scale bar = 0.5 μm.

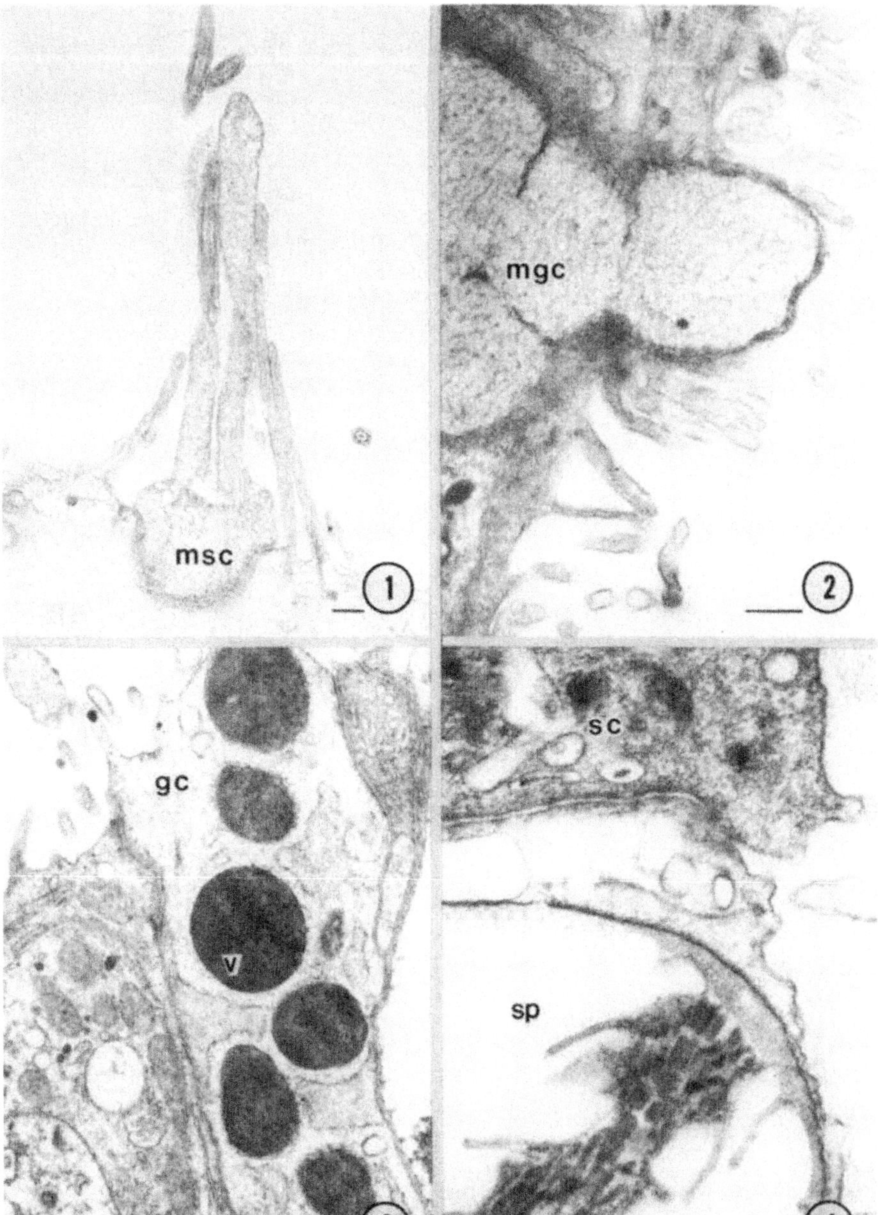
msc
1
mgc
2
gc
v
3
sc
sp
4

B. Distribution of bound BSM-gold at the cell surface

Bound BSM-gold seems to be randomly distributed along that portion of the plasma membrane in contact with seawater (Figs. 7,8).

C. Translocation of BSM-gold within the supporting cell

Bound BSM-gold is internalized into a network of tubular vesicles in the apex of the supporting cell (Figs. 7,8). These vesicles have smooth (uncoated) membranes and an average diameter of 0.25 μm (Fig. 5). Morphologically similar vesicles have been called receptosomes (Pastan and Willingham, 1981) or endosomes (Neutra et al., 1985) in other systems. The term "endosome" seems to have wider usage in the literature and, therefore, will be used here.

FIG. 5. Longitudinal section of a nematocyte containing a microbasic p-mastigophore (mpm) nematocyst at the cell surface. The adjacent supporting cell (sc) has stereocilia (st) at the cell surface and numerous endosomes (e) beneath the cell surface. Ruthenium red was included with the osmium tetroxide. Scale bar = 0.5 μm.

FIG. 6. Longitudinal section of a nematocyte containing a basitrichous isorhiza (bs) nematocyst. A supporting cell (sc) is visible on each side of the nematocyte. Scale bar = 0.5 μm.

FIG. 7. Cross section of a tentacle exposed to colloidal gold coated with the glycoprotein, bovine submaxillary mucin (BSM-gold) for one hour and then processed cytochemically for acid phosphatase activity. BSM-gold is located at the surface, endosome (e, arrowhead) and multivesicular body (mvb) of the supporting cells. A lysosome (ly) is visible in the vicinity of the mvb. The lysosome is positive for acid phosphatase activity as indicated by the electron-dense precipitate, whereas the mvb lacks such precipitate. Scale bar = 0.5 μm.

FIG. 8. Longitudinal section of a nematocyte containing a microbasic p-mastigophore nematocyst at its apex. This tissue was exposed to BSM-gold for one hour prior to fixation. BSM-gold is located on a supporting cell (sc) stereocilium (st). BSM-gold also is visible in a supporting cell endosome (e, arrowhead). Scale bar = 0.5 μm.

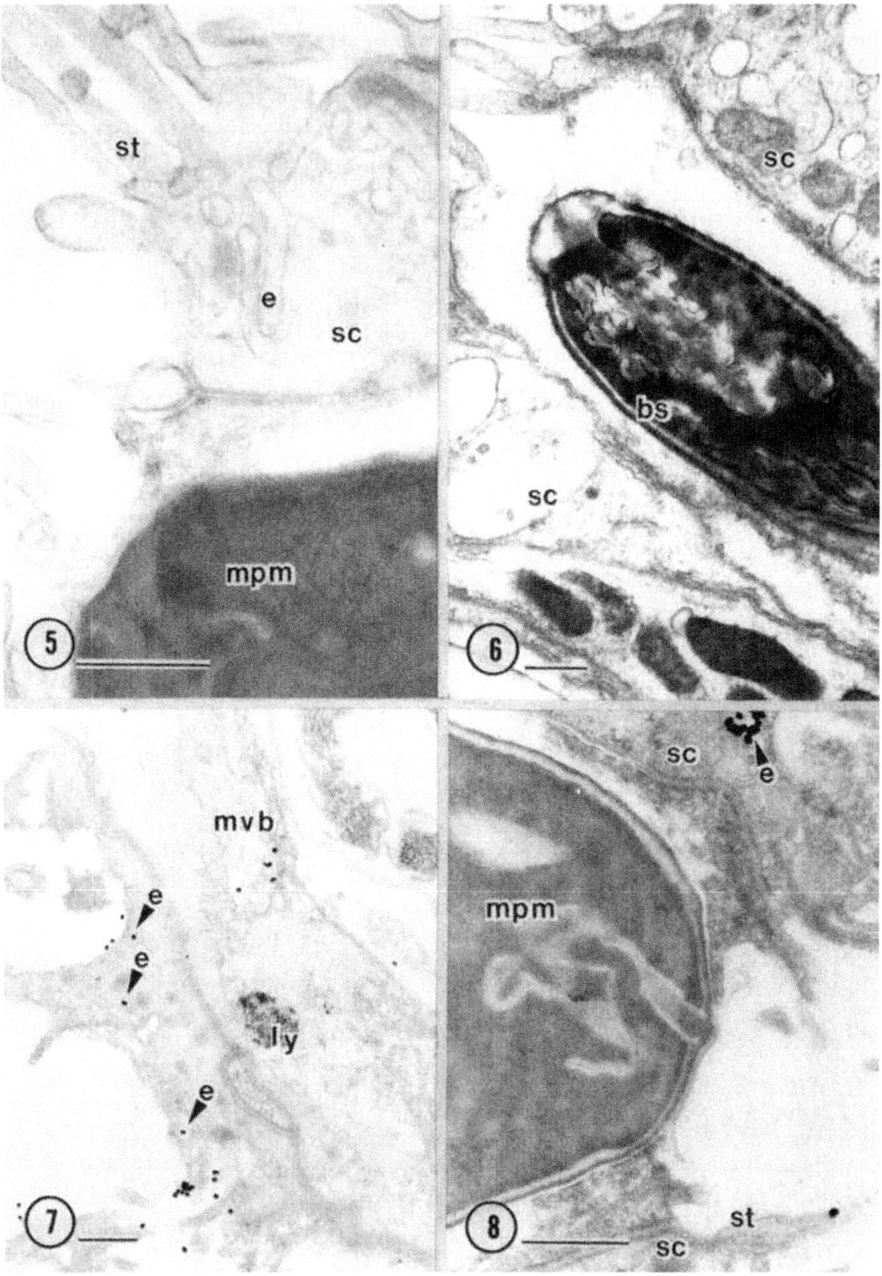
st
e
sc
mpm
5
sc
bs
sc
6
mvb
e
e
ly
e
7
sc
e
mpm
8
st
sc

BSM-gold also appears in multivesicular bodies, each consisting of a large vesicle containing one or more smaller vesicles (Figs. 7,9). It is presumed that several endosomal vesicles fuse to form a multivesicular body (Pastan and Willingham, 1985; Livne and Oliver, 1986). After one hour of exposure to BSM-gold, cells have numerous BSM-gold particles in the MVB, indicating that the MVB can accumulate BSM-gold.

In addition, BSM-gold appears in electron-dense vesicles (Fig. 9) that stain cytochemically for acid phosphatase activity (Fig. 7) and thus, are lysosomes. Given sufficient time before fixation, BSM-gold appears in secondary lysosomes (Fig. 10). The time-course of BSM-gold translocation through intracellular compartments in the supporting cell is complex and will be presented elsewhere (Watson and Hessinger, in prep).

## D. Quantitative analyses of BSM-gold binding and translocation

Each cnidocyte in the tentacle ectoderm is surrounded by two or more supporting cells. "Nearest neighbor" cnidocytes normally are separated by at least two supporting cells (i.e., two cnidocytes are not adjacent to the same supporting cell). It appears that the cnidocyte and its complement of adjacent supporting cells form a multicellular complex. Such complexes may have some degree of functional autotomy. One might predict that complexes consisting of the same type of cnidocyte would exhibit the same or a similar sensitivity to a given molecule. On the other hand, complexes formed by different types of cnidocyte might exhibit different sensitivities to a given molecule. It is conceivable that such differences would be reflected in differences in binding and/or the translocation of chemosensitizers through intracellular compartments.

FIG. 9. Longitudinal section of a supporting cell showing BSM-gold in a multivesicular body (mvb). An electron-dense vesicle (v), probably a lysosome, is in the vicinity of the mvb. This tissue was exposed to BSM-gold for one hour prior to fixation. Scale bar = 0.5 μm.

FIG. 10. Longitudinal section of a supporting cell exposed to BSM-gold for three hours prior to fixation. BSM-gold is located in a secondary lysosome (sl); which can be identified by the membranous inclusions. Scale bar = 0.5 μm.

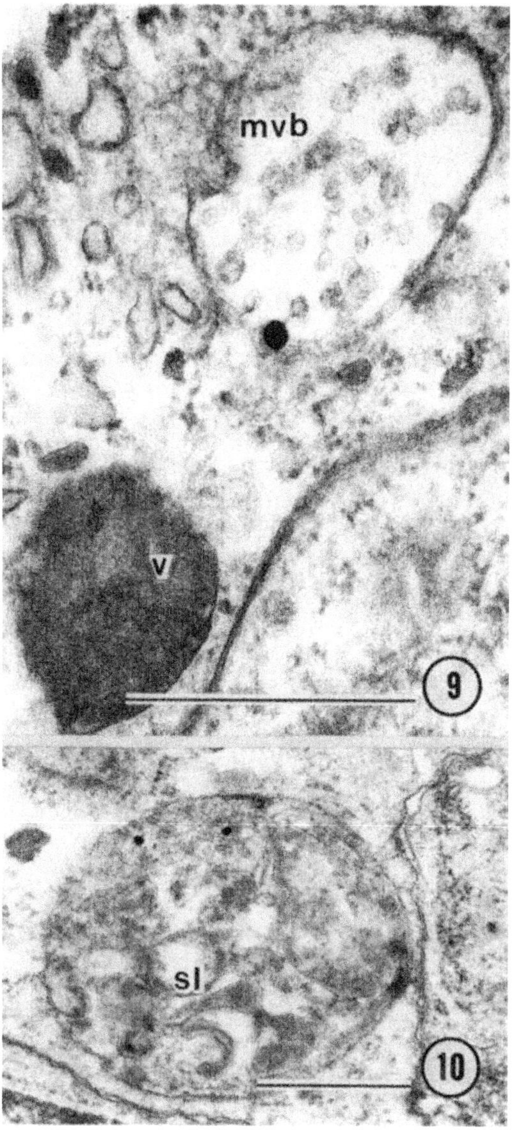
mvb
v
9
sl
10

1. Tentacle tip

We formulated the hypothesis that supporting cells adjacent to one type of cnidocyte bind and/or translocate different quantities of BSM-gold than supporting cells adjacent to another type of cnidocyte. Animals were exposed to BSM-gold for one hour and then processed for electron microscopy. Thin sections of one hundred cells were observed for each of three categories of supporting cell as follows, (1) supporting cells adjacent to spirocytes (sp-supporting cells), (2) supporting cells adjacent to microbasic p-mastigophore nematocytes (mpm-supporting cells) and (3) supporting cells adjacent to cell-types other than cnidocytes, such as other supporting cells and gland cells (free-supporting cells).

To determine if populations of supporting cell adjacent to different types of cnidocyte bind and/or translocate BSM-gold differently, we compared the number of cells (per 100 cells) having one or more BSM-gold particles at the cell surface, endosome and MVB compartments. The results from three identical experiments were combined and tested for heterogeneity using analysis of variance (Sokal and Rohlf, 1969). We find that the three categories of supporting cell are comparable in the numbers of cells (per 100 cells) having one or more BSM-gold particles at the cell surface, endosome and MVB compartments, respectively (Fig. 11). This observation indicates that the three categories of supporting cell have a similar proportion of cells that bind and translocate BSM-gold through intracellular compartments.

To determine if different categories of supporting cell bind and/or translocate different amounts of BSM-gold, we determined the numbers of BSM-gold particles at the cell surface, endosome and MVB compartments (per 100 cells). The results from three identical experiments were combined and tested for heterogeneity using analysis of variance (Sokal and Rohlf, 1969). The relative amount of BSM-gold at the cell surface is comparable among the different categories of supporting cell (Fig. 11, Table II). However, significantly less BSM-gold is present at the endosome of mpm-supporting cells than at the endosome of free-supporting cells ($P=0.01$, Table II). In addition, significantly less BSM-gold is present at the MVB of mpm-supporting cells than at the MVB of sp-supporting cells ($P=0.01$, Table II), or at the MVB of the free-supporting cells ($P=0.01$, Table II). No significant differences were observed in BSM-gold binding or in BSM-gold translocation between free-supporting cells and sp-supporting cells (Fig. 11, Table II).

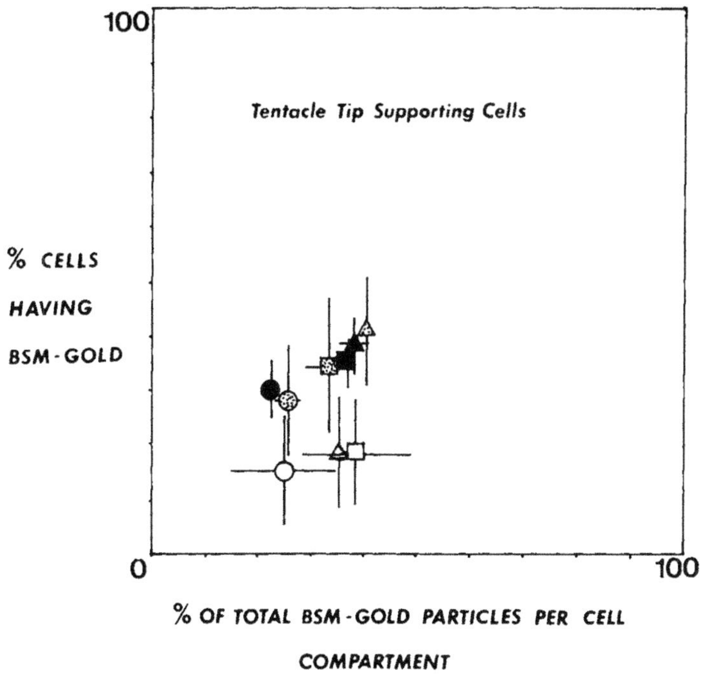


FIG. 11. Comparison of BSM-gold distribution among three categories of supporting cell at the tentacle tip. These are: mpm-supporting cells (circles), sp-supporting cells (squares) and free-supporting cells (triangles). Data were analyzed separately for the cell surface (white figures), endosome (stippled figures) and multivesicular body (black figures). Average values from three replicate experiments ± standard deviation are depicted. The ordinate represents the percent cells having BSM-gold at a particular cell compartment and the abscissa represents the percent of the total BSM-gold particles counted at a particular cell compartment.

Table II. Differences in Numbers of BSM-gold Particles per Cell Compartment

| | SURFACE | | ENDOSOME | | MVB | |
|---|---|---|---|---|---|---|
| MPM-SC | ns | ---- | ns | ---- | 0.01 | ---- |
| FREE-SC | ns | ns | ns | 0.01 | ns | 0.01 |
| | SP-SC | MPM-SC | SP-SC | MPM-SC | SP-SC | MPM-SC |

(MPM-SC, FREE-SC, SP-SC, supporting cell adjacent to a MPM-nematocyte, non-cnidocyte, and spirocyte, respectively; ns, not significant; 0.01, significant at the 0.01 level. Results based upon data shown in Fig. 11)

It is interesting that while the three categories of supporting cell bind comparable amounts of BSM-gold, the mpm-supporting cell translocates less BSM-gold to the MVB than either the sp-supporting cell or the free-supporting cell. These data support the hypothesis that supporting cells adjacent to different types of cnidocyte translocate BSM-gold differently.

Some recent experiments using the anemone Aiptasia pallida indicate that MPM nematocytes are more responsive to BSM chemosensitization (i.e., as measured by discharge) than are spirocytes (Thorington and Hessinger, this volume).

2. Tentacle tip versus tentacle base

To determine whether categories of supporting cell located at opposite ends of the tentacle bind and/or translocate BSM-gold in similar or different fashions, we compared (1) sp-supporting cells located at the tip of the tentacle with sp-supporting cells located at the base of the tentacle and (2) free-supporting cells located at the tip of the tentacle with free-supporting cells located at the base of the tentacle. Thus, four subcategories of supporting cell were formed: sp-supporting cells at the tip and at the base of the tentacle, and free-supporting cells at the tip and at the base of the tentacle. Comparisons were made in (1) the number of cells (per 100 cells) having one or more BSM-gold particles at the cell surface, endosome and MVB compartments, and (2) the numbers of BSM-gold particles at the cell surface, endosome and MVB compartments (per 100 cells). Data from three identical experiments were combined and tested for heterogeneity using analysis of variance.

We find that the number of cells having BSM-gold at the cell surface is similar among the subcategories of supporting cell (Fig. 12, Table III). Likewise, the number of cells having BSM-gold at the endosome compartment is similar among the supporting cells (Fig. 12, Table III). In addition, the relative amount of BSM-gold at the cell surface is comparable among the four subcategories of supporting cell (Fig. 12, Table IV). Likewise, the relative amount of BSM-gold at the endosome compartment is comparable among the four subcategories of supporting cell (Fig. 12, Table IV).

On the other hand, significantly more free-supporting cells at the tentacle tip have BSM-gold at the MVB compartment than free-supporting cells located at the tentacle base (P=0.01, Fig. 12, Table III). Likewise, significantly more sp-supporting cells at the tentacle tip have BSM-gold at the MVB compartment than sp-supporting cells located at the tentacle base (P=0.05, Fig. 12, Table IV).

Significantly more BSM-gold is present at the MVB of free-supporting cells at the tip of the tentacle than in the MVB of free-supporting cells at the tentacle base (P=0.01, Fig. 12, Table IV). In contrast, sp-supporting cells at the tentacle tip have comparable amounts of BSM-gold at the MVB as sp-supporting cells located at the base of the tentacle (Table IV).

Apparently, the translocation of BSM-gold is more likely to be arrested at the endosome in supporting cells located at the base of the tentacle than at the tip. This phenomenon is more clear cut for the free-supporting cells than for the sp-supporting cells because data for free-supporting cells are significant at a lower P value (P=0.01) than data for sp-supporting cells (P=0.05).

## II. CONCLUSIONS

We have described the location, distribution and mode of intracellular translocation of a chemoreceptor involved in triggering cnida discharge in sea anemones. The chemoreceptor is restricted to a single type of cell in the tentacle epithelium, the supporting cell. Thus, in anemones, it appears that the chemosensitization leading to the discharge of cnidae involves at least two types of cell, the cnidocyte and the supporting cell. Whereas supporting cells adjacent to different types of cnidocyte bind comparable amounts of ligand-coated gold, the tendency to internalize ligand-coated gold into intracellular compartments, and the rate of translocation between intracellular compartments, differs among these supporting cells. At the present time, we cannot distinguish between the possibility that the type of cnidocyte affects receptor-ligand translocation through the adjacent supporting cell, and that intrinsically different categories of supporting cell are present in the tentacle. Supporting cells adjacent to the same type of cnidocyte, but located at different points along the tentacle length translocate ligand-coated gold in differing fashions. Thus, it also is possible that the position of the supporting cell along the length of the tentacle affects the translocation of receptor-ligand complexes through intracellular compartments.

ACKNOWLEDGEMENTS

We thank Dr. Robert Schultz and Mr. Ernest Whitter and the Department of Anatomy, Loma Linda University for the use of electron microscopic facilities.

FIG. 12. Comparison of BSM-gold distribution within supporting cell categories: Tentacle tip versus tentacle base. Data were pooled for four subcategories of supporting cell: sp-supporting cells at the tentacle tip and base, and free-supporting cells at the tentacle tip and base. Data were analyzed separately for the cell surface (white figures), endosome (stippled figures) and multivesicular body (black figures). A. Sp-supporting cells at the tentacle tip (squares) are compared with sp-supporting cells at the tentacle base (circles). B. Free-supporting cells at the tentacle tip (triangles) are compared with free-supporting cells at the tentacle base (hexagons). Data are represented as for Figure 11.

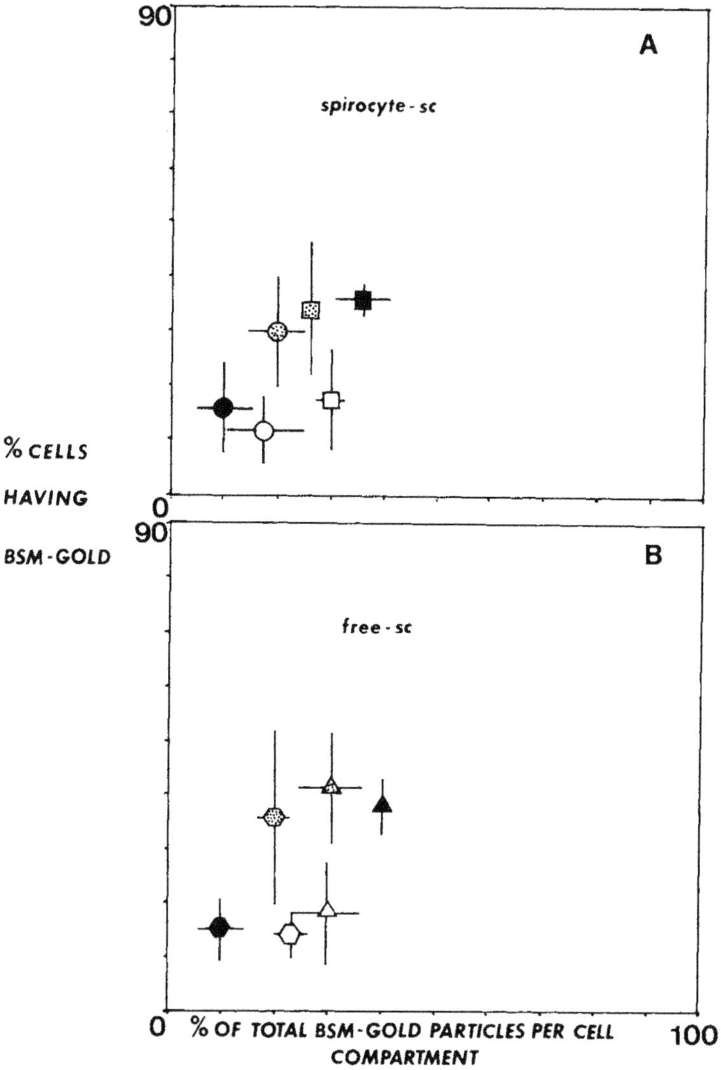
90
A
spirocyte - sc
% CELLS
HAVING
BSM - GOLD
0
90
B
free - sc
0
% OF TOTAL BSM - GOLD PARTICLES PER CELL COMPARTMENT
100

Table III. Differences in Numbers of Cells with BSM-gold at a Particular Cell Compartment: Tentacle Tip vs Tentacle Base

| | SURFACE | | ENDOSOME | | MVB | |
|---|---|---|---|---|---|---|
| | SP-SC BASE | FREE-SC BASE | SP-SC BASE | FREE-SC BASE | SP-SC BASE | FREE-SC BASE |
| TIP SP-SC | ns | ---- | ns | ---- | 0.05 | ---- |
| TIP FREE-SC | ---- | ns | ---- | ns | ---- | 0.01 |

(SP-SC, FREE-SC, supporting cell adjacent to a spirocyte and non-cnidocyte, respectively; ns, not significant; 0.05, significant at the 0.05 level; 0.01, significant at the 0.01 level. Results are based upon data shown in Fig. 12)

Table IV. Differences in Numbers of BSM-gold Particles per Cell Compartment: Tentacle Tip vs Tentacle Base

| | SURFACE | | ENDOSOME | | MVB | |
|---|---|---|---|---|---|---|
| | SP-SC BASE | FREE-SC BASE | SP-SC BASE | FREE-SC BASE | SP-SC BASE | FREE-SC BASE |
| TIP SP-SC | ns | ---- | ns | ---- | ns | ---- |
| TIP FREE-SC | ---- | ns | ---- | ns | ---- | 0.01 |

SP-SC, FREE-SC, supporting cell adjacent to a spirocyte and non-cnidocyte, respectively; ns, not significant; 0.01, significant at the 0.01 level. Based upon data shown in Fig. 12)

## REFERENCES


Bettelheim, F. A., and S. K. Dey. 1965. Molecular parameters of submaxillary mucins. Arch. Biochem. Biophys. 109: 259-265.

Carré, D., and C. Carré. 1980. On triggering and control of cnidocyst discharge. Mar. Behav. Physiol. 7: 109-117.

Chapman, G. B., and L. G. Tilney. 1959. Cytological studies of the nematocysts of Hydra. I. desmonemes, isorhizas, cnidocils, and supporting structures. J. Biophysic. Biochem. Cytol. 5: 69-88.

Cormier, S. M., and D. A. Hessinger. 1980. Cnidocil apparatus: Sensory receptor of Physalia nematocytes. J. Ultrastruct. Res. 72: 13-19.

Ewer, R. F. 1947. On the functions and mode of action of the nematocysts of hydra. Proc. Zool. Soc. London 117: 365-376.

Francis, L. 1973. Intraspecific aggression and its effect on the distribution of Anthopleura elegantissima and some related anemones. Biol. Bull. 144: 73-92.

Hausmann, K., and T. Holstein. 1985. Bilateral symmetry in the cnidocil-nematocyst complex of the freshwater medusa Craspedacusta sowerbii Lankester (Hydrozoa, Limnomedusae). J. Ultrastruct. Res. 90: 89-104.

Horisberger, M., and J. Rosset. 1977. Colloidal gold, a useful marker for transmission and scanning electron microscopy. J. Histochem. Cytochem. 25: 295-305.

Hyman, L. H. 1940. The Invertebrates: Protozoa through Ctenophora, vol I. McGraw-Hill, New York.

Livne, E., and C. Oliver. 1986. Internalization of cationized ferritin by isolated pancreatic acinar cells. J. Histochem. Cytochem. 34: 167-176.

Mariscal, R. N. 1972. The nature of adhesion to shells of the symbiotic sea anemone Calliactis tricolor Leseur. J. Exp. Mar. Biol. Ecol. 8: 217-224.

Mariscal, R. N., E. J. Conklin, and C. H. Bigger. 1978. The putative sensory receptors associated with the cnidae of cnidarians. Scanning Electron Microsc. 2: 959-966.

Neutra, M. R., A. Ciechanover, L. S. Owen and H. F. Lodish. 1985. Intracellular transport of transferrin- and asialoorosomucoid-colloidal gold conjugates to lysosomes after receptor-mediated endocytosis. J. Histochem. Cytochem. 33: 1134-1144.

Pantin, C. F. A. 1942. The excitation of nematocysts. J. Exp. Biol. 19: 294-310.

Parker, G. H., and M. A. Van Alstyne. 1932. The control and discharge of nematocysts, especially in Metridium and Physalia. J. Exp. Zool. 63: 329-344.

Pastan, I. H., and Willingham, M. C. 1981. Journey to the center of the cell: Role of the receptosome. Science 214: 504-509.

Pastan, I. H., and Willingham, M. C. 1985. The pathway of endocytosis. In I. Pastan, and M. C. Willingham (eds.) Endocytosis, pp. 1-44. Plenum Press, New York and London.

Peteya, D. J. 1975. The ciliary-cone sensory cell of anemones and cerianthids. Tiss. Cell 2: 243-252.

Slautterback, D. B. 1967. The cnidoblast-musculoepithelial cell complex in the tentacles of Hydra. Z. Zellforsch. 79: 296-318.

Sokal, R. R., and F. J. Rohlf. 1969. Biometry. W. H. Freeman and Company, San Francisco.

Thorington, G., and D. A. Hessinger. 1984. Identification and partial characterization of cnidocyte chemoreceptors on the sea anemone, Aiptasia pallida. J. Cell Biol. 99: 221a.

Thorington, G., and D. A. Hessinger. 1987. The characterization of cnidocyte chemoreceptors in the sea anemone, Aiptasia pallida. Biol. Bull. (submitted).

Watson, G. M., and D. A. Hessinger. (in prep). Receptor-mediated endocytosis of a purported chemoreceptor associated with cnida discharge in a sea anemone tentacle.

Watson, G. M., and R. N. Mariscal. 1983. The development of a sea anemone tentacle specialized for aggression: Morphogenesis and regression of the catch tentacle of Haliplanella luciae (Cnidaria, Anthozoa). Biol. Bull. 164: 506-517.

Watson, G. M., and R. N. Mariscal. 1985. Ultrastructure of nematocyst discharge in catch tentacles of the sea anemone Haliplanella luciae (Cnidaria: Anthozoa). Tiss. Cell 17: 199-213.

Westfall, J. A. 1965. Nematocysts of the sea anemone Metridium. Amer. Zool. 5: 377-393.

Westfall, J. A., 1970. The nematocyte complex in a hydromedusan, Gonionemus vertens. Z. Zellforsch. 110, 457-470.

15

# ON THE PREPARATION AND PROPERTIES OF ISOLATED CNIDOCYTES AND CNIDAE[1]

M. Craig McKay
Peter A. V. Anderson

C. V. Whitney Laboratory and
Departments of Physiology and Neurosciences
University of Florida
St. Augustine, Florida, USA

## SUMMARY

The complexity of tissues bearing cnidocytes has made it very difficult to assess the steps by which a contact-chemical stimulus leads to discharge of the cnidae. This reason, among others, has led investigators to develop methods for isolating cnidae. The methods used to produce these preparations are reviewed and compared, with emphasis on the role of the cnida membrane in regulating the discharge of the isolated cnida. Also discussed is a recently developed method for isolating intact cnidocytes thereby allowing the study of the physiology of these cells. The feasibility of making electrophysiological recordings from single cnidocytes both in situ and isolated from a variety of cnidarians is demonstrated. These studies show that there are physiological features common to cnidocytes from different classes of cnidarians.

[1] Supported by N.S.F. Grant BNS 85-06193 to P.A.V.A.

# I. INTRODUCTION

The study of the "nettling response" of cnidarians has enjoyed a long history, capturing the attention of successive generations of zoologists for over a century (for review, see preparations (i.e. whole animals or isolated tentacles). Cnidocyte-bearing tissues are invariably complex and contain, in addition to cnidocytes, a variety of cell types, including myoepithelial cells, neurons, sensory, and secretory cells. Given the potential for interaction between many of these cell types, a reliance on intact tissue has made it exceedingly difficult to identify the specific stimulus necessary to trigger the cnidocytes and more significantly, to identify the pathways and sequence of events by which the detection of stimuli gives rise to cnida discharge. These problems, compounded by the generally small size of cnidarian cells, provided the motivation for the development of methods for isolating cnidae. Whereas work with preparations of isolated cnidae has provided a great deal of useful information on certain aspects of discharge, the absence of the cell membrane in these preparations precludes their use for studies of the control of discharge at the tissue and cellular level.

This paper reviews some of the methods used to prepare isolated cnidae and compares them with a method (McKay and Anderson, 1985) whereby intact, functional cnidocytes can be prepared from a variety of sources, thus avoiding many of the problems associated with isolated cnidae and intact tissues. We compare the properties of isolated cnidae as reported by various workers, and finally we demonstrate the feasibility of making electrophysiological recording from both isolated cnidocytes and single cnidocytes *in situ*.

# II. HISTORY OF CNIDA ISOLATION

The techniques used over the years to isolate cnidae fall into three basic classes: mechanical methods, extrusion following chemical treatment, and autolysis of the surrounding tissue. As will be evident later, each technique has its own advantages and drawbacks.

Mechanical techniques such as maceration and homogenization have been used by a variety of investigators. This approach was first used by Glaser and Sparrow (1909) who used all three techniques to isolate cnidae from *Metridium* and

Physalia. Cnidae released by autolysis from Physalia were purified and concentrated by centrifugation. These authors reported also that the high specific gravity of undischarged cnidae from the acontia of Metridium caused them to sink to the bottom of the vessel while discharged ones remained near the surface. Phillips (1956) obtained large numbers of cnidae from Metridium by homogenizing whole animals in a solution of 1M sucrose in sea water and filtering and centrifuging the homogenate to purify the cnidae. A similar procedure, with homogenization in sea water only, has been used with tentacles of Chrysaora (Burnett, et al., 1968). In this case, four types of cnidae were identified but their relative proportions were not reported. There is some doubt as to the efficacy of this technique because in subsequent work (Burnett and Goldner, 1970) it was reported that extensive rupture of the cnidae occurred during homogenization.

A mechanical technique used to isolate cnidae from the cubomedusa Chironex fleckeri (Endean et al., 1969) consisted of freezing the tentacles on dry ice and then, after thawing the tissue, grinding it in a mortar. The resultant suspensions were allowed to settle and were purified by repeated centrifugation. As with Chrysaora tentacles, this procedure provided a mixture of all cnida types present in the tentacle; it was later reported (Endean and Rifkin, 1975) that these could be separated by discontinuous density gradient centrifugation. Finally, Wilby (1976) reported the recovery of undischarged cnidae by filtering whole hydra that had been sonicated in 2M sucrose. Unfortunately, neither the properties nor the purity of the isolated cnidae were described.

The most popular technique used for isolating cnidae from sea anemones is probably extrusion, whereby the cnidocyte-rich acontia are placed in a particular medium and, by some as yet not understood process, the cnidae are extruded intact from the tissue. According to Mariscal (1974), extrusion is not part of the normal discharge process. A variety of media have been used and, as will be seen later, there are major differences between the properties of cnidae isolated in the different instances. The extrusion process was first reported by Abric (1904) who found that cnidae from a variety of actinians were expelled undischarged when the tissues were immersed in a solution of thionine (3,7-diaminopheno-athiazonium chloride). Subsequently, both Glaser and Sparrow (1909) and Parker and Van Alstyne (1932) found that when the acontia of Metridium were placed in a sea water saturated with ether, their cnidae were extruded intact.

The first detailed study of the conditions under which cnidae could be extruded was made by Yanagita (1951, 1959b) working with the acontia of Diadumene (= Haliplanella) luciae. He found that undischarged cnidae could be obtained when acontia were placed in distilled water, dilute salt solutions, or 1M glycerol. Based on these findings, Blanquet (1968) introduced a method for liberating large numbers of undischarged cnidae from the acontia of Aiptasia pallida. Acontia were placed into 1M sodium citrate and the extruded cnidae separated from other debris by filtration and centrifugation. This technique has now become standard for preparing large numbers of cnidae but has certain drawbacks; specifically, it works with the acontia of certain species only and does not work with cnidae from anemone tentacles (Blanquet, 1970).

In some cases, a combination of extrusion and mechanical methods have been applied. Picken and Skaer (1966) removed individual cnidae from the tentacles of Corynactis by placing the tentacles in liquid paraffin. The cnidae were extruded and collected individually with micropipettes. A similar method was employed by Lubbock and Amos (1981) to isolate cnidae from the mesenteric filaments of Rhodactis rhodostoma. While this approach can provide sufficient material for small scale analyses such as freezing point determination (Pickens and Skear, 1966; Lubbock and Amos, 1981), it is impractical for producing large numbers of isolated cnidae.

The third technique, autolysis, takes advantage of the presence of the tough cnida wall. In this method the surrounding tissues essentially rot away leaving the cnidae intact. This technique was first employed by Glaser & Sparrow (1909) to obtain undischarged cnidae from Physalia. Subsequently, Lane and Dodge (1958) modified this technique for preparing large numbers of cnidae from Physalia. They placed the tentacles in sea water at $4^{o}C$ for one to two days; the resultant autolysate was filtered and the cnidae washed and concentrated by repeated centrifugation and resuspension. The same technique has been used with tentacles from Chrysaora (Burnett et al., 1968) although in this case the tentacles were apparently more resistant to autolysis than those from Physalia, requiring seven days to disintegrate. Finally, Salleo, et al., (1983) reported a method for isolating large numbers of cnidae from the scyphozoan Pelagia noctiluca. They found that when tentacles were placed in distilled water for several hours, the surrounding tissues disintegrated releasing a suspension of undischarged cnidae and debris, the former being purified by centrifugation.

## III. ISOLATING INTACT CNIDOCYTES

The above methods purify isolated cnidae which, as described later, retain the ability to discharge to some degree or another. Experiments with whole animals have made it clear, however, that the normal stimuli to discharge have both chemical and mechanical components and that these presumably must be transduced by some structure in contact with the external environment. Presumably that structure is on the cell membrane of either the cnidocyte itself or some adjacent cell. Consequently, it is difficult to evaluate many of the experiments with isolated cnidae in a physiologically meaningful way. A solution to this problem is to use preparations of functional, intact cnidocytes; to date, however, such preparations have rarely been obtained. Gierer, et al., (1972) described a technique for dissociating intact hydra into individual cells. Animals were placed in a high salt bathing medium (70 mOsM) and repeatedly triturated with a pipette. Unfortunately, because the authors' interests lay in morphogenisis and cell aggregation they were not particularly concerned with cnidocytes, but did report that over 15% of the recognizable isolated cells were nematocytes. Another technique commonly used with hydra is to shake whole animals in a solution of glycerol, acetic acid and water (David, 1973). This technique dissociates the animals into the component cells, amongst which nematocytes are recognizable, but the composition of the medium precludes the recovery of living cells.

We have now developed a rapid method for isolating intact cnidocytes from various cnidarians (McKay and Anderson, 1985). Briefly, the method, as applied to sea anemones, involves anesthetizing animals in 1:1 isotonic $MgCl_2$ and sea water followed by removal of the tentacles and acrorhagi. The tissue is added to a volume of anesthetizing medium containing the enzyme papain (1.5 mg/ml; 2.5 mM Dithiothreatol; pH 7.5). After several hours on an orbit shaker the tissues dissociate liberating large numbers of undischarged cnidocytes. The cnidocytes are separated from the rest of the dissociated tissue by centrifugation in Percoll (0.72 ml Percoll + 0.48 ml ASW) made isotonic with sea water by the addition of a concentrated artificial sea water (NaCl, 1.12 M; KCl, 22 mM; $CaCl_2$, 20 mM; $MgCl_2$, 65 mM; $NaHCO_3$, 2 mM). The separation of the cnidocytes from the other cells is based upon their high density compared with the other tissues present. All of the different cnidocyte types sank to the bottom of the Percoll solution (density = 1.12 g/ml). Because of the difficulties in creating higher density solutions we did not attempt to separate the different types of cnidocytes released by the enzymatic digestion of the tentacles. It should be noted that a homogeneous population of cnidocytes can be isolated by using a tissue such as a catch tentacle or acrorhagus which is composed almost exclusively of one type of cnidocyte.

We have tested the efficacy of the enzymatic method of isolation on various cnidarians. To date, it has been applied successfully to tentacles and acrorhagi of *Anthopleura elegantisssima*, tentacles of *Callactis tricolor*, *Aiptasia* sp., and *Bunodosoma cavernosa*, all anthozoans, and *Chrysaora quinquecirrha*, a scyphozoan. We have found that, depending on the type and amount of tissue being dissociated, it was necessary to employ more or less enzyme, vary pH, and digestion time. The percentage of Percoll necessary to separate the cnidocytes from the rest of the digestion products in the centrifugation step is also variable.

While this method is versatile and appears to be applicable to a variety of cnidarian tissues bearing cnidocytes, there are some limitations. We have applied our procedure to the siphonophore, *Physalia physalis* but isolated cnidae, not cnidocytes. Likewise, enzymatically treated acontia of *Aiptasia* sp. yielded only cnidae.

## A. Physical appearance of isolated cnidocytes

Suspensions of isolated cnidocytes from two species of anemones are shown in Figure 1 a,b. The bulk of the material present is intact, isolated cnidocytes, contaminated by a small amount of debris. The plasma membranes of the cnidocytes appear intact. The cnida is clearly visible within each cell and completely dominates the cell's interior. The appearance of these cnidocytes agrees well with that of cnidocytes in tentacles of *Calliactis* (Robson, 1973) but is quite different from that of cnidae isolated from the acontia of *Aiptasia* using 1M sodium citrate (Blanquet, 1970). The intactness of the isolated cnidocytes was further verified by scanning and transmission electron microscopy. The ultrastructure of isolated cnidocytes is consistent with published data on the ultrastructure of *in situ* cnidocytes (Westfall, 1965; Mariscal, *et al.*, 1976). The isolated cells are surrounded by a continuous cell membrane and the cytoplasm had the normal complement of organelles plus the undischarged cnida.

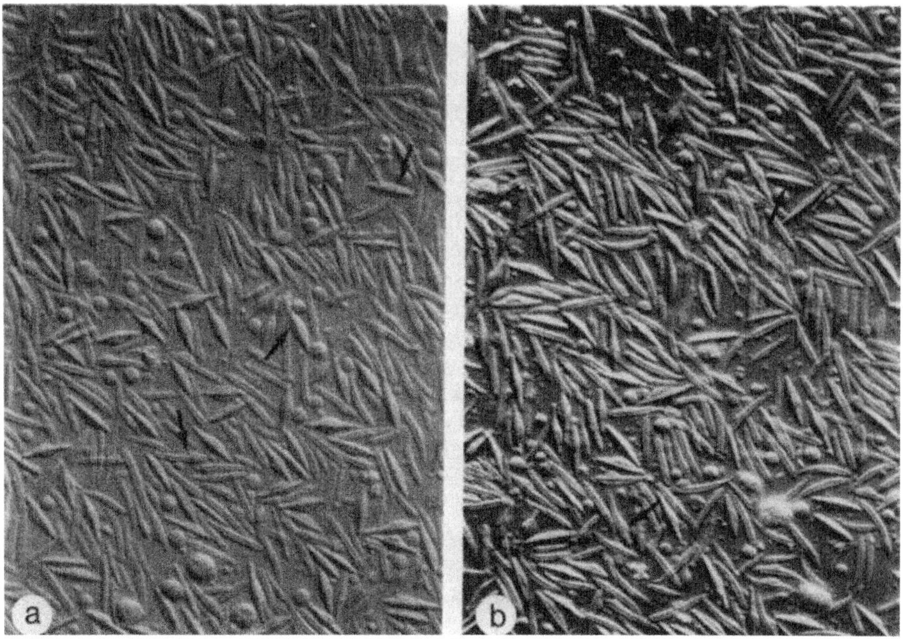

*Fig. 1. Nomarski micrographs of representative aliquots of cnidocytes isolated from the tentacles of (a) Anthopleura elegantissima and (b) Bunodosoma cavernosa. The majority of cells are spirocytes with smaller numbers of other types of cnidocytes, some of which are marked by arrows. Plasma membranes can be clearly seen surrounding the enclosed cnidae.*

## IV. PROPERTIES OF ISOLATED CNIDAE

The requirements for cnida discharge in intact tissue have been reviewed (Picken and Skaer, 1966; Mariscal, 1974). Likewise, the discharge of cnidae isolated from acontia (Glaser and Sparrow, 1909; Yanagita, 1951, 1959a,b; Blanquet, 1970;) and tentacles (Lubbock, et al., 1981) of sea anemones as well as jellyfish tentacles (Burnett, et al., 1968; Salleo, et al., 1983,1984) has been studied.

The conditions under which cnidae, isolated in various ways, have been found to discharge are summarized in Table I. Examination of this table reveals that although the isolated cnidae retain the ability to discharge in different media, many of the results appear contradictory. While some of these differences may reflect species specific differences, some undoubtedly reflect differences in the isolation procedures used and, in particular, differences in the chemical composition of the various media used in their preparation. For instance, cnidae isolated from the acontia of Calliactis by squashing them in 1M sodium citrate (Hidaka and Mariscal, 1985) discharged when placed in $Ca^{++}$-free artificial sea water, 50 mM $CaCl_2$ or distilled water, while cnidae isolated by squashing Calliactis acontia in artificial sea water discharged only in distilled water or with 5mM EGTA. Clearly, the medium in which the cnidae were isolated affected their ability to discharge.

The same is true, perhaps to a greater degree, for cnidae isolated by extrusion. For instance, Yanagita (1959a) found that Diadumene acontial cnidae extruded in 1M glycerol discharged when placed in any of a variety of isotonic salt solutions; sodium citrate being particularly effective. Discharge failed to occur only when those solutions were diluted 10-fold or more with 1M glycerol. In contrast, Blanquet (1970) found that cnidae isolated by extrusion with 1M sodium citrate would discharge only when the medium was diluted to 0.65M citrate or less. Nine different salts, isotonic with sea water failed to discharge more than 20% of the cnidae although one-half isotonic concentrations of KCl, $Na_2SO_4$ and $K_2SO_4$ discharged more than 50% of the cnidae. The notable exceptions were salts of calcium and magnesium which failed to evoke significant discharge even when diluted to 35% of their isotonic strength.

Conceivably the isolation of cnidae in glycerol sensitized them to the presence of ionic media while isolation in citrate produced the opposite effect. The stabilizing effects of glycerol on protein structure have been known for a long time, but only recently has the mechanism become known. Biophysical studies with model systems have shown that glycerol increases the amount of water at a solvent-protein interface (Gekko and Timasheff, 1981). The addition of salts to a glycerol-water mixture effects the balance of water-glycerol and water-protein interactions, because the salts themselves compete with the other system components for hydration. Thus, those salts which are the most strongly hydrated will have the greatest effect on the solvent properties of the solution, causing a relatively higher proportion of the glycerol to interact with the protein surface. This could lead to a protein conformational change in the cnida wall, operculum or cnida contents of the type that has been hypothesized to play a role in the events immediately preceding discharge (Blanquet, 1970; Lubbock and Amos, 1981; Lubbock, et al., 1981).

TABLE I. Discharge Properties of Isolated Cnidae. + discharge; - no discharge

| Source | Method of isolation | Test substances | Effects | Reference |
|---|---|---|---|---|
| *Metridium* whole animal | homogenization/ 1M sucrose in SW | DW | + | Phillips (1956) |
| | | dilute acid or base | + | |
| | | methylene blue | + | |
| *Metridium* tentacles & acontia | maceration/SW with chlorotone | SW | - | Glaser & Sparrow (1909) |
| | | DW | + | |
| | | saturated solutions: $NaCl$, $K_2CO_3$, $Na_2CO_3$ $SrCl$, $MgSO_4$, $ZnSO_4$, sugar | - | |
| | | 95% Ethanol | - | |
| | | chloroform | - | |
| *Metridium* acontia | extrusion/ether saturated SW | SW saturated with ether | - | Parker & Van Alstyne (1932) |
| *Diadumene* acontia | extrusion/DW | 1M Glycerol | - | Yanagita (1951) |
| | extrusion/1M Glycerol | DW | - | |
| | | SW diluted 8x with 1M Glycerol | + | Yanagita (1959a) |
| | | 0.16M $NaCl$ | +* | |
| | | 0.25M $KCl$ | +* | |
| | | 0.062M $CaCl_2$ | +* | |
| | | 0.062M $NaSCN$ | +* | |
| | | 0.025M $NaHCO_3$ | +* | |
| | | 0.125M $CH_3CO_2Na$ | +* | |
| | | 75% Ethanol | - | |

TABLE I (Continued)

| Source | Method of isolation | Test substances | Effects | Reference |
|---|---|---|---|---|
| Aiptasia acontia | extrusion/ 1M sodium citrate | 0.65M sodium citrate | + | Blanquet (1970) |
| | | DW | + | |
| | | $CaCl_2$ then DW | +[a] | |
| | | SW | - | |
| | | 1M Glycerol | + | |
| | | SW pH2, pH10 | + | |
| | | isotonic salt solutions: $K_2SO_4$, $NaNO_3$, $MgCl_2$, $Ca(NO_3)_2$, KCl, $KNO_3$, $Na_2SO_4$, $CaCl_2$ | -[b] | |
| | | 1/2 isotonic solutions: KCl, $K_2SO_4$, NaCl, $Na_2SO_4$ | +[c] | |
| | | 1/2 isotonic solutions: $KNO_3$, $NaNO_3$, $MgCl_2$, $CaCl_2$, $Ca(NO_3)_2$ | -[d] | |
| Calliactis acontia | extrusion/DW | 5mM EGTA | +[e] | Hidaka & Mariscal (1985) |
| | | Ca-free ASW or 50mM $CaCl_2$ | - | |
| | extrusion/1M sucrose | 5mM EGTA | +[e] | |
| | | Ca-free ASW or 50mM $CaCl_2$ | - | |

TABLE I (Continued).

| Source | Method of isolation | Test substances | Effects | Reference |
|---|---|---|---|---|
| Calliactis acontia | Squashing/1M sodium citrate | 5mM EGTA, Ca-free ASW or DW | + | |
| | Squashing/ASW | Ca-free ASW, 50mM $CaCl_2$ | - | |
| | | 5mM EGTA or DW | + | |
| Rhodactis mesentarial filaments | dissection in parafin oil | 0.01M potassium citrate & imidazole, pH7.0 | + | Lubbock & Amos (1981) |
| | | 0.01M Imidazole | - | |
| | | 0.01M potassium citrate & imidazole + 0.05M $CaCl_2$ | - | |
| | | 0.01M potassium citrate & imidazole + 50 mM NaCl, KCl, or $MgCl_2$ | - | |
| | | 0.01M EGTA & imidazole, pH7.0 | + | |

TABLE I (Continued).

| Source | Method of isolation | Test substances | Effects | Reference |
|---|---|---|---|---|
| *Chrysaora*[†] tentacles | autolysis or homogenization | DW | + | Burnett, *et al* (1968) |
| | | dilute acid or base | + | |
| | | deoxycholate | + | |
| | | SW (pH4) | + | |
| | | SW (pH9) | - | |
| | | SW (pH13) | + | |
| | | 0.5M KCl | -[b] | |
| | | 0.3M $MgCl_2$ | - | |
| | | 0.3M $CaCl_2$ | -[d] | |
| *Pelagia* tentacles | autolysis/DW | Sodium citrate | + | Salleo, *et al* (1984) |
| | | 0.53M KCl, NaCl, LiCl or CsCl, choline chloride, $Na_2SO_4$, $K_2SO_4$, $(NH_4)_2SO_4$ | -[b] | |
| | | 0.53M KI or NaI | +[e] | |
| | | 0.53M KI + 0.01M $CaCl_2$ or 0.01M $BaCl_2$ | - | |

*Minimum concentration necessary to induce >50% discharge. (Solutions made isotonic with sea water by addition of glycerol.); †Averaged results for mixed types of cnidae; [a]diminished; [b]<25% discharge; [c]<60% discharge; [d]<40% discharge; [f]after latency [e]>50% discharge.

Abbreviations. ASW - artificial sea water; DW - distilled water; SW - sea water.

The ability of various salts to discharge cnidae extruded in citrate is more complex. Citrate is a chelating agent for divalent cations which in turn have been shown to have important effects in stabilizing protein structure (Arakawa and Timasheff, 1984). In the absence of divalent cations, the stabilizing effects of monovalent cations have a more pronounced role, but the mechanism is different, and for a given concentration the stabilizing effect is less than that of divalent cations (Arakawa and Timasheff, 1985). This may explain why cnidae isolated in citrate are stable in many different isotonic salt solutions, but very sensitive to dilution except in the case of calcium and magnesium salts which strongly interact directly with proteins, stabilizing their structure.

Cnidae isolated from *Chrysaora* by autolysis (Burnett, *et al.*, 1968) were not discharged in any quantity by the chloride salts of potassium, calcium, and magnesium at isotonic concentrations, whereas distilled water, dilute acids and bases, deoxycholate, and highly acid or alkaline sea water did induce discharge. Holotrichous isorhizas isolated from the tentacles of *Pelagia* by autolysis in distilled water (Salleo, *et al.*, 1984), failed to discharge in significant numbers when placed in a variety of isotonic salts solutions. The only exceptions were the iodine salts of sodium and potassium. These findings are complicated by the fact that even potassium iodide produced only slightly more than 50% discharge and this only after 15 minutes. Interestingly, the effects of the iodine salts could be blocked by the addition of calcium, barium or magnesium chloride. They also reported (Salleo, *et al.*, 1983) that the sodium salts of perchlorate and thiocyanate were more effective than the iodine salts in causing discharge. They proposed that the effectiveness of these salts was related to their lyotrophic properties as previously noted by Yanagita (1959b). Unfortunately, all of their tests were made after the isolated cnidae had been previously frozen ($-20^{o}C$) for unspecified lengths of time. The effect freezing may have had on altering or diminishing the responsiveness was not investigated.

In drawing conclusions from the foregoing one must keep in mind not only the highly variable results of several workers but variability of the preparations themselves. When a percentage of cnidae do or do not discharge in response to some treatment does it mean that the remainder are defective in some way or that there are subclasses of cnidae sensitive to different sets of stimuli? For instance, what percentage of acontial cnidae were not extruded when treated and do these differ from cnidae which were extruded? These questions apply to all types of preparatory methods.

## A. Role of the Cnida Membrane in Regulating Discharge

Because the cnida is a membrane-bound product of the Golgi apparatus, any successful theory of cnidocyte discharge will have to account somehow for the role of the cnida membrane and that of the cnidocyte plasma membrane, which regulates the enclosed cnida's environment. One suggestion is that the events preceeding cnida discharge are essentially those of an exocytotic process (Lubbock, et al., 1981; Holstein and Tardent, 1984). The sequence of steps in cellular exocytosis are well documented (for review see De Lisle and Williams, 1986), although the precise mechanism is not understood. What is clear is that for exocytosis to occur there must be apposition and fusion of the exocytotic vesicular membrane (i.e. the cnida membrane) with the plasma membrane, followed by release of the vesicular contents to the outside. Although it has been demonstrated that a Paramecium trichocyst, a so-called nematocyst analogue, exits the cell via an exocytotic process (Plattner, et al., 1982), this process has never been rigorously demonstrated for true cnidocytes. Interestingly, there are similarities between the ultrastructure of the Paramecium membrane with its in situ trichocysts and the ultrastructure of the cnidocyte plasma membrane and cnida's apical end.

Our interest in the role of the cnida membrane in the regulation of discharge was aroused by considering the methods with which acontial cnidae have been isolated. We hypothesized that the differences in the responsiveness of cnidae isolated by different methods could be explained by the condition of the cnida membrane; i.e., was it perturbed by chemical treatment and was it completely intact? For instance, glycerol is known to permeabilize membranes; citrate may have other effects. To test this hypothesis we isolated acontial cnidae from a local Aiptasia species by three different methods: extrusion in 1M glycerol, extrusion in 1M sodium citrate, and enzymatically with papain. In the first two cases, all the isolated cnidae rapidly discharged when transferred to a $Ca^{++}$-free artificial sea water containing EGTA. On the other hand, the enzymatically treated cnidae were isolated in $Ca^{++}$-free artifical sea water that contained EGTA, yet were recovered undischarged; however, when the calcium ionophore A23187 was added to the medium all the cnidae discharged within a few minutes. These results are consistent with the view that the enzymatically isolated cnidae were enveloped in a functional membrane whereas, the membranes of cnidae isolated by the citrate and glycerol methods had been altered. To further explore this possibility we examined the ultrastructure of the cnidae isolated by the three methods but observed no obvious difference between the three types of cnida. However, a permeability change of the type suggested by these experiments could well occur in the absence of any detectable morphological change.

## V. ELECTROPHYSIOLOGY OF CNIDOCYTES

Cnidaria have attracted the attention of electrophysiologists since the 1950's and with few exceptions, the results of those studies have been tantalizing but at the same time frustrating (for review see Anderson & Schwab, 1982). The problems encountered with these animals stem largely from the small size of the animals' cells and the relative complexity of their tissues, many of which are spontaneously active.

One area of cnidarian electrophysiology that has received little attention is that of cnidocytes, and this despite several findings that suggested that electrical phenomena might play a major role in their discharge or at least in its regulation (Lubbock and Shelton, 1981). Evidence that certain cnidocytes in hydra might be under some form of nervous control was obtained by Spangenberg and Ham (1960) who found that nerve fibers stained with methylene blue appeared to terminate on battery cells. This finding was confirmed electronmicroscopically in other cnidarians (Westfall, 1969, 1970) with the description of chemical synapses onto battery cells. More recently, Hufnagel, et al., (1985) reported that battery cells in the tentacles of hydra are always associated with two types of neuron. Histochemical evidence that the nervous system and battery cells of hydra contained common enzymes was presented by Lentz and Barrnett (1961, 1962) and led them to suggest that cnidocyte discharge was, at least partially, under the control of the nervous system. Finally, several behavioral studies have demonstrated that the physiological state of the animal can influence the responsiveness of cnidocytes to discharge (for review see Shelton, 1982).

Reasons for the absence of electrophysiological studies on cnidocytes are that they are often found in muscularized tissue that is frequently spontaneously active and by the fact that the bulk of these cells is occupied by the hard capsule wall which provides very little room for the insertion of microelectrodes. We are aware of only one instance of intracellular recordings being attempted in cnidocyte-rich tissue. Yanagita (1973) reported that a microelectrode could be inserted into an excised acontium and that a stable negative potential was measured (-25 mV). Definative proof as to the location of the electrode was, however, not provided.

We initiated an electrophysiological study of cnidocytes primarily to determine whether or not the cnidocytes we isolated were indeed intact cells. Structural and ultrastructural studies clearly confirm the presence of a cell membrane on these cells but provide no information on the integrity of that membrane. However, evidence that the isolated cells maintained a negative resting potential and supported voltage-activated electrical events, would clearly indicate that the cells' membranes were intact and functional. To this end we have examined the electrophysiology of isolated cnidocytes using whole-cell patch clamp recording techniques (Sakmann and Neher, 1983). We complemented these recordings with others on cnidocytes *in situ*, using conventional microelectrode recording techniques. This was done to provide control data with which to identify changes in electrophysiological properties resulting from the isolation procedure.

Recordings have now been obtained from cnidocytes isolated from tentacles of hydra, acrorhagi of *Anthopleura*, and tentacles of *Chrysaora* and from *in situ* cnidocytes in *Physalia* tentacles and the battery cells of hydra. A voltage clamp recording made from an acrorhagial cnidocyte is presented in Figure 2a. This recording demonstrated both inward and outward currents but because of the scarcity of recordings, their ionic dependencies were not studied. With cnidocytes obtained from other tissues good recordings were obtained routinely, especially with the atrichous isorhiza cnidocytes from *Chrysaora*. When depolarized by current injection these cells show very pronounced delayed rectification (Fig. 2b). Voltage clamp recordings indicate that this delayed rectifier has the same origin as the delayed rectifier in other cell types, namely a delayed, steady state, outward potassium current ($I_K$). In some cells there is also a transient outward current, presumably an A-current ($I_A$). No other obvious ionic currents were recorded from these cells. Current- and voltage-clamp recordings from

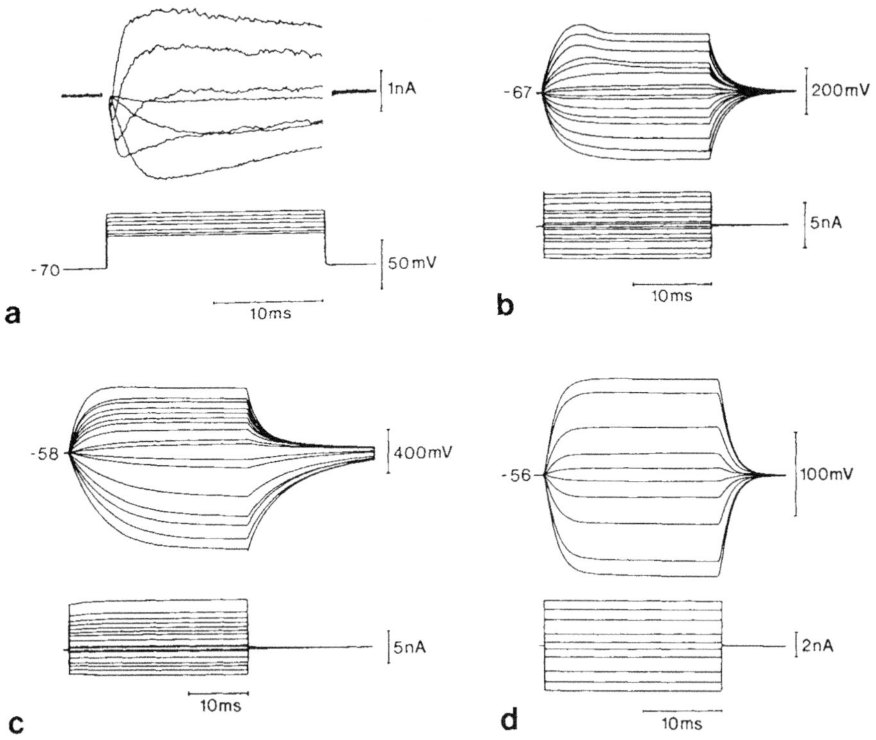


*Fig. 2. Electrophysiological recordings from cnidocytes. (a) A voltage clamp recording from a cnidocyte isolated from an acroraghus of Anthopleura. The cell was clamped at a holding potential of -70 mV and step-depolarized in the manner shown on the lower trace. Both inward and outward currents were generated by this regime. (b) A current-clamp recording from an atrichous anisorhiza isolated from Chrysaora. Current (lower trace) was injected into the cell through the recording pipette, a patch electrode, producing the voltage responses shown in the upper trace. The asymmetrical nature of these responses is indicative of delayed rectification. (c) The same from a battery cell isolated from hydra. (d) A microelectrode recording from an in situ cnidocyte in a Physalia tentacle. Note the absence of delayed rectification.*

battery cells isolated from hydra indicate that these cells too possess a delayed rectifier (Fig. 2c).

Intracellular recordings from cnidocytes *in situ* suggest that the isolated cnidocytes are essentially the same as those *in situ*. For reasons outlined above we have been able to obtain intracellular recordings from only a few types of *in situ* cnidocytes and frequently those are of limited duration because of movement of the preparation. The most consistent recordings were obtained from tentacles of *Physalia* (Fig. 2d). These cnidocytes have resting potentials in the range -52 to -67 mV (mean = -58.8 mV) and their input impedances, determined from conventional current/voltage plots are in the range 34 to 39 Mohm. Interestingly, however, these cells were the only cnidocyte examined not to display delayed rectification (Fig. 2d). One problem frequently encountered while attempting to impale small cells *in situ* is the identity of the impaled cell but, in this case, the problem was circumvented by filling the impaled cell with the fluorescent dye, lucifer yellow. The dye was invariably restricted to the impaled cell indicating that under the conditions employed, it was not dye-coupled to adjacent cells.

One very consistent and important observation that came out of these several recordings was that changes in membrane potential alone are not sufficient to trigger discharge. This finding applies to both isolated and *in situ* cells. Out of the approximately 100 cells we have recorded from in one fashion or another only two have discharged during the recording, despite the fact that the cells sometimes were depolarized or hyperpolarized by as much as 150 mV. This finding indicates that discharge is not the result of a change in membrane potential and suggests that external stimuli to the cell must be transduced into alternative signals such as second messengers in order to effect discharge. The role of the animals' nervous systems in this process remains unknown.

## REFERENCES

Abric, P. 1904. Sur le fonctionnement des nematocystes des Coelenteres. *C. R. Soc. Biol. Paris.* 56:1008-1010.

Anderson, P.A.V. and W.E. Schwab. 1982. Recent advances and model systems in coelenterate neurobiology. *Prog. Neurobiol.* 19:213-236.

Arakawa, T. and S.N. Timasheff. 1984. Mechanism of protein salting in and out by divalent cation salts: balance between hydration and salt binding. Biochem. 23:5912-5923.

Blanquet, R. 1968. Properties and composition of the nematocyst toxin of the sea anemone, Aiptasia Pallida. Comp. Biochem. Physiol. 25:893-902.

Blanquet, R. 1970. Ionic effects on discharge of the isolated and in situ nematocysts of the sea anemone, Aiptasia pallida: a possible role of calcium. Comp. Biochem. Physiol. 35:451-461.

Burnett, J.W., J.H. Stone, L.H. Pierce, D.G. Cargo, E.C. Layne, and J.S. Sutton. 1968. A physical and chemical study of sea nettle nematocysts and their toxin. J. Invest. Dermatology. 51:330-336.

Burnett, J.W. and R. Goldner. 1970. Partial purification of sea nettle toxin (Chrysaora quinquecirrha) nematocyst toxin. Proc. Soc. Exp. Biol. Med. 138:978-981.

Burnett, J.W. and G.J. Calton. 1977. The chemistry and toxicology of some venomous pelagic coelenterates. Toxicon. 15:177-196.

Carré, C. and D. Carré. 1973. Etude du cnidome et de la cnidogenese chez Apolemia uvaria (Lesueur, 1811) (Siphonophore Physonecte). Exp. Cell Res. 81:237-249.

David, C.N. 1973. A quantitative method for maccration of Hydra tissue. Wilhelm Roux Archiv. 171:259-268.

De Lisle, R.C. and J.A. Williams. 1986. ReguLation of membrane fusion in secretory exocytosis. Ann. Rev. Physiol. 48:225-238.

Endean, R., C. Duchemin, D. McColm, and E. H. Fraser. 1969. A study of toxic material derived from nematocysts of the cubomedusan Chironix fleckeri. Toxicon. 6:179-204.

Endean, R. and J. Rifkin. 1975. Isolation of different types of nematocyst from the cubomedusan Chironix fleckeri. Toxicon. 13:375-376.

Gekko, K. and S.N. Timasheff. 1981. Mechanism of protein stabilization by glycerol: preferential hydration in glycerol-water mixtures. Biochem. 20:4667-4676.

Gierer, A., S. Berking, H. Bode, C.N. David, K. Flick, G. Hansmann, H. Schaller, and E. Trenkner. 1972. Regeneration of hydra from reaggregated cells. Nature New Biology. 239:98-101.

Glaser, O.C. and C.M. Sparrow. 1909. The physiology of nematocysts. J. Exp. Zool. 6:361-382.

Hidaka, M. and R.N. Mariscal. 1985. Effects of ions on nematocysts isolated from acontia of a sea anemone by different methods. Am. Zool. 25:111A.

Holstein, T. and P. Tardent. 1984. An ultrahigh-speed analysis of exocytosis: nematocyst discharge. Science. 223:830-832.

Hufnagel, L.A., G. Kass-Simon, and M. K. Lyon. 1985. Functional organization of battery cell complexes in tentacles of Hydra attenuata. J. Morph. 184:323-341.

Lane, C.E. and E. Dodge. 1958. The toxicology of Physalia nematocysts. Biol. Bull. 115:219-226.

Lentz, T.L. and R.J. Barrnett. 1961. Enzyme histochemistry of Hydra nematocysts. Anat. Rec. 142:315.

Lentz, T.L. and R.J. Barnett. 1962. The effect of enzyme substrates and pharmacological agents on nematocyst discharge. J. Exp. Zool. 149:33-38.

Lubbock, R. and W.B. Amos. 1981. Removal of bound calcium from nematocyst contents causes discharge. Nature. 290:500-501.

Lubbock, R. and B.A.B. Shelton. 1981. Electrical activity following cellular recognition of self and non-self in a sea anemone. Nature. 289:59-60.

Lubbock, R., B.L. Gupta and T.A. Hall. 1981. Novel role of calcium in exocytosis: mechanism of nematocyst discharge as shown by x-ray microanalysis. Proc. Nat. Acad. Sci., USA. 78:3624-3628.

Mariscal, R.N. 1974. Nematocysts, in Coelenterate Biology Reviews and New Perspectives ed by Leonard Muscatine and Howard M. Lenhoff. Academic Press, New York.

Mariscal, R.N., C.H. Bigger and R.B. McLean. 1976. The form and function of cnidarian spirocysts. 1. Ultra-structure of the capsule exterior and relationship to the tentacle sensory surface. Cell Tiss. Res. 168:465-474.

McKay, M.C. and P.A.V. Anderson. 1985. Properties of single cnidocytes isolated from sea anemones. Amer. Zool. 25:31A.

Parker, G.H. and M.A. Van Alstyne. 1932. The control and discharge of nematocysts in Metridium and Physalia. J. Exp. Zool. 63:329-344.

Phillips, J.H. 1956. Isolation of active nematocysts of Metridium senile and their chemical composition. Nature. 178:932.

Picken, L.E.R. and R.J. Skaer. 1966. A review of researches on nematocysts. J. Symp. Zool. Soc. Lond. 16:19-50.

Plattner, H., C. Westphal, and R. Tiggemann. 1982. Cytoskeleton-secretory vesicle interactions during the docking of secretory vesicles at the cell membrane in Paramecium tetraurelia. J. Cell Biol. 92:368-377.

Robson, E.A. 1973. The discharge of nematocysts in relationship to properties of the capsule. Proc. 2nd Int. Symp. Cnidaria. Seto. Mar. Biol. Lab. 20:653-673.

Sakmann, B. and E. Nehr (eds.). 1983. Single channel recording. Plenum Press, New York.
Salleo, A., M. Alfa, and G. LaSpada. 1983. Blockage of trypsin-induced discharge of nematocyst of Pelagia noctiluca by $Ca^{2+}$. Mol. Physiol. 3:89-97.
Salleo, A., G. LaSpada, G. Falzea, and M. Gabriella. 1983. Discharging effectiveness of lyotrophic anions on nematocysts of Pelagia Noctiluca. Mol. Physiol. 6:19-26.
Salleo, A., G. La Spada, G. Falzea, and D.M. Gabriella. 1984. Discharging effect of anions and inhibitory effect of divalent cations on isolated nematocysts of Pelagia Noctiluca. Mol Physiol. 5:25-34.
Shelton, G.A.B. 1982. Anthozoa, in Electrical Conduction and Behaviour in Simple Invertebrates ed by G.A.B. Shelton. p.203-242, Oxford Univ. Press, New York.
Skaer, R.J., and L.E.R. Picken. 1965. The structure of the nematocyst thread and the geometry of discharge in Corynactis viridis. Phil. Trans. Roy. Soc. Lond. 250:131-164.
Spangengerg, D.B. and R.G. Ham. 1960. The epidermal nerve net of Hydra. J. Exp. Zool. 143:195-201.
Westfall, J.A. 1965. Nematocysts of the sea anemone Metridium. Am. Zool. 5:377-393.
Westfall, J.A. 1969. Nervous control of nematocyst discharge: chemical synapses. Amer. Zool. 9:1141.
Westfall, J.A. 1970. Ultrastructure of synapses in a primitive coelenterate. 1970. J. Ultrastruct. Res. 32:237-246.
Wilby, O.K. 1976. Nematocyst discharge and the effects of antibiotics on feeding in Hydra. Nature. 262:387-388.
Yanagita, T.M. 1951. The influences of immersion media on the "longevity" of isolated nematocysts of sea-anemone. Nat. Sci. Rep., Ochanomizu Univ. 2:117-123.
Yanagita, T.M. and T. Wada. 1953. Discharge inducing concentrations of acids and bases for the nematocysts of sea anemone. Nat. Sci. Rep., Ochanomizu Univ. 4:112-118.
Yanagita, T.M. 1959a. Physiological mechanism of nematocyst responses in sea-anemone. II Effects of electrolyte ions upon the isolated cnidae. J. Fac. Sci., Univ. Tokyo. Sect. 4. 8:381-400.
Yanagita, T.M. 1959b. Physiological mechanism of nematocyst responses in sea-anemone. VII extrusion of resting cnidae-its nature and its possible bearing on the normal nettling response. J. Exp. Zool. 36:478-494.
Yanagita, T.M. 1973. The "cnidoblast" as an excitable system. Proc. 2nd. Int. Symp. Cnidaria. Seto Mar. Biol. Lab. 20:675-693.

# 16

# THE ROLE OF NEMATOCYSTS IN ANTHOZOAN AGGRESSION

Charles H. Bigger

Department of Biological Sciences
Florida International University
Miami, Florida

## SUMMARY

There are four well documented anthozoan behaviors that fulfill the criteria of aggression, those associated with: (1) sea anemone acrorhagi, (2) sea anemone catch tentacles, (3) coral sweeper tentacles and (4) coral mesenteric filaments. Nematocysts are certainly the major effector of damage to the target animal in the first three cases and may play a role in the fourth. The necrosis that results from acrorhagial, catch tentacle, and sweeper tentacle interactions can be caused by either the penetration of the nematocyst tubules, the actions of the nematocyst venom, or the shearing action of the "anchored" aggressor tissue. The behaviors associated with these responses and the final stage of nematocyst discharge are stimulated by currently unknown molecules on the cell surface of the target animal, a conspecific in many cases. Some evidence suggests the acrorhagial behavior and the discharge of acrorhagial nematocysts may involve different stimulating molecules. Hypotheses concerning the receptors associated with the acrorhagial response are now being suggested. Whereas the involvement of these behaviors and the nematocyst effectors do not seem to fit the standard criteria for immunological systems, the same or similar types of molecular recognition mechanisms may play a role. However, more investigations at the molecular and genetic levels will be required to substantiate this concept.

## I. INTRODUCTION

Activities associated with prey capture or defense from predators have been studied for a long time but the aggressive behaviors of some anthozoans against conspecifics ("clonal aggression") have only recently been recognized. These behaviors, associated with the acrorhagi and catch tentacles of some sea anemones and the sweeper tentacles and mesenterial filaments of some corals, are now well documented examples of directed behaviors that are capable of harming the target animals and that may result in the aggressor limiting non-clonemate access to space (e.g., Francis, 1973b; Bigger, 1980; Purcell, 1977; Richardson, et al., 1979; Lang, 1971). The cnidae of acrorhagi (Bigger, 1980), of catch tentacles (Watson and Mariscal, 1983a) and of sweeper tentacles (Hidaka and Yamazato, 1984) differ significantly from those of the respective feeding tentacles. The concerted discharge of large numbers of these nematocysts constitutes the means of damaging the target animal. These responses with the associated discharge are generally limited to contact with other coelenterates, with the strongest responses toward conspecifics or other anthozoans (e.g. Bigger, 1980). At least some of the receptors for these systems would appear to be associated with the nematocysts so that nematocysts may serve not only in an effector but also a receptor role. While some ecological or behavioral aspects of these responses will be mentioned in this review, the focus is on the role of the nematocyst in these aggressive behaviors.

## II. THE AGGRESSIVE RESPONSES

### A. Acrorhagi

Acrorhagi are "hollow" vesicles that occur at the base of the tentacles of some endomyarium sea anemones. Acrorhagi differ from anemone feeding tentacles, pseudoacrorhagi, fronds, verrucae, and column vesicles in terms of their morphology, cnidae, and function (Doumenc, 1972; Bigger, 1982). While some basitrichous isorhizas (basitrichs) and spirocysts may occur in acrorhagi, the predominant nematocyst type is that of a large holotrichous isorhiza (holotrich) bearing "T" shaped spines that is not found in the other aforementioned structures (Schmidt 1969, 1974; Bigger, 1982).

Even when acrorhagi are expanded, the surface density of holotrichs is still 2-3 times the density of nematocysts found in the feeding tentacles of the same anemone (Bigger, 1982). Based on the presence of spirocysts, acrorhagi appear to be derived from the tentacles rather than other column structures such as verrucae, which lack spirocysts (Bigger, 1982).

Similar acrorhagial behaviors have been described in seven species of sea anemones (Abel, 1954; Bonnin, 1964; Francis, 1973b; Ottaway, 1978; Williams, 1978; Brace et al., 1979; Bigger, 1980; Brace, 1981; Sebens, 1984). When the anemone touches certain conspecifics or other coelenterates with its tentacles the acrorhagi in the area of contact swell and elongate. The attacking anemone then leans over and places the expanded acrorhagi on the target animal (Fig. 1). After a short time the acrorhagi are withdrawn and the process is repeated. Contact with the target animal causes massive, relatively synchronous discharge of the acrorhagial cnidae. The holotrich tubules anchor the acrorhagial ectoderm to the target tissue so that when the acrorhagus is withdrawn, up to 90% of the ectoderm on the acrorhagial tip remains attached to target in what has been termed a "peel" (Fig. 2). Necrotic lesions occur on the target animal at the site of contact and can even lead to the target animal's death. In Anthopleura krebsi the lost ectodermal region is fully restored to the acrorhagus by twenty-four hours (Bigger, 1982). The "loser" of such an encounter usually contrives to "avoid" the winner by moving out of contact or out of the area, although recently Sebens (1984) has reported what appears to be habituation of the response in some cases with Anthopleura xanthogrammica, so that some otherwise antagonistic anemones can remain in contact.

Although acrorhagial aggression was originally viewed (e.g., Francis, 1973a) solely in terms of a direct competition for space between genetically distinct individuals or groups (clones), other possible functions are now considered. These include: the effect on juvenile settlement and recruitment (Ayre, 1982; Sebens, 1984); serving as a form of information transfer communicating the position of neighbors (Sebens, 1984); and, contributing to population polymorphism and general fitness (Bigger, 1980; Brace and Quicke, 1986). Although most individuals of an appropriate species bear acrorhagi, Francis (1976) noted that clonemates on the periphery of Anthopleura elegantissima clones possess more and larger acrorhagi, presumably because of their interactions with other conspecifics while protecting the integrity of the group.

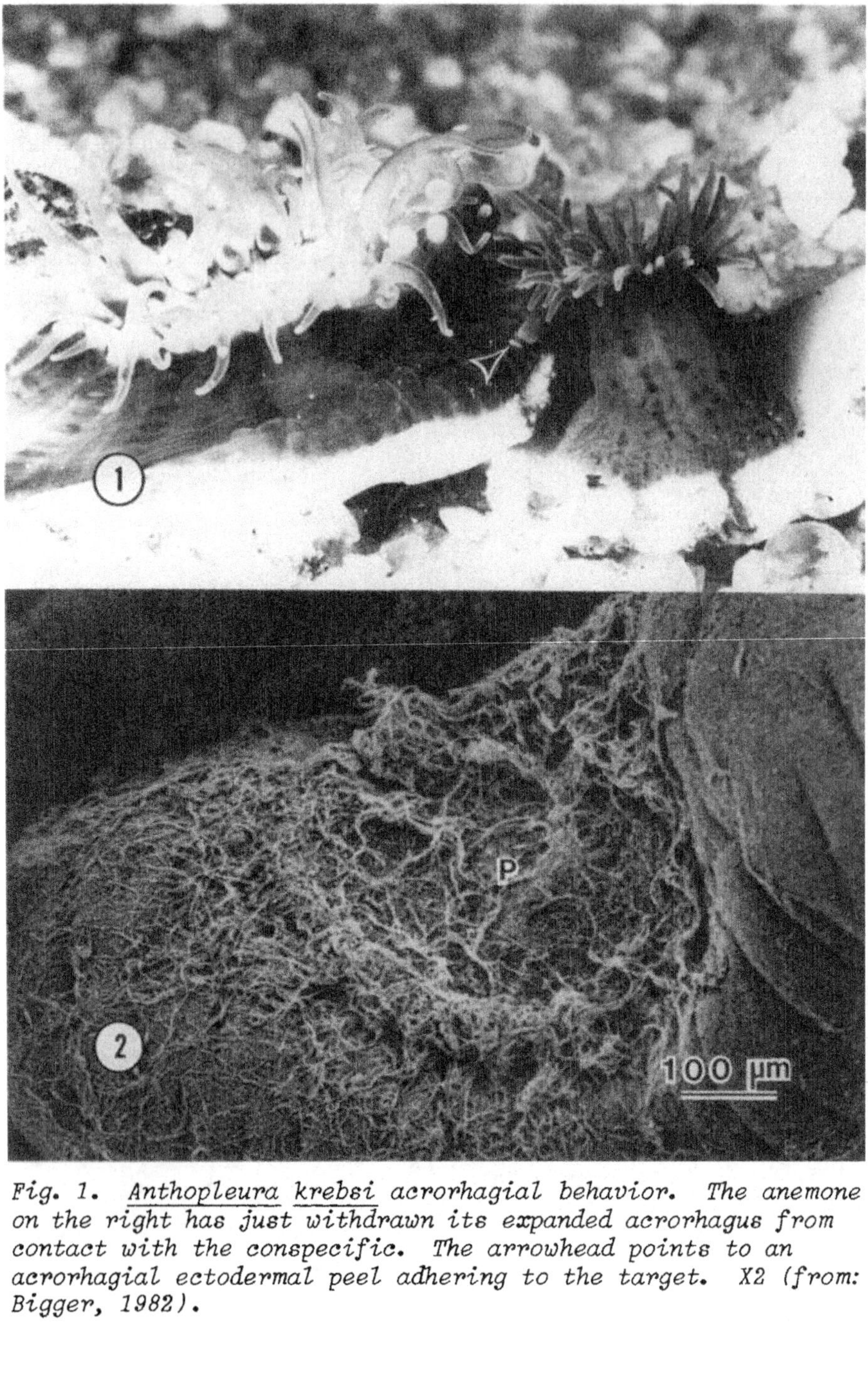


*Fig. 1. Anthopleura krebsi acrorhagial behavior. The anemone on the right has just withdrawn its expanded acrorhagus from contact with the conspecific. The arrowhead points to an acrorhagial ectodermal peel adhering to the target. X2 (from: Bigger, 1982).*

B. Catch Tentacles

Some acontiate sea anemones develop typically opaque, blunt-tipped "catch" tentacles about twice as wide as their relatively slender feeding tentacles. Only approximately ten years ago did Williams (1975) first recognize and then Purcell (1977) fully describe the aggressive function of these special tentacles. Although some aspect of catch tentacle morphogenesis or function has been mentioned for more than six sea anemone species (Williams, 1975; den Hartog, 1977; Purcell, 1977; Watson and Mariscal, 1983a) those that are best studied are found in Metridium senile (Purcell, 1977; Purcell and Kitting, 1982) and Haliplanella luciae (Watson and Mariscal, 1983a & b, 1984, 1985). During a period of three (den Hartog, 1977) to nine weeks (Purcell, 1977) of contact with appropriate conspecifics or other anemone species, some feeding tentacles change their morphology and cnidome to become catch tentacles. Feeding tentacles contain predominantly spirocysts and microbasic p- (H. luciae) or b- (M. senile) mastigophores while catch tentacles of H. luciae contain large numbers of large and small holotrichs that do not have the T-shaped spines of acrorhagial holotrichs (Watson and Mariscal, 1983a & b) and catch tentacles of M. senile contain holotrichs and atrichs (Purcell, 1977). After contacting the appropriate anemone, the catch tentacles of the attacking anemone inflate to approximately three to four times their length and the tips touch the upper region of the target anemone. Contact stimulates massive nematocyst discharge in the catch tentacle tip, thereby anchoring it to the target. As the catch tentacle withdraws the tip autotomizes. A new tip then reforms, apparently through a process of cellular migration and remodeling (Watson and Mariscal, 1983b).

As was the case with the acrorhagial response the original concept of catch tentacles being used in a simple competition for space (e.g., Purcell, 1977) has required further thought in the light of new data (Purcell and Kitting, 1982; Kaplan, 1983). While catch tentacles are certainly used to maintain the space between some clones of M. senile, some clonal aggregates can contain anemones of mixed genotypes (Purcell and Kitting, 1982). The decrease in fighting within mixed clones is not due to fatigue (or depletion of cnidae)

*Fig. 2. An A. krebsi acrorhagial peel on an Anemonia sargassensis anemone tentacle. Note the numerous holotrich nematocyst tubules, from the peel, and the contraction of the tentacle tip (from: Bigger, 1982).*

since newly introduced anemones of different genotypes are attacked (Purcell and Kitting, 1982). This demonstrates that M. senile can differentiate between conspecifics, in a manner analogous to the induction of specific immunological tolerance.

C. Sweeper tentacles

Some scleractinian corals possess extremely long tentacles called sweeper tentacles. Lewis and Price (1975) described a feeding function for these tentacles. It now appears that, in addition to feeding, sweeper tentacles are also used for aggression (Richardson, et al., 1979; Sheppard, 1981; Chornesky, 1983; Hidaka and Miyazaki, 1984; Hidaka and Yamazato, 1984; Logan, 1984). The acrospheres at the tips of Montastrea cavernosa sweeper tentacles have many (63%) large holotrichs and few (3.5%) spirocysts (den Hartog, 1977) while the acrospheres of Galaxea fascicularis sweeper tentacles have approximately 50% large microbasic b-mastigophores and almost 40% spirocysts (Hidaka and Yamazato, 1984). This differs markedly from the feeding tentacles where the cnidae are composed of approximately 80% spirocysts and the remainder, microbasic p-mastigophores, holotrichs, and microbasic b-mastigophores (den Hartog, 1977; Hidaka and Yamazato, 1984). Chornesky (1983) has reported that Agaricia agaricites sweeper tentacles are induced over a 30 day period following contact with the tissues of other corals, or a gorgonian, or a zooanthid.

The sweeper tentacles extend at night (e.g. Richardson, et al. 1979) and move in a waving or "flailing" motion. Nematocysts discharge when the tip contacts an appropriate target. A cluster of discharged nematocysts are left penetrating target tissue but the ecodermal surface of the acrosphere does not remain on the target (see Fig. 1, Hidaka and Miyazaki, 1984). Galaxea fascicularis responds to a wide range of corals, including Millepora sp., a sponge, a bryozoan, mollusks, an arthropod, and an echinoderm (Hidaka and Miyazaki, 1984). Sweeper tentacles may also be used against conspecifics (Hidaka and Yamazato, 1984). The relative importance of mesenteric filaments, sweeper tentacles, or other forms of direct competition in coral ecology is currently a matter of debate (e.g. Sheppard, 1982; Chornesky, 1983; Logan 1984).

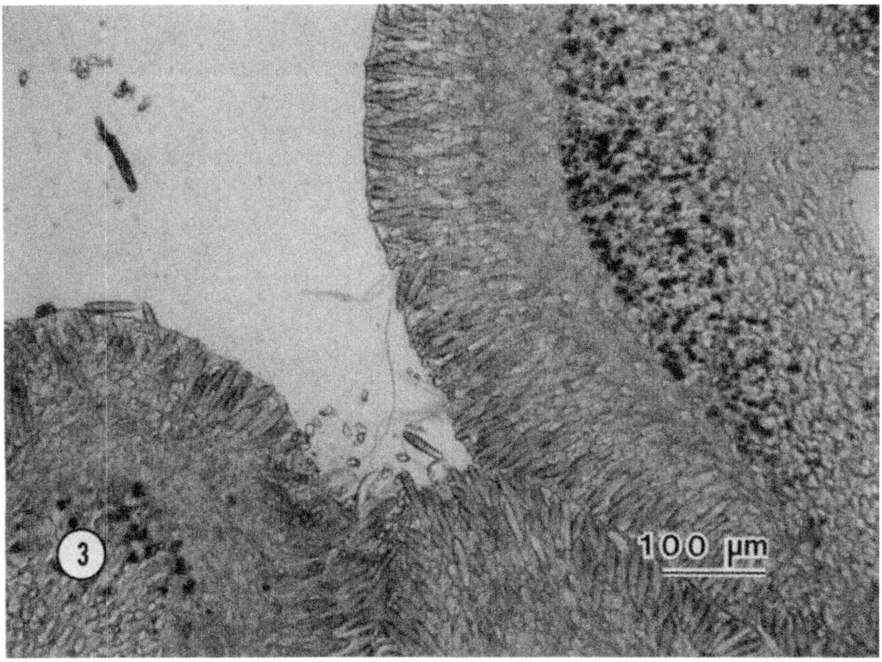


*Fig. 3. A mesenterial filament "squash" of the coral Fungia scutaria. Note the numerous nematocysts.*

D. Mesenterial Filaments

Regardless of any role of coral mesenterial filaments in digestion, these tissues are also used in aggressive behavior. Mesenterial filaments can be extruded through the polyp wall onto adjacent corals; in less than a day the tissues of the target animal may be digested away in the zone of contact (Lang, 1971, 1973). Although digestion seems to be the major source of damage to the target animal, coral mesenterial filaments contain a large number of nematocysts (Fig. 3) a fact appreciated in the last century by Bourne (1887) who considered them to be acontia. Accordingly, the nematocysts of the mesenterial filaments may contribute to the damage inflicted on the target coral. Although digestion caused by mesenterial filaments is usually thought to be interspecific competition, it may also occur intraspecifically (Sheppard, 1982).

### E. Other Possibilities

Additional possible examples of anthozoan tissues employed in aggression include: the marginal tentacles of *Rhodactis sanctithomae* (den Hartog, 1977), the "sweeper polyps" of the coral *Goniopora* spp. (Sheppard, 1982), and the feeding tentacles of certain anemones such as *Bartholomea annulata* (Sebens, 1976).

## III. NEMATOCYSTS AS EFFECTORS

The target tissue lesions associated with the aforementioned anthozoan aggressions are caused by several mechanisms. First, in all cases the entry of large numbers of nematocyst tubules into the target tissue causes mechanical disruption (see Figs. 30 and 31 in Bigger, 1982). Secondly, although there is not really much data on the effects of anthozoan toxin on other anthozoans, the nematocyst venom probably contributes to the necrosis. As a result of acrorhagial peels on tentacles, the target animal often retracts the affected tentacle and may keep it that way for a day or more (Bigger, 1980). Additionally, the tissues of the autotomized catch tentacle tip and acrorhagial peel hold the discharged nematocyst capsules together. Where the acrorhagial peel covers the edge of two tentacles/acrorhagi it can bind them together and cause an impediment to their use. Additionally, with the nematocyst tubules anchored in the target tissue and bound together any local movement of the target tissue will produce a shearing force through the tubules and disrupt the target tissue (Bigger, 1982).

## IV. ELICITING/INFLUENCING FACTORS

As discussed above, acrorhagial and catch tentacle responses are elicited by conspecifics or closely related anthozoans. In corals, mesentarial filaments are elicited by other corals, whereas sweeper tentacle nematocysts are stimulated by a wider variety of animals. Bonnin (1964) suggests that the acrorhagial response is elicited by nematocyst discharge from the target animal, but Lubbock (1980) and Bigger (1976), using nematocyst inactivated tissue and nematocyst-lacking tissue, were unable to elicit acrorhagial nematocyst discharge. A number of factors have been proposed to effect some of these responses with the loser often not even responding aggressively: (1) residency -- the resident anemone always won the acrorhagial encounters (Ottaway, 1978); (2) size -- the larger anemone seems to attack first in acrorhagial encounters (Brace and Pavey, 1978), with some evidence to the contrary (Bigger, 1980); and (3) sex -- catch tentacles are used against conspecifics of the same sex (Kaplan, 1983), with data to the contrary in acrorhagial responses (Ayre, 1982; Sebens, 1984). The major variable that every study of conspecific interactions has confirmed is the genotype combination. Syngeneic animals are not attacked and will not elicit nematocyst discharge. The response is initiated and the strength is governed by specific molecules on the surface of target animal cells. Bigger (1976) demonstrated that for the acrorhagial response these molecules are found on ectoderm from all regions and the endoderm of the target animal. Some target tissues have elicited the behavioral portion of the response but not the nematocyst discharge suggesting that there is either a quantitative or qualitative difference in the molecules perceived by the receptors associated with the behavior and the nematocysts (Bigger, 1976 & 1980). To date, no one has isolated or defined the nature of the eliciting molecules for any of the responses.

## V. ACRORHAGIAL NEMATOCYST RECEPTORS

Bonnin (1964) and Bigger (1982) demonstrated that inflation of the acrorhagus was required before the acrorhagial nematocysts would discharge. Based on this data and on ultrastructural evidence Bigger (1982) proposed a hypothesis to explain the change in sensitivity (threshold) of the acrorhagial nematocysts. The nematocysts require both a chemical and a mechanical stimulus for discharge. The cnidocyte that contains the holotrich has a long cilium and a number of shorter microvilli. It was assumed, that similar to other systems, the cilium acts as the mechanoreceptor and the chemoreceptors reside in the microvilli or other apical surface of the nematocyte (cf., Watson, this volume about chemoreceptors on other cells). In the unexpanded acrorhagus, the majority of the apical surface and the microvilli are folded into a pit with a small opening that surrounds the base of the cilium. As diagrammed in Figure 4, when the acrorhagus expands, the apical region of the nematocyte expands, opening

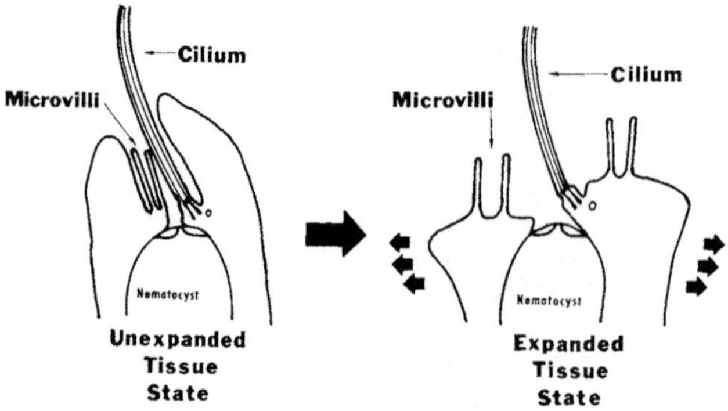


*Fig. 4. A schematic representation of the shape changes in the apical region of a holotrich-containing nematocyte as the acrorhagus expands. Such a morphological change of shape could be responsible for the coordinated nematocyst discharge threshold change that occurs in sea anemone acrorhagi (from: Bigger, 1982).*

the pit and exposing more of the apical surface and the microvilli. At that time the chemoreceptor would then be in a position to contact the eliciting factor on the target tissue. Lubbock and Shelton (1981) suggested that the eliciting molecules (on the target tissue surface) binding to the receptors of "cells of the tip of the acrorhagus" would induce an electrical excitation that would in summation correspond to 10μV impulses that they measured prior to the nematocyst discharge. These ideas await further experimentation (Note: McKay, this volume finds no evidence for the cnidocyte being excitable). It should be noted that Watson and Mariscal (1983b) did not find ciliary pits at the apex of the catch tentacle nematocytes and that no hypotheses have been formulated concerning the mechanisms of sweeper tentacles or mesenterial filament operation.

## VI. RECOGNITION SYSTEMS

Hildemann, et al. (1975) specifically included the "acute intraspecific and interspecific aggression" of this discussion in a continuum of reactions that they felt involved cellular mechanisms of recognition similar to those involved in transplant rejection among higher animals. Those views were later changed, particularly as regards specific sensitization in acrorhagial responses (cf., Bigger, 1980). Especially in the case of acrorhagi and catch tentacles, these aggressive responses all involve self/not-self recognition. Although such recognition is the cornerstone of immune responses, these responses do not meet the other criteria generally considered to be part of an immune response (cf., Hildemann, et al., 1979). That does not preclude the same or similar recognition molecules, stimulating factors and/or receptors, from being involved in these aggressive reactions and transplantation immunological responses but, this remains speculation until the molecules and genes involved are characterized and compared.

## VII. CONCLUSIONS

It is obvious from this brief review that the basic parameters of the acrorhagial response, and to a lesser degree, the catch tentacle, sweeper tentacle and mesenterial filament responses, have been delineated. Some hypotheses concerning the underlying cellular, chemical, and genetic bases of these responses are being formulated; but, it is only now that we are beginning to seriously investigate at these levels. Hopefully in the future we will discover the underlying mechanisms of these and other, yet undiscovered, anthozoan aggressive responses.

## VII. ACKNOWLEDGMENTS

I thank the N.I.H. for partial support under grant # RR08205 and would especially like to acknowledge many conversations on this topic with Drs. R.N. Mariscal and W.H. Hildemann.

## VII. REFERENCES


Abel, E.F. 1954. Ein Beitrag zur Giftwirkung der Aktinien und Funktion der Randsackchen. Zoologischer Anzeiger 153:259-268.

Ayre, D.J. 1982. Inter-genotype aggression in the solitary sea anemone Actinia tenebrosa. Marine Biology 68:199-205.

Bigger, C.H. 1976. The acrorhagial response in Anthopleura krebsi: intraspecific and interspecific recognition. In G.O. Mackie (ed.), Coelenterate Ecology and Behavior, pp. 127-136. Plenum Press, New York.

Bigger, C.H. 1980. Interspecific and intraspecific acrorhagial aggressive behavior among sea anemones: a recognition of self and not-self. Biol. Bull. 159:117-134.

Bigger, C.H. 1982. The cellular basis of the aggressive acrorhagial response of sea anemones. J. Morph. 173:259-278.

Bonnin, J.P. 1964. Recherches sur la "reaction d'agression" et sur le fonctionnement der acrorrhages d'Actinia equina L. Bull. Biol. 98:225-250.

Bourne, G.C. 1887. The anatomy of the madreporarian coral Fungia. Q. J. Micr. Sci. 27:293-324.

Brace, R.C. 1981. Intraspecific aggression in the color morphs of the anemone Phymactis clematis from Chile. Marine Biology 64:85-93.

Brace, R.C., Pavey, J. and D.L.J. Quicke. 1979. Intraspecific aggression in the color morphs of the anemones Actinia equina: the `convention' governing dominance ranking. Anim. Behav. 27:553-561.

Brace, R.C. and J. Pavey. 1978. Size-dependent dominance hierarchy in the anemone Actinia equina. Nature 273:752-753.

Brace, R.C. and D.L.J. Quicke. 1986. Seasonal changes in dispersion within an aggregation of the anemone, Actinia equina, with a reappraisal of the role of intraspecific aggression. J. mar. biol. Ass. U.K. 66:49-70.

Chornesky, E.A. 1983. Induced development of sweeper tentacles on the reef coral Agaricia agaricites: a response to direct competition. Biol. Bull. 165:569-581.

Doumenc, D. 1972. Adaptation morphologique de l'acrorrhage chez Actinia equina L. Z. Zellforsch. Microsk. Anat. 129:386-394.

Francis, L. 1973a. Clone specific segregation in the sea anemone Anthopleura elegantissima. Biol. Bull. 144:64-72.

Francis, L. 1973b. Intraspecific aggression and its effect on the distribution of Anthopleura elegantissima and some related sea anemone. Biol. Bull. 144:73-92.

Francis, L. 1976. Social organization within clones of the sea anemone *Anthopleura elegantissima*. *Biol. Bull.* 150:361-376.

den Hartog, J.C. 1977. The marginal tentacles of *Rhodactis sanctithomae* (Corallimorphia) and the sweeper tentacles of *Montastrea cavernosa* (Scleractinia); their cnidom and possible function. *Proceedings, Third International Coral Reef Symposium*, pp. 463-469, U. of Miami, Miami.

Hidaka, M. and I. Miyazaki. 1984. Nematocyst discharge and surface structure of the ordinary and sweeper tentacles of a scleractinian coral, *Galaxea fascicularis*. *Galaxea* 3:119-130.

Hidaka, M. and K. Yamazato. 1984. Intraspecific interactions in a scleractinian coral, *Galaxea fascicularis*: induced formation of sweeper tentacles. *Coral Reefs* 3:77-85.

Hildemann, W.H., Linthicum, D.S. and D.C. Vann. 1975. Immunoincompatibility reactions in corals (Coelenterata). *In* W.H. Hildemann and A.A. Benedict (eds.), *Immunologic Phylogeny*, pp. 105-114. Plenum Press, New York.

Hildemann, W.H., Bigger, C.H. and I.S. Johnston. 1979. Histocompatibility reactions and allogeneic polymorphism among invertebrates. *Transplantation Proc.* 11:1136-1141.

Kaplan, S.W. 1983. Intrasexual aggression in *Metridium senile*. *Biol. Bull.* 165:416-418.

Lang, J. 1971. Interspecific aggression by scleractinian corals. 1. The rediscovery of *Scolymia cubensis* (Milne Edwards & Haime). *Bull. Mar. Sci.* 21:952-959.

Lang, J. 1973. Interspecific aggression by scleractinian corals. 2. Why the race is not only to the swift. *Bull. Mar. Sci.* 23:260-279.

Lewis, J.B. and W.S. Price. 1975. Feeding mechanisms and feeding strategies of Atlantic reef corals. *J. Zool. Lond.* 176:527-545.

Logan, A. 1984. Interspecific aggression in hermatypic corals from Bermuda. *Coral Reefs* 3:131-138.

Lubbock, R. 1980. Clone-specific cellular recognition in a sea anemone. *Proc. Natl. Acad. Sci. USA* 77:6667-6669.

Lubbock, R. and G.A.B. Shelton. 1981. Electrical activity following cellular recognition of self and non-self in a sea anemone. *Nature* 289:59-60.

Ottaway, J.R. 1978. Population ecology of the intertidal anemone *Actinia tenebrosa*. I. Pedal locomotion and intraspecific aggression. *Aust. J. Mar. Freshwater Res.* 29:787-802.

Purcell, J.E. 1977. Aggressive function and induced development of catch tentacles in the sea anemone *Metridium senile* (Coelenterata, Actiniaria). *Biol. Bull.* 153:356-368.

Purcell, J.E. & C.L. Kitting. 1982. Intraspecific aggression and population distributions of the sea anemone Metridium senile. Biol. Bull. 162:345-359.
Richardson, C.A., Dustan, P. and J.C. Lang. 1979. Maintenance of living space by sweeper tentacles of Montastrea cavernosa, a Caribbean reef coral. Marine Biology 55:181-186.
Schmidt, H. 1969. Die Nesselkapseln der Aktinien und ihre differential-diagnostische Bedeutung. Helgolander Meeresunters 19:284-317.
Schmidt, H. 1974. On evolution in the Anthozoa. Proc. Second Int. Coral Reef Symp. 1:533-560.
Sebens, K.P. 1976. The ecology of Caribbean sea anemones in Panama: utilization of space on a coral reef. In G.O. Mackie (ed.) Coelenterate Ecology and Behavior, pp. 67-77. Plenum Press, New York.
Sebens, K.P. 1984. Agonistic behavior in the intertidal sea anemone Anthopleura xanthogrammica. Biol. Bull. 166:457-472.
Sheppard, C.R.C. 1981. "Reach" of aggressively interacting corals, and relative importance of interactions at different depths. Proc. 4th Int. Coral Reef Symp. 2:363-368.
Sheppard, C.R.C. 1982. Coral populations on reef slopes and their major controls. Mar. Ecol. Prog. Ser. 7:83-115.
Watson, G.M. and R.N. Mariscal. 1983a. The development of a sea anemone tentacle specialized for aggression: morphogenesis and regression of the catch tentacle of Haliplanella luciae (Cnidaria, Anthozoa). Biol. Bull. 164:506-517.
Watson, G.M. and R.N. Mariscal. 1983b. Comparative ultrastructure of catch tentacles and feeding tentacles in the sea anemone Haliplanella. Tissue & Cell 15:939-959.
Watson, G.M. and R.N. Mariscal. 1984. Ultrastructure and sulfur cytochemistry of nematocyst development in catch tentacles of the sea anemone Haliplanella luciae (Cnidaria:Anthozoa). J. Ultrastruct. Res. 87:159-171.
Watson, G.M. and R.N. Mariscal. 1985. Ultrastructure of nematocyst discharge in catch tentacles of the sea anemone Haliplanella luciae (Cnidaira:Anthozoa). Tissue & Cell 17:199-213.
Williams, R.B. 1975. Catch tentacles in sea anemones: Occurrence in Haliplanella luciae (Verill) and a review of current knowledge. J. Nat. Hist. 9:241-248.
Williams, R.B. 1978. Some recent observations on the acrorhagi of sea anemones. J. mar. biol. Ass. U.K. 58:787-788.

17

# HISTORY AND CURRENT STATE OF KNOWLEDGE CONCERNING DISCHARGE OF CNIDAE[1]

*Pierre Tardent*

Zoological Institute
University of Zürich-Irchel
8057 Zürich, Switzerland

SUMMARY

This paper presents a short overview of the morphodynamic and energetic aspects of nematocyst discharge in cnidarians. This process which includes the eversion of the tubular content of the cyst and the release of its soluble constituents is one of the fastest biomechanical processes occurring at the level of cellular organoids. According to a high-speed cinematographical analysis (40,000 frames per second) the whole exocytotic process takes less than 3 milliseconds during which certain components of the ejected tubule accelerate in the order of 40 000 x g. The forces required for achieving this performance appear to be of dual origin: through osmotically generated increase of the intracapsular hydrostatic pressure and by release of intrinsical spring-like energies stored in the walls of both the cyst and the tubule.

[1]*Generously supported by the Swiss National Science Foundation (Grant Nr. 3.072-0.84).*

## I. INTRODUCTION

The explosive discharge of cnidarian nematocysts is one of the most spectacular events occurring at the level of an individual cell. No wonder that both the morphodynamic and energetic aspects of this process have challenged many cellular biologists and biophysicists since these sophisticated secretory products of the Golgi apparatus were first seen by Leewenhoek (1704), Baker (1743) and Trembley (1744) in the tentacles of the fresh water hydra. Later Corda (1836, 1837) found that there was not only one but at least two different types of cysts he considered as being "weapons of defense" operated by the polyp. Ehrenberg (1836) concluded from his observations that the nematocysts were "fishing hooks" serving hydra for grabbing and catching its prey, mostly small crustaceans and insect larvae.

According to Laurent (1842), who probably was the first person to observe the actual discharge of cysts, this process consists in the extrusion of a "magma" which was not defined further. Almost simultaneously Doyère (1842) reported that what the discharging cyst releases is not a solid "thread" but a slender tubule which is extruded from the capsule by a rather complicated process of eversion by which the inside of the tubule is turned out like the fingers of an inverted glove.

Doyère's (1842) interpretation proved to be accurate. In fact the structural content of all the numerous kinds of cnidarian nematocysts (Weill, 1930; Werner, 1965; Mariscal, 1974) was recognized as being a tubule, the dimensions of which vary from one type to another (Fig. 1). The same holds true for the accessory structures of the tubule such as spines, stylets, lamellae and barbs. In the resting cyst they are located within the lumen of the undischarged tubule (Fig. 3), while upon discharge they finally decorate its outer surface (Fig. 4).

The entire process of cyst discharge occurs at such an extraordinary high speed that a thorough analysis of its subsequent phases and their chronology had been out of reach. According to Reisinger's (1937) calculations the eversion time is in the order of milliseconds. Various authors (Möbius, 1866; Ewald, 1915; Schulze 1917, 1921/1922; Kepner *et al.*, 1951)

*Fig. 1. Plate from an early work by Möbius (1866) about the morphology and discharge of cnidarian nematocyst. Certain drawings (5, 6, 10, 12) represent intermediate stages of cyst discharge by eversion.*

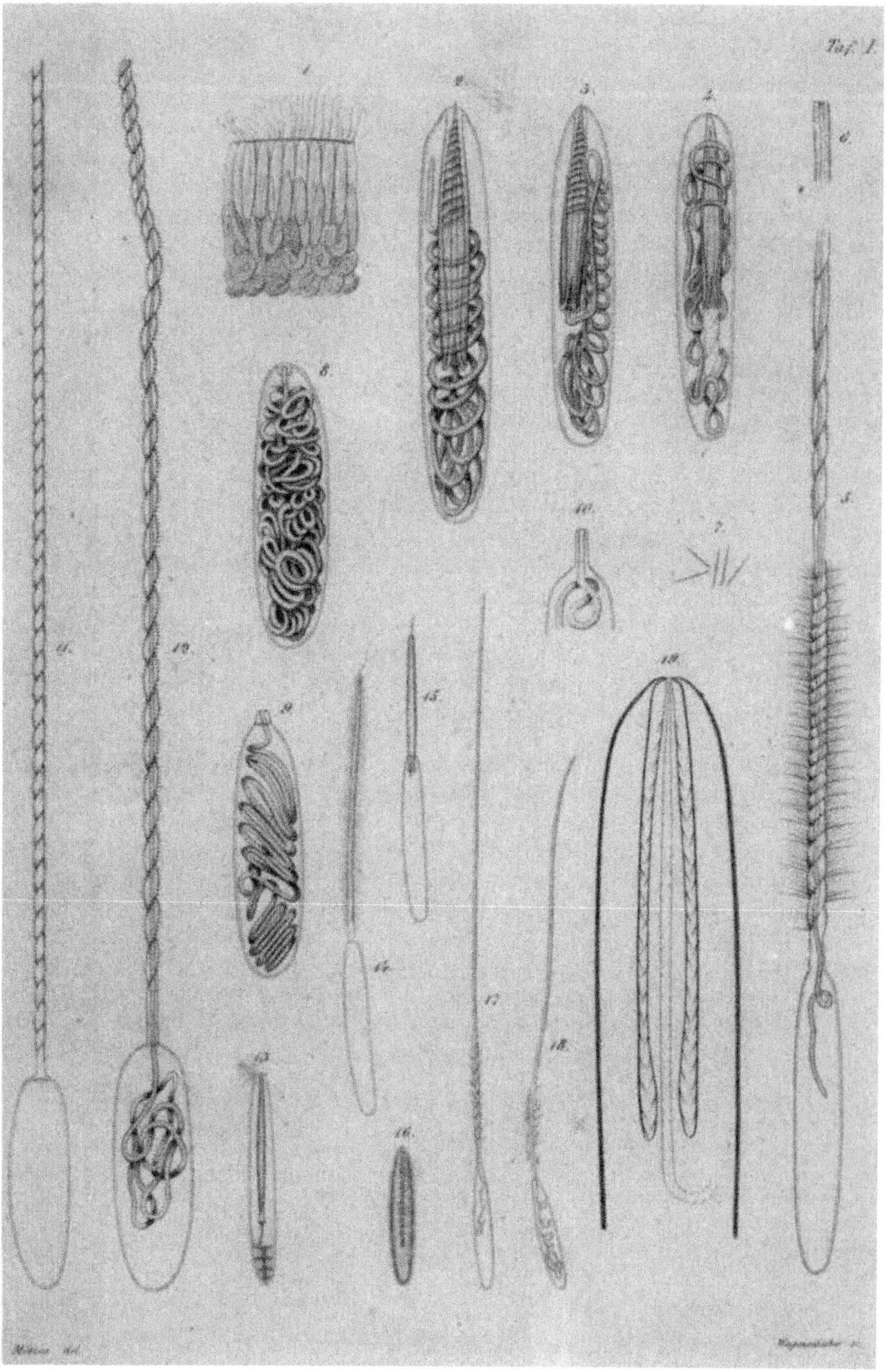

however have described intermediate stages of the process which can be found now and then when a discharging cyst hits an obstacle that blocks the course of events (Fig. 1).

We exploited this opportunity systematically by discharging electrically triggered stenoteles (Fig. 6) of hydra into gels (Gelatine, agar, N-alginate, polyacrylamide gels) of varying degrees of viscosity (Tardent and Holstein, 1982). The higher the viscosity of the medium was, the earlier the process of discharge could be blocked. This method allowed us to reconstruct the course of morphodynamic events including the associated behavior of the stylets and lamellae (Figs. 7-9). It also provided the opportunity for subjecting the intermediate stages to ultrastructural examination (Figs. 10-12).

This procedure, of course, was inadequate for gaining a much needed insight as to the speed at which discharge occurs and as to the times required for the completion of the individual phases of the event. This information is essential not only for the characterization of the nature of forces involved and estimations concerning their magnitude, but also for the understanding of how triggered cysts act upon their natural targets (Figs. 16, 17). With the technical assistance of the "Bundesstaatliche Hauptstelle für wissenschaftliche Kinematographie" in Vienna we subjected stenotele discharge to a successful cinematographical analysis for the first time using a high-speed camera which yields 40 000 frames per second (Holstein and Tardent, 1984). This study not only confirmed Reisinger's (1937) earlier calculations about the duration of the entire process of cyst discharge but yielded valuable informations about the speed of the various phases of the event. Additionally the cinematographical documents provided the opportunity for a biometrical survey of the event (Fig. 15), which gave valuable data in connection with the existing theories and speculations as to the nature and sources of forces which are required for enacting the process and accomplishing it successfully.

This survey about the present state of knowledge concerning cyst discharge considers nothing but the cyst (Figs. 2-4), its behavior during discharge and its action upon natural and artificial targets. Associated problems such as to the kind of external stimuli that induce the nematocyte to discharge its projectile and by what pathways the external signal is transmitted to the effector are discussed elsewhere in this volume.

The relevant investigations which were and are dedicated to the discharge problem are necessarily limited to a relatively small number of cnidarian species and to a few of the numerous types of cysts. Consequently, the facts presented here are valid for the investigated objects only and should be extrapolated to the entire family of nematocysts only with great caution. It is highly improbable though that such a marvellous cellular gadgetry should have polyphyletic origins and that cyst discharge could depend on more than one set of mechanical and energetical principles.

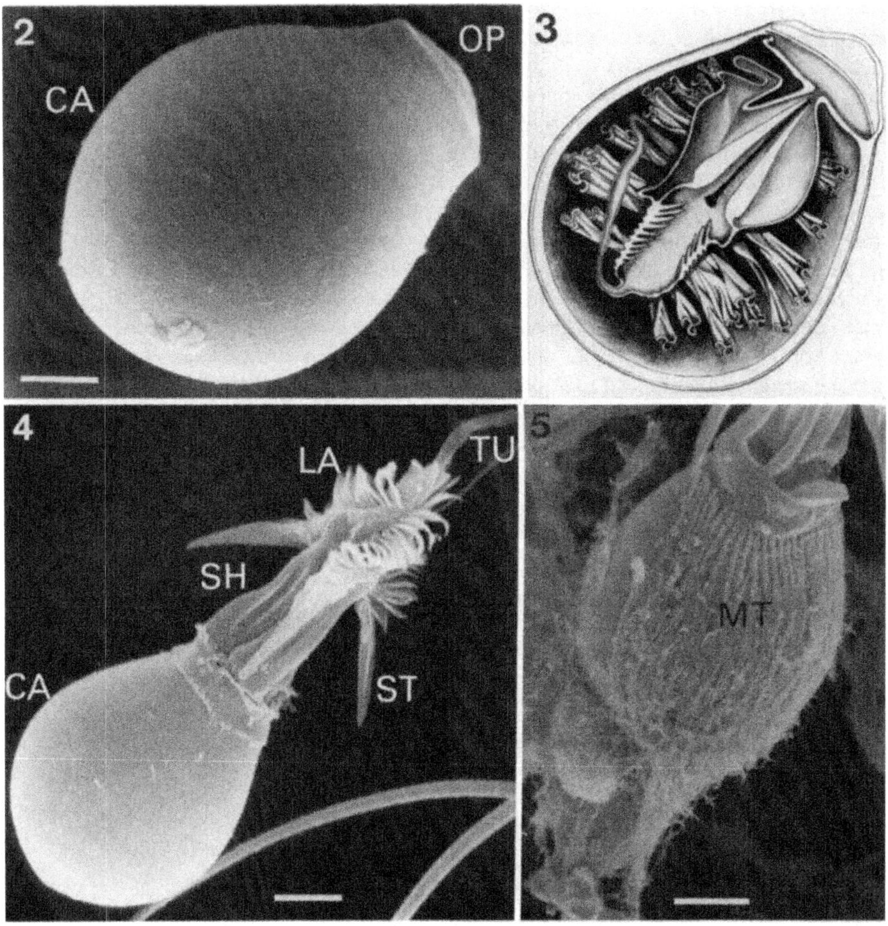


*Figs. 2-5. Stenoteles of Hydra attenuata. Fig. 2: isolated undischarged cyst from which all extracapsular components had been removed (bar: 3 μm); Fig. 3: longitudinal section through an undischarged stenotele as reconstructed from TEM documents; Fig. 4: fully discharged stenotele with only a short section of the long tubule (bar: 3 μm). Fig. 5: Discharged stenotele from which most cytoplasmatic components except the microtubular envelope and the nucleus have been removed (bar: 3 μm). (CA = capsule, LA = lamellae, N = nucleus, MT = microtubules, OP = operculum, SH = shaft, ST = stylet, TU = tubule).*

## II. THE MORPHODYNAMICS AND CHRONOLOGY OF NEMATOCYST DISCHARGE

An essential methodological prerequisite for any study dealing with nematocyst discharge consists in the possibility of controlling the onset of the process. Even if the latter occurs at such a prohibitive speed that makes it inaccessable to any visual analysis, the microscopic examination of individual cysts before and after their discharge yields valuable information as to changes of structural properties and dimensions associated with discharge.

When dealing with isolated live tentacles of polyps or medusae, an electrical pulse is the most appropriate stimulus for discharging at least a certain number of cyst types at will. In hydra tentacles, stenoteles (Figs. 2-5) and desmonemes readily respond to an electrical shock of 24 V D.C., while the isorhizas do not. It must be stressed that not every stenotele and desmoneme will respond. Those nematocytes which have not reached their final position within the tentacular epithelial cell and those which are still migrating do not discharge.

For triggering individual functional stenoteles isolated tentacles or entire polyps are placed in a drop of culture

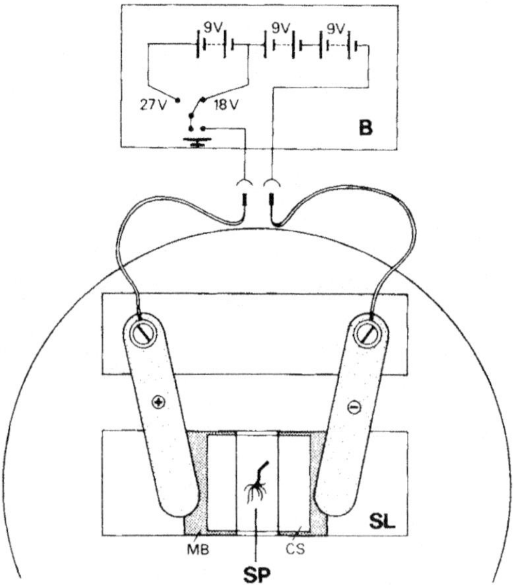


*Fig. 6. Set up for the electrical triggering of stenoteles under the microscope (B = battery; CS = cover slip; MB = adhesive metal band, ST = slide, SP = specimen).*

solution (Loomis and Lenhoff, 1956) between two electrodes which are about 1 cm apart (Fig. 6). The latter are mounted on a slide by coating the surface of the marginal areas of the slide with chromium (150 A°), palladium (150 A°) and gold (200 A°). The electrodes are then reinforced by galvanization with a 30 µm thick layer of elemental gold (Holstein and Tardent, 1984). More simple electrodes manifactured with aluminium foil or silver paint will serve the purpose equally well. The object to be examined is then covered with a cover slip that is large enough to bridge the two electrodes. This set up was used for the high-speed cinematographical recordings of stenotele discharge (see below).

Isolated stenoteles of hydra (Fig. 2) which were cleaned from other cellular components by freezing (-20°C) and subsequent thawing, homogenizing and Percoll-centrifugation (10 min, 3000 x g) (Weber *et al.*, this volume) do not react to electrical stimulation anymore. However, they readily discharge their content when exposed to disulfide reducing agents such as DTE (dithioerythritol) or mercaptoethanol.

Discharge occurs in a normal manner although somehow slower than when the cyst is part of a live cell and is electrically triggered. We have not yet made a cinematographical analysis of discharge performed by isolated cysts, therefore exact data about the difference between the two velocities are not available.

Not even complete dehydration deprives the isolated hydra stenotele cysts from their ability to perform normal discharge. The oddly shaped dehydrated stenoteles quickly take up water and after regaining their normal appearance (Fig. 13) are ready again to be triggered by means of reducing agents. This surprising finding demonstrated that the withdrawal of water does not cause any irreversable damages to those components of the cyst which are essential for enacting and sustaining the normal process of discharge. The same is true for dried nematocysts of other cnidarian species (Amos, personal communication).

Many early authors (Möbius, 1866; Schulze, 1921/1922; Kepner *et al.*, 1951) have described and magnificiently depicted (Fig. 1) intermediate stages of the discharge of various kinds of cysts. Such fortunate cases can be found where the tubule of a discharging cyst encounters a solid or highly viscuous obstacle which inhibits the course of further events. This phenomenon can be used to assemble whole series of intermediate stages which provide an insight into structural changes accompaning the process of cyst discharge.

For this purpose we have discharged hydra stenoteles electrically into gels (gelatine, agar-agar, polyacrylamide, Na-alginate) of varying degrees of viscosity (Tardent and

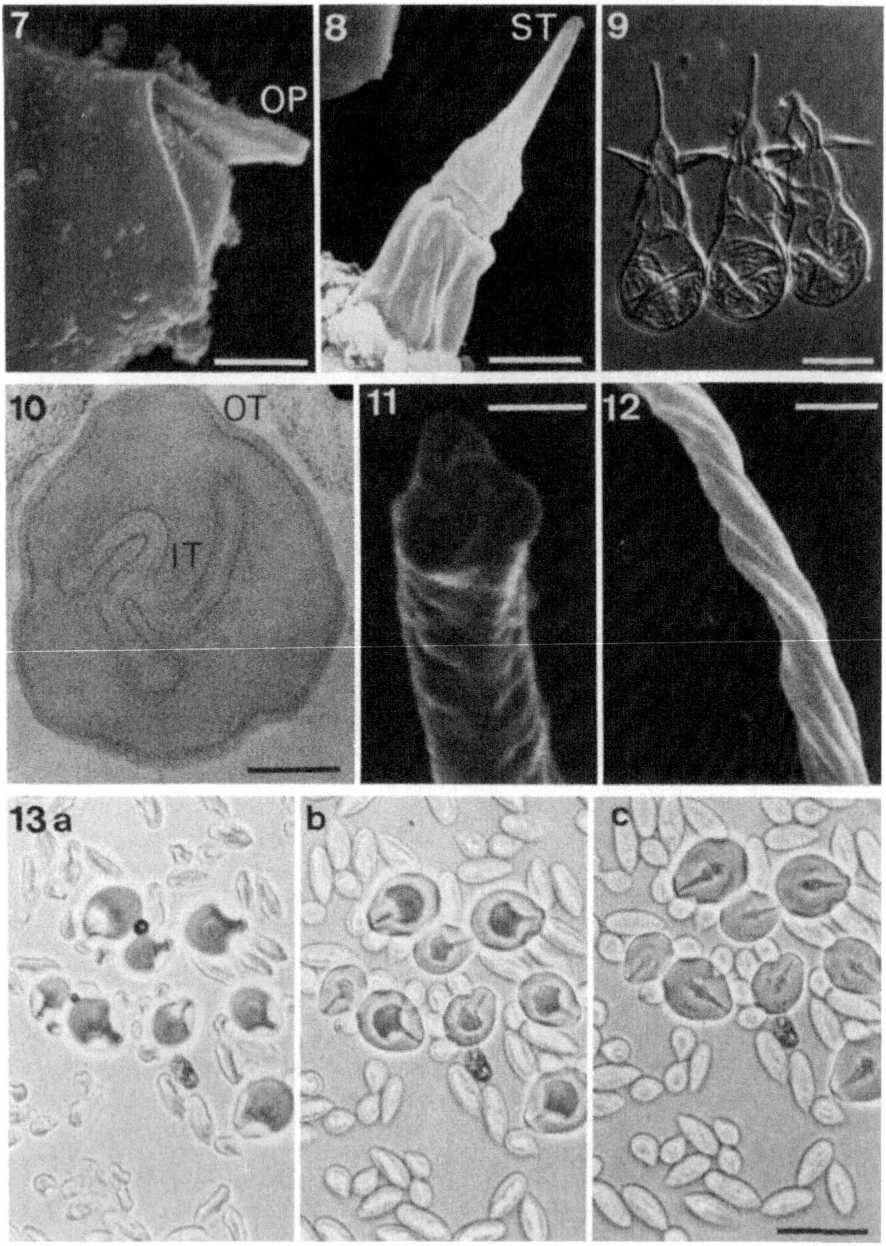


*Figs. 7-13. Stages of stenotele discharge. Fig. 7: half-way open operculum (phase b, SEM, bar: 2 µm); Fig. 8: arrow head formed by the three stylets (phase b, SEM, bar: 5 µm);*

Holstein, 1982). Figures 7-9 show some examples of such arrested intermediate situations.

A break-through in the efforts aiming at a better understanding of the chronology of cyst discharge was possible thanks to high-speed cinematography which in combination with electrical triggering of individual cysts (Fig. 6) was for the first time used by Holstein and Tardent (1984). The Hycam (16 mm/120 m) film camera allowed to record a total of 20 stenotele discharges at a speed of 40 000 frames per second (for technical details see Holstein and Tardent, 1984). Their analysis yielded most valuable information about the time required for the cyst to discharge completely, about the velocity of particular steps of the event and about the behavioral and dimensional changes to which the various structural components of the exploding cyst are subjected. The triggered stenoteles were allowed to discharge freely into a liquid medium without being confronted with an obstacle. So far technical difficulties have prohibited an analogous cinematographical study of exploding cysts hitting a solid natural or artificial target as depicted in Figure 16.

The following description of the various phases of stenotele discharge is based upon both, the cinematographical recordings as well as SEM and TEM pictures taken from intermediate stages of the evagination process.

The total time required for the discharge of a stenotele cyst including the completion of the evagination of the long (470 $\pm$ 77 m) and slender (diameter: 0.8-0.9 m) tubule (Fig. 12) was about 3 milliseconds) thus approximately confirming Reisinger's (1937) calculation. One of the surprising discoveries was that all 20 recordings matched perfectly not only with respect to the total time required for discharge but also regarding the length of each subsequent phase of the process.

A certain degree of variability (0.025 to 2.5 msec) was observed as to the delay between the onset of the electrical

*Fig. 9: cysts with flipped back stylets and partly evaginated tubule (phase d, phase contrast, bar: 10 µm; Fig. 10: TEM cross section through a partly evaginated tubule. Note the inversion of the two layers of the tubular wall (bar: 200 nm); Fig. 11: SEM of the tip of a partly evaginated tubule (phase d, bar: 500 nm); Fig. 12: short segment of a fully evaginated tubule (SEM, bar: 1 µm). (IT = wall of the inner tubule; OP = operculum; OT = wall of the outer tubule; ST = stylets, TU = tubule) (Figs. 9-12 from Tardent and Holstein, 1982); Fig. 13: Protocoll of the recovery of dehydrated and accordingly odd-shaped nematocysts of hydra. 40 sec (13a), 4 min (13b) and 14 min (13c) after having been reimmersed in water (bar: 20 µm).*

pulse (duration 30 µsec) and the first apparent sign of the cysts reaction to the stimulus (phase a). This variation is most probably due to variations in the time required for the contacts of the relay to close.

Phase b of the process starts with the first detectable reaction of the cyst characterized by the capsule's triangular operculum flipping open (Fig. 7). Since this event also takes place in isolated and chemically triggered cysts (see above) we must assume that in hydra stenoteles no structures other than the naked cyst (Fig. 2) are functionally involved in this event. The opening of the operculum is immediately followed by the extrusion of the basal part of the tubule including the short broad shaft and the 3 stylets which at this stage are still joined thus forming a pointed arrow head (Figs. 8, 16, 19). Both steps, the opening of the cyst and the ejection of the stylet apparatus require considerably less than 10 µsec. Within this period the tip of the stylets cover a distance of about 20 µm and, accordingly, travel at a speed of 2m/sec. To produce this average velocity over a 10 µsec period from a standing start a constant acceleration of about 40 000 x g is required. This finding is important because this sudden deployment of energy functionally coincides with the crucial moment when the stenotele hits its target and punches a hole into it (Figs. 16, 18, 19).

At this stage (phase c) discharge is arrested for about 150 µsec. During this period the stylets must be withdrawn from the opening they have produced in the prey's integument in order to clear the way for the evagination of the rest of the tubule (Fig. 16). How this backward movement is achieved is not fully understood. It is presumably caused by the eversion of the shaft's conical section, the lamellae of which may force the joined stylets to flip backward. Investigations using artificial targets such as araldite films (Leemann and Tardent, unpublished) have shown that if for one reason or another the stylets remain stuck in the hole (Figs. 18, 19) they had produced the eversion of the rest of the tubule does not occur.

The last phase (d) of the process is characterized by the eversion of the long and slender terminal section of the tubule, which - in the undischarged stenotele capsule - is more or less regularly coiled around the centrally located shaft (Fig. 3), while in other types of cysts it appears to be rather randomly packed within the capsule (Fig. 1).

The cross-section of the tubule features the shape of a three-bladed propeller with sideward bent blades (Figs. 10, 14d). In addition the tubule is twisted around its longitudinal axis (Fig. 12). Since in hydra stenoteles both the propeller shaped cross-section and the longitudinal twisting of the tubule exist not only before (Fig. 3) but also after the completion of eversion this particular structural pattern must be considered as expressing the resting state of the tubule (see below). However, in the course of eversion the fraction of the tube which has already been extruded through the capsule's opening becomes temporarily cyclindrical and untwisted (Figs. 9, 10). Thanks to this particular situation the friction between the wall of the already evaginated section and that of the part enclosed by the latter is considerably reduced (Fig. 14). The transition from one kind of cross-section to the other occurs at the tip of the evaginating tubule and is achieved as shown in Fig. 11.

The question whether or not the length of the tubule and the degree of its twisting are subjected to changes during evagination remains unanswered for most types of cysts because of the difficulties which are met in measuring the tubule

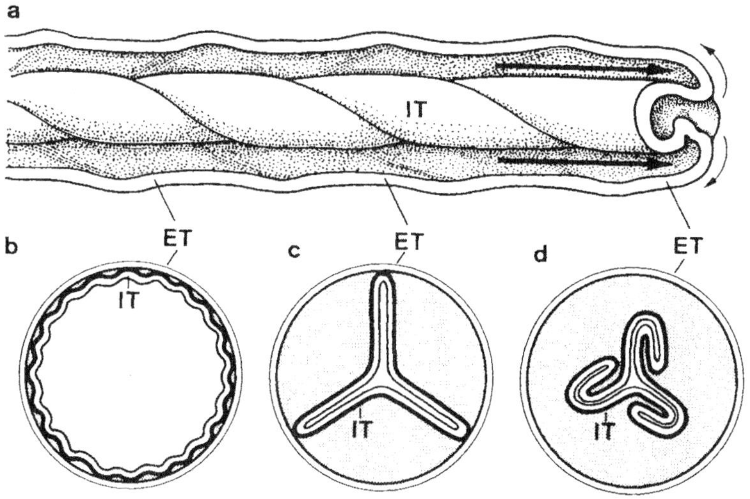


*Fig. 14. Schematical representation of tubule eversion of a hydra stenotele. Fig. 14a: longitudinal section through the tip of the tubule (cfr. Fig. 11). The large arrows indicate the direction of forces necessary for bringing about eversion. Fig. 14b-d: Theoretical representation of three possible cross sections of the inner tubule (IT) as compared to the already everted outer part (ET). The examples b and c would involve considerable friction between the inner and external sections of the tubule, while d representing the true situation (cfr. Fig. 10) minimizes friction (from Tardent and Holstein, 1982).*

while it is still inside the capsule (Figs. 1,3). Measurements performed by Godknecht (1985) on B-rhabdoids of *Anemonia sulcata* (Anthozoa), where inside the capsule the coils of the tubule are arranged in a most regular fashion, have shown, that the everted tubule ( ∿ 500 µm) is about twice as long as compared to its regularly coiled state within the cyst ( ∿ 257 µm). This increase in length following eversion has proved to be the consequence of an accordion-like relaxation of the degree of twisting through which the number of spiral turns per unit length is reduced from 4.1 to 2.1. This finding indicates that in this case the tubule is endowed with intrinsically stored mechanical forces which at least in part support the process of eversion (see below).

Much attention has been paid to the question whether the discharge of the cyst is accompanied by changes of its shape and/or volume. In most types of cysts examined so far by Ewald (1915), Will (1919) and others the changes of shape are less dramatic than those of the nematocyst's volume. Careful measurements performed by Will (1919) on cysts of numerous species indicate that discharge always leads to drastic reduction of the cyst's volume. The magnitude of this reduction depended on the type of cyst, but was always in the order of 50% or less.

Our own measurements (Tardent and Holstein, 1982, Holstein and Tardent, 1984) carried out on numerous undischarged and

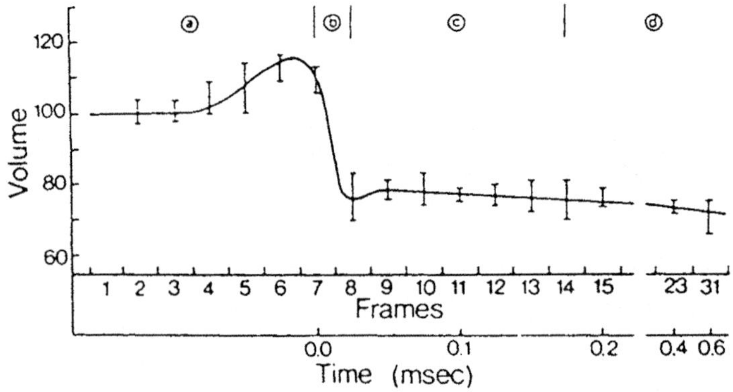


*Fig. 15. Average values (n = 5) of volumetric changes (expressed as percentages of the initial values) of the stenotele cyst after triggering and during the subsequent phases of the process of discharge. The calculations are based on the high-speed film sequences (Holstein and Tardent, 1984).*

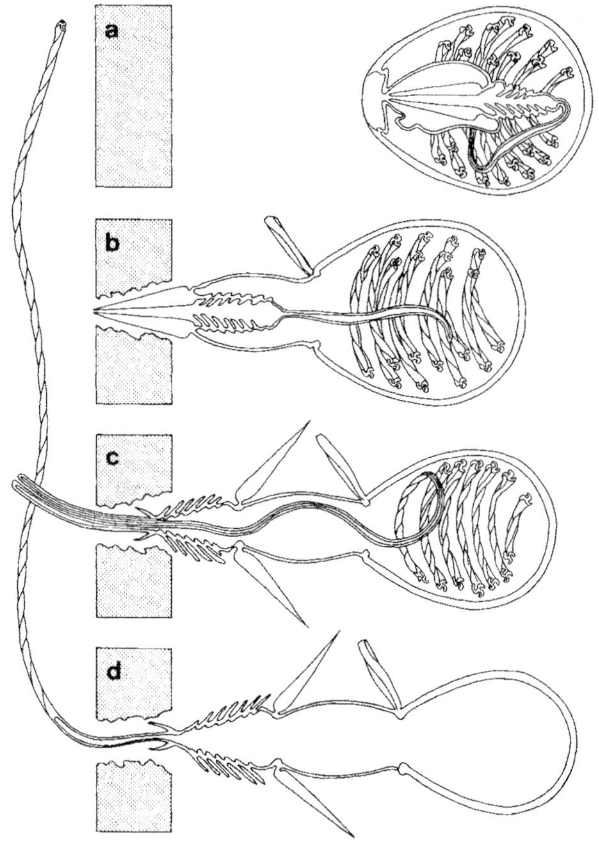


*Fig. 16. Four stages of the discharge of a hydra stenotele hitting and perforating the cuticule of a prey and everting its tubule into the target (from Tardent and Holstein, 1982).*

discharged hydra stenoteles confirmed these earlier findings. While the average volume of the undischarged stenoteles of *Hydra attenuata* amounted to 1283.6 ± 126.5 $\mu m^3$ (n = 250) it dropped to 652 ± 110.3 $\mu m^3$ (n = 250) following complete discharge. This ratio was confirmed by data obtained from individually discharged specimens and from the high-speed film sequences.

These sequences have shown (Fig. 15) that this reduction in volume of the cyst is not continuous. A first sharp drop to 75% of the original value is associated with the ejection of the basal part of the shaft including the stylet apparatus (phase b, see above). However, during the relatively slow eversion of the tubule the decrease proceeds rather continuously down to 50%.

## III. THE ENERGETICAL PROBLEMS OF CYST DISCHARGE

Ever since people became aware of the complexity of the morphodynamics of nematocyst discharge speculations about the nature and magnitude of forces involved in this process have found their way into the literature. The questions of when and how the energies required for enacting and sustaining this cellular explosion are generated are still open, because so far the smallness of the objects has prevented the performance of direct measurements and recordings. Therefore, we depend largely on pieces of indirect evidence. Nevertheless, it is reasonable to expect that inspite of the numerous types of cnidarian nematocysts the energetic principles will be the same for all of them.

The various concepts which have been expressed in relation to this pending problem have been reviewed and discussed by various authors (Weill, 1934; Hyman, 1940; Picken, 1953; Skaer and Picken, 1965; Picken and Skaer, 1966; Mariscal, 1974). There are the 3 main hypothesis to be considered:

a) the "contractile hypothesis" (von Lendenfeld, 1887; Toppe, 1910; Jones, 1947; Mueller, 1950; Chapman and Tilney, 1959a, 1959b; Yanagita and Wada, 1954, 1959; Mattern *et al.*, 1965; Westfall, 1965, 1970; Cormier and Hessinger, 1980) which states that the pressure required for opening the operculum and for the eversion of shaft and tubule is generated by the contraction of either contractile elements of the cell's cytoskeleton or of the cyst's wall.

In the case of hydra stenoteles, at least, the enaction of such mechanisms appears to be unlikely for the following reasons: Although in the live cell the surface of the cyst is covered by a unit membrane and a dense "basket of microtubules (Fig. 5), no actin filaments could be detected surrounding the cyst (Stidwill, unpublished). Furthermore, as mentioned above, isolated and pure cysts (Fig. 2) have preserved their ability to perform an absolutely normal process of tubular evagination without the help of other cellular constituents.

b) The "intrinsic forces hypothesis" ("tension hypothesis") supported by Dujardin (1845), Jones (1947) and Carré (1980) postulates that the energy deployed during eversion stems from spring-like intrinsic forces which are built up during cnidogenesis and stored within the walls of the cyst and the tubule. At discharge they would be released like the energies stored in a tightened spring.

Our observations on stenoteles and B-rhabdoids suggest that this source may indeed cover at least part of the energy requirements. In fact, when considering the exceptional speed at which the basal part of the shaft and the stylets are ejected by the capsule it is reasonable to assume that this phase of the process depends on such "tension forces". Carré (1980) who has surgically injured cysts of siphonophores concludes from the dynamic behavior of the tubule that the latter must be under considerable tension forces when packed within the capsule. We have recently observed the same phenomenon in injured stenoteles of hydra where the tubule squirts out of an artificially produced opening in the cyst wall without everting properly. Figure 18 is a SEM close up of a hydra stenotele the stylets of which have remained stuck within the target. The tubule which for this reason was prevented from evaginating has succeeded in rupturing the wall of the shaft. Whether this spring-like behavior is due to a longitudinal compression of the twisted tubule is not clear yet. However, Godknecht's data (1985) obtained from B-rhabdoids (see above) speak in favour of this interpretation. The question of course remains how this mechanical force is built up during cnidogenesis and how the tubule is kept in this state of permanent tension inside the undischarged cyst.

c. The "osmotic theory" advocated by a considerable number of authors (Iwanzoff, 1896; Schneider, 1900; Glaser and Sparrow, 1909; Will, 1909, 1914, 1919; Weill, 1934; Picken, 1953, 1957; Robson, 1953; Slautterback, 1963; Picken and Skaer, 1966; Lubbock and Amos, 1981; Lubbock *et al.*, 1981; Gupta and Hall, 1984; Holstein and Tardent, 1984) claims that the intracapsular pressure increases immediately before and/or during the process of discharge by swelling of the cyst's matrix. A certain degree of intracapsular pressure undoubtedly exists already in the resting cyst. With the aid of a micromanipulator we have succeeded in introducing micropipettes into isolated stenoteles of hydra that are stained with toluidine-blue (Fig. 21). When the pipet hits the capsule at a favourable angle and pierces its wall the dye can be seen rushing into the lumen of the pipet (Flückiger and Tardent, unpublished).

However, there is no doubt whatsoever that this intracapsular pressure increases further just before and/or during the opening of the cyst's operculum flips by artificial triggering. Volumetric measurements performed on the high-speed film sequences (Holstein and Tardent, 1984) clearly show (Fig. 15) that the volume of the triggered capsule increases by the order of 10%. This swelling of the cyst which starts about 60 μsec before the cyst's cover flips open is undoubtedly due to water rushing into the capsule.

Although most authors supporting this theory agree that the increase in intracapsular pressure is osmotically generated, many different ideas have been expressed as to how this is achieved. Gosse (1858) proposed that the contraction of the stimulated nematocyte would produce heat, which in its turn would cause a swelling of the matrix. Iwanzoff (1896) proposed that the content of the cyst was gelatinous and that acids secreted by the cell would lead to a swelling of this material. Jacobson (1912) believed that the intracapsular pressure was maintained by means of a continuous secretion of water by the cell into the cyst.

The fact that naked cysts (Fig. 2) can explode even after a prolonged storage indicates that its intracapsular pressure is independent of a controlling activity of the cyst's membrane and of other cytoplasmatic components.

The actual influx of water upon triggering can be visualized as follows: while undischarged cysts of a live tentacle hardly take up any dye when immersed in a methylene-blue solution the latter readily penetrates into an electrically triggered capsule. This phenomenon, however, is only true for cysts which are still within their live cells. Isolated and cleaned undischarged capsules do take up the dye. Therefore, the uptake of water is in this particular situation not actively mediated and controlled by the wall of the cyst, but presumably by the cyst's membranous envelope.

By exposing naked cysts of stenoteles to dyes of varying molecular weights we found that their wall was freely permeable to molecules of sizes up to 700 daltons. The cyst wall is therefore to be considered as a sieve which lacks active regulatory functions. It is still a matter of speculation which components actually fulfill this task. Since discharge is possible without the cyst's membrane the triggering system must be looked for within the matrix of the capsule.

It has been shown that in the nematocysts of sea anemones (Anthozoa) the onset of discharge is accompanied by the release of intracapsular calcium and an influx of water (Lubbock and Amos, 1981; Lubbock *et al.*, 1981; Gupta and Hall, 1984, Gupta, this volume). Indeed cysts of all types examined so far contain extraordinary high concentration of $Ca^{++}$ (Johnson and Lenhoff, 1958; Lubbock and Amos, 1981, Mariscal, 1984; Tardent *et al.*, 1985). Our own x-ray spectrophotometrical (EDAX) analysis performed on isolated cysts of members of various genera (*Hydra*, *Podocoryne*, *Physalia*, *Chrysaora*, *Cassiopea*, *Aurelia*, *Anemonia*) have not only confirmed this finding but have also revealed the presence of equal or even larger amounts of Mg-ions (Weber *et al.*, this volume). The sum of these divalent ions contained in the capsule exceeds the value of 1 molar as revealed by chemical analysis.

Inspite of the fact that the capsular wall is permeable to small molecules (see above) isolated, membrane-free cysts (Fig. 2) can be washed thoroughly without loosing significant amounts of $Ca^{++}$ and $Mg^{++}$. Consequently, the latter must be bound to and controlled by proteins, peptides or glycoproteins. How this system operates and how it is functionally linked to the triggering mechanism is not yet known. Another still open question concerns magnesium. Are these cations like $Ca^{++}$ also involved in cyst discharge and, if so, are $Ca^{++}$ and $Mg^{++}$ acting in a synergistic or antagonistic manner?

In our opinion there is no doubt that an osmotically generated intracapsular hydrostatic pressure together with "tension forces" (see above) is one of the main force which enacts and sustains cyst discharge. What the respective contribution of these two forces is and what their concerted actions are remain the subjects of further investigations.

## IV. CONCLUSIONS

The present knowledge about the spectacular process of discharge of nematocysts is still fragmentary. While the ultrastructure and the morphodynamic events taking place in the course of discharge are fairly well investigated in the case of particular types of cysts, little information exists about the speed of the event and its accurate chronology. For this purpose the high-speed film camera has already proved its usefulness (Holstein and Tardent, 1984). It should therefore be used systematically for analyzing the explosion of cyst types other than stenoteles and for comparing the modes of discharge of isolated capsules and cysts which still are a functional part of the living cell. It would also be of interest to subject the action of exploding cysts on natural and artificial targets to a cinematographical analysis. Although the high-speed film sequences have permitted insight into the chronology of stenotele discharge, the ejection of the stylets (phase b), which occurs within the time of one single exposure, could be analyzed only with an even faster camera.

Nematocyst discharge undoubtedly is one of the fastest processes occurring at the cellular level. Whether or not all types of cysts match the velocity of hydra stenoteles remains to be investigated. The speed, as recorded in hydra, is

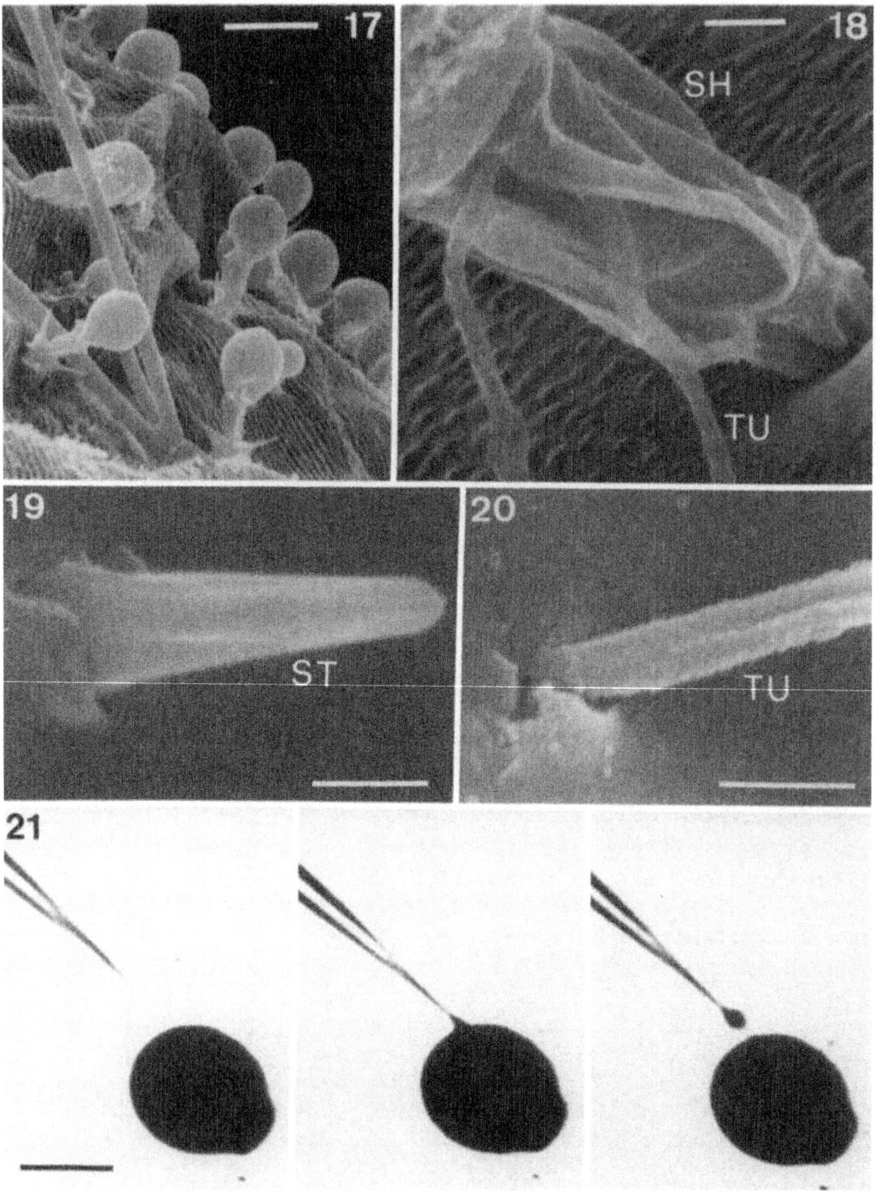


*Figs. 17-21. Stenoteles of hydra. Fig. 17: SEM view of the cuticula of a mosquito larva (Aedes aegyptii) which had been exposed to the tentacles of hydra. Numerous discharged cysts are sticking within the prey's integument (bar: 20 μm). Fig. 18: SEM-close-up of a discharged stenotele that has hit*

certainly required for the fulfillment of its principal function, that of breaking the integument of the prey by mechanically punching a hole into it (Figs. 16, 18, 19). It is still not known, how the arrow-head of the stenotele consisting of 3 joined stylets is withdrawn from the opening it has produced. If this crucial movement does not take place (Fig. 19) the rest of the shaft and the tubule are prevented from everting into the prey's body. Another unsolved problem concerns the behavior of the everting tubule within its target and the ways and means it releases the toxins. The stenotele tubule does not act like the needle of a syringe which releases its toxins at its tip once it is completely evaginated. Should it act in this manner a considerably shorter tubule whould meet the requirements of the "syringe principle" and an additional force would be needed for pressing the liquid from the capsule into the "needle", i.e. into the tubule.

Experiments with stained and subsequently discharged stenoteles have provided evidence that the cyst's liquid content is more or less continuously released all along the everting tubule. Yet we do not know which one of the two principles which are schematically depicted in Figure 22 is actually adopted. Nevertheless, this fast mechanical spreading of the toxins by the tubule through the prey's body appears to be much more efficient than a passive and relatively slow diffusion from the tip of the fully evaginated tubule (Tardent *et al.*, 1985).

According to our observations a relatively small number of successful "stenotele-shots" are required for paralyzing a larva of *Artemia* or that of a mosquito (*Aedes*) larva. Yet great numbers of cysts are triggered (Fig. 17) when such a prey organism collides with a tentacle of hydra. However, there is always a considerable number of failures (Fig. 18, 19, Leemann and Tardent, unpublished).

The challenging problems concerning the energetic aspects of cyst discharge are still far from being solved. In our opinion the "contraction hypothesis" can now be discarded for the reasons explained in this paper. It seems that there is no

*an araldite film. For unknown reasons the stylets (ST) did not withdraw from the hole they have produced, meanwhile the tubule (TU) has ruptured the shaft (SH, bar: 2 μm). Fig. 19: Backside of an araldite film with the stylets (ST) having punched a hole into the artificial target (bar: 1 μm). Fig. 20: Stenotele tubule (TU) emerging through an opening successfully produced in an araldite film by the stylets of a stenotele (bar: 1 μm). Fig. 21: Micropipet operated by a micromanipulator entering in and withdrawing from a toluidine-blue stained, isolated (cfr. Fig. 2) stenotele (bar: 10 μm).*

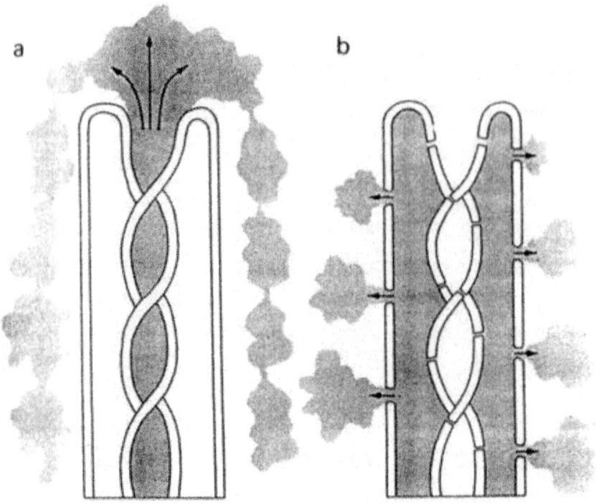


*Fig. 22. Schematic drawings of two possible ways by which toxins (stippled areas) could be released by the everting tubule (from Tardent et al., 1985).*

need for seeking a decision as to whether the "tension theory" or the "osmotic theory" corresponds to the truth. In fact both principles seem to be involved in producing the energies required for expelling the long tubule from the cyst by eversion. Further biochemical investigations about the cyst's matrix and the chemical and physical properties of the structural components as well as direct measurements of the intracapsular pressure before and during discharge may one day allow to make much more precise statements about the nature, origin and magnitude of forces involved in nematocyst explosion.

ACKNOWLEDGEMENTS

The author thanks his collaborators and students in particular Drs. Th. Holstein, J. Weber, R. Stidwill and M. Klug for their assistance and critical comments and R. Tardent for carefully reading and typing the manuscript.

## REFERENCES


Baker, H. 1743. An attempt towards a natural history of the Polyp: In a letter to Martin Folkes, Esg. President of the Royal Society, Dodsley, London.

Carré, D. 1980. Hypothesis on the mechanism of cnidocyst discharge. Eur. J. Cell Biol. 20:265-271.

Chapman, G.B. and L.G. Tilney. 1959a. Cytological studies of the nematocysts of *Hydra*. I. Desmonemes, isorhizas, cnidocils and supporting structures. J. Biophys. Biochem. Cytol. 5:68-78.

Chapman, G.B. and L.G. Tilney. 1959b. Cytological studies of the nematocysts of *Hydra*. The stenoteles. J. Biophys. Biochem. Cytol. 5:79-84.

Corda, A.J. 1836. Anatome Hydrae fuscae. Nov. Act. Acad. Caes. Leopold 18:302.

Corda, A.J. 1837. Anatome Hydra fuscae. Ann. Sc. natur. 2 sér. t.8.

Cormier, S.M. and D.A. Hessinger. 1980. Cnidocil apparatus: sensory receptor of *Physalia physalis* nematocytes. J. Ultrastruct. Res. 72:13-19.

Doyère. 1842. Notes sur quelques points de l'anatomie des Hydres d'eau douce. C.R. Acad. Sci. Paris 15:429-432.

Dujardin, F. 1845. Mémoire sur le développement des Polypes hydraires et Méduses. Ann. Sc. Nat. 4:257-281.

Ehrenberg, C. 1836. Beobachtungen über die Organisation der Armpolypen. Mitt. Ges. naturf. Freunde Berlin, 2. Quartal 27-29.

Ewald, A. 1915. Ueber den Bau, die Entladung und die Entwicklung der Nesselkapseln von *Hydra* und *Porpita mediterranea* nebst einigen histologischen Bemerkungen über die letztere Form. Verhandl. naturwiss. und med. Verein Heidelberg, N.F. 13:303-354.

Glaser, D.C. and C.M. Sparrow. 1909. The physiology of nematocysts. J. Exp. Zool. 6:361-382.

Godknecht, A. 1985. Die Entladung der B-Rhabdoiden der Tentakel von *Anemonia sulcata* Pennant (Cnidaria, Anthozoa. Diploma thesis, University of Zürich.

Gosse, R. 1858. Researches on the poison apparatus in the Actiniidae. Proc. Roy. Soc. 9:125-128.

Gupta, B.L. and Th.A.Hall. 1984. Role of high concentrations of Ca, Cu and Zn in the maturation and discharge in situ of sea anemone nematocysts as shown by x-ray microanalysis of cryosection. *In* L. Bolis, J. Zadunaisky and R. Gilles (eds.), *Toxins, Drugs and Pollutants in Marine Animals*, pp. 77-95, Springer Verlag, Heidelberg.

Holstein, Th. 1987. Cnidocil-nematocyst complex in *Craspedacusta*. This volume.

Holstein, Th. and P. Tardent. 1984. An Ultrahigh-Speed Analysis of Exocytosis: Nematocyst Discharge. Science 223: 830-833.

Hyman, L. 1940. *The Invertebrates: Protozoa through Ctenophora*. Vol. I, pp. 1-726, McGraw Hill, New York.

Jacobson, A. 1912. Die Nesselzellen. Arch. Naturgesch. 78A:111-144.

Johnson, F.B. and H.M. Lenhoff. 1958. Histochemical study of purified hydra nematocysts. J. Histochem. Cytochem. 6: 394.

Jones, C.S. 1947. The control and discharge of nematocysts in hydra. J. Exp. Zool. 105:25-60.

Iwanzoff, N. 1896. Ueber den Bau, die Wirkungsweise und die Entwicklung der Nesselkapseln der Coelenteraten. Bull. Imp. Soc. Nat. Moscou 10:95.

Kepner, W.A., B.D. Reynolds, L. Goldstein, G. Britt, E. Atcheson, E.Q.Zielinsky, M. Brickert Rhodes. 1951. The discharge of nematocysts of *Hydra* with special reference to the penetrant. J. Morphol. 88:23-47.

Laurent, L. 1842. Recherches sur trois sortes reproducteurs, l'anatomie de l'Hydra. C.R.Acad. Sci. Paris 12.

Leewenhoek, A.V. 1704. Concerning green weeds growing in water and some animalcula found about them. Phil. Trans. Roy. Soc. London 23:1304-1311.

Lendenfeld, R. von 1887. The function of nettle cells. J. Roy. Micr. Soc. 7:247-248.

Loomis, W.F. and H.M. Lenhoff. 1956. Growth and sexual differentiation of *Hydra* in mass culture. J. Exp. Zool. 132:555-574.

Lubbock, R. and W.B. Amos. 1981. Removal of bound calcium from nematocysts causes discharge. Nature 290:500-501.

Lubbock, R., B.L. Gupta and T.A. Hall. 1981. Novel role of calcium in exocytosis: Mechanism of nematocyst discharge as shown by x-ray microanalysis. Proc. Natl. Acad. Sci. USA 78:3624-3628.

Mariscal, R.N. 1974. Nematocysts. *In* L. Muscatine and H.M. Lenhoff (eds.). *Coelenterate Biology, Reviews and Perspectives*, pp. 129-178, Academic Press, New York.

Mariscal, R.N. 1984. Cnidaria: Cnidae. *In* J. Breiter-Hahn, A.G. Matoltsy, K.S. Richards (eds.). *Biology of the Integument*, Vol. 1, Invertebrates, pp. 57-68, Springer Verlag, Berlin.

Mattern, C.F.T., H.D. Park and W.A. Daniel. 1965. Electron microscope observations on the structure and discharge of the stenotele of hydra. J. Cell Biol. 27:621-638.

Möbius, C. 1866. Ueber den Bau, den Mechanismus und die Entwicklung der Nesselkapseln einiger Polypen und Quallen. Hamburg, Naturwiss. Verein Abhandl. V:1-22.
Picken, L.E.R. 1953. A note on the nematocysts of *Corynactis viridis*. Quart. J. Microsc. Sci. (N.S.) 94:203-227.
Picken, L.E.R. 1957. Stinging capsules and designing nature New Biol. 22:56-71.
Picken, L.E.R. and R.J. Skaer. 1966. A review of researches on nematocysts. Symp. Zool. Soc. London 16:19-50.
Reisinger, E. 1937. Der Entladungsvorgang der Nesselkapseln. Zool. Anz. Suppl. 10: 311-315.
Robson, E.A. 1953. Nematocysts of *Corynactis:* The activity of the filament during discharge. Quart. J. Micr. Sci. (N.S.) 94:229-235.
Schneider, K.C. 1900. Mitteilungen über Siphonophoren. V. Nesselzellen. Arb. Zool. Inst. Wien 12:133-242.
Schulze, P. 1917. Neue Beiträge zu einer Monographie der Gattung Hydra. Arch. f. Biontol. IV, 2:33-119.
Schulze, P. 1921/1922. Der Bau und die Entladung der Penetranten von *Hydra attenuata* Pallas. Arch. Zellforsch. 16:383-438.
Skaer, R.J. and L.E.R. Picken. 1965. The structure of the nematocyst thread and the geometry of discharge in *Corynactis viridis* Allman. Phil. Trans. Roy. Soc. London 250:131-164.
Slautterback, D.B. 1963. Cytoplasmic microtubules. I. Hydra. J. Cell Biol. 18:367-388.
Tardent, P. and Th. Holstein. 1982. Morphology and morphodynamics of the stenotele nematocyst of *Hydra attenuata* Pall. (Hydrozoa, Cnidaria). Cell Tissue Res. 224:269-290.
Tardent, P., Th. Holstein, J. Weber and M. Klug. 1985. The morphodynamics and actions of stenotele nematocysts in *Hydra*. Arch. Sc. Genève 38:401-418.
Toppe, O. 1910. Untersuchungen über Bau und Funktion der Nesselzellen der Cnidarier. Zool. Jahrb. Abt. Anat. 29:191-280.
Trembley, A. 1744. Mémoires pour servir à l'histoire d'un genre de polypes d'eau douce, à bras en forme de cornes. Verbeek, Leyde.
Weber, J. 1987. Chemistry of hydra nematocysts. This volume.
Weill, R. 1930. Essai d'une classification des nématocystes des Cnidaires. Bull. Biol. Fr. Belg. 64:141-153.
Weill, R. 1934. Contribution à l'étude des cnidaires et de leurs nématocystes. Trav. Stat. Zool. Wiméreux 10/11: 1-701.

Werner, B. 1965. Die Nesselkapseln der Cnidaria mit besonderer Berücksichtigung der Hydroida. I. Klassifikation und Bedeutung für die Systematik und Evolution. Helgoländer wiss. Meeresuntersuch. 12:1-39.

Westfall, J.A. 1965. Nematocysts of the sea anemone *Metridium*. Amer. Zool. 5:377-393.

Westfall, J.A. 1970. The nematocyte complex in a Hydromedusan *Gonionemus vertens*. Z. Zellforsch. mikrosk. Anat. 110: 457-470.

Will, L. 1909. Die Klebekapseln der Aktinien und der Mechanismus ihrer Entladung. Sitz.ber. Naturf. Ges. Rostock, Abhandl. N.F. 1:65-103.

Will, L. 1914. Kolloidale Substanz als Energiequelle für die mikroskopischen Schusswaffen der Coelenteraten. Abhandl. Preuss. Akad. Wiss. Berlin 1:1-28.

Will, L. 1919. Die Volumenreduktion der Nesselkapseln bei der Explosion und infolge "Alterns". Anat. Hefte, Abt. I, 57:485-549.

Yanagita, T.M. and T. Wada. 1954. Effects of trypsin and thiocollate upon the nematocysts of the sea anemone. Nature (London), 173:171.

Yanagita, T.M. and T. Wada. 1959. Physiological mechanism of nematocyst responses in sea anemone. VI. A note on the microscopical structure of acontium with special reference to the situation of cnidae within its surface. Cytologia 24:81-97.

18

# NEMATOCYST VENOMS AND TOXINS

David A. Hessinger

Department of Physiology and Pharmacology
School of Medicine
Loma Linda University
Loma Linda, California 92350

## SUMMARY

Although many people are stung each year by marine cnidarians, proportionally few stingings are fatal. Nonetheless, nematocyst venoms are among the most potent of all known venoms. Detailed information about nematocyst venoms and toxins is presently confined to only a few species owing in part to the use of less than optimum procedures for collecting and preparing the venom. The relative merits and liabilities of isolating nematocyst venoms and toxins by four different procedures are discussed. Currently the venoms from the sea anemone (Aiptasia pallida) and the Portuguese Man-of-War (Physalia physalis) are the most completely characterized biochemically and in terms of their cellular and molecular mechanisms of action. Unusually steep mortality curves, very short lethality times, and similar pharmacological effects for venoms from at least five species representing all three classes of cnidarians suggest similar overall mechanisms of action. Hydrozoan and scyphozoan venoms, in particular, seem to share common features, with each characterized primarily by a major, single-component stoichiometric type of cytolysin. Anthozoan venoms, on the other hand, are considerably different and characterized by a multi-component catalytic cytolysin plus neurotoxins. Evidence is presented suggesting that the full potency of the physiological effects of nematocyst venoms are the result of cooperative and synergistic interactions between various components of the venom.

## I. INTRODUCTION

In this paper I evaluate the findings most relevant to understanding the nature\and mechanism of action of nematocyst venoms and toxins at the cellular and molecular levels. I especially attempt to identify unresolved questions and underlying mechanisms, and at the risk of grossly over-simplifying, to propose how nematocyst venoms differ among different cnidarian taxa. The first section of this review is introductory and deals with aspects of nematocyst structure and function, public health of nematocyst stings, and the terminology and general principles of toxinology. The second section reviews selected aspects of the literature on nematocyst venoms and toxins, including methods of obtaining venoms and toxins, their physiological effects, physico-chemical and enzymic properties, lethality and cellular mechanisms of action. The third section attempts to distill current knowledge into general principles and mechanisms of action, and proposes characteristics of nematocyst venoms from different taxa.

### A. Structure and Function of Venomous Nematocysts

The cnidarians, which include the jellyfish, sea anemones, corals, and hydroids, are the most venomous phylum. Each of the approximately 10,000 species in this phylum possess tiny eversible and venomous organelles called nematocysts, which are mainly used to capture prey. There are more than 24 different morphological types of nematocysts (Mariscal, 1974). The morphological differences that distinguish different types of nematocysts imply specific functions related to capture of prey or defense, such as penetrating, holding, adhering, and entangling (Purcell and Mills, 1988). It is generally assumed that only the penetrant nematocysts are venomous and capable of stinging. For example, the fully everted tubules of holotrichous isorhizas from <u>Physalia physalis</u> (Portuguese Man-of-War) extend more than fifty times the diameter of the nematocyst capsule (Cormier and Hessinger, 1980a) and these nematocysts (Ionnides and Davis,1965) and those from the "sea wasp" (<u>Chironex fleckeri</u>; Kingston and Southcott, 1960) can penetrate into vascularized tissues of human flesh.

It is not yet known in which intranematocyst compartment of the undischarged nematocyst the venom is stored or by which route the venom is released upon discharge. Isolated Man-of-War nematocysts discharged into 1% glutaraldehyde and immunohistochemically stained with polyclonal antibodies to the Man-of-War venom show antigenic reactivity along the length of the everted tubule (Jewell and Hessinger, in preparation) and a similar pattern of release is indicated by discharging nematocysts of other species into red blood cell suspensions (Klug et al., 1988). In addition, the barbs on the predominant holotrichous isorhizas of Man-of-War fishing tentacles appear to be hollow (Hessinger and Ford, 1988). On the other hand, when isolated mastigophores from the acontia tissue of the sea anemone Aiptasia pallida are osmotically discharged into acidic medium the soluble contents of discharging nematocysts are observed to precipitate at the tip of the everted tubule (Blanquet, 1968). Hence, both modes of venom delivery may occur depending upon type of nematocyst.

## B. Public Health Aspects

Our knowledge of cnidarian stinging incidents on humans on a global scale is meager and is based on scattered reports provided by lifeguards or published as clinical case studies. Most of these reports are from American and Australian workers dealing with the hydrozoan, the Portuguese Man-of-War, and certain scyphozoan "jellyfish," particularly the sea wasp. The information coming from the Southeastern Atlantic and Gulf coasts of the United States is extremely deficient owing to the fact that such stings are non-reportable public health illnesses.

On the basis of published reports Halstead (1965) notes that there have been 502 reported stingings with 46 fatalities world-wide from the mid-1800's to 1960. This gives a crude fatality rate of approximately 9%. However, the results of Southcott (1952, 1959) and Barnes (1960) indicate that only a small fraction of the stingings are ever published in the medical science literature. Many deaths listed as drownings may in fact be due to stings from either the Man-of-War or the sea wasp. To my knowledge there have been no more than six reported fatalities by cnidarians on U.S. coasts since 1950, and two of these accounts were only reported in local newspapers. All of these fatalities have occurred in the southeastern U.S. and have been caused by Physalia, believed to be the most harmful cnidarian in the western hemisphere.

The stings of Physalia are not usually fatal; nonetheless, many hundreds of people are stung annually on U.S. Gulf of Mexico and Atlantic beaches. Such non-fatal encounters produce excruciating pain for periods lasting from 15 minutes to several hours with pain rapidly becoming more generalized. In severe cases the acute symptoms include difficulty in breathing, apparently due to tetany of abdominal muscles and diaphragm, convulsions and cardiac irregularities (Russell, 1966). Acute renal failure along with systemic hemolysis and attendent hyperkalemia have been reported (Spielman et al., 1982; Guess et al., 1982). In addition, delalyed cutaneous eruptions both at (M'ansson et al., 1985) and remote from (Matosow, 1980) the site of envenomation have been reported. Consequently, individuals previously exposed to the Physalia sting, those with histories of cardiovascular disorders and those who are generally immunologically hypersensitive (e.g., asthmatics) are likely to be at the greatest risk.

Current therapy for victims is entirely supportive; no antidotes or antisera are available. Neither are there adequate preventative methods other than avoidance for individuals likely to be exposed to a high risk of Physalia encounter, such as professional divers, environmental scientists, lifeguards and commercial fishermen.

While the sting from the Pacific "bluebottle" (Physalia utriculus) is considered less dangerous than the larger Atlantic Man-of-War (P. physalis), the Indo-Pacific "sea wasp" (C. fleckerii) is considered to inflict the most dangerous cnidarian sting (Halstead, 1978). At least 60 deaths are attributable to the sea wasp near Brisbane, Australia during a recent 25 year period (Cleland and Southcott, 1965). As with Man-of-War most stings from the sea wasp painful but minor injuries. Such stings produce a "weal and flare" cutaneous reaction. There is considerable edema and subsequent subcutaneous necrosis, slow healing and long-term discoloration or permanent scarring, with susceptibility to infection for years (Barnes, 1966). With fatal stings the victim collapses within 20 minutes and death appears to be due to cardiac arrest. Maintainance of respiratory exchange and circulation by artificial means have resusitated some severly stung victoms (Barnes, 1966). A commercial sea wasp antivenin (Commonwealth Serum Laboratories, Parkville, Victoria, Australia) is available and is indicated whenever a victim has received first aid and continues to have difficulty breathing, swallowing or continues in severe pain (Halstead, 1978).

## C. Terminology and General Principles

Mariscal, in his 1974 review, noted that most of the reports on nematocyst venoms or toxins are difficult to harmonize with others in this field and often it is not certain whether the toxins are of the nematocyst origin. Even in those cases claiming that the toxins were extracted from pure nematocysts, there is some doubt concerning the purity and stability of these preparations. Perhaps this ambiguity would be resolved if (i) a definition of the nematocyst venom and toxins were agreed upon, and (ii) reliable methods were available to obtain and to study these venoms.

## 1. Definitions and Terms

Classically, venoms are defined as toxic secretions intended for injection into other animals for use in either capturing food, aggression, or defense. In the case of cnidarians the venom refers to soluble material originally contained within undischarged penetrant nematocysts. It does not include or refer to cytoplasmic substances that might also be toxic. Neither does the term, nematocyst venom, necessarily refer to any single toxic substance within the nematocysts, but to all the soluble materials extruded from the nematocyst at discharge regardless of whether the individual components are toxic alone or not.

The term, toxin, must also be used precisely in order to distinquish it from poison. In general, toxins are defined as poisonous substances of biological origin. Toxins confer adaptive advantage to the organisms employing them by exerting adverse biological effects on other organisms. In the context of this review, the term "toxin" implies a specific anatomical site of origin, a unique mode of deployment, and a specific biological function. With respect to studying cnidarian toxins it is essential that care be taken to distinguish between those toxins that originate within nematocysts as constituents of the nematocyst venom, and those poisonous substances that originate outside the nematocysts as constituents of the tissues. Whether poisonous substances of extra-nematocyst origin should be termed "toxins" or simply "poisons" may be difficult to determine since it depends upon whether these substances assist in capture prey or defense or whether they exist as normal tissue components which are only incidentally harmful to other organisms.

Finally, it is helpful to use such terms as potency and specificity when comparing different nematocyst toxins with each other or with toxins of other origins. Potency is a quantitative term that refers to effective toxicity or lethality. Lethality is commonly expressed in terms of $LD_{50}$ values in standardized animals, such as white mice, and is expressed in units of μg/kg body weight. Analogous units may be used to express toxicity in terms of physiological or biochemical effects. Specificity is a qualitative term that is not expressed in terms of defined units but refers to relative restriction of function and/or site of action and is, therefore, difficult to objectively measure.

## 2. General Principles

Before proceeding with an evaluation of nematocyst toxins some additional generalizations regarding toxins may be of use.

a. Most Toxins Target Plasma Membranes. There are several reasons why most toxins have as their primary target sites the plasma membranes of susceptible cells. (i) The plasma membrane is the most accessible part of a target cell, especially if the toxin is macromolecular and/or hydrophilic. (ii) Relatively less toxin is needed to saturate binding sites at a given density on a two-dimensional cellular surface than at the same concentration within the three-dimensional cytoplasmic space enclosed by that cellular surface. Thus, the amount of toxin needed to exert its effect is theoretically reduced when it acts on the surface of the target cell. (iii) Many vital functions which are mediated by plasma membranes serve as potential targets. Such functions include excitability, regulation of metabolism, maintenance of cell integrity, transport, and recognition. Of these functions excitability, regulation, and cell integrity are most often targeted by toxins since inhibition or disruption of these functions is most likely to have immediate deleterious or lethal effects.

b. Most Toxins are Proteinaceous. This is true for several reasons. (In cases where the toxin is not a protein the target site is likely to be a protein.) (i) Protein toxins are two to five orders of magnitude more lethal than non-proteinaceous poisons. (ii) Proteins can be designed to interact specifically with any other kind of molecular site. (iii) Some protein toxins may possess enzymatic activity in order to amplify the extent of their biological effects due to a turnover of target-derived substrates. (iv) Some protein toxins, due to their large size and three-dimensionsal structure, may be able to span the plasma membrane to affect membrane permeability and/or cytoplasmic processes. (v) Some proteins are capable of undergoing functional and conformational changes from inactive to active states in response to changes in extrinsic conditions. Thus, a toxin could be inactive but stable within the nematocyst, and become active and labile upon release or upon contact with a target membrane. Venom toxins, in particular, are almost always proteins. Furthermore, venom toxins are not likely to harm the predator when ingested with envenomated prey since protein toxins are not likely to survive the action of digestive proteases.

c. Venom Toxins Act Rapidly. They are designed either to have an immediate defensive effect to stun or induce pain in a potential predator or to have an immediate offensive effect to kill or paralyse prey.

## II. NEMATOCYST VENOMS AND TOXINS

The following evaluation is presented in terms of the two difficulties that have most consistently hampered research of nematocyst venoms: (i) development of methods for obtaining nematocyst venom free of cytoplasmic components, and (ii) development of suitable assays from which a systematic study of the mechanism of toxicity could be directed.

### A. Methods of Obtaining Nematocyst Venom

Four general methods have been used in attempts to obtain nematocyst venoms and toxins.

#### 1. Extracts of Whole Animals and Tentacles

Possibly because mature nematocysts are fragile structures and very prone to discharge by slight mechanical shock workers have often attempted to study nematocyst venoms by extracting toxic substances from the whole cnidarian tissue. Richet (1902; 1903) and Richet and Portier (1936) isolated three pharmacologically active extracts from whole tissues of cnidarians: (i) an aqueous extract termed "hypnotoxin," which caused cardiac arrest; (ii) an alcohol extract termed "thalassin," which caused the release of histamine; and (iii) a glycerin extract termed "congestin," which induced anaphylaxis. None of the active components of these extracts have been chemically identified.

Several quarternary ammonium bases have been identified in extracts of whole sea anemone (*Actinia equina*; Ackermann *et al*., 1923; 1924; 1953). Welsh and Prock (1958) listed many identifiable quaternary amines in tissue extracts from a variety of cnidarians, but found tetramine, which has toxic properties, only in hydra. In addition, Welsh (1956; 1960) reported finding 5-hydroxy-tryptamine (5-HT or serotonin), a powerful pain producer, histamine releaser, and neurotransmitter, in all cnidarians tested. The highest levels of 5-HT were found in so-called nematocyst-rich tissues (acontia and tentacles). Mathias *et al*., (1960), however, found that the levels of 5-HT in the tissues devoid of nematocysts, in *Calliactis parasitica*, were as high as that found in the tentacles. These workers also found a histamine-releasing substance from a sea anemone while Unvas (1960) described a similar substance from a jellyfish (*Cyanea*).

In addition to the many low molecular weight chemicals which have been identified in cnidarian extracts, researchers have more recently found toxic activity associated with various protein fractions. The earliest work of this type (Kline and Waravdekar, 1960; Shapiro, 1968a) followed the growing conviction now held by most workers in the field that the nematocyst venoms are proteinaceous whereas the low molecular weight pharmacologically active compounds of the tissue extracts represent materials from outside the nematocysts. By far, most of the work with proteinaceous extracts of whole tissues has involved sea anemones. This body of work is reviewed elsewhere in this volume by Kem (1988).

Shapiro pioneered the work on extracted, protein neurotoxins. He began with an acetone extract of the tentacles of the sea anemone, *Condylactis gigantea*, and purified a single, homogenous, low-molecular weight, basic protein (1968a) which transformed action potentials in lobster ventral nerve cord giant fibers into cardiac-like potentials (1968a,b). The neurophysiological action of the protein on nerve membranes was later elucidated by voltage clamp experiments (Narahashi *et al*., 1969). In spite of the toxic and neurophysiological activity associated with the extracted protein, it has never been shown to be a component of nematocyst venom. At best, the extraction method can isolate only one nematocyst toxin at a time. At worst, the extraction method can yield pharmacologically active substances of non-nematocyst origin (Mathias *et al*., 1960).

## 2. "Milked" Venom

By this method, live isolated tentacles of *Chironex* (Barnes, 1967) are electrically stimulated to discharge their nematocysts into and through a specially treated amniotic membrane. Nematocyst tubules which fully penetrate the membrane emit their venom on the opposite side of the membrane from which it is collected by rinsing. This procedure simulates the stinging reaction of the cnidarian against its prey, but in this case only small quantities of emitted venom can be collected and analyzed.

The venom collected in this manner is very toxic and stable (Barnes, 1967) but it is quite conceivable that some of the biologically active, components of the venom are removed by adsorption to biological moieties on the surface of the membrane. It is widely known that toxic components of many venoms, including nematocyst venoms (Hessinger and Lenhoff, 1973b; Lim and Hessinger, 1979), bind to cell surfaces.

### 3. Extracts of Isolated Nematocysts

Some recent workers have endeavored to avoid the ambiguities associated with crude extracts of tentacles and whole animals, and have sought to isolate intact nematocysts from which the toxic contents could be extracted. In general, the nematocysts are isolated from disrupted nematocyst-bearing tissue and then disintegrated to release their venomous contents. Although these methods ensure that the nematocyst contents are included in the final nematocyst extract, the toxicites of many extracts have tended to be low and unstable. Possible reasons for this are described below.

a. Several Types of Nematocysts Are Obtained (Table I). Endean *et al*. (1969) reported that isolated nematocyst preparations from the sea wasp consist of: 43% microbasic mastigophores, 42% atrichous isorhizas, 13% microbasic euryteles, and 2% holotrichous isorhizas. However, histological studies of the skin of human victims indicate that only the microbasic mastigophores are likely to be of any clinical significance (Kingston and Southcott, 1960). Burnett *et al*. (1968) and Blanquet (1970) have also reported four different types of nematocysts in their preparations of nematocysts from the tentacles of *Chrysaora quinquecirrha* (the sea nettle) and Lane (1960) reported that *Physalia* tentacles yield two distinct sizes of holotrichous isorhizas. Recently Burnett *et al*. (1986) separated the two sizes of *Physalia* nematocysts by flow cytometry, but the method is not suitable for obtaining preparative quantities of nematocysts. It is, however, relatively easy to separate large quantities of the two sizes of isorhizas in *Physalia* by differential centrifugation (Tamkun and Hessinger, 1981).

b. Immature Nematocysts May Be Isolated Preferentially. Mature nematocysts are fragile and notoriously susceptible to discharge. Methods which disrupt nematocyst-bearing tissues by techniques such as homogenization (Phillips and Abbott, 1957) or pressing the tissue through mesh screens (Lane and Dodge, 1958; Burnett *et al*., 1968; Banquet, 1970) may discharge mature nematocysts while yielding mostly immature nematocysts which contain "unripe" venoms (Barnes, 1967).

c. Isolated Nematocysts Are Often Contaminated. Nematocysts are often isolated with adhering cytoplasmic materials. Uncontaminated nematocysts when sedimented, however, are pure white in appearance. Lane (1960) noted that preparations of isolated Physalia nematocysts have the appearance of "grey putty." Such nematocysts, when veiwed by phase-contrast microscopy or by scanning electron microscopy (Cormier and Hessinger, 1980) are contaminated with "fibrillar baskets" which are attached to the capsules. These can be removed with proteolytic enzymes.

Lane (1961) found his extract of isolated Physalia nematocysts to contain a heterogenous mixture of small peptides. This extract also exhibited substantial proteolytic activity and showed rapid loss of toxicity (1967). Nematocysts prepared from Stomolophus are contaminated with pigmented cytoplasmic debris (Toom and Chan, 1972a) and yield extracts possessing proteolytic activity (Toom and Chan, 1972b). Bodansky and Rose (1922) have shown that both the nematocyst-bearing and nematocyst-free tissues of Stomolophus possess proteolytic activity. Cytoplasmic contaminants (Goldner et al., 1969) and proteolytic activity (Burnett and Calton, 1974a; Lal et al., 1981a; Calton and Burnett, 1982; 1983) characterize nematocysts prepared by conventional methods from both the sea nettle Chrysoara and from the Man-of-War (Lal et al., 1981b).

We have not been able to detect proteolytic activity in venom obtained from the uncontaminated nematocysts from the sea anemone, Aiptasia pallida (unpublished observations) nor in toxic extracts of nematocysts from Physalia after they have had their fibrillar baskets removed (Tamkun and Hessinger, 1981). We, therefore, believe that reports of proteolytic activity associated with venomous nematocyst contents actually reflect the presence of contaminating substances of non-nematocyst origin. As a general rule, venom preparations exhibiting proteolytic activity possess low toxicity (Table I).

d. Disintegration Techniques Solubilize Contaminants. The disintegration techniques used to release the soluble, venomous contents of the isolated nematocysts also solubilize some of the proteins from the nematocyst capsules, including any adhering contaminants. We have detected this with SDS-polyacrylamide electrophoresis (unpublished observations). Using nematocyst capsules from Aiptasia that were discharged osmotically (see next section) and washed extensively to free them of all venom, we monitored the total amount and variety

TABLE I. Lethalities of nematocyst venoms

| Genus (cnida type)[a] | $LD_{50}$ (µg/kg) | Preparations[b] | References |
|---|---|---|---|
| *Stomolophus* (ME) | | | Toom and Chan, 1972a |
| | 5,300 | NE | Toom and Chan, 1972a |
| *Chrysaora* (ME/HI/HA; 55:25:20) | | | Burnett *et al.*, 1968 |
| | 2,000 | NE | Burnett and Goldner, 1970 |
| | 2,500 | NE | Burnett and Gould, 1971 |
| | 1,400 | NE | Burnett and Calton, 1973 |
| | 2,780 | NE | Cobbs *et al.*, 1983 |
| *Chironex* (MM/ZI/ME/HI; 43:42:13:2) | | | Endean *et al.*, 1969 |
| | 250-400 | MV | Olsen *et al.*, 1984 |
| | 12.5 | TE | Crone and Keen, 1971 |
| | 11 | TE | Calton and Burnett, 1986 |
| *Aiptasia* (MM) | | | Blanquet, 1968 |
| | 136 | OD | Hessinger and Grove, 1979 |
| | 300[c] | OD | Hessinger *et al.*, 1973 |
| Neurotoxin-II | 11[c] | purified | Hessinger *et al.*, 1973 |
| *Physalia* (HI) | | | Cormier and Hessinger, 1981 |
| | 2,000 | NE | Lane, 1960 |
| | 750 | NE | Garriott and Lane, 1969 |
| | 700 | NE | Calton and Burnett, 1973 |
| | 700 | NE | Burnett *et al.*, 1986 |
| | 145 | NE | Tamkun and Hessinger, 1981 |
| | 70 | NE | Flowers and Hessinger, 1981 |
| | 75 | NE | Loredo *et al.*, 1985 |
| | 32 | NE | Hessinger, 1988 |
| Physalitoxin | 200 | purified | Tamkun and Hessinger, 1981 |

[a]AI, atrichous isorhiza; HA, homotrichous anisorhiza; HI, holotrichous isorhiza; ME, microbasic euryteles; MM, microbasic mastigophore.

[b]MV, milked venom; NE, nematocyst extract; MV, milked venom; OD, osmotically discharged venom; TE, tissue extracts

[c]Minimum lethal dose ($LD_{100}$) in *Uca*.

of materials released from the capsules following disintegration by different techniques. We found that brief sonication solubilized the capsular materials most extensively; less effective in declining order were French pressing at 16,000 psi, grinding with mortar and pestle, and motor-driven homogenization. Hence, caution should be used in evaluating venoms prepared in these ways.

## 4. Venom Obtained by Osmotic Discharge

The acontiate sea anemones possess acontia threads which contain high densities of nematocysts. Exposure of excised acontia threads to 1 M glycerol or sodium citrate causes the undischarged nematocysts to be extruded from the threads (Yanagita, 1959; Blanquet, 1968). The threads are removed by filtration through fine nylon netting and the remaining suspension of nematocysts are washed repeatedly in the cold by centrifugation. These can then be caused to discharge by placing them in a low ionic strength medium. The released venom is collected as the supernatant phase by sedimenting the discharged capsules. The venom collected in this manner is quite toxic (Table I) and stable (Blanquet, 1968; Hessinger *et al*., 1973). Unfortunately, this method can only be used to purify nematocysts from acontia.

As a model system for characterizing all of the components of a nematocyst venom the nematocysts from the acontia offer several advantages over other sources of nematocysts. (i) Acontia threads from *Aiptasia* possess only the large microbasic mastigophores, which contain the venom, and the much smaller holotrichous isorhizas, which contain no detectable venom (unpublished observations). These nematocysts are easily separated by either differential (Hessinger and Lenhoff, 1973a) or sucrose density gradient centrifugation. (ii) Nematocysts extruded from the acontia are free of adhering cytoplasmic materials. (iii) All of the isolated nematocysts can be discharged by hypotonic solutions indicating that only functionally mature nematocysts are extruded. This method of releasing the venom is gentle and closely simulates the natural mechanism of discharge, thereby ensuring that no capsular components are solubilized. (iv) Many acontiate anemones, such as *A. pallida*, can be cultivated and asexually cloned *en masse* in the laboratory under optimal and controlled conditions thereby ensuring a convenient and continual supply of fresh venom (Hessinger and Hessinger, 1981).

## B. Physiological Effects

A wide range of physiological and pharmacological effects have been reported for various nematocyst venoms. Because so many different physiological systems, experimental animals and methods of preparing the toxic materials and venoms have been used it is difficult to identify generalized underlying mechanisms of action common to any taxonomic group of cnidarians. To simplify comparisons the physiological effects reported here will be grouped under general categories: cardiovascular effects, neuromuscular effects, and effects on transport. Most of these effects, however, if not all, can be explained in terms of molecular effects on membrane or function (Section II. F.).

### 1. Cardiovascular Effects

Chrysaora venom depolarizes frog ventricles (Shyrock and Bianchi,1983) and dog Perkinje fibers (Kleinhaus et al., 1973). Burnett and Goldner (1969) have not only observed irregularities in cardiac conduction, but also an increase in coronary vascular resistance and arterial blood pressure. In addition, Burnett et al. (1985) have demonstrated that the venom can increase contractile force of guinea pig atria. Reports on Chironex venom, however, are conflicting. Whereas Endean (1987) has observed a venom-induced increase in contractile force of guinea pig atria, others have shown that the venom can decrease contractile force of whole rat (Endean and Noble, 1971) and guinea pig hearts (Turner and Freeman, 1969). Other parasympathomimetic effects, such as a decrease in heart rate, coronary vasoconstriction (Turner and Freeman, 1969), progressive heart failure (Endean and Henderson, 1969) and atrio-ventricular block (Endean and Noble, 1971), have also been seen. Illustrating its dual effect on cardio-vascular function, Endean and Noble (1971) reported that the venom initially increases blood pressure but later profoundly decreases it. Freeman (1974) suggested that the atrial depolarization and reduction in transmembrane action potentials observed by her is linked to an alteration of ion permeability.

Physalia venom has also produced conflicting results. Hastings et al. (1967) observed an increase in cardiac output and a strong increase in blood pressure in the dog, both sympathomimetic effects. However, mixed responses, such as an increase in ventricular repolarization and subsequent atrioventricular block in the rat (Larsen and Lane, 1966), and an early depression of atrial strip contractile force followed by later enhancement (Burnett et al., 1985), have also been shown. Sympatholytic effects include a decrease in arterial pressure in rats (Garriott and Lane, 1969), occasional generalized hypotension and accompanying respiratory distress (Burnett et al., 1975) as well as skeletal muscle vasodilation in the dog (Loredo et al., 1985). This latter effect is due to venom-induced relaxation of rabbit arterial rings precontracted with norepinephrine (Loredo et al., 1986).

### 2. Neuromuscular Effects

Chrysaora venom decreases the action potentials of rat ventral caudal nerve and skeletal muscle (Burnett and Goldner, 1970) as well as of frog sartorius muscle (Shryock and Bianchi, 1983). Chironex venom blocks nerve conduction on sciatic nerve (Endean and Noble, 1971) and semi-purified, high molecular weight fractions of the venom elicit contraction in rat diaphragm and guinea pig ileum and vas deferens smooth muscle (Endean, 1987). Physalia venom also blocks nerve conduction on frog sciatic nerve (Larsen and Lane, 1970) and contracts isolated guinea pig ileum (Garriott and Lane, 1969).

### 3. Effects on Membrane Transport

Chrysaora venom non-competitively inhibits glucose transport in brushborder membranes from guinea pig small intestine (Watrous and Walsh, 1976) and decreases uptake and binding of calcium into isolated mitochondria (Calton et al., 1973) and into skeletal muscle sarcoplasmic reticulum (Calton and Burnett, 1973a). Chironex venom also inhibites calcium uptake in sacroplasmic reticulum from mouse striated muscle (Endean and Henderson, 1974) as does Physalia venom (Calton and Burnett, 1973a). Physalia venom also increases the rate of sodium influx into cells of frog skin (Larsen and Lane, 1970b).

## C. Physical and Chemical Properties of Purified Toxins

There is fairly uniform agreement in the literature since 1960 that the toxins of nematocyst venoms are proteinaceous (Mariscal, 1974). Of the several published attempts to fractionate and characterize nematocyst venoms only a few have been successful.

Physalia-derived toxins have been studied longer than other cnidarian toxins beginning with toxic tissue extracts

(Portier and Richet, 1902) from which the discovery of anaphylaxis was made. Lane et al. (1961) first attempted to fractionate Physalia nematocyst extracts by ascending paper chromatography and reported the presence of eight peptides. The presence of peptides is not surprising since he later found his venom extracts to have high proteolytic activity (Lane, 1967a). Attempts to fractionate Physalia venom by gel-filtration on G-200 (Calton and Burnett, 1973a) and by preparative gel electrophoresis (Burnett and Calton, 1974b) yielded eighteen and nine lethal fractions, respectively; none of which were hemolytic. More recently, a single, potently hemolytic toxin in Physalia venom was identified, purified and characterized although other, apparently non-toxic, components were also present (Tamkun and Hessinger, 1981). A monoclonal antibody to the lethal activity of Physalia venom has been prepared (Gaur et al., 1982). While it has been used to purify toxic fractions from Chironex venom (Olsen et al., 1984), from the venoms of Chrysaora and from two other species of scyphozoans, from brown recluse spider venom, and from cholera toxin (Olsen et al., 1985), it apparently has not yet been used to purify toxic components of Physalia venom.

Extracts of Chrysoara nematocysts yielded 25 lethal fractions by gel-filtration (Burnett and Calton, 1973), while immunochromatography yielded a fraction containing two major proteins having molecular weights of 100,000 and 190,000 exhibiting cardiotoxic, neurotoxic and enhanced lethal activities (Cobbs et al., 1983).

Attempts to isolate toxins from Chironex have started with "milked" venom, venom extracted from isolated nematocysts, and extracts of tentacles. Fractionations using "milked" venom have toxic activities associated with lower molecular weight fractions (10,000-30,000, Baxter and Marr, 1969; 20,000-70,000, Olson et al., 1984), while extracts of isolated nematocysts (50,000 and 150,000, Calton and Burnett, 1986; 150,000 and 600,000, Endean, 1987) and extracts of tentacles (a hemolytic fraction of 70,000 and a toxic fraction of 140,000, Crone and Keen, 1969; 1971) have toxicity associated with higher molecular weight fractions. These results strongly suggest that the biologically active, high molecular weight components found in tentacle and nematocyst extracts may be lost during the "milking" procedure.

Nearly 80% of the constituent amino acids of Aiptasia venom are glutamic and aspartic acid (Blanquet, 1968; Phelan and Blanquet, 1985). Blanquet (1968) reported finding two proteins in Aiptasia venom; one toxic to fiddler crabs (Uca) with a molecular weight of 130,000, and the other not toxic with a molecular weight of 30,000. More recent studies reveal that Aiptasia venom is more complex. At the present we have been able to identify five toxic constituents of *Aiptasia* venom of which four have been isolated and partially characterized. These five proteins account for all of the known toxic effects of the venom. Currently, *Aiptasia* venom is physically and chemically the best characterized nematocyst venom and the venoms from *Aiptasia* and *Physalia* are the only two from which individual toxins have been purified and characterized. These are listed along with some of their physical and chemical properties on Table II.

### D. Non-toxic Proteins and Enzymic Activities

Several non-toxic proteins have been found in association with venoms extracted from isolated nematocysts while most of these non-toxic proteins have been identified on the basis of enzymatic activity. One protein has been identified on the basis of the absence of aromatic amino acids (Weber *et al*., 1988). Although several non-toxic enzymatic activities have been listed only a few of these enzymes have been isolated, and these all constitute small portions of the venoms from which they were obtained. This low occurrance coupled with the fact that they were all extracted from nematocysts by ultrasonic disintegration and were isolated from a heterogeneous population of more than one type of nematocyst raises questions as to their origin and biological role.

Neeman *et al*. (1980) isolated an endonuclease with a molecular weight of 75,000 that constituted 0.8% of the protein of *Physalia* venom. Lal *et al*. purified a collagenase both from extracted *Physalia* (1981b) and *Chrysaora* (1981a) nematocysts having molecular weights of 25,000 and 34,000, and constituting 0.6% and 0.4% of the total venom protein, respectively. In addition, acid and alkaline proteases have been found in association with *Chrysaora* venom having molecular weights of approximately 120,000-150,000 (Calton and Burnett, 1982) and 100,000 (Calton and Burnett, 1983), respectively. Although the acidic protease constitutes an estimated 1.6% of the total venom protein it was reported that the specific activity of the enzyme is ten-fold higher in the float, where there are few nematocysts, than with the isolated nematocysts (Calton and Burnett, 1982). For the alkaline protease it is not possible to calculate the percent of the total protein from the published data.

Is it possible that these enzymes are of extranematocyst origin or components of non-penetrant nematocysts that have been isolated along with penetrants? If they originate from penetrant nematocysts do they contribute to the toxic action of the venom, or might they be "left-overs" from nematocyst assembly that impliment post-ribosomal modifications

TABLE II. Physical and chemical properties of toxic proteins purified from nematocyst venoms.

| Genus/Protein | Mol. Wt. | Charge/pI | Sed. Coef. | Quat. Struct. | CHO | Shape | % Venom |
|---|---|---|---|---|---|---|---|
| A. *Aiptasia*: | | | | | | | |
| Co-lytic factor[a] | 98,000 | 4.5 | 4.5 S | monomer | + | prolate | 1.3 |
| Phospholipase $A_2$[a,b] | | | | | | | |
| α | 45,000 | - | - | monomer | ? | - | 1.9 |
| β | 43,000 | 8.8 | - | monomer | + | - | 3.0 |
| Direct lytic fraction[b] | 30,000 | basic | - | - | ? | - | - |
| Neurotoxin:[a,c] | | | | | | | |
| I | 15,000 | basic | - | monomer | ? | - | 12 |
| II | 12,500 | basic | - | monomer | ? | - | 10 |
| B. *Physalia*: | | | | | | | |
| physalitoxin[d] | 240,000 | subunits: 5.5 8.2 8.3 | 7.8 S | heterotrimer | + | prolate | 28 |

References: (a) Grotendorst and Hessinger, 1988; (b) Hessinger and Lenhoff, 1976; (c) Hessinger, Lenhoff and Kahan, 1973; (d) Tamkun and Hessinger, 1981.

of nematocyst precursors during development as suggested by Blanquet (1988) or might they be components of a system of nematocyst proteins that function to cause eversion of the tubule. Until proof of origin and evidence of function can be determined it is best if proposed roles for these materials be viewed with caution.

## E. Lethality

Nematocyst extracts and venom are commonly assayed for lethality on mice and/or crabs (e.g. fiddler crab, *Uca pugilator*). Toxicities are usually reported as the amount of venom per unit body weight needed to kill 50% of the animals within a specified period and are expressed as $LD_{50}$ values. Such $LD_{50}$ values for the more extensively studied nematocyst venoms are listed on Table I.

### 1. Mortality Curves

The $LD_{50}$ of *Aiptasia* venom on mice is similar to the more toxic species of cobras (family Elapidae) such as the kraits (e.g. *Bungarus*; Fischer and Kabara, 1967) and the mambas (e.g. *Dendroaspis*; Schwick and Dickgeisser, 1963) and the most toxic of the pit vipers (family Crotalidae) including the hundred pace snake (*Arcistrodon acutus*; Ouyang, 1957). Consequently, the venoms from *Physalia* and *Chironex*, which are the more potent nematocyst venoms, are among the most toxic of all known venoms. Several features of the mortality curves and of the lethal effects in injected test animals are common to nematocyst venoms from all three classes of cnidarians.

The mortality curves of nematocyst venoms from *Physalia* (Tamkun and Hessinger, 1981), *Aiptasia* (Hessinger and Grove, 1979), *Chironex* (Baxter and Marr, 1975) and *Chrysaora* (Cobbs *et al*., 1983) display extreme vertical slopes. This is due to the narrow dose range between the maximum non-lethal doses and minimal lethal doses. Such steep mortality curves are in contrast to the broad mortality curves that have been published for various microbial toxins and reptilian venoms (Ipsen, 1951). I suggest that broad mortality curves are characteristic of toxins which act on widely distributed sites having a broad range of binding affinities whereas the narrow and steep mortality curves of the nematocyst venoms are due to toxic effects on limited or specialized target sites having a narrow range of binding affinities.

The mortality curves of nematocyst venoms also exhibit a threshold effect in which a minimal dose must be exceeded before any lethal effect occurs. This suggests that either cooperativity occurs between the toxin molecules on the target cells to form lethal lesions or a critical number of lesions must be formed per target cell in order to produce the lethal effect.

### 2. Rate of Lethality

The time required to produce lethality by lethal doses of intravenously injected venom is fairly short for nematocyst venoms. Mice injected with $2LD_{50}$ doses of *Aiptasia* venom die within minutes and at five or more $LD_{50}$ doses they die within seconds (Hessinger and Grove, 1979). Multilethal doses of *Chironex* venom kill within minutes (Baxter and Marr, 1975). On the other hand, while Burnett *et al*. (1985) have reported that $3LD_{50}$ doses of *Physalia* venom requires an average of almost two hours to kill, we (Flowers and Hessinger, unpublished observations) find that one $LD_{50}$ dose of *Physalia* venom kills half the injected mice in an average time of approximately 30 minutes and $2LD_{50}$ doses kill the mice within seconds. Microbial exotoxins, in contrast, require on the order of days to kill mice injected with minimal lethal doses (Carpenter, 1965). The rapidity of the lethal effects of intravenously injected nematocyst venoms along with the observation that mice can tolerate up to $100LD_{50}$ doses of intradermally or subcutaneously injected *Chironex* venom without ill effect (Keen, 1970) suggests that the preferred target sites for the lethal factor(s) are on the cell surfaces of cellular components of the circulatory system.

### 3. Symptoms

Many of the symptoms exhibited by lethally injected mice are common to the different nematocyst venoms. Mice injected with $2LD_{50}$ or less of *Aiptasia* venom become lethargic and breathe irregularly. This is followed by partial paralysis of limb, neck and jaw muscles. Immediately before death violent convulsions and spasms of major skeletal muscles occur, sometimes causing animals to leap or jump vertically (Hessinger and Grove, 1979). Mice injected with *Chironex* venom also become lethargic and develop a progressive impairment of breathing movements followed by spasmodic jerking of the legs before death (Endean *et al*., 1969). They also show alterations in rate and depth of breathing and die in respiratory arrest (Keen, 1970) after showing terminal jumping and leaping (Baxter *et al*., 1972). At low doses of *Physalia* venom mice show no immediate symptoms but develop respiratory distress and abdominal convulsions within minutes. Mice then become quiescent for a dose-dependent period of time ranging from a few minutes to an hour. Following this interval, the mice either completely recover or expire beginning with a series of spasmodic convulsions involving the hind legs, causing them to hop uncontrollably. A bloody urine and bloody nasal discharge is often observed (Jewell and Hessinger, in preparation).

In the fiddler crab (Uca) Aiptasia venom induces quivering of the legs, violent motor activity with subsequent rigid paralysis, extensive autotomy (loss of legs), and progressive insensitivity to external stimuli (Blanquet, 1968; Hessinger et al., 1973). With Chrysaora venom there is also a severe quivering of the legs with legs becoming hyperextended, but no autotomy (Blanquet, 1972). With Physalia venom there is a vigorous contraction of the leg extensors leading to paralysis, but no autotomy (Lane and Dodge, 1958). Common to these three venoms is the quivering of the legs whereas violent motor activity and autotomy are only seen with the sea anemone venom. The anemone venom is unique among these venoms in that it contains at least two neuro-active proteins; one responsible for causing increased motor activity, and one for autotomy (Hessinger et al., 1973). The toxin causing increased motor activity is responsible for for than two-thirds of the lethality of the venom in fiddler crabs (Table II) and is representative of the typical sea anemone neurotoxin that delays "sodium" inactivation (Kem, 1988). The higher molecular weight cytolytic proteins in Aiptasia venom, on the other hand, only produce quivering of the legs and a subsequent slow, "sleep-like" death in crabs (Hessinger et al., 1973), whereas they are responsible for most of the lethality in mice (Hessinger and Grove, 1979).

### 4. Synergistic Actions

Some of the constituent nematocyst proteins from Aiptasia and Physalia venoms act synergistically to produce lethal effects. For Aiptasia venom synergy is discussed in detail in the next section (II. F. 1. a.) with regard to the mechanism of venom-induced hemolysis. For Physalia venom only one lethal toxin, physalitoxin PTX), has been detected and purified (Tamkun and Hessinger, 1981) and this toxin is both a major glycoprotein of the venom and an extremely potent hemolysin. During purification of PTX it was noted that while the specific hemolytic activity was increased 3.6-fold the specific toxicity (i.e. reciprocal of $LD_{50}$) was decreased by 28% (Table II). This discrepency suggests that the venom possesses a non-lethal component, in addition to

PTX, that somehow enhances the lethality of PTX. This possibility is supported further by the observation that the lethality of the venom is largely lost upon incubating the venom at 30°C for 30 minutes while the hemolytic activity is not affected until the incubation temperature is increased to 40°C (Flowers and Hessinger, unpublished). From the examples of these two venoms, each of which originate from different classes of cnidarians (Anthozoa and Hydrozoa) and different morphological types of nematocysts (Table I), I suggest that the constituent proteins of most nematocyst venoms exhibit an economy of action based upon synergistic interactions. In such an economy the nematocyst venom, rather than being merely a simple mixture of independently acting proteins, some of which are toxic, consists instead of a system of interacting proteins. On the one hand, these interacting proteins would stabilize each other while contained within the capsule and, on the other hand, enhance, and possibly diversify, the actions of the toxins to a variety of target tissues when released from the capsule.

In summary, nematocyst venoms are lethal and very potent, having steep mortality curves that exhibit threshold effects. The onset of symptoms and lethality is rapid. Common symptoms shared by the different nematocyst venoms in mice and crabs suggest that different nematocyst venoms act via generally similar mechanisms. Lethality is likely to involve synergistic actions of different venom components acting primarily on elements of the circulatory system.

### F. Cellular Mechanisms of Action

Most toxins, and in particular protein toxins, act by disrupting either the structural or functional features of biological membranes. For reasons discussed previously (section I. C.) membrane-acting toxins affect primarily cell integrity, cell excitability, or membrane-mediated regulation of cell metabolism. I believe that the major physiological and lethal effects of nematocyst venoms can be explained in terms of one or more of these three classes of toxic effects.

#### 1. The Cytolysins

The cytolysins alter cell membrane permeability and/or disrupt cell integrity. Cytolysis, or in the case of red blood cells, hemolysis, is one of the most commonly reported actions by nematocyst venoms since first reported in *Aiptasia* venom (Hessinger and Lenhoff, 1968). Hemolytic activity has been reported for nematocyst venoms from *Physalia* (Garriott and Lane, 1969), the sea wasp, Chironex (Crone and Keen, 1969), the sea nettle, Chrysaora (Burnett and Goldner, 1971), and hydra (Weber *et al.*, 1988).

In general, there are two types of cytolysins based upon differences in mechanisms of action: catalytic and stoichiometric. Both catalytic and stoichiometric cytolytic actions have been reported for nematocyst venoms, but in no case are both known to occur in the same venom.

a. Catalytic Cytolysins. The catalytic cytolysins (or "catalysins") alter membrane permeability and/or structure by chemically changing membrane structure. These cytolysins are enzymes or, more specifically, phospholipases that are capable of hydrolysing many times their number of phospholipid substrates. They act as either direct or indirect catalysins. The direct catalysins alter cell membranes by hydrolysing membrane phospholipids involved in maintaining the integrity of the plasma membrane and are typified by the phospholipases from insect and elapid snake venoms. The indirect catalysins hydrolyse extracellular phospholipids into detergent and membrane-altering products and are typified by the phosplipase $A_2$ of viperid snake venom. Lysis induced by direct catalysins is characterized by (i) time-courses of lysis that reach 100% regardless of venom dose, (ii) rectangularly hyperbolic dose-response curves of the rate of lysis (Hessinger and Lenhoff, 1973; Grotendorst and Hessinger, submitted), and (iii) the presence of phospholipase activity (Hessinger and Lenhoff, 1976; Tu, 1977).

At the present the cytolytic agents of Aiptasia, Physalia and Chironex venoms are the best characterized cytolysins of nematocyst origin. Of these only the nematocyst venom from the sea anemone A. pallida is known to possess a catalytic cytolysin, while the others possess stoichiometric cytolysins. Although it has been reported that Physalia venom possesses phospholipase activity (Stillway and Lane, 1971) the data is unconvincing and others have been unable to confirm this finding (Burnett and Calton, 1974a; Tamkun and Hessinger, 1981).

The phospholipase from Aiptasia venom, is of the $A_2$ type (Hessinger and Lenhoff, 1974) and can hydrolyse the phospholipids of osmotically ruptured red cell "ghosts" but cannot act on the phospholipids of an intact red cell membrane unless a second venom protein termed direct lytic factor (DLF) is present (Hessinger and Lenhoff, 1976). The mechanism of Aiptasia venom-induced hemolysis is rather complicated, involving the sequential interaction of three venom proteins and calcium with the red cell membrane as follows: (i) Initially, the DLF and phospholipase (Table II) bind tightly but independently to the red cell surface (Hessinger

and Lenhoff, 1973b) via trypsin- and neuraminidase-insensitive sites (unpublished).
(ii) The DLF alters the cell membrane so that phospholipids of the membrane become available to enzymatic attack by the phospholipase (Hessinger and Lenhoff, 1976).
(iii) Because the Km values for enzyme substrate and calcium are similar (Grotendorst and Hessinger, in preparation) we believe that the true substrate for the venom phospholipase is a phospholipid/calcium complex.
(iv) The co-lytic factor (CLF; Table II) interacts reversibly with the cell membrane (Ka = $3.9 \times 10^{-9}$ M) to remove lysophospholipids from the membrane as products of phospholipid hydrolysis ($4 \times 10^{4}$ molecules per molecule of CLF) following the action of the venom phospholipase (Grotendorst and Hessinger, submitted).
(v) It is believed that removal of the products of phospholipid hydrolysis produces local destabilization of the membrane around each DLF/phospholipase complex and causes the altered locus of membrane to become permeable to hydrated ions.

_Aiptasia_ venom is the only known direct lytic venom whose phospholipase is regulated; that is, requires an accessary protein (DLF) in order to hydrolyse phospholipids of intact membranes. Furthermore, other known venom phospholipases are notoriously stable, have broad pH optima for activity, and are fairly low molecular weight, basic proteins, whereas, the _Aiptasia_ venom phospholipase, is labile, has a narrow pH optimum, and is an acidic protein with a molecular weight of 43-45 kd (Table II). Such functional and physical differences characterize the differences between the _exocrine_ phospholipases of pancreatic and snake and insect venom origins, on the one hand, and the _regulated_ phospholipases that regulate endogenous ecosinoid biosynthesis in many tissues and also occur in _Aiptasia_ venom (Brockenhoff and Jensen, 1974).

b. Stoichiometric Cytolysins. The stoichiometric cytolysins interact with membranes and physically change membrane structure so as to alter membrane permeability. Such cytolysins may form size- and/or charge-specific channels, or less specific and larger membrane pores, or may disrupt membrane integrity by micellizing membrane components. Stoichiometric cytolysins characteristically exhibit (i) graded (i.e. dose-dependent) final extents of lysis and (ii) sigmoidal dose-response curves that exhibit threshold effects at low concentrations of lysin (Tamkun and Hessinger, 1981).

Stoichiometric cytolysins have been found in several nematocyst venoms, all of which seem to lack phospholipase activity, including venom from *Physalia* (Burnett and Calton, 1974a; Tamkun and Hessinger, 1981), *Chironex* (Keen and Crone, 1969; Crone, 1976), and hydra (Weber *et al*., 1988). In contrast, very little is known about the hemolytic activity of *Chrysaora* venom other than it is reported to be "very weak" (Burnett and Calton, 1977) if not negligible (Kelman *et al*., 1984). It has been shown, however, that the depolarizing effect of *Chrysaora* venom on excitable tissues can be explained by the demonstrated ability to form monovalent cation channels in black lipid membranes (Shryock and Bianchi, 1983). Thus, the best studied of these stoichiometric cytolysins are from *Physalia* and *Chironex*, but only the hemolysin from *Physalia* has been purified and characterized (Tamkun and Hessinger, 1981; Table II).

Physalitoxin (PTX) is an heterotrimeric glycoprotein and hemolytic toxin that constitutes about 25% of the Man-of-War venom protein. It is labile to heat and dithiothreitol (DTT). The susceptibility of PTX to DTT is one reason why Klug *et al*. (1988) did not find hemolytic activity in *Physalia* nematocysts discharged in DTT (another being that *Physalia* fishing tentacles do not possess eurytele nematocysts; Cormier and Hessinger, 1981; Hessinger and Ford, 1988). PTX is able to lyse 50% of a washed rat red cell suspension at $10^{-11}$ M, making it one of the most potent hemolysins known. Several lines of evidence point to PTX interacting with red cell membrane integral proteins (Lin and Hessinger, 1979). While PTX will not directly interact with membrane phospholipids it does bind free fatty acids, lysophospholipids and various nonionic detergents at sub-critical micellar concentrations (unpublished) indicating that PTX is a lipophilic protein (Tanford, 1980). In terms of its affinity for small amphiphiles and its hydrodynamic shape PTX resembles CLF from *Aiptasia* venom (see previous section), yet CLF is not lytic by itself and PTX is not able to substitute for CLF in combination with *Aiptasia* DLF and phospholipase (Grotendorst and Hessinger, in preparation).

The hemolysin from "milked" *Chironex* venom and extracts of *Chironex* tentacles have molecular weights of about 70,000 (Crone and Keen, 1969). Hemolytic activity co-chromatographs with dermonecrotic and lethal activities (Keen and Crone, 1969b). As with PTX, lytic activity is labile above 30°C and below pH 6.0 (Keen and Crone, 1969a). The hemolysin is not surface-active and does not interact with monolayers of either phospholipid or cholesterol (Keen, 1972), but does interact with monolayers of gangliosides (Keen, 1973). It is sensitive to dithiothreitol and is neutralized by ganglioside and N-acetylneuraminic acid (Crone, 1976). Unfortunately, like much of the published work with *Chironex* these results have not been expressed in units of protein concentration, thus it is not possible to estimate the hemolytic potency or lethality of the hemolysin.

2. The Neurotoxins

Toxins which preferentially act on excitable membranes may be referred to as neurotoxins. Since neurotoxins usually affect mechanisms associated with axonal conduction or synaptic transmission three classes of membrane-acting neurotoxins can be described: axonal; pre-synaptic; and post-synaptic. The axonal neurotoxins consist of two types: the _channel blockers_, as typified by tetrodotoxin and saxitoxin; and the _gate inhibitors_, as typified by sea anemone and scorpion neurotoxins. There are also two types of synaptic neurotoxins: pre-synaptic and post-synaptic. The pre-synaptic neurotoxins include two types: _blockers of neuro-transmittor release_ (e.g. botulinum toxin) and _stimulators of release_ (e.g. black widow spider venom neurotoxin). Post-synaptic neurotoxins generally block transmittor receptors as typified by the curarimimetic $\alpha$-bungarotoxin.

Several excitable tissue and organ preparations have been affected by nematocyst venoms from several species (section II. B.). In the absence of contrary data, all of these effects, with one exception, can be explained in terms of the known cytolytic or membrane permeability effects of these venoms. That one exception is a basic protein of low molecular weight (10-15,000) from _Aiptasia_ venom. This neurotoxin is a gate inhibitor which causes violent motor activity in crabs and specifically prolongs the early, transient inward ("sodium") current on isolated crustacean ventral nerve cord (Hessinger _et al_., 1973). Qualitatively indistinguishable effects are produced by protein neurotoxins prepared from whole animal and tissue extracts of other sea anemones (Kem, 1988).

3. The Regulatory Toxins

Regulatory toxins either stimulate or inhibit vital metabolic processes. Classical examples of such toxins are cholera toxin, which stimulates adenyl cyclase activity via covalent alteration of transduction G-proteins in cells bearing the appropriate ganglioside surface receptors, and diphtheria toxin, which inhibites protein synthesis by covalently altering elongation factor tu.

At the present, the only reported candidate among nematocyst venoms for such an effect is the vasodilatory effect of Man-of-War venom on arterial musculature (Loredo _et al._, 1985). This effect appears to be due to the production of a vasodilatory ecosinoid (Loredo _et al._, 1986) of endothelial origin (Gonzalez and Hessinger, in prepration) due to stimulation by the venom of an endogenous, regulatory phospholipase $A_2$ (Shier, 1980). This effect, however, has not yet been attributed to a purified venom component and, therefore, may be ascribable to a cytolytic permeabilizing effect.

## III. CONCLUDING REMARKS

Although some nematocyst venoms are among the most potent venoms known (especially Man-of-War and sea wasp venoms; Table I), proportionally few stingings of humans are fatal. This is probably because of the relatively small amounts of venom that are injected. At the present there is no known specific antidote for nematocyst stings and symptomatic treatment is all that is available in most cases. A commercial antivenin is available for life-threatening sea wasp stings.

All known nematocyst toxins are proteins that act on the plasma membranes of target cells. Common to all nematocyst venoms are cytolytic/hemolytic effects caused by either of two types of cytolysins: catalytic or stoichiometric. Hydrozoan and scyphozoan venoms possess single-component stoichiometric cytolysins. This type of cytolysin is typified by physalitoxin (PTX) from the Man-of-War venom. PTX is a high molecular weight heterotrimeric glycoprotein that is an extremely potent hemolysin. It is the only known toxic component in Man-of-War venom although evidence suggests that other venom components synergistically act with PTX to exert the full lethal potency of the venom.

Anthozoan venoms, on the other hand, possess multi-component catalytic cytolysins. The only anthozoan venom to be extensively studied is from the acontia of the sea anemone, _A. pallida_. The hemolytic system from this anemone consists of three synergistically interacting proteins including an unusual regulated phospholipase $A_2$. It is important to realize that although stoichiometric cytolysins have been purified from whole tissue extracts of sea anemones (Kem, 1988), these cytolysins have never been found in association with or localized in sea anemone nematocysts and, therefore, may represent proteins of extra-nematocyst origin.

In addition, anthozoans venoms are the only nematocyst venoms known to possess true neurotoxins. These neurotoxins primarily interfere with the normal inactivation of "sodium" currents in excitable membranes. Hydrozoan and scyphozoan venoms do not possess this type or any other known type of true neurotoxin. Although the hydro-scyphozoan venoms have mainly depolarizing effects on excitable tissues these effects can at the present be explained in terms of the known cytolyic and membrane permeabilizing components of these venoms. Thus, it appears that hydro-scyphozoan venoms more generally resemble each other in that they possess potent stoichiometric cytolysins but no neurotoxin, whereas anthozoan venoms possess both multi-component catalytic cytolysins plus neurotoxins.

It is noteworthy that microbasic mastigophore nematocysts both from an anthozoan, the sea anemone (A. pallida), and from a scyphozoan, the sea wasp (C. fleckeri), possess very different kinds of venom. Thus, a single morphological type of nematocyst can contain very different types of venoms. The considerable differences in the constituent toxins of these two mastigophore venoms probably reflect the intended use of these nematocysts against dissimilar prey. The preferred prey for the sea anemone are small crustaceans, against which the anemone neurotoxins are especially potent. The preferred prey for the sea wasp, however, are fish in which the cardiovascular, myotoxic and hemolytic effects of the venom would be especially effective in disabling.

The complete biochemical and functional characterization of venom components from individual types of nematocysts is within the capablities of current technologies. This has not yet been achieved for any specific nematocyst venom. Nonetheless, the picture that emerges from the information that is currently available suggests that venom components synergistically interact in functionally specific ways. In addition to having functional significance during toxin attack on target cells following nematocyst discharge, such molecular interactions are also likely to be important before nematocyst discharge occurs. During development and maturation of the nematocyst specific molecular interactions likely guide the labile toxic components of the venom to proper intra-nematocyst compartments for future deployment while also maintaining them in stable configurations.

ACKNOWLEDGEMENTS

I wish to thank Drs. Ramon R. Gonzalez, Jr., Ian M. Fraser, Howard M. Lenhoff, and Glen M. Watson for their valuable suggestions on this manuscript. Supported in part by funds from NIH grant ESO 2182-01.

## REFERENCES

Ackermann, D., F. Holtz, and H. Reinwein. 1923. Reindarstellung und konstitutionsermittelung des tetramines, eines giftes aus Aktinia equina. Zeit. Biol. 79: 113-120.
Ackermann, D., F. Holtz, and H. Reinwein. 1924. Uber die extrakstoffe von Aktinia equina. Zeit. Biol. 81: 61-64.
Ackermann, D. and R. Janka. 1953. Konstitution und synthese des anemonins. Hoppe-Seylers Zeit. Physiol. Chem. 294: 93-97.
Barnes, J. H. 1960. Observations on jellyfish stingings in north Queensland. Med. J. Australia 2: 993-999.
Barnes, J. H. In W.J. Rees, (ed.), The Cnidaria and Their Evolution. Symp. Zool. Soc. Lond. 16: 307-332, Academic Press, London.
Barnes, J. H. 1967. Extraction of cnidarian venom from living tentacle. In F.E. Russell and P.R. Saunders (eds.), Animal Toxins, pp. 115-129. Pergamon Press, New York.
Baxter, E. H. and A. G. M. Marr. 1969. Sea wasp (Chironex fleckeri) venom: lethal, hemolytic and dermonecrotic properties. Toxicon 7: 195-210.
Baxter, E. H. and A. G. Marr. 1975. Sea wasp toxoid: an immunizing agent against the venom of the box jelly fish, Chironex fleckeri. Toxicon 13: 423-435.
Baxter, E. H., N. B. Walden, C. E. Liefman, and A. G. Marr. 1972. Fatal intoxication of rabbits, sheep and monkeys by the venom of sea wasp (Chironex fleckeri). Toxicon 10: 653-655.
Blanquet, R. 1969. Properties and composition of the nematocyst toxin of the sea anemone, Aiptasia pallida. Comp. Biochem. Physiol. 25B: 893-902.
Blanquet, R. 1970. Ionic effects on discharge of the isolated and in situ nematocysts of the sea anemone Aiptasia pallida. Comp. Biochem. Physiol. 35: 451-461.
Blanquet, R. 1972. A toxic protein from the nematocysts of the scyphozoan medusa, Chrysaora quinquecirrha. Toxicon 10: 103-109.
Blanquet, R. 1988. The chemistry of cnidae. In D.A. Hessinger and H.M. Lenhoff (eds.), The Biology of Nematocysts. Academic Press, San Diego.
Bodansky, M. and W. C. Rose. 1922. Comparative studies of digestion. I. The digestive enzymes of coelenterates. Amer. J. Physiol. 62: 473-481.
Brockenhoff, H. and R. G. Jensen. 1974. Lipolytic Enzymes.

Academic Press, New York.
Burnett, J. W. and G. J. Calton. 1973. Purification of sea nettle nematocyst toxins by gel diffusion. Toxicon 11: 243-247.
Burnett, J. W. and G. J. Calton. 1974a. The enzymatic content of the venoms of the sea nettle and the Portuguese man-o'-war. Comp. Biochem. Physiol. 47B: 815-820.
Burnett, J. W. and G. J. Calton. 1974b. Sea nettle and man-o'-war venoms: a chemical comparison of their venoms and studies on the pathogenesis of the sting. J. Invest. Derm. 62: 372-377.
Burnett, J. W. and G. J. Calton. 1977. The chemistry and toxicology of some venomous pelagic coelenterates. Toxicon 15: 177-196.
Burnett, J. W., G. J. Calton, H. L. Meir and A. P. Kaplan. 1975. A study on the kinin activity of three venomous jellyfish. Comp. Biochem. Physiol. 51C: 153-159.
Burnett, J. W., C. J. Gean, J. W. Calton and J. E. Warnick. 1985. The effect of verapamil on the cardiotoxic activity of Portuguese man-o'-war (Physalia physalis) and sea nettle (Chrysaora quinquecirrha) venoms. Toxicon 23: 681-689.
Burnett, J. W. and R. Goldner. 1969. Effects of Chrysaora quinquecirrha (sea nettle) toxin on the rat cardiovascular system. Proc. Soc. Exp. Biol. Med. 132: 353-356.
Burnett, J. W. and R. Goldner. 1970. Effect of Chrysaora quinquecirrha (sea nettle) toxin on rat nerve and muscle. Toxicon 8: 179-181.
Burnett, J. W. and R. Goldner. 1971. Some immunological aspects of sea nettle toxins. Toxicon 9: 271-277.
Burnett, J. W. and W. M. Gould. 1971. Further studies on the purification and physiological actions of sea nettle toxin. Proc. Soc. Exp. Biol. Med. 138: 759-762.
Burnett, J. W., J. V. Ordonez and G. J. Calton. 1986. Differential toxicity of Physalia physalis (Portuguese man-o'-war) nematocysts separated by flow cytometry. Toxicon 24: 514-518.
Burnett, J. W., L. H. Pierce, U. Nawachinda, and J. H. Stone. 1968. Studies on sea nettle stings. Arch. Dermatol. 98: 587-589.
Calton, G. J. and J. W. Burnett. 1973a. The effect of two jellyfish toxins on calcium ion transport. Toxicon 11: 357-360.
Calton, G. J. and J. W. Burnett. 1973b. The purification of Portuguese Man-of-War nematocyst venom by gel diffusion. Comp. Gen. Pharmac. 4: 267-270.
Calton, G. J. and J. W. Burnett. 1982. Partial purification and characterization of the acid protease of sea nettle

(Chrysaora quinquecirrha) nematocysts. Comp. Biochem. Physiol. 72B: 93-97.

Calton, G. J. and J. W. Burnett. 1983. Partial purification and characterization of the alkaline protease of sea nettle (Chrysaora quinquecirrha) nematocyst venom. Comp. Biochem. Physiol. 74C: 361-364.

Calton, G. J. and J. W. Burnett. 1986. Partial purification of Chironex fleckeri (sea wasp) venom by immunochromatography with antivenom. Toxicon 24: 416-420.

Calton, G. J., J. W. Burnett, J. Garbus, and S. R. Marx. 1973. The effects of Chrysaora and Physalia venoms on mitochondrial structure and function. Proc. Soc. Exp. Biol. Med. 143: 971-977.

Carpenter, P. L. 1965. Immunology and Serology, pp. 276, 277. W. B. Saunders, Philadelphia.

Cleland, J. B. and R. V. Southcott. 1965. Injuries and diseases in Australia attributable to animals. Series VI. Dept. of Health, Australia.

Cobbs, C. S., P. K. Gaur, A. J. Russo, J. E. Warnick, G. J. Calton, and J. W. Burnett. 1983. Immunosorbent chromatography of sea nettle (Chrysaora quinquicirrha) venom and characterization of toxins. Toxicon 21: 385-391.

Cormier, S. M. and D. A. Hessinger. 1980. Cellular basis for tentacle adherence in the Portuguese Man-of-war (Physalia physalis). Tissue Cell 12: 713-721.

Crone, H. D. 1976. On the inactivation by gangliosides of the haemolytic protein toxin from the sea wasp (Chironex fleckeri). Toxicon 14: 494-498.

Crone, H. D. and T. E. B. Keen. 1969. Chromatographic properties of the hemolysin from the cnidarian Chironex fleckeri. Toxicon 7: 79-87.

Crone, H. D. and T. E. B. Keen. 1971. Further studies on the biochemistry of the toxins from the sea wasp Chironex fleckeri. Toxicon 9: 145-151.

Endean, R. 1987. Separation of two myotoxins from nematocysts of the box jellyfish (Chironex fleckeri). Toxicon 25: 483-492.

Endean, R., C. Duchemin, D. McColm, and E. H. Fraser. 1969. A study of the biological activity of toxic material derived from nematocyst of the cubomedusan Chironex fleckeri. Toxicon 6: 179-204.

Endean, R. and L. Henderson. 1969. Further studies of toxic material from nematocysts of the cubomedusan Chironex fleckeri Southcott. Toxicon 7: 303-314.

Endean, R. and M. Noble. 1971. Toxic material from the tentacles of the cubomedusan Chironex fleckeri. Toxicon 9: 255-264.

Fischer, F. G. and J. J. Kabara. 1967. In F.E. Russell and P.R. Saunders (eds.), Animal Toxins, pp. 283-292.

Pergamon Press, New York.

Flowers, A. L. and D. A. Hessinger. 1981. Mast cell histamine release induced by Portuguese Man-of-War (Physalia) venom. Biochem. Biophys. Res. Comm. 103: 1083-1091.

Freeman, S. E. 1974. Actions of Chironex fleckeri toxins on cardiac transmembrane potentials. Toxicon 12: 395-404.

Garriott, J. C. and C. E. Lane. 1969. Some autonomic effects of Physalia toxin. Toxicon 6: 281-286.

Gaur, P. K., R. L. Anthony, G. J. Calton, and J. W. Burnett. 1982. Isolation of hybridomas secreting monoclonal antibodies against Physalia physalis (Portuguese man-o'-war) nematocyst venom. Toxicon 20: 419-425.

Goldner, R., J. W. Burnett, J. S. Stone, and M. S. Dilaimy. 1969. The chemical composition of sea nettle nematocysts. Proc. Soc. Expl. Biol. Med. 131: 1386-1388.

Guess, H. A., P. L. Saviteer, and C. R. Morris. 1982. Hemolysis and acute renal failure following a Portuguese Man-of-War sting. Pediatrics 70: 979-981.

Halstead, B. W. 1965. Poisonous and Venomous Marine Animals of the World, Vol. I--Invertebrates, U.S. Gov't. Printing Office, Washington, D.C.

Halstead, B. W. 1978. Poisonous and Venomous Marine Animals of the World. Darwin Press, Princeton.

Hastings, S. G., J. B. Larsen, and C. E. Lane. 1967. Effects of nematocyst toxin of Physalia physalis (Portuguese Man-of-War) on the canine cardiovascular system. Proc. Soc. Exp. Biol. Med. 125: 41-45.

Hessinger, D. A. and M. T. Ford. 1988. Ultrastructure of the small cnidocyte of the Portuguese Man-of-war (Physalia physalia) tentacles. In D.A. Hessinger and H.M. Lenhoff (eds.), The Biology of Nematocysts. Academic Press, San Diego.

Hessinger, D. A. and R. I. Grove. 1979. Antibodies to sea anemone nematocyst venom: II. Neutralization of the hemolytic, phospholipase $A_2$ and lethal activities by purified antibodies produced in response to attenuated venom. Toxicon. 17: 109-120.

Hessinger, D. A. and J. A. Hessinger. 1981. Methods for rearing sea anemones in the laboratory. In Committee on Marine Invertebrates (eds.), Marine Invertebrates, pp. 153-179. National Academy Press, Washington, D.C.

Hessinger, D. A. and H. M. Lenhoff. 1968. Effects of purified sea anemone toxin on red blood cells. Proc. Internl. Union Physiol. Sci. 7: 191.

Hessinger, D. A. and H. M. Lenhoff. 1973a. Assay and properties of the hemolysis activity of pure venom from the nematocysts of the acontia of the sea anemone Aiptasia pallida. Arch. Biochem. Biophys. 159: 629-638.

Hessinger, D. A. and H. M. Lenhoff. 1973b. Binding of active and inactive hemolytic factor of sea anemone nematocyst venom to red blood cells. Biochem. Biophys. Res. Comm. 53: 475-481.

Hessinger, D. A. and H. M. Lenhoff. 1974. Degradation of red cell membrane phospholipids by sea anemone venom. Toxicon 12: 379-383.

Hessinger, D. A. and H. M. Lenhoff. 1976. Mechanism of hemolysis induced by nematocyst venom: roles of phospholipase A and direct lytic factor. Arch. Biochem. Biophys. 173: 603-613.

Hessinger, D. A., H. M. Lenhoff, and L. Kahn. 1973. Haemolytic, phospholipase A, and nerve-affecting activities of sea anemone nematocyst venom. Nature New Biol. 241: 125-127.

Ioannides, G. and J. H. Davis. 1965. Portuguese Man-of-War stinging. Arch. Derm. 91: 448-451.

Ipsen, J. 1951. The effect of environmental temperature on the reaction of mice to tetanus toxin. J. Immunol 66: 687-694.

Keen, T. E. B. 1970. Recent investigations on sea-wasp stingings in Australia. Med. J. Aust. (Feb. 7): 266-270.

Keen, T. E. B. 1972. Surface properties of the hemolytic fraction derived from tentacle extracts of _Chironex fleckeri_. Toxicon 10: 587-596.

Keen, T. E. B. 1973. Interaction of the hemolysin of _Chironex fleckeri_ tentacle extracts with lipid monolayers. Toxicon 11: 293-299.

Keen, T. E. B. and H. D. Crone. 1969a. The hemolytic properties of extracts of tentacles from the cnidarian _Chironex fleckeri_. Toxicon 7: 55-63.

Keen, T. E. B. and H. D. Crone. 1969b. Dermatonecrotic properties of extracts from the tentacles of the cnidarian _Chironex fleckeri_. Toxicon 7: 173-180.

Kelman, S. N., G. J. Calton, and J. W. Burnett. 1984. Isolation and partial characterization of a lethal sea nettle (_Chrysaora quinquecirrha_) mesenteric toxin. Toxicon 22: 139-144.

Kem, W. R. 1988. Sea anemone toxins: structure and action. In D.A. Hessinger and H.M. Lenhoff (eds.), _The Biology of Nematocysts_. Academic Press, San Diego.

Kingston, C. W. and R. V. Southcott. 1960. Skin histopathology in fatal jellyfish stinging. Trans. Roy. Soc. Trop. Med. Hyg. 54: 373-384.

Kleinhaus, A. L., P. F. Cranefield, and J. W. Burnett. 1973. The effects on canine cardiac Purkinje fibers of _Chrysaora quinquecirrha_ (sea nettle) toxin. Toxicon 11: 341-349.

Kline, E. S. and V. S. Waravdekar. 1960. Inhibitor of

succinoxidase activity from Hyrdra littoralis. J. Biol. Chem. 235: 1803-1808.

Klug, M., J. Weber, and P. Tardent. 1988. Direct observation of hemolytic activity associated with single nematocysts. In D.A. Hessinger and H.M. Lenhoff (eds.), The Biology of Nematocysts. Academic Press, San Diego.

Lal, D. M., G. J. Calton, I. Neeman, and J. W. Burnett. 1981a. Characterization of Chrysaora quinquicirrha (sea nettle) nematocyst venom collegenase. Comp. Biochem. Physiol. 69B: 529-533.

Lal, D.M., G. J. Calton, I. Neeman, and J. W. Burnett. 1981b. Characterization of Physalia physalis (Portuguese man-o'-war) nematocyst venom collagenase. Comp. Biochem. Physiol. 70B: 635-638.

Lane, C. E. 1960. The toxin of Physalia nematocysts. Ann. N.Y. Acad. Sci. 90: 742-750.

Lane, C. E. 1961. Physalia nematocysts and their toxin. In H.M. Lenhoff and W.F. Loomis (eds.), The Biology of Nematocysts, pp. 169-178. Univ. of Miami Press, Coral Gables.

Lane, C. E. 1967a. Pharmacologic action of Physalia toxin. Fed. Proc. 26: 1225-1226.

Lane, C. E. 1967b. In F.E. Russell and P.R. Saunders (eds.) Animal Toxins, pp. 131-136. Pergamon Press, Oxford.

Lane, C. E. and E. Dodge. 1958. Toxicity of Physalis nematocysts. Biol. Bull. 115: 219-225.

Lane, C. E., B. W. Coursen, and K. Hines. 1961. Biologically active peptide in Physalia nematocysts. Proc. Soc. Exp. Biol. Med. 107: 670-672.

Larsen, J. B. and C. E. Lane. 1966. Some effects of Physalia physalis toxin on the cardiovascular system of the rat. Toxicon 4: 199-203.

Larsen, J. B. and C. E. Lane. 1970a. Direct action of Physalia toxin on frog nerve and muscle. Toxicon 8: 21-23.

Larsen, J. B. and C. E. Lane. 1970b. Some effects of Physalia physalis toxin on active $Na^+$ transport across frog skin. Comp. Biochem. Physiol. 34: 333-338.

Lin, D. C. and D. A. Hessinger. 1979. Possible involvement of red cell membrane proteins in the hemolytic action of Portuguese Man-of-War venom. Bioch. Biophys. Res. Comm. 91: 761-769.

Loredo, J. S., R. R. Gonzalez, Jr., and D. A. Hessinger. 1985. Vascular effects of Physalia physalis venom on the skeletal muscle of the dog. J. Pharmacol. Exp. Ther. 232: 301-304.

Loredo, J. S., R. R. Gonzalez, Jr., and D. A. Hessinger. 1986. Effects of Portuguese Man-of-War venom on isolated vascular segments. J. Pharmacol. Exp. Ther. 236: 140-

143.
M'ansson, T., H. W. Randle, R. M. Mandojana, G. J. Calton, and J. W. Burnett. Recurrent cutaneous jellyfish eruptions without envenomation. 1985. Acta Derm. Venereol. (Stockh) 65: 72-75.
Mariscal, R. N. 1974. In L. Muscatine and H.M. Lenhoff (eds.), Coelenterate Biology, pp. 129-178. Academic Press, New York.
Mathias, A. P., D. M. Ross, and M. Schachter. 1960. The distribution of 5-hydroxytryptamine, tetramethylammonium, homarine, and other substances in sea anemones. J. Physiol. 151: 296-311.
Matusow, R. J. 1980. Oral inflammatory response to a sting from a Portuguese Man-of-War. J. Am. Dent. Assoc. 100(1): 73-75.
Narahashi, T., J. W. Moore, and B. I. Shapiro. 1969. Condylactis toxin: interaction with nerve membrane ionic conductances. Science 163: 680-681.
Neeman, I., G. J. Calton, and J. W. Burnett. 1980. An ultrastructural study of the cytotoxic effect of the venoms from the sea nettle (Chrysaora quinquecirrha) and Portuguese Man-of-War (Physalia physalis) on cultured Chinese hamster ovary K-1 cells. Toxicon 18: 495-501.
Olson, C. E., E. E. Pockl, G. J. Calton, and J. W. Burnett. 1984. Immunochromatographic purification of a nematocyst toxin from the cnidarian Chironex fleckeri (sea wasp). Toxicon 22: 733-742.
Olson, C. E., D. G. Cargo, G. J. Calton, and J. W. Burnett. 1985. Immunochromatography and cardiotoxicity of sea nettle (Chrysaora quinquecirrha) polyps and cysts. Toxicon 23: 127-133.
Ouyang, C. J. 1957. The effects of Formosan snake venoms on blood coagulation in vitro. J. Med. Assoc. Formosa 56: 435-438.
Phelan, M. A. and R. S. Blanquet. 1985. Characterization of nematocyst proteins from the sea anemone Aiptasia pallida and Pachycerianthus torrey: (Cnidaria: Anthozoa). Comp. Biochem. Physiol. 81B: 661-666.
Phillips, J. H. and D. P. Abbott. 1957. Isolation and assay of the nematocyst toxin of Metridium senile. Biol. Bull. 113: 296-301.
Portier, P. and C. Richet, C. 1902. Sur les effects physiologiques du poison des filaments pecheurs et des tentacules des Coelenteres (hypnotoxine). Compt. Rend. Soc. Biol. 134: 247-248. Compt. Rend. Soc. Biol. 54: 548-551.
Purcell, J. E. and C. Mills. 1988. The correlation between nematocyst types and diets in pelagic hydrozoa. In D.A. Hessinger and H.M. Lenhoff (eds.), The Biology of

Nematocysts. Academic Press, San Diego.
Richet, C. 1902. On poison pruritogene et urticant cantenu dans les tentacules d'Actinies. C. R. Soc. Biol., Paris. 54: 1438-1440.
Richet, C. 1903. Des poisons contenus dan les tentacules des Actinies, congestine et thallasine. C. R. Soc. Biol., Paris. 55: 246-248.
Richet, C. and D. Portier. 1936. Reserches sur la toxine des coelenteres et le phenomene d'anaphylaxie. Res. Comp. Sci., Monaco. 95: 3-24.
Russell, F. E. 1966. Physalia stings: a report of two cases. Toxicon 4: 65-67.
Schwick, G. and F. Dickgeisser. 1963. Probleme der antigen- und fermentanalyse in zusammenhang mit herstellung polyvalenter schlangen giftseren. In Die Giftschlangen der Erde, pp. 35-36. Behringwerke-Mitt, Sonderband.
Shapiro, B. I. 1968a. Purification of a toxin from tentacles of the anemone Condylactis gigantea. Toxicon 5: 253-259.
Shapiro, B. I. 1968b. A site of action of toxin from the anemone Condylactis gigantea. Comp. Biochem. Physiol. 27: 519-531.
Shier, W. T. 1980. Activation of self-destruction as a mechanism of action for cytolytic toxins. In D. Eaker and T. Wadstrom (eds.), Natural Toxins. Pergamon Press, New York.
Shryock, J. C. and C. P. Bianchi. 1983. Sea nettle (Chrysaora quinquecirrha) nematocyst venom: mechanism of action on muscle. Toxicon 21: 81-95.
Southcott, R. V. 1952. Fatal stings to north Queensland bathers. Med. J. Australia 1: 272-273.
Southcott, R. V. 1959. Tropical jellyfish and other marine stingings. Military Med. 124: 569-579.
Spielman, F. J., E. A. Bowe, C. B. Watson, and E. F. Klein, Jr. 1982. Acute renal failure as a result of Physalia physalis sting. South. Med. J. 75: 1425-1426.
Stillway, L. W. and C. E. Lane. 1971. Phospholipase in the nematocyst toxin of Physalia physalis. Toxicon 9: 193-195.
Tamkun, M. M. and D. A. Hessinger. 1981. Isolation and partial characterization of a hemolytic and toxic protein from the nematocyst venom of the Portuguese Man-of-War, Physalia physalis. Biochim. Biophys. Acta 667(1): 87-98.
Tanford, C. 1980. The Hydrophobic Effect, pp. 162-163. John Wiley and Sons, New York.
Toom, P. M. and D. S. Chan. 1972a. Preliminary studies of nematocysts from the jellyfish Stomolophus meleagris. Toxicon 10: 605-610.
Toom, P. M. and D. S. Chan. 1972b. Enzymatic activities of

venom from the jellyfish Stomolophus meleagris. Comp. Biochem. Physiol. 43: 435-441.

Tu, A. T. 1977. Venoms: Chemistry and Molecular Biology. John Wiley and Sons, New York.

Turner, R. J. and S. E. Freeman. 1969. Effects of Chironex fleckeri toxin on the isolated perfused guinea pig heart. Toxicon 7: 277-286.

Uvnas, B. 1960. Mechanism of action of a histamine-liberating principle in jellyfish (Cyanea capillata). Ann. N.Y. Acad. Sci. 90: 751-759.

Weber, J., M. Klug, and P. Tardent. 1988. Chemistry of hydra nematocysts. In D.A. Hessinger and H.M. Lenhoff (eds.), The Biology of Nematocysts. Academic Press, San Diego.

Welsh, J. H. P. B. and Prock. 1958. Quarternary ammonium bases in the coelenterates. Bio. Bull. 115: 551-561.

Welsh, J. H. 1956. On the nature and action of coelenterate toxins. Deep Sea. Res. Suppl. 3: 287-299.

Welsh, J. H. 1960. 5-Hydroxytryptamine in coelenterates. Nature 186: 811-812.

Yanagita, T. M. 1959. Physiological mechanisms of nematocyst responses in the sea anemone. I. Effects of trypsin and thioglycolate upon isolated nematocysts. Jap. J. Zool. 12: 361-375.

# 19

## CHARACTERIZATION OF NEMATOCYST VENOMS

Gary J. Calton
Joseph W. Burnett

Division of Dermatology
University of Maryland School of Medicine
Baltimore, Maryland

The chemical and pharmacological characterization of nematocyst venoms is fraught with major difficulties. Even in relatively recent literature, we find a great deal of inaccurate chemical and especially pharmacological data. The reason for this is that most of the isolated venoms were impure and, since the pharmacological activity of many of the components is extremely potent, a single activity was not being assayed. Thus, the pharmacological activities were not reproducible, as the compounds present were quite different. In addition, chemical analyses were often based on degraded components. This presentation centers on the methods of isolation that our group at the University of Maryland has used and the conclusions that can be drawn concerning nematocyst venoms from the toxin isolations we have carried out.

Early studies with Chrysaora quinquecirrha, the Sea Nettle, and Physalia physalis, the Portuguese Man-O'War, showed a loss of activity upon storage as well as upon purification in regard to mouse lethality, dermonecrosis and hemolysis (Burnett et al., 1968;Burnett and Golder, 1970;Calton and Burnett, 1973). Part of this loss was probably due to the methods used for nematocyst venom preparation. Autolysis (Lane and Dodge, 1958) undoubtedly resulted in contamination from musculature components and nematocyst venom degradation due to proteolysis (Calton and Burnett, 1973; Endean and Noble, 1971;Burnett and Calton, 1973; Burnett and Calton, 1974). Only the hardiest of nematocysts withstood the high osmotic pressures involved in Yanagita's procedure (Yanagita, 1959) since chrysaora nematocysts were virtually all ruptured but some Physalia nematocysts survived, however, residual tentacular debris was always present (Calton and Burnett, 1973). This method worked well only in anthozoan acontia. (Hessinger and Lenhoff 1973) In the early efforts on nematocyst venom component isolation, the investigators were faced with an even more complex problem than existed in the nematocyst itself. This problem was shown to occur in the Chironex fleckeri, the Sea Wasp, venom purification where toxic materials were shown to exist outside of the nematocyst such as in the tentacular muscle. (Endean and Noble, 1971)

The hydrophobic amino acids present in Chrysaora and Physalia venoms also led to adsorption of venom components on Sephadex (Calton and Burnett, 1973; Burnett and Calton 1973) thus confusing the venom character further.

Most coelenterate venoms, as opposed to snake and wasp venoms are extremely labile at all ranges of temperature above $0^{\circ}C$, and decompose on repeated freezing and thawing (Burnett and Goldner, 1970). No improvement in stability was imparted to the venoms of Chrysaora and Physalia by the addition of agents such as dithiotheitol or ion chelation reagents. Furthermore, stability appears to be decreased by purification (Calton and Burnett, 1973). There are a number of low molecular weight coelenterate toxins obtained from nematocysts of the sea anemones Bolacera tuediae, Anemonia sulcata, and Anthopleura xanthogrammica, which are stable to heat (60-80° for 30 minutes) (Beress et al., 1977;Wanderer and Eulitz, 1978;Beress and Zwick, 1980). The structure of these polypeptides has been elucidated. There is considerable homology between them (Beress et al., 1977). The molecular weight of these polypeptides is quite low (~ 4,000) when compared to those in the venom from Chrysaora, Physalia, or Chironex (20,000-250,000).

The presence in venoms of acid and alkaline proteases (Burnett and Calton, 1974), collagenases (Lal et al., 1981 a, b) and other deleterious enzymes suggested a similarity between the various coelenterate venoms (Toom and Chan, 1972, Burnett and Calton, 1974;Cobbs et al., 1983a) and the enzymatic profile of snake venoms. Since Chrysaora, Physalia and Stomolophus venoms have both alkaline and acid proteases present, it may be assumed that autolysis plays an important role in limiting stability. Protein folding may also contribute to this stability, since sulfhydryl protection and metal chelation do not greatly affect the stability of the pharmacological activity of these venoms.

Our current interest in characterizing these venoms lies in the use of antibodies to separate components in venom, analyze component degradation and to test for cross reactivity between species. The lability of these venoms makes the isolation of pure fractions difficult. The use of immobilized antibodies of low affinity provides a method for the rapid isolation of specific components under mild conditions (Calton, 1984b).

Polyclonal antisera against a number of nematocyst venoms have been available for several years (Calton and Burnett, 1973, Baxter and Marr, 1974). These antisera neutralize lethal activities, hemolytic activities and pain factors (Calton and Burnett, 1986; Baxter and Marr, 1974; Williamson 1986). However, their lack of specificity for any single component does not provide a method of separation of individual components. The introduction of monoclonal antibodies (Kohler and

Milstein, 1975) has provided the means for an elegant, yet simple, method for the isolation of a single component from a complex mixture (Hamman, and Calton, 1985; Calton, 1984a).

We have prepared monoclonal antibodies (MAB) against Chrysaora and Physalia venom components (Gaur et al., 1981; Gaur et al., 1982). Using hybridomas generated by mice, which had been immunized against crude Chrysaora venom, we were able to select a MAB which neutralized one of the lethal factors, a cardiotoxin, in the venom (Gaur et. al, 1981;Cobbs et al., 1983b). Only one $LD_{100}$ of crude venom could be neutralized, possibly due to the large number of deleterious enzymes present which might cause death. Alternatively, additional cardiotoxic factors may be present which are not neutralized by the single MAB (Kelman et al., 1984). The same effect was seen with an MAB obtained in the same manner against crude Physalia venom which neutralized Physalia cardiotoxins (Gaur, et al., 1982; Cobbs et al, 1983b; Kelman et al., 1984). For use as immunosorbent chromatography reagents, we immobilized each of these MAB with CNBr activated Sepharose (Cobbs et al. 1983b; Burnett et al 1983; Kelman et al. 1984). This immobilization procedure is not the best method for isolation of pure components, however, since CNBr provides an unstable isourea linkage (Peng et al., 1986b). Approximately 10ug of murine protein/ml of eluant are found as contaminants in the eluant of an immunosorbent column prepared in this manner (Peng et al, 1986b). Alternative methods of immobilization are recommended. The most stable and active of these is oxidation of vicinal glycols on either a Sepharose or cellulose matrix to provide aldehyde groups (Peng 1986b). Reaction of the aldehydes with primary amines, available primarily as the epsilon amino group of lysine, form Schiff bases. The reduction of the Schiff bases with an appropriate borohydride provides a stable linkage for the immunosorbent (Peng et al., 1986a). Sepharose$^R$ cellulose and Trisacryl$^R$ provide a matrix which have low non-specific adsorption of other proteins in the mixtures to be separated.

Antibodies to be used in immunosorbent chromatography must be selected on the basis of affinity. High affinity antibodies may not release the antigen under reasonable dissociation conditions while low affinity antibodies will not retain the antigen when a washing step is employed. In addition, the antigen-antibody couplex must be dissociated by a buffer in which the antigen is stable or in which pharmacological/enzymatic activity is not lost. An ELISA screen for this selection has been described (Calton, 1984a).

We determined by ELISA with the Chrysaora MAB that a reaction took place not only with Chrysaora venom but also with Physalia venom (Russo et al. 1983a). Serendipitously, we also found this MAB cross reacted with a hornet venom (Russo et al., 1983b). Based on this information, we looked at a number of the venoms and found that the Chrysaora MAB, C6, reacted with a rattlesnake venom (Crotalus durissus terrificus), but not with ATX I or ATX II from Anemonia sulcata, APA or APC from Anthopleura xanthogrammica or Bolacera tuediae crude venom (Russo et al, 1983a). We were able to use this antibody to isolate cardiotoxins from the venoms which reacted (Russo et al., 1983a,b; Calton and Burnett, 1986. Based on this data we proposed that this antibody mimicked a common receptor site on cardiac tissue. The MAB provided a specific reagent for probing other venoms for cardiotoxins and the subsequent isolation of these cardiotoxins from these venoms (Russo et al 1983b; Burnett et. al. 1983; Kelman et al. 1984; Olson et al. 1984; Olson et al. 1985). We do not believe that the recognition site in the isolated cardiotoxins necessarily has the same chemical structure, although the data to support this speculation remains to be developed. Rather, we believe that the conformation of these proteins is expected to fit in the receptor site. Thus, this antibody, and others that may be developed against the numerous pharmacological activities of coelenterate venoms, could provide receptors for the isolation of different agents having a similar biological activity.

In this respect, a number of future directions in our lab might be of interest. The active site of these venoms is of high interest insofar as definition of molecular structure and action is concerned. The immunological activity of the degraded protein may be followed by immune blot. This allows determination of the most appropriate proteolysis scheme for component degradation. Recently, it has been shown that proteolytic degradation of an antigen-antibody couplex allows preservation of the binding site (Jemmerson and Paterson, 1986). Using a MAB which mimics the receptor site, it should be possible to identify and sequence the lowest molecular weight immunologically or pharmacologically active subunit of a toxin and, thus, to simplify its structural elucidation.

The study of nematocyst venoms with MAB probes provides access to the isolation of specific venom components, to the isolation of specific pharmacological classes from these venoms, and to the isolation and chemical identification of active sites and subunits.

## REFERENCES

Baxter, E.H. and A.G.M. Marr 1974. Sea Wasp (Chironex Fleckeri) antivenene. Neutralizing potency against the venom of the other jellyfish. Toxicon 12:223-229.

Beress, L., G. Wunderer, and E. Wachterer 1977 Amino acid sequence of toxin III from Anemonia Sulcata. Hoppe Seylers Z. Physiol. Chem. 358:985-988.

Beress, L., and J. Zwick. 1980 Purification of two crab-paralyzing polypeptides from the sea anemone Bolocera tuediae. Mar. Chem. 8:333-338.

Burnett, J.W., L. H. Pierce, U. Nawachinda and J.H. Stone. 1968. Studies on sea nettle stings. Arch. Derm. 98: 587-589.

Burnett, J.W. and G.J. Calton. 1973. The purification of sea nettle nematocyst toxins by gel diffusion. Toxicon 11: 243-246.

Burnett, J. W. and G.J. Calton. 1974. The enzymatic content of the venoms of the sea nettle and Portuguese an O'War. Comp. Biochem. Physiol. 47B:815-820.

Burnett, J.W., G.J. Calton, C.S. Cobbs, and S.N. Kelman. 1983. Sea nettle and Portuguese Man O'War nematocyst venoms. Studies with monoclonal antibodies and affinity chromatography. Toxicon Suppl. 3:49-52.

Burnett, J.W. and R. Golder. 1970. Partial purification of sea nettle (Chrysaora quinquecirrha) nematocyst toxin. Proc. Soc. Exp. Biol. Med. 133:978-981.

Calton, G.J. 1984a. Immunosorbent separations in W.B. Jacoby (ed.), Methods In Enzymology, Vol. 104, pp381-387. Academic Press, New York.

Calton, G.J. 1984b. System design for industrial scale purification of high value proteins by immunosorbent chromatography in I.M. Chaiken, M. Wilchek & I. Parikh (eds.), 5th Int'l Symposium on Affinity Chromatography, pp 383-392. Academic Press, N.Y.

Calton, G.J. and J. W. Burnett. 1986. Partial purification of Chironex fleckeri (sea wasp) venom by immunochromatography with antivenom. Toxicon 24:416-420.

Calton, G.J. and J.W. Burnett. 1973. The purification of Portuguese Man O'War nematocyst toxins by gel diffusion. Comp. Gen. Pharm. 4:267-270.

Cobbs, C.S., P. Gold, G.J. Calton and J. W. Burnett. 1983a. Sea nettle (Chrysaora quinquecirrha) nematocysts venom hemagglutinins. Comp. Biochem. Physiol. 74C:225-228.

Cobbs, C.S., P.K. Gaur, A.J. Russo, J.E. Warnick, G.J. Calton and J. W. Burnett. Immunosorbent chromatography of sea nettle, Chrysaora quinquecirrha, venom and chracterization of toxins. Toxicon 21:385-391.

Endean R. and M. Noble. 1971. Toxic material from the tentacles of the cubomedusan Chironex fleckeri. Toxicon. 9:255-264.

Gaur, P.K., R.L. Anthony, G.J. Calton and J.W. Burnett. 1982. Isolation of hybridomas secreting monoclonal antibodies against Physalia physalis (Portuguese Man O'War) nematocyst venom. Toxicon 20:419-425.

Gaur, P.K., R.L. Anthony, T.S. Cody, G.J. Calton, and J.W. Burnett. 1981. Production of a monoclonal antibody against sea nettle venom mouse lethal factor. Proc. Soc. Exp. Biol. Med. 167:374-377.

Hamman, J.P., and G.J. Calton. 1985. Immunosorbent chromatography for recovery of protein products in Purification of Fermentation Products. ACS Symposium Series. 271, Am. Chem. Soc., Washington, D.C. pp105-112.

Hessinger, D.A. and H.M. Lenhoff. 1973. Assay and Properties of the hemolysis activity of pure venom from the nematocysts of the acontia of the sea anemone Aiptasia Pallida. Arch. Biochem. Biophys. 159:629-638.

Jemmerson, R. and Y. Paterson. 1986. Mapping epitopes on a protein antigen by the proteolysis of antigen-antibody complexes. Science 232:1001-1004.

Kelman, S.N., G.J. Calton and J.W. Burnett. 1984. Isolation and partial characterization of a lethal sea nettle (Chrysaora quinquecirrha) mesenteric toxin. Toxicon 22:139-144.

Kohler, G. and C. Milstein. 1975. Continuous cultures of fused cells secreting antibody of predetermined specificity. Nature 256:495.

Lal, D.M., G.J. Calton, I. Neeman and J.W. Burnett. 1981a. Characterization of Chrysaora quinquecirrha (Portuguese Man O'War) venom collagenase. Comp. Biochem. Physiol. 69B:529-533.

Lal,D.M., G.J.Calton, I. Neeman and J.W.Burnett. 1981b. Characterization of Physalia physalis (Sea Nettle) venom collagenase. Comp. Biochem. Physiol. 70B:635-638.

Lane, C.E. and E. Dodge. 1958. The toxicity of Physalia nematocysts. Biol. Bull. 115:219-226.

Olson, C.E., M.G. Heard, G.J. Calton and J.W. Burnett, J.W. 1985. Interrelationships between toxins: studies on the cross-reactivity between bacterial or animal toxins and monoclonal antibodies to two jellyfish venoms. Toxicon 23:307-316.

Olson, C.E., E.E. Pockl, G.J. Calton and J.W. Burnett, J.W. 1984. Immunochromatographic purification of a nematocyst toxin from the cnidarian Chironex fleckeri (sea wasp). Toxicon 22:733-742.

Peng, L., J.W. Burnett and G.J. Calton. 1986a. Effect of borohydride reduction on antibodies. Appl. Biochem. Biotech. (in press)

Peng, L., G.J. Calton and J.W. Burnett. 1986b. Stability of Antibody attachment in immunosorbent chromatography. Enzyme Microbial Tech. 8:681-685.

Russo, A.J., C.S. Cobbs, G.J. Calton and J.W. Burnett. 1983a. Detection of common antigenic sites in lethal proteins of non-related animal venom. Toxicon 21:433-437.

Russo,A.J. C.S. Cobbs, J.S. Ishay, G.J. Calton, and J.W. Burnett. 1983b. Isolation of lethal factor from venom of Vespa orientalis (oriental hornet) by affinity chromatography using cross reactive monoclonal antibody. Toxicon 21:166-169.

Toom, P.M. and D.S. Chan. 1972. Enzymatic Activities of venom from jellyfish Stomolophus meleagris. Comp. Biochem. Physiol. 43B:436-441.

Wunderer, G. and M. Eulitz. 1978. Amino acid sequence of toxin I from Anemonia sulcata. Eur. J. Biochem. 89:11-17.

Williamson, Jack. 1986. Personal communication

Yanagita, T.M. 1959. Physiological mechanism of nematocyst responses in sea anemone. II. Effects of electrolyte ions upon the isolated cnidae. J. Fac. Sci. Univ. Tokyo 8:381-400.

# 20

# SEA ANEMONE TOXINS: STRUCTURE AND ACTION

William R. Kem

Department of Pharmacology and Therapeutics
University of Florida
College of Medicine
Box J-267 JHMHC
Gainesville, Florida 32610

## SUMMARY

Sea anemones apparently produce only polypeptide and protein toxins. All of the anemone neurotoxins thus far reported possess a single polypeptide chain crosslinked by several disulfide bonds. These polypeptides usually consist of 46-49 amino acid residues, but two shortor neurotoxins (27 and 31 residues) have also been isolated. The initial sequences reported for the *Anthopleura xanthogrammica* and *Anemonia sulcata* polypeptides suggested that sea anemone "long" toxins possessed a highly conservative primary structure. During the past two years, the sequences of three stichodactylid neurotoxins were elucidated and found to differ considerably from these actiniid "long" toxin sequences, particularly near the N- and C-termini. On structural grounds, the long toxins are divided into two types. Electrophysiological and pharmacological studies with type 1 (*Anthopleura* and *Anemonia*) toxins indicate that they act in a membrane voltage-dependent manner upon the same receptor site 3 on the Na channel as do scorpion alpha toxins. It is not yet clear if the stichodactylid toxins (type 2) also act upon this site; an initial report suggests that they act upon a separate sodium channel site. The structures of the two actiniid short toxins reveal little sequence similarity with the long toxins; further experiments are needed to establish whether they bind to site 3 or another site on the Na channel.

 
 

Sea anemone cytolysins have only begun to be characterized. The only cytolysin which has been characterized in any detail is *Stichodactyla* (previously *Stoichactis*) *helianthus* cytolysin III. It is a single chain polypeptide of 153 residues which lacks half-cystine residues. This toxin aggregates in the cell membrane to form a moderately large pore for both cations and anions, depending on the pH. The toxin acts much more readily upon sphingomyelin-containing membranes; several molecules of this lipid apparently act as a membrane acceptor site. Several other species of actiniarian sea anemones seem to contain very similar homologous proteins. It is proposed that those actiniarian pore-forming cytolysins homologous with the *S. helianthus* cytolysin be called "actinoporins."

## I. INTRODUCTION

The phylum Cnidaria probably offers the largest and most diverse group of toxic animals for investigation: approximately 10,000 species, most probably possessing one or more types of toxin. Although a few sea anemone toxins have been extensively investigated during the past ten years, the overall status of knowledge concerning cnidarian toxins must still be considered to be somewhat rudimentary; even with the best characterized toxins, little is yet known about their tertiary structure, mechanisms of cardiac inotropism, biosynthesis, tissue localization, and biological roles. In this review, I shall focus upon sea anemone polypeptide toxins affecting sodium channels and protein cytolysins affecting cell membranes. The major emphasis will be upon the laboratory characterization of toxin structures and mechanisms of action; the knowledge thus obtained should ultimately be useful in understanding the biological functions of the toxins. Sodium channel polypeptide and cytolytic protein toxins recently isolated from the caribbean Sun Anemone *Stichodactyla* (formerly *Stoichactis*) *helianthus*. At the end, I suggest some possible ecological implications of the laboratory results for understanding how sea anemones interact with other animals under natural conditions.

## II. STRATEGIES FOR PURIFYING SEA ANEMONE TOXINS

From the very beginning, the purification procedure selected should exploit the known tissue localization of the toxin, in order to minimize contaminants. The milking of snake or other venoms is an example of what might be called a "biological" purification step. In the case of certain cnidarians, it is possible to obtain undischarged nematocysts and then osmotically or otherwise shock them to obtain practically pure venom. Such a procedure has been attempted with sea anemones several times, but has only been successful for the acontial nematocysts of a few species, including the anemones *Metridium senile* (Phillips and Abbott, 1957) and *Aiptasia pallida* (Blanquet, 1968). A preparative method for obtaining nematocysts from the fishing tentacles of sea anemones would be very useful for isolating toxins; perhaps the isolation methods described by MacKay and Anderson and Weber *et al*. at this conference can be exploited for this purpose. Although less appealing, agitation of whole animals in 1% acetic acid for several minutes, a method (Kem, 1976) which is very useful for obtaining protein toxins from the integuments of heteronemertine worms may also be useful for certain cnidarians when nematocyst isolation methods are not successful.

The method most commonly used for extracting the sea anemone polypeptide toxins involves homogenizing whole animals in an equal volume of ethanol, heating to 60°C to denature enzymes and other protein contaminants, pelleting insoluble impurities by centrifugation, and finally, batch adsorption to ion exchangers (Béress, 1978; Schweitz *et al*., 1981). This protocol seems fairly efficient if only polypeptide toxins are to be isolated. However, many species also possess larger protein toxins which could be denatured by this rather harsh procedure. In purifying *Stichodactyla helianthus* toxins, we used a milder procedure: toxin exudation into cold water after two freeze-thaw cycles, followed by centrifugal removal of insoluble components, and ammonium sulfate precipitation of toxic proteins in the supernatant. This procedure avoids introducing the tissues of the entire animal into the purification. It was previously used to obtain hemolysins, neurotoxins, and and proteinase inhibitors from three other stichodactylid anemone species (Mebs and Gebauer, 1980; Mebs *et al*., 1983). Regrettably, I have found that it is ineffective in extracting polypeptide toxins from *Anthopleura xanthogrammica* and *A. elegentissima*.

Before proceeding to the chemical and biological characterization of a purified toxin, it is necessary to assess its purity. The degree of purity required varies with the experiment planned. For determining the amino acid sequence, lesser purity may suffice in comparison with a pharmacological experiment where one has to remove even trace amounts of another type of interfering toxin or enzyme (example: phospholipase). Unfortunately, some investigators have relied almost exclusively upon the use of SDS polyacrylamide gel electrophoresis under reducing, denaturing conditions to assess the purity of their toxin samples. The most difficult part of any protein purification is generally that of separating the variants of that protein, as they usually have nearly identical molecular sizes and may differ only slightly in ionization behavior. Since SDS gel electrophoresis separates polypeptide chains almost entirely on the basis of molecular size, a sample may appear quite homogeneous by this method and yet be quite impure. This was the case for *Stichodactyla helianthus* cytolysin (Bernheimer and Avigad, 1976; Kem *et al.*, 1982). The assessment of protein purity should thus be made by several methods. Analytical isoelectric focusing on polyacrylamide gels is an excellent method, as even small differences in isoelectric point are sufficient for resolving most isotoxins. N-terminal sequencing is probably the most discerning method for evaluating protein homogeneity.

## III. BIOASSAYS

The selection of appropriate bioassays is obviously quite important when purifying toxins from a new organism. We have generally relied upon whole animal (mouse and crab or crayfish) paralysis and hemolysis assays for detecting and quantitating toxins specifically affecting excitable and other cell membranes. An advantage of the acute whole animal assays is that they respond to a variety of toxins affecting different physiological systems. This nonspecificity allows the detection of active substances which might not be noticed with more selective bioassays. A major advantage of *in vitro* cell preparations is that their responsiveness is more specific and less influenced by pharmacokinetic factors such as blood:brain barriers and proteolytic inactivation. For instance, isolated mammalian heart preparations have been very useful for detecting the mammalian-active sea anemone polypeptide toxins (Shibata *et al.*, 1974). Hemolysis assays are extremely simple, quantitative procedures for detecting cytolysins which act by making the cell membrane leaky to ions. One can, by simple visual inspection, determine if the proper range of toxin dilutions has been selected in order to obtain a complete concentration-hemolysis curve; if not, additional dilutions can be quickly prepared. The sensitivity of hemolysis assays is however, contingent upon several factors: erythrocyte source species, the erythrocyte cell count (hematocrit) used, and the incubation conditions. Some investiga-

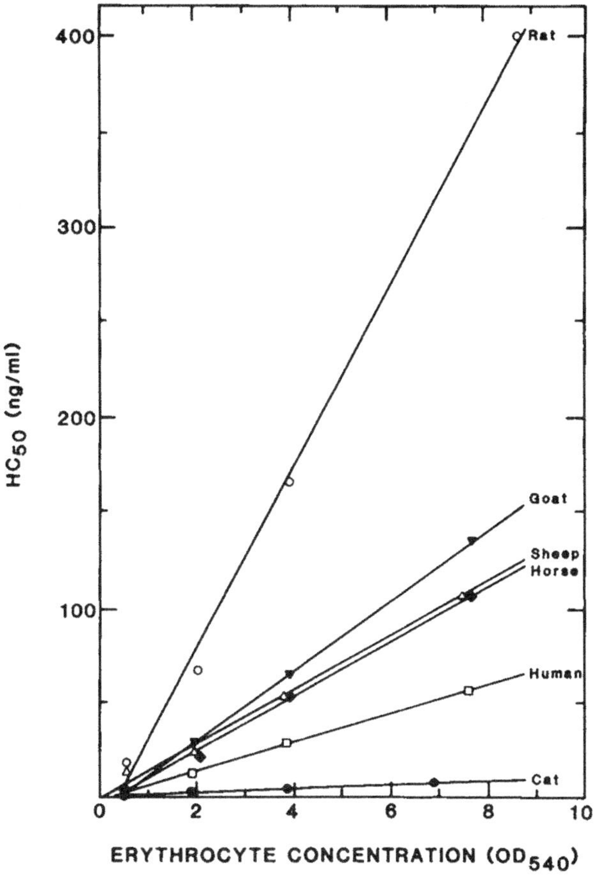


Fig. 1. The dependence of the Stichodactyla helianthus cytolysin III median hemolytic concentration upon erythrocyte cell count (measured by total hemoglobin release; a density of 1.0 means that total lysis yielded a 540 nm absorbance of 1.0). Each point is the average of 3 estimates.

tors have attempted to compare toxin hemolysis data obtained with erythrocytes of one mammalian species with data for other toxins obtained with erythrocytes from other mammalian species, without recognizing the importance of erythrocyte source. Bernheimer and Avigad (1976, 1978) have reported erythrocyte species comparisons for Stichodactyla helianthus and *Metridium senile* toxins which show that species differences in sensitivity are often quite large. The erythrocyte cell count is an extremely critical factor since most cytolysins are tightly bound to the cell membrane within a short time after mixing: the final free concentration of the lysin is thus affected by the hematocrit and is only a small fraction of the initial free concentration. When considering a non-enzymatic cytolysin, it can be assumed that lysis only occurs when a critical amount of toxin is bound to each cell. The greater the cell count, the less likely is this critical density of bound toxin per cell to be reached for a given initial toxin concentration. Thus, the median hemolytic concentration ($HC_{50}$) of a toxin will increase in a linear fashion as the cell count increases (Fig. 1; also Thron, 1964). For convenience, we have used a cell count which gives an absorbance of 1.0 at 540 nm upon complete lysis with the detergent TX-100 (Kem and Blumenthal, 1978).

## IV. POLYPEPTIDE TOXINS

### A. Primary Structure

Eleven years ago, the first anemone toxin (*Anemonia sulcata* toxin II or As-II) amino acid sequence was reported (Wunderer *et al.*, 1976). Sequences of several other *Anemonia* and *Anthopleura* isotoxins have subsequently been elucidated. Wunderer *et al.* (1978) assigned the disulfide crosslinks in As-II. The sequences of these actiniid toxins are quite conservative--at equivalent positions one usually observes the same amino acid.

However, during the past two years three different laboratories reported sequences for stichodactylid polypeptides (a family of large, predominantly Indopacific anemones, including *Heteractis* species) that are quite different from the initially studied toxins, but which closely resemble each other. All complete sequences of sea anemone "long" polypeptide toxins are shown in Table I. The stichodactylid toxins differ from the actiniid toxins by: (1) lacking a single residue at the N-terminus, (2) possessing a very basic C-terminus consisting of the Arg-Lys-Lys-Lys sequence, (3) occurrence of a Lys residue between the first two half-cystines, (4) invariability of only one Trp residue, at position 30, and (5) occurrence of Tyr residues at positions 35 and 36, in at least two of the toxins. The C-terminal half of the sequence originally reported for *Heteractis* (formerly *Radianthus*) *paumotensis* II was incon-

TABLE I. Amino acid sequences of sea anemone long polypeptide toxins affecting sodium channels. References: As-I, Wunderer and Eulitz, 1978; As-II, Wunderer *et al.*, 1976; As-V, Scheffler *et al.*, 1982; Ax-I (anthopleurin A), Tanaka *et al.*, 1977; Ax-II (anthopleurin B), Reimer *et al.*, 1985; Sh-I, Kem *et al.*, 1986; Hm-III, Zykova *et al.*, 1985; Hp-II, Schweitz *et al.*, 1985.

| | 1 | | | | 5 | | | | | 10 | | | | | 15 | | | | | 20 | | | | | | 25 | | | | | 30 | | | | | 35 | | | | | 40 | | | | | 45 | | |
|---|---|---|---|---|---|---|---|---|---|---|---|---|---|---|---|---|---|---|---|---|---|---|---|---|---|---|---|---|---|---|---|---|---|---|---|---|---|---|---|---|---|---|---|---|---|---|---|---|---|
| As-I | G | A | P | C | L | C | K | S | D | G | P | N | T | R | G | N | S | M | S | G | T | I | W | V | | | F | G | C | P | S | G | W | N | N | C | E | G | R | A | | I | I | G | Y | C | C | K | Q |
| As-II | G | V | P | C | L | C | D | S | D | G | P | S | V | R | G | N | T | L | S | G | I | I | W | L | | | A | G | C | P | S | G | W | H | N | C | K | K | H | G | P | T | I | G | W | C | C | K | Q |
| As-V | G | V | P | C | L | C | D | S | D | G | P | S | V | R | G | N | T | L | S | G | I | L | W | L | | | A | G | C | P | S | G | W | H | N | C | K | K | H | K | P | T | I | G | W | C | C | K | |
| Ax-I | G | V | S | C | L | C | D | S | D | G | P | S | V | R | G | N | T | L | S | G | T | L | W | L | Y | P | S | G | C | P | S | G | W | H | N | C | K | A | H | G | P | T | I | G | W | C | C | K | Q |
| Ax-II | G | V | P | C | L | C | D | S | D | G | P | R | P | R | G | N | T | L | S | G | I | L | W | F | Y | P | S | G | C | P | S | G | W | H | N | C | K | A | H | G | P | N | I | G | W | C | C | K | K |

| | 1 | | | | 5 | | | | | 10 | | | | | 15 | | | | | 20 | | | | | | | 25 | | | | | 30 | | | | | 35 | | | | | 40 | | | | | 45 | | | |
|---|---|---|---|---|---|---|---|---|---|---|---|---|---|---|---|---|---|---|---|---|---|---|---|---|---|---|---|---|---|---|---|---|---|---|---|---|---|---|---|---|---|---|---|---|---|---|---|---|---|---|
| Sh-I | A | A | C | K | C | D | D | E | G | P | D | I | R | T | A | P | L | T | G | T | V | D | L | | | G | S | C | N | A | G | W | E | K | C | A | S | Y | Y | T | I | I | A | D | C | C | R | K | K | K |
| Hm-III | G | N | C | K | C | D | D | E | G | P | Y | V | R | T | A | P | L | T | G | Y | V | D | L | | | G | Y | C | N | E | G | W | E | K | C | A | S | Y | Y | S | P | I | A | E | C | C | R | K | K | K |
| Hp-II | A | S | C | K | C | D | D | D | G | P | D | V | R | S | A | T | F | T | G | T | V | D | F | | | W | N | C | N | E | G | W | E | K | C | T | A | V | Y | T | P | V | A | S | C | C | R | K | K | K |

Abbreviations: A = alanine, C = half-cystine, D = aspartic acid, E = glutamic acid, F = phenylalanine, G = glycine, H = histidine, I = isoleucine, K = lysine, L = leucine, N = asparagine, P = proline, Q = asparagine, R = arginine, S = serine, T = threonine, V = valine, W = tryptophan, Y = tyrosine.

sistent with its NMR spectral analysis, but the revised sequence (Wemmer *et al.*, 1986) is very similar to the sequences of the other two stichodactylid toxins. The *H. paumotensis* toxins were immunologically distinct from *A. sulcata* toxins (Schweitz *et al.*, 1985). Structurally, one can thus separate the anemone "long" polypeptide toxins into two types. The *Anemonia* and *Anthopleura* polypeptides are called type 1 toxins since they were isolated and characterized first. The three stichodactylid polypeptides are designated as type 2 toxins. As other anemone long toxin sequences become available, it will be interesting to see if the distributions of these two polypeptide types strictly follows anemone family systematics. Do both types co-exist in any species? Although there are many differences between the two toxin types, it is equally obvious that the six half-cystines are always in the same places, as well as are a few other residues such as Arg 14 (13 in type 2 toxins). Thus, it seems highly probable that these two types are evolutionary descendants of a common ancestral gene.

Other anemone polypeptides currently being sequenced include *Bolocera tuediae* II (see Table 3 of Béress, 1986), *Condylactis gigantea* toxin III (Kem *et al.*, 1986), *Phyllactis flosculifera* I (Kem *et al.*, unpublished results), and a toxin from the acontiate anemone *Calliactis parasitica* (L. Cariello, personal communication). The *Bolocera* toxin appears to have a unique sequence dissimilar to either of the two types of long toxins discussed above, even though it also delays the closing of the sodium channel in a similar way. The *Condylactis* and *Phyllactis* toxin sequences indicate that they are type 1 toxins. One very interesting aspect of their structures is the occurrence of trans-4-hydroxy-L-proline at position 3 (Kem *et al.*, 1986); this hydroxyproline has also been recently found in polypeptide toxins from two other invertebrate phyla, the nemertines (Blumenthal *et al.*, 1981) and molluscs (Stone and Gray, 1982). The influence of hydroxyproline upon the toxicity of these polypeptides has not yet been studied.

Only two "short" polypeptide toxins have been reported. The major polypeptide toxin of the common Indopacific actiniid *Entacmea quadricolor* (formerly named *Parasicyonis actinostoloides*) possesses only 31 residues, but eight are half-cystines. About half of its sequence closely resembles that of *Anemonia sulcata* III (Nishida *et al.*, 1985). Only a trio (Gly 32-Cys 33-Pro 34) of residues in these two toxins is alignable with those of the long toxins.

The toxicities of the purified sea anemone polypeptide neurotoxins are listed in Table II. Although all of the toxins are at least moderately toxic to crabs, only a few

TABLE II. Pharmacological properties of purified sea anemone long polypeptide toxins affecting sodium channels. $LD_{50}$'s are based on intrahaemocoelic (crab) or intraperitoneal injections; $K_D$ = equilibrium dissociation constant; $EC_{50}$ = median effective concentration increasing sodium fluxes. References: 1) Cg, Pf, Sh (Kem, submitted); 2) As, Ax, Sg (Schweitz et al., 1981); 3) Hm (Zykova et al., 1985); 4) Hp (Schweitz et al., 1985).

| | $LD_{50}$ (µg/Kg) | | Rat Tissue Response (nM) | |
|---|---|---|---|---|
| Toxin | Crab | Mouse | Brain $K_D$ | Heart $EC_{50}$ |
| Type 1: | | | | |
| Cg-II | 0.2 | >50,000 | ------ | >1,000 |
| Pf-I | 0.4 | >20,000 | ------ | >1,000 |
| As-I | 2 | 4,000 | 7,000 | ----- |
| As-II | 2 | 100 | 150 | 15 |
| As-V | 5 | 19 | 50 | 2 |
| Ax-I | 11 | 66 | 120 | 3 |
| Ax-II | 39 | 8 | 35 | 2 |
| Type 2: | | | | |
| Sh-I | 0.3 | >15,000 | ------ | >8,000 |
| Sg-I | 7 | > 2,000 | > 10,000 | ----- |
| Hp-III | 10 | 53 | 300 | 4,000 |
| Hp-II | 15 | 4,200 | >100,000 | 5,000 |
| Hp-I | 36 | 145 | 900 | 3,000 |
| Hp-IV | 90 | 40 | 10,000 | 1,300 |
| Hm-III | 820 | 20 | ------ | ----- |

As = *Anemonia sulcata*, Ax = *Anthopleura xanthogrammica*, Cg = *Condylactis gigantea*, Hm = *Heteractis macrodactylus*, Hp = *Heteractis paumotensis*, Pf = *Phyllactis flosculifera*, Sg = *Stichodactyla giganteum*, Sh = *Stichodactyla helianthus*.

seem to be very active upon whole mice or rat brain synaptosome sodium fluxes. Toxins with extremely low $LD_{50}$ estimates for one group generally displayed a high $LD_{50}$ for the other.

## B. Secondary and Tertiary Structure

There is considerable interest in how these polypeptides bind to their receptor sites on sodium channels. A detailed three-dimensional toxin structure is needed to serve as a conceptual guide for further experimentation. Several crystallographic laboratories are currently pursuing this problem. Two-dimensional proton magnetic resonance spectroscopy now offers an alternative approach for obtaining the solution conformation of polypeptides, although in less detail. Gooley and Norton (1986) have used this approach to locate four short B strands in Ax-I at residues 2-4, 20-24, 33-37, and 44-48. The large loop consisting of residues 7-28 was shown to be partially linked to other regions by several antiparallel β-sheet hydrogen bonds. Secondary structures previously inferred from other spectroscopic measurments and Chou-Fasman predictions (Ishizaka _et al._, 1979; Nabiullin _et al._, 1982) only roughly resembled the PMR-deduced secondary structure. Many more interproton distance estimates are needed before even an approximate solution conformation can be deduced by the PMR method, but this seems feasible in the near future. The type 2 toxins display CD spectra nearly identical with the type 1 toxins, so one may expect very similar secondary and tertiary structures for these toxin types (Nabiullin _et al._, 1982; Kem, unpublished results).

### C. Structure-Activity Relationships

In view of the potential uses of sea anemone toxin-mimetic compounds as cardiac inotropic drugs or as insecticides, a challenging research problem is that of identifying the "receptor-binding domain" on the surface of the polypeptide toxin. This knowledge, if detailed enough, might conceivably allow the molecular design of simpler compounds which could selectively control sodium channel function. Two types of information are needed: (1) a high resolution 3-dimensional structure by x-ray crystallographic analysis, and (2) knowledge of the importance of individual amino acid sidechains for binding and stimulating sodium channels obtained by both chemical modification and analog synthesis approaches. The chemical modification approach has several limitations, the major ones being limited selectivity and types of substitutions. In principle, the first weakness can be overcome by careful optimization of reaction conditions and rigorous purification and description of the ensuing products. Barhanin _et al._ (1981) have carried out the most extensive anemone toxin modification experiments to date, using As-II. Only the most interesting results will be discussed here. Cyclohexanedione addition to Arg 14 abolished crab toxicity. Kolkenbrock _et al._ (1983), using the same reagent failed to observe an effect of modifying Arg 14 in As-II. Phenylglyoxal modification of Ax-I failed to affect its inotropic activity (Newcomb _et al._, 1980), but this may have been due to conversion back to the native polypeptide at alkaline pH. Kozlovskaya et al. (1982) modified Arg 13 and Arg 45 in Heteractis macrodactylus I by converting the guanidyl moiety into a pyrimidine ring; the modified toxin possessed less than 5% of the native toxin's activity. Since Arg 14 (13 in type 2 toxins) seems practically invariant among the long toxins, it is likely that it is part of the toxin's receptor binding domain. Further experiments are required to clarify the importance of this particular amino acid sidechain for receptor binding and activation.

The carbodiimide-mediated addition of glycine ethyl ester to the three carboxyl-containing sidechains of As-II was also investigated by Barhanin et al. (1981). Two monosubstituted toxic derivatives were isolated from the reaction mixture but were not further characterized. However, a derivative resulting from the modification of all three carboxyls was reported to act as a competitive antagonist capable of preventing the action of the native toxin. This extremely interesting result could not be confirmed by Kolkenbrock et al. (1983). Gruen and Norton (1985) recently reported that when this reaction modified both Asp 7 and Asp 9 in Ax-I, its conformation was greatly altered and its biological activity lost.

We recently succeeded in the solid phase synthesis of Sh-I, after unsuccessfully attempting to obtain analogs of this toxin by semisynthesis for some time. The crab toxicity, spectra (CD, fluorescence, NMR), amino acid composition, and isoelectric point of the synthetic toxin were identical with that of the native toxin. Total solid phase synthesis now offers a versatile and well defined means of obtaining toxin analogs useful for investigating structure-activity relationships of sea anemone polypeptide toxins. (Pennington et al., in press).

## D. Mechanisms of Action

Space does not permit an extensive review of the many recent papers on the physiological effects of the sea anemone polypeptide toxins; rather I will attempt to touch on some of the more interesting aspects of their actions. The sea anemone polypeptides, like tetrodotoxin and saxitoxin, have a remarkably uncomplicated spectrum of actions: they selectively alter the function of the sodium channel present in electrically-excitable tissues and all of their pharmacological effects seem explainable on this basis. The major effect of these polypeptides upon the sodium channel is to delay the inactivation process (Narahashi *et al.*, 1969). Inactivation involves a transition from an activated, open channel state to a closed, but not readily activatable state:

$$\underline{R}\text{esting} \rightleftharpoons \underline{O}\text{pen} \rightleftharpoons \underline{I}\text{nactivated}$$

Normally, membrane depolarization triggers the transition of the sodium channel from R ⟶ O ⟶ I. After exposure to these polypeptides, the probability of the O ⟶ I transition is reduced considerably, so that some sodium channels never seem to inactivate (Ulbricht and Schmidtmayer, 1981). This causes greatly prolonged action potentials, leading to an organism's hyperexcitability, at least in the earlier stages of intoxication. Hartung and Rathmayer (1985) recently reported that the *Anemonia* polypeptides also affect the activation (R ⟶ O) process in a crayfish neuronal soma preparation, though their action is primarily upon the inactivation process.

An intriguing aspect of the mechanism of action of the sea anemone and scorpion α-toxin polypeptides is the way in which membrane potential controls their interaction with sodium channels. As the membrane is depolarized, the affinity of the polypeptide for its receptor site is reduced; the voltage-dependence of polypeptide toxin binding most likely reflects the preferential binding of the toxin to the activated state of the channel. An alternative hypothesis involving the influence of the membrane's electrical field upon the toxin's orientation near the receptor seems less plausible, partly because various anemone and scorpion toxin variants acting upon this site show similar voltage dependencies, though their isoelectric points differ markedly. One possible exception is the short toxin isolated from *Entacmea quadricolor*. The ability of this polypeptide to prolong the duration of the action potential (crayfish giant axon) was affected by much smaller depolarizations than were necessary to affect *Anemonia* and *Leiurus* (scorpion α-toxin) polypeptide toxin actions (Warashina *et al.*, 1981).

The actions of the vertebrate-active polypeptides upon several neuromuscular preparations were reported recently. *A. sulcata* II enhanced the contractility of both amphibian and mammalian skeletal muscle preparations. Since the toxin enhanced rat diaphragm contractions elicited by electrical stimulation of either nerve or muscle, the toxin seems to act upon sodium channels in both types of cell (Alsen *et al.*, 1981). Recent experiments on frog muscle are also consistent with this interpretation (Erxleben and Rathmayer, 1984; Khan *et al.*, 1986). Erxleben and Rathmayer also showed that enhancement of transmitter release by As-II was due to a selective increase in quantal content (the number of quantal units of transmitter released per nerve impulse). The effects of Ax-II (anthopleurin B) upon frog spinal cord and guinea pig smooth muscle preparations seem to be entirely explainable in terms of toxin enhancement of neurotransmitter release (Kudo and Shibata, 1980; Ohizumi and Shibata, 1981; Norton et al., 1981). Denervated skeletal muscle apparently possesses the same form of sodium channel as in the heart, based upon its high sensitivity to As-II and low sensitivity to tetrodotoxin (Alsen et al., 1981; Frelin et al., 1984).

A potentially therapeutic property of some sea anemone polypeptides is their ability to selectively stimulate cardiac contractility in intact mammals without affecting neuromuscular function (Scriabne et al., 1979). This effect is directly upon a myocardial sodium channel which is unusually sensitive to sea anemone but not scorpion toxins (Catterall and Coppersmith, 1981). The myocardial channel differs from brain sodium channels in being about 500 times less sensitive to tetrodotoxin (for a recent review, see Catterall, 1986). A similar but less antigenic substance displaying this selectivity and mode of action upon heart sodium channels might be a very useful drug for treating patients with congestive heart failure (Shibata and Norton, 1982). Development of such a drug is an exciting but long term goal of several laboratories investigating the sea anemone polypeptide toxins. Besides the challenge of identifying the receptor-binding domain on the polypeptide and designing simpler, possibly non-peptide analogs, it will also be necessary to minimize the predominant toxic effect of these substances: their ability to induce extra-systolic contractions resulting from the delayed inactivation of cardiac sodium channels (Scriabne et al., 1979). Clinical use of the naturally-occurring polypeptides seems unlikely, because of their antigenicity and their presumed poor bioavailability when administered orally.

At this time it isn't clear how the polypeptides actually increase cardiac contractility, although their direct effect is to stabilize the open state of the sodium channel, thereby enhancing sodium influx during the heart beat. The duration of the action potential, normally controlled by calcium and potassium permeability changes, can be increased by keeping the sodium channels open for several hundred milliseconds. The augmented sodium influx into the myocardial cell is expected to considerably increase the intracellular sodium ion concentration during each heart beat. This toxin action was initially thought to indirectly cause an increase in intracellular calcium concentration by stimulating a sodium-calcium exchange mechanism located in the plasma membrane. However, the polypeptide-mediated positive inotropic effect was reported to be relatively insensitive to the external calcium concentration. It was also antagonized by drugs (dantrolene and ryanodine) which inhibit the release of calcium from intracellular stores (Bailey _et al_., 1980; Kodama _et al_., 1981). Perhaps the increased intracellular sodium concentration may in some way enhance excitation-contraction coupling at the sarcoplasmic reticulum membrane surface, rather than at the plasma membrane. Elucidation of the inotropic mechanism of action of these polypeptides could thus provide an important contribution to our understanding of the underlying processes involved in excitation-concentration coupling in muscle.

In a recent paper Renaud _et al_. (1986) compared the cardiac actions of several sea anemone (and scorpion) toxins. The _Heteractis paumotensis_ (type 2) long toxins produced only small, transient increases in contractility relative to the _Anemonia_ and _Anthopleura_ type 1 toxins. The actions of all anemone polypeptides were biphasic, the initial positive inotropic effect being followed by a decrease in contractility below the control level. Since adrenergic and cholinergic antagonists were not present during these experiments, it is possible that some of the polypeptide effects upon isolated rat atria were mediated through its autonomic innervation. The positive inotropic effects of _Goniopora_ (coral) toxin were similar but not identical to the effects of the type 1 toxins (Fujiwara _et al_., 1979; Noda _et al_., 1985).

## E. Binding to Sodium Channels

The type 1 long toxins have been most intensively investigated. Catterall and Béress (1978) first demonstrated the saturable binding of As-II to rat brain synaptosomal and neuroblastoma sodium channels and the ability of As-II to inhibit the specific binding of a scorpion (_Leiurus_) toxin. The affinity of this site for As-II was about one-hundred times less than it's affinity for _Leiurus_ toxin; As-II binding was inhibited by membrane depolarization but less than for _Leiurus_ toxin. Another group also investigated the binding of As-II to rat brain synaptosomes but their results differed in rather important ways: (1) the number of As-II binding sites was about ten times greater than the number of scorpion (_Androctonus_) toxin binding sites, and (2) As-II binding was not voltage-dependent. They have interpreted these results as an indication of sodium channel heterogeneity--only a small fraction of sodium channels bind the scorpion α-toxin. Relative to the number of tetrodotoxin binding binding sites, they found about twice as many As-II sites and about 1/5 as many scorpion toxin sites (Vincent *et al.*, 1980). Further experiments are needed to resolve these important differences.

The binding of the type 2 long toxins to the sodium channel has not yet been extensively investigated. Schweitz *et al.* (1985) tested the ability of several type 2 long toxins from *H. paumotensis* to displace the binding of labelled *Androctonus* (scorpion) toxin and As-V to rat brain synaptosomes. The *H. paumotensis* toxins were able to compete for the *Androctonus* toxin binding sites, but not the As-V binding sites. These results indicate that the *Heteractis* toxins bind to the scorpion α-toxin binding site, whereas the binding site for As-V is separate. Hp-II, a crustacean-selective isotoxin, was unable to displace either bound AS-V or *Androctonus* toxin. These results lead to the tentative conclusion that the type 2 toxins bind to a site on the sodium channel which is different from the receptor site for the type 1 toxins. It is now known that there are two separate sites for scorpion toxins; only the α-toxins display voltage-dependent binding (Jover *et al.*, 1980). That the action of one type 2 (*H. macrodactylus*) toxin is voltage dependent (Kryshtal *et al.*, 1983), is consistent with the conclusion of Schweitz *et al.*, (1985), that *H. paumotensis* toxins bind to the scorpion α-toxin site.

Fujita *et al.* (1983) recently reported on the binding of iodinated *Entacmea quadricolor* toxin (a short toxin) to intact crayfish leg nerves. The equilibrium dissociation constant for binding was 310 x $10^{-9}$ M. Toxin binding was inhibited by membrane depolarization, by *Leiurus* toxin, and by As-II. These results, though not very extensive (Also, specific binding could not be demonstrated with nerve vesicle preparations), suggest that this short toxin also binds to the type 1 long toxin binding site (Catterall's site 3). *Goniopora* toxin, a 9,700 dalton polypeptide isolated from a coral, did not affect the binding of *Leiurus* toxin to mouse neuroblastoma cells, so it must interact with a separate site on the sodium channel.

The relationships between the different polypeptide binding sites on the sodium channel are yet to be deciphered. The recently published DNA sequences of the electric eel and rat brain sodium channel α-subunits indicate that each consists of four repeating sequences which are homologous but not identical (Noda *et al.*, 1986). It is possible that each of these four homologous sequences possesses one or more non-equivalent polypeptide binding sites which bind different types of sea anemone, coral, and scorpion toxins.

## V. PROTEIN CYTOLYSINS

### A. Initial Comments

These substances, so named because of their readily bioassayed ability to lyse cells, are the most widespread type of toxin found in nature. They generally act upon the cell membrane, upsetting cell electrolyte balance by making the membrane abnormally permeable to inorganic ions. All the anthozoan cytolysins so far described are proteins. (This may soon change: hemolytic lipids were recently found in a hydrozoan by Fusetani *et al.*, 1986.) Under natural conditions the cytolysins probably produce their effects at sublytic concentrations by modulating nervous, cardiovascular, gastrointestinal, and inflammatory systems in envenomated animals.

The major cytolysin of *Stichodactyla helianthus* is the only sea anemone cytolysin which has been extensively investigated by both protein structural and membrane mechanistic approaches. In this section, we shall focus upon recent work on this type of anthozoan toxin and at the end briefly consider the few other types of cytolysins which have been investigated.

### B. Purification and Characterization of *Stichodactyla* Cytolysins

Initial experiments carried out with Tahitian *Stichodactyla kenti* (Dunn, 1981, found this species synonomous with *S. gigantea*) demonstrated that extracts of this Indo-pacific species possessed antineoplastic and mouse central nervous system stimulatory activities (Norton and Kashiwagi, 1972; Turlapaty *et al.*, 1973). Unfortunately, the substances responsible for these activities were not sufficiently purified and characterized to permit their comparison with the pure cytolysins recently isolated from *S. helianthus*. Bernheimer and Avigad (1976) isolated a potent 16,000 dalton basic hemolytic protein from *S. helianthus* which contained some carbohydrate. Their protein samples, isolated by preparative isoelectric focusing, were later found to be inhomogeneous by gel or ion exchange chromatography and by analytical isoelectric focusing on polyacrylamide gels. Eventually, four separate protein isotoxins were obtained from *S. helianthus* by CM cellulose gradient chromatography in the presence of 4 M urea to solubilize the toxins and prevent the formation of mixed isotoxin complexes. There were two major toxins (II, III) and two minor toxins (I, IV); toxin III contributed about 70% of the total hemolytic activity, II about 28%, and the remaining ones less than 2% of the total activity. The molecular weight (19,000) of toxin IV exceeded those of the other toxins by about 2,000 daltons. The N-terminal amino acid sequences of the isotoxin are shown in Table III. It is possible that isotoxins IV is an incompletely processed form of toxin III; further analysis of its sequence should answer this question. The N-terminal sequence of *Actinia equina* toxin (Ferlan and Jackson, 1983; Table III) clearly resembles the sequences of the *S. helianthus* toxins. In contrast with the results of Bernheimer and Avigad (1976), we failed to detect significant amounts of half-cystine or carbohydrate in these toxins; presumably these were due to contaminants. The hemolytic potency of purified Sh-CIII for human erythrocytes was about 15 times greater than was previously reported (Kem and Dunn, submitted).

Table III. Amino terminal sequences of *Stichodactyla helianthus* cytolysins and *Actinia equina* cytolysin. References: 1) Sh cytolysins (Kem and Dunn, submitted); 2) Ae cytolysin (Ferlan and Jackson, 1983).

| | |
|---|---|
| Sh-CI: | SER-GLU-LEU-ALA-GLY-THR-ILE-ILE-ASP- |
| Sh-CII: | ALA-GLU-LEU-ALA-GLY-THR-ILE-ILE-ASP- |
| Sh-CIII: | ALA-LEU-ALA-GLY-THR-ILE-ILE-ALA- |
| Sh-CIV: | ASP-GLN-LYS-GLU-GLN-THR-ARG-ALA-LEU-ALA-GLY-THR-ILE-ILE-ALA- |
| Ae-CI: | SER-ALA-ASP-VAL-ALA-GLY-ALA-VAL-ILE-ASP- |

The amino acid sequence of Sh-CIII (Blumenthal and Kem, 1983) possesses several interesting features. The first 17 N-terminal residues are quite hydrophobic and, therefore, it is possible that this portion may span the entire lipid bilayer, particularly if they occur as a B-sheet. The polypeptide chain in the C-terminal one-third of this protein is quite polar, suggesting that this portion remains on the outside surface of the target cell membrane. From denaturation experiments one can conclude that the toxin has a rather stable secondary and tertiary structure despite lacking disulfide crosslinks. This seems largely due to the existence of a very high (about 65%) proportion of antiparallel β-pleated sheet in this protein. Although the Chou-Fasman method predicted 16% α-helix and 31% β-sheet, we could not detect any α-helix using either circular dichroism or Raman spectroscopic techniques (Kem and Williams, in preparation). The current dogma in membrane protein research is that α-helices are used to span the lipid bilayer interior of the membrane. Sh-CIII however, like the bacterial porins, seems to exploit a different, as yet unknown, molecular strategy to penetrate through the lipid bilayer. Crystals of Sh-CIII have recently been obtained (R. Honzatko, personal communication); once the 3-dimensional structure of this protein becomes available, it should be possible to formulate and test hypotheses regarding the surfaces of this protein are involved in self-association, lipid binding, and pore formation.

## C. Lipid Dependence of *Stichodactyla* Cytolysin Action

Bernheimer and Avigad (1976) made the very interesting observation that preincubation of *S. helianthus* cytolysin with microgram amounts of a particular phospholipid, sphingomyelin, inhibited its ability to lyse erythrocytes. The only other lipid inhibiting this cytolysin was ceramide, which results from the removal of choline from sphingomyelin. Isoelectric focusing of the purified toxin following preincubation with sphingomyelin caused it to migrate to a neutral pH region with the lipid, while exposure to lecithin failed to influence the toxin's migration to its isoelectric pH (9.8). Using the electron microscope, ferritin-labelled toxin was shown to be bound to sphingomyelin-containing membranes, even after they were washed (Linder *et al.*, 1977).

The ability of Sh cytolysin to release radioactive glucose from sphingomyelin-containing multilamellar liposomes was inhibited by previous incubation of the liposomes with staphylococcal sphingomyelinase. Liposomes prepared in the same way with ceramide replacing sphingomyelin were insensitive to the toxin. These experimental results suggest that sphingomyelin acts as a membrane acceptor for the toxin (Linder and Bernheimer, 1978).

We have corroborated and extended many of these observations using phospholipid monolayers, unilamellar liposomes, and red cell ghosts as well as intact red cells. Somewhat surprising was the similarity of Sh-CIII effects upon dipalmitoylphosphatidylcholine or sphingomyelin monolayers initially adjusted to a surface pressure (40 dynes/cm) where the lipid packing is comparable to that in a bilayer. The toxin was able to reduce the surface pressure in monolayers containing either lipid regardless of the initial surface pressure. We also found the surface activity of Sh-CIII increased as the pH approached the toxin's isoelectric point (Doyle, Kem and Villalonga, submitted).

In order to investigate more quantitatively the toxin's binding to small unilamellar liposomes, we measured the interaction of $^{3}$H-labelled Sh-CIII with $^{14}$C-labelled sphingomyelin-containing liposomes. Liposomes were prepared with varying proportions (0-20%, on a molar basis) of $^{14}$C-labelled sphingomyelin mixed with dimyristoylphosphatidylcholine (a synthetic lecithin). The Sh-CIII was reductively methylated with $^{3}$H-formaldehyde and the least labelled toxin derivative (2 methyls per toxin) was isolated from the reaction mixture by CM cellulose chromatography. It retained nearly full cytolytic activity. After preincubating a fixed amount of $^{3}$H-toxin with liposomes (lipid:protein molar ratio of 400) for 1 hour, the mixture was separated on a Sepharose 4B gel column. The liposomes bound toxin eluted with the void volume whereas the unbound toxin eluted later in the included volume. As the proportion of sphingomyelin increased, the fraction of the toxin eluting with the liposomes increased in a linear manner, until all of the active $^{3}$H-cytolysin was bound to lipid vesicles. It was calculated that 15-20 sphingomyelin molecules were required per toxin monomer to allow maximum binding to the liposomes. If the same gel-lipid binding experiment was carried out using a lower (50:1) lipid:protein molar ratio, a new "micellar" peak with an intermediate elution volume containing both lipid and protein was observed. Each micelle was calculated to contain 3-4 toxin monomers and, thus may represent a single ion channel complex (Kem and Doyle, unpublished results).

The above-mentioned ability of Sh-CIII to nonspecifically bind to lipid monolayers seems to have its counterpart in lipid bilayers. Shin et al. (1979), using relatively high concentrations of cytolysin, showed that multilamellar liposomes containing other phospholipids could also be attacked; this interaction apparently included a detergent-like action, as some lipid was actually removed from the liposomes. Liposomes containing lecithins with long chain fatty acyl groups were more readily attacked by the toxin than those containing shorter acyl chains; Shin et al. suggested that the toxin's apparent preference for sphingomyelin and other phospholipids containing long chained fatty acyl groups reflects an optimum lipid bilayer thickness for toxin insertion.

### D. Stichodactyla Cytolysin Ion Channels

Two laboratories have demonstrated that Sh cytolysin can form ion channels in lipid bilayer membranes (Michaels, 1979; Varanda and Finkelstein, 1980). The average lifetime of a single channel was several seconds. The conductance increase caused by the formation of many such ion channels was found to be proportional to (total toxin concentration)$^n$, where $n = 3$ or 4. Such results suggest that a single ion channel consists of a complex of 3-4 toxin monomers. The ionic selectivity of the channel was low at pH 5 or below, but above this pH the channel was cation selective, as if an anionic site with $pK_a$ of about 5 occurs near the outer surface of the pore. The diameter of the ion channel was estimated to be about 10 Å, based upon its ability to pass sugars of different molecular sizes. Varanda and Finkelstein found that the presence of only small proportions (2%) sphingomyelin enhanced the conductance increase 10,000-fold and suggested that this could be explained by a 10-fold increase in the probability that the toxin monomer would insert into the membrane. Assuming that the ion channel complex contains four toxin monomers, the probability of channel formation would be enhanced by a factor of $10^4$. Both laboratories reported that sphingomyelin is not an absolute requirement for ion channel formation.

### E. Stichodactyla Cytolysin Binding and Action upon Erythrocyte Membranes

The equilibrium binding of Sh-CIII to mammalian erythrocytes was investigated by indirect (Thron, 1964) and direct ($^3$H-toxin) binding methods. Among the six species examined, cat and rat erythrocytes were most and least sensitive, respectively (Fig. 1). The median hemolytic concentration was proportional to the red cell density (hematocrit).

Although on theoretical grounds one would expect at least a low free toxin concentration to be observed as a y-axis intercept at an infinitely low red cell count, this could not be measured with the indirect method, due to the toxin's extremely high affinity for the erythrocyte membrane. The number of toxin monomers bound per red cell, under conditions where 50% lysis occurred, were 680 for the cat and about 18,000 for the rat. The indirect and direct binding estimates for all six mammalian species were in good agreement. There was no obvious relationship between the sphingomyelin content of a red cell and its sensitivity to the cytolysin. This is not too surprising. First, the red cells of all six mammalian species possess reasonably high concentrations of this lipid. The rat erythrocyte membrane, which possesses the least sphingomyelin, should contain 20-30% sphingomyelin in it's external lipid layer. Second, membranes such as the ruminant erythrocytes which possess very high proportions of sphingomyelin are expected to be very tightly packed and ordered; this would tend to inhibit toxin binding and aggregation.

We found that binding of $^{3}$H-Toxin to red cells was extremely rapid, reaching a steady state within about 10 seconds regardless of the toxin concentration used. The increase in ion permeability (measured by release of radioactive rubidium) preceded the release of hemoglobin; rubidium ion release always occurred after a definite time lag, whose minimum value was only about 10 seconds at room temperature. Cell lysis depended upon medium osmolarity as would be expected for colloid osmotic lysis (Kem and Doyle, in preparation).

## VI. OTHER CYTOLYSINS

Bernheimer and colleagues have surveyed several sea anemone species and found many cytolysins which resemble the *Stichodactyla* toxins in molecular size, isoelectric point, and sphingomyelin inhibitability (Table IV). It is likely most of these toxins are homologous; that is, they possess a common primary structure as a result of evolutionary divergence from a common ancestral toxin gene. With the concurrence of Dr. Bernheimer (personal communication), I suggest the name "actinoporin" be hereafter used as a general name for those actiniarian pore-forming toxins which can be shown by partial protein sequencing to have a primary sequence related to *Stichodactyla helianthus* cytolysin III. Use of such a general name will encourage comparison and possible synthesis of data derived from the various isotoxins and also emphasize their evolutionary relationship.

TABLE IV. Sphingomyelin-inhibitable cytolysins in sea anemones resembling *Stichodactyla helianthus* cytolysin.

| Species | Isoelectric Point | Molecular Size | Reference |
|---|---|---|---|
| *Actinia equina* | 12.5 | 20,000 | Ferlan and Lebez (1974) |
| *Anthopleura xanthogrammica* | 9.4 | <10,000 | Bernheimer and Avigad (1982) |
| *Entacmea quadricolor* | 7.9 | 18,000 | Shimomi *et al.* (1985) |
| *Epiactis prolifera* | 8.3 | <10,000 | Bernheimer and Avigad (1982) |
| *Heteractis macrodactylus* | ---- | 15,000 | Rudnev *et al.* (1984) |
| *Stichodactyla helianthus* | 9.8 | 16,000 | Bernheimer and Avigad (1976) |
| *Stichodactyla gigantea* | 9.0 | 18,000 | Bernheimer and Lai (1985) |
| *Pseudactinia varia* | 9.8 | 19,500 | Bernheimer *et al.* (1984) |

Specific actinoporins can then be designated by the genus and species abbreviations followed by a number assigned to the isotoxin (for example, Sh-CIII) which is determined by its basicity relative to the other cytolysins of that species, in order of increasing isoelectric point. Bernheimer (1987) prefers to refer to the particular cytolysins by species (helianthin, kentin, etc.).

In addition to the actinoporin-like toxins, Bernheimer and Avigad (1978) reported their discovery of a cholesterol-inhibitable protein cytolysin in the acontiate anemone *Metridium senile*. This protein has much greater hemolytic activity as the pH is decreased, even to pH 5. In addition, it is absent from the tentacles but present in the rest of the body (Bernheimer and Avigad, 1978). In my experiments with this species, I also failed to find it in the tentacles, but rather found it localized in the mesenterial filaments (Kem, unpublished results). For these reasons, I surmise that this cytolysin, by enhancing cell and tissue breakdown, aids digestion.

## VII. BIOLOGICAL ROLES OF SEA ANEMONE TOXINS

The occurrence of a substance within the tissues of an organism which is active in toxicity bioassays does not necessarily mean that the substance biologically functions as a toxin. For instance, neurotransmitters or digestive enzymes (including phospholipases) could conceivably be detected by the standard toxicity assays we earlier discussed. Consequently, it is necessary to satisfy certain anatomical, physiological, and other criteria before a presumed biological role for a toxin can be considered as established; the problem is very similar to establishing that a putative neurotransmitter is the real neurotransmitter. For cnidarians, it is desirable that a toxic substance be shown to be localized within nematocysts (or other integumentary glands that might be involved in a chemical defense). None of the toxins I have discussed in this article have yet been unequivocally localized in nematocysts. The *Aiptasia palida* polypeptide affecting action potentials is the only anthozoan neurotoxin yet shown to originate from nematocysts (Hessinger *et al*., 1973). The chemical similarity of *Aiptasia* neurotoxin with those shown in Table 1 is not known at this time since no sequence information on this particular toxin is yet available. Immunohistochemical as well as nematocyst purification methods should allow further delineation of the biological functions of these toxins in the near future. More information regarding the tissue distributions of both toxins and the nematocyst types is obviously desirable to better understand the biological functions of cnidarian toxins.

The remarkable differences in the relative mammalian and crustacean toxicities of some of the sodium channel polypeptide toxins (Table II) is provocative. Most of these polypeptide toxins have high crustacean toxicity and only a relatively small number possess significant mammalian toxicity. Species (*Anemonia sulcata* and *Anthopleura xanthogrammica*) which have vertebrate-toxic polypeptides seem to lack or possess relatively small amounts of cytolysin activity. Species with potent crustacean-active polypeptides (*e.g.*, *Stichodactyla helianthus* and *Condylactis gigantea*) seem to generally possess cytolysins. Possibly these two types of toxins co-evolved to provide toxicity against a broad phylogenetic spectrum of potential prey or predatory species. The plausibility of this ecological speculation can only be evaluated after a wider variety of sea anemones have been investigated toxinologically.

## ACKNOWLEDGEMENTS

I wish to thank my collaborators, William Doyle, Ben Dunn, Kenneth Blumenthal, Roy King, David Price, Bennie Parten, and Mike Pennington, for their assistance in the sea anemone toxin research, Daphne Fautin for identifying and naming sea anemones, Lazlo Béress for providing a copy of his manuscript for Halstead's book, and Judy Adams and Jeanette Janney for typing the manuscript.

## REFERENCES

Alsen, C., Harris, J.B., and Tesseraux, I. (1981). Mechanical and electrophysiological effects of sea anemone (*Anemonia sulcata*) toxins on rat innervated and denervated skeletal muscle. Brit. J. Pharmacol. 74, 64.

Bailey, L.E., Shibata, S., Seriguchi, D., and Dresel, P.E. (1980). Inhibition of the positive inotropic effect of Anthopleurin-A (AP-A) by dantrolene. Life Sci. 26, 1061.

Barhanin, J., Hughes, M., Schweitz, H., Vincent, J-P., and Lazdunski, M. (1981). Structure-function relationships of sea anemone toxin II from *Anemonia sulcata*. J. Biol. Chem. 256, 5764.

Béress, L., Béress, R., and G. Wunderer, G. (1975). Isolation and characterization of three polypeptides with neurotoxic activity from *Anemonia sulcata*. FEBS Lett. 50, 311.

Béress, L. (1986). Biologically active compounds from coelenterates. *In* "Poisonous and Venonous Marine Animals of the World" (Halstead, B., ed.), in press.

Bernheimer, A.W. (1987). Proposed nomenclature for cytolytic toxins of sea anemones. Toxicon 24, 1031.

Bernheimer, A.W., and Avigad, L.S. (1976). Properties of a toxin from the sea anemone *Stoichactic helianthus*, including specific binding to sphingomyelin. Proc. Nat. Acad. Sci. USA 73, 467.

Bernheimer, A.W., and Avigad, L.S. (1978). A cholesterol-inhibitable cytolytic protein from the sea anemone Metridium senile. Biochem. Biophys. Acta 541, 96.
Bernheimer, A.W., and Avigad, L.S. (1981). New cytolysins in sea anemones from the west coast of the United States. Toxicon 19, 529.
Bernheimer, A.W., and Avigad, L.S. (1982). Toxins of the sea anemone Epiactis prolifera. Arch. Biochem. Biophys. 217, 174.
Bernheimer, A.W., Avigad, L.S., Branch, G., Dowdle, E., and C.Y. Lai, C.Y. (1984). Purification and properties of a toxin from the South African sea anemone Pseudactinia varia. Toxicon 22, 183.
Bernheimer, A.W., and Lai, C.Y. (1985). Properties of a cytolytic toxin from the sea anemone, Stoichactis kenti. Toxicon 23, 791.
Blanquet, R. (1968). Properties and composition of the nematocyst toxin of the sea anemone, Aiptasia pallida. Comp. Biochem. Physiol. 25, 893.
Blumenthal, K.M., Keim, P.S., Heinrikson, R.L., and Kem, W.R. (1981). Amino acid sequence of Cerebratulus lacteus toxin B-II and revised structure of toxin B-IV. J. Biol. Chem. 256, 9063.
Blumenthal, K.M., and Kem, W.R. (1983). Primary structure of Stoichactis helianthus cytolysin III. J. Biol. Chem. 258, 5574.
Catterall, W.A., and Béress, L. (1978). Sea anemone toxin and scorpion toxin share a common receptor site associated with the action potential sodium iontophore. J. Biol. Chem. 253, 393.
Catterall, W.A., and Coppersmith, J. (1981). Pharmacological properties of sodium channels in cultured rat heart cells. Mol. Pharmacol. 20, 533.
Catterall, W.A. (1986). Molecular properties of voltage-sensitive sodium channels. Ann. Rev. Pharmacol. 55, 953.
Dunn, D. (1981). The clownfish sea anemones: Stichodactylidae (Coelenterata: Actinaria) and other sea anemones symbiotic with pomacentrid fishes. Trans. Amer. Phil. Soc. 71, 3.
Erxleben, C., and Rathmayer, W. (1984). Effects of the sea anemone Anemonia sulcata toxin II on skeletal muscle and on neuromuscular transmission. Toxicon 22, 387.
Ferlan, I., and Lebez, D. (1974). Equinatoxin, a lethal protein from Actinia equina-I Purification and Characterization. Toxicon 12, 57.
Ferlan, I., and Jackson, K.W. (1983). Partial amino acid sequence of equina-toxin. Toxicon, Suppl. 3, 141.

Frelin, C., P. Vigne, Schweitz, H., and Lazdunski, M. (1984). The interaction of sea anemone and scorpion neurotoxins with tetrodotoxin-resistant $Na^+$ channels in rat myoblasts. A comparison with $Na^+$ channels in other excitable and non-excitable cells. Mol. Pharmacol. 26, 70.

Fujita, S., Warashina, A., and Satake, M. (1983). Binding characteristics of a sea anemone toxin from Parasicyonis actinostolodes with crayfish leg nerves. Comp. Biochem. Physiol. 76C, 25.

Fujiwara, W., Muramatsu, I., Hidaka, H., Ikushima, S., and Ashida, K. (1979). Effects of Goniopora toxin, a polypeptide isolated from coral on electromechanical properties of rabbit myocardium. J. Pharm. Exper. Ther. 210, 153.

Fusetani, N., Yasukawa, K., Matsunaga, S., and Hashimoto, K. (1986). Isolation and identification of three hemolysins from the hydroid Solanderia secunda (Inaba). Comp. Biochem. Physiol. 8313, 511.

Gonoi, T., Ashida, K., Feller, D., Schmidt, J., Fujiwara, M., and Catterall, W.A. (1986). Mechanism of action of a polypeptide neurotoxin from the coral Goniopora on sodium channels in mouse neuroblastoma cells. Mol. Pharmacol. 29, 347.

Gooley, P.R., Blunt, J.W., and Norton, R.S. (1984). Conformational heterogeneity in polypeptide cardiac stimulants from sea anemones. FEBS Lett. 174, 15.

Gooley, P.R., and Norton, R.S. (1986). Secondary structure in sea anemone polypeptides: Proton nuclear magnetic resonance study. Biochem. 25, 2349.

Gross, G.J., Warltier, D.C., Hardman, H.F., and Shibata, S. (1985). Cardiotonic effects of anthopleurin-A (AP-A), a polypeptide from a sea anemone, in dogs with a coronary artery stenosis. Eur. J. Pharmacol. 110, 271.

Gruen, L.C., and Norton, R.S., (1985). Role of aspartate residues in the cardiac stimulatory activity of anthopleurin-A. Biochem. Intern. 11, 69.

Hartung, K., and Rathmayer, W., (1985). Anemonia sulcata toxins modify activation and inactivation of $Na^+$ currents in a crayfish neurone. Pflug. Arch. 404, 119.

Hessinger, D.A., Lenhoff, H.M., and Kahan, L.B. (1973). Haemolytic, phospholipase A and nerve-affecting activities of sea anemone nematocyst venom. Nature New Biol. 241, 125.

Ishizaki, H., McKay, R.H., Norton, T.R., Yasunobu, K.T., Lee, J., and Tu, A.T. (1979). Conformational studies of peptide heart stimulant anthopleurin a. J. Bio. Chem. 254, 9651.

Jover, E., Couraud, F., and Rochat, H. (1980). Two types of scorpion neurotoxins characterized by their binding to two separate receptor sites on rat brain synaptosomes. Biochem. Biophys. Res. Comm. 95, 1607.

Kem, W.R. (1976). Purification and characterization of a new family of polypeptide neurotoxins from the heteronemertine Cerebratulus lacteus (Leidy). J. Biol. Chem. 251, 4184.

Kem, W.R., and Blumenthal, K.M., (1978). Purification and characterization of the cytotoxic Cerebratulus A toxins. J. Biol. Chem. 253, 5752.

Kem, W.R., Blumenthal, K.M., and Doyle, J.W. (1980). Cytotoxins of some marine invertebrates: In "Natural Toxins" (D. Eaker and T. Wadstrom, eds.), Pergamon Press, New York, p. 157.

Kem, W.R., Dunn, B.M., Parten, B.F., King, R.W., and Price, D.A. (1986). Mass spectrometric identification of hydroxyprolyl phenylthiohydantoins during protein sequencing. Proc. 6th Intern. Conf. Meth. Prot. Seq. Analysis (Abstract D55).

Kem, W.R., Dunn, B., Parten, B., Pennington, M., and Price, D. (1986). The unconventional amino acid sequence of the sea anemone (Stoichactis helianthus) polypeptide neurotoxin. Fed. Proc. 45, 1795 (abstract).

Khan, A.R., Lemeignan, M., and J. Molgo, J. (1986). Effects of toxin II from the sea anemone Anemonia sulcata on contractile and electrical responses of frog skeletal muscle fibers. Toxicon 24, 373.

Kodama, I., Shibata, S., Toyama, J., and Yamada, K. 1981. Electromechanical effects of anthopleurin-A (AP-A) on rabbit ventricular muscle: influence of driving frequency, calcium antagonists, tetrodotoxin, lidocaine, and ryanodine. Brit. J. Pharmacol. 74, 29.

Kodama, I., Toyama, J., Shibata, S., and Norton, T.R. (1981). Electrical and mechanical effects of anthopleurin-A, a polypeptide from a sea anemone, on isolated rabbit ventricular muscle under conditions of hypoxia and glucose-free media. J. Cardiov. Pharmacol. 3, 75.

Kolkenbrock, H.J., Alsen, C., Asmus, R., Béress, L., and Tschesche, H. (1983). On the biological activity of chemically modified ATX II (Anemonia sulcata). Proc. 5th Europ. Symp. Animal, Plant, and Microbial Toxins (D. Mebs and G. Habermehl, eds), p. 72 (abstract).

Kozlovskaya, E., Vozhova, H., and Elyakov, G. Structure activity studies on neurotoxin I from sea anemone Radianthus macrodactylus. In "Chem. Pept. Prot." 1, 379. (W. Voelter et al, eds) de Gruyter and Co., Berlin.

Kudo, Y., and Shibata, S. (1980). The potent excitatory effect of a novel polypeptide, anthopleurin-B, isolated from a sea anemone (Anthopleura xanthogrammica) on the frog spinal cord. J. Pharm. Exper. Ther. 214, 443.

Linder, R., Bernheimer, A.W., and Kim, K-S (1977). The interaction between sphingomyelin and a cytolysin form the sea anemone Stoichactis helianthus. Biochem. Biophys. Acta 467, 290.

Linder, R., and Bernheimer, A.W. (1978). Effect on sphingomyelin-containing liposomes of phospholipase D. from Corynebacterium ovis and the cytolysin from Stoichactis helianthus. Biochim. Biophys. Acta 530, 236.

Mebs, D., and Gebauer, E. (1980). Isolation of proteinase inhibitory, toxin, and hemolytic polypeptides from a sea anemone, Stoichactis sp. Toxicon 18, 97.

Mebs, D., Liebrich, M., Reul, A., and Samejima, Y. (1983). Hemolysins and proteinase inhibitors from sea anemones of the Gulf of Aqaba. Toxicon 21, 257.

Michaels, D.W. (1979). Membrane damage by a toxin from the sea anemone Stoichactis helianthus. I. Formation of transmembrane channels in lipid bilayers. Biochem. Biophys. Acta 555, 67.

Nabiullin, A.A., Odinokov, S.E., Kozlovskaya, E.P., and Elyakov, G.B. (1982). Secondary structure of sea anemone toxins. FEBS Lett. 141,124.

Narahashi, T., Moore, J.W., and Shapiro, B.I. (1969). Condylactis toxin: interaction with nerve membrane ionic conductances. Science 163, 680.

Neumcke, B., Schwarz, W., and Stampfli, R. (1985). Comparison of the effects of Anemonia toxin II on sodium and gating currents in frog myelinated nerve. Biochem. Biophys. Acta 814, 111.

Newcomb, R., Yasunobu, K.T., Seriguchi, D., and Norton, T.R. Effects of chemical modification of anthopleurin A, a peptide heart stimulant. In "Frontiers in Protein Chemistry" (Liu, Mamiya, and Yasunobu, eds.), p. 539.

Nishida, S., Fujita, S., Warashina, A., Satake, M., and Tamiya, N. (1985). Amino acid sequence of a sea anemone toxin from Parasicyonis actinostoloides. Eur. J. Biochem. 150, 171.

Noda, M., Muramatsu, I., Fujiwara, M., and Ashida, K. (1985). Effects of goniopora toxin on bullfrog atrial muscle are frequency-dependent. N.S. Arch. Pharmacol. 330, 59.

Noda, M., Ikeda, T., Kayano, T., Suzuki, H., Takeshima, H., Kurasaki, M., Takahashi, H., and Muna, S. (1986). Existence of distinct sodium channel messenger RNAs in rat brain. Nature 320, 188.

Norton, T.R., and Kashiwagi, M. (1972). Purification of a potent antitumor agent from a Tahitian sea anemone and methods of administration studies with Ehrlich ascites tumor in mice. J. Pharmaceut. Sci. 61, 1814.

Ohizumi, Y., and S. Shibata, S. (1981). Possible mechanism of the dual action of the new polypeptide (Anthopleurin-B) from sea anemone in the isolated ileum and taenia caeci of the guinea-pig. Brit. J. Pharmacol. 72, 239.

Pauron, D., Barhanin, J., and M. Lazdunski, M. (1985). The voltage-dependent $Na^+$ channel of insect nervous system identified by receptor sites for tetrodotoxin, and scorpion and sea anemone toxins. Biochem. Biophys. Res. Commun. 131, 1226.

Pennington, M.W., Kem, W.R., and Dunn, B.M. Purification, sequence determination, and chemical synthesis of the major neurotoxin in the sea anemone *Stichodactyla helianthus*. *In* "Macromolecular Sequencing and Synthesis" (Schlesinger, D.H., ed.) Alan Liss, Inc., New York (in press.

Phillips, J.H., and Abbott, D.P. (1957). Isolation and assay of the nematocyst toxin of *Metridium senile fimbriatum*. Biol. Bull. 113, 296.

Reimer, N.S., Yasunobu, C.L., Yasunobu, K.T., and Norton, T.R. (1985). Amino acid sequence of the *Anthopleura xanthogrammica* heart stimulant, anthopleurin-B. J. Biol. Chem. 260, 8690.

Renaud, J-F., Fosset, M., Schweitz, H., and Lazdunski, M., (1986). The interaction of polypeptide neurotoxins with tetrodotoxin resistant $Na^+$ channels in mammalian cardiac cells. Correlation with inotropic and arrhythmic effects. Eur. J. Pharmacol. 120, 161.

Rudnev, B.S., Likhatskaya, G.N., Kozlovskaya, E.P., Monastyrnaya, M.M., and Elyakov, G.B. (1984). Effect of hemolysin from the sea anemone *Radianthus macrodactylus* on lipid membrane conductivity (in Russian). Biol. Membr. 1, 1019.

Scheffler, J-J., Tsugita, A., Linden, G., Schweitz, H., and Lazdunski, M. (1982). The amino acid sequence of toxin V from *Anemonia sulcata*. Biochem. Biophys. Res. Commun. 107, 272.

Schweitz, H., Vincent, J-P., Barhanin, J., Frelin, C., Linden, G., Hugues, M., and Lazdunski, M. (1981). Purification and pharmacological properties of eight sea anemone toxins from *Anemonia sulcata*, *Anthopleura xanthogrammica*, *Stoichactis giganteus*, and *Actinodendron plumosum*. Biochem. 20, 5245.

Schweitz, H., Bidard, J-N., Frelin, C., Pauron, D., Vijverberg, H.P.M., Mahasneh, D.M., Lazdunski, M.,

Vilbois, F., and Tsugita, A. (1985). Purification, sequence, and pharmacological properties of sea anemone toxins from Radianthus paumotensis. A new class of sea anemone toxins acting on the sodium channel. Biochem. 24, 3554.

Scriabne, A., Van Arman, C.G., Morgan, G., Morris, A.A., Bennett, C.D., and Bohidar, N.R. (1979). Cardiotonic effects of anthopleurin-A, a polypeptide from a sea anemone. J. Cardiov. Pharmacol. 1, 571.

Shibata, S., and Norton, T.R. (1982). Potent cardiotonic action of polypeptides isolated from sea anemone. In "Recent Developments in Cardiac Muscle Pharmacology" (S. Shibata and L.E. Bailey, eds.). Igaku-Shoin, Tokyo and New York. p. 13.

Shin, M.L., Michaels, D.W., and Mayer, M.M. (1979). Membrane damage by a toxin from the sea anemone Stoichactis helianthus II. Effect of membrane lipid composition in a liposome system. Biochim. Biophys. Acta 555, 79.

Shiomi, K., Tanaka, E., Yamanaka, H., and Kikuchi, T. (1985). Isolation and characterization of a lethal hemolysin in the sea anemone Parasicyonis actinostoloides. Toxicon 23, 865.

Stengelin, S., Rathmayer, W., Wunderer, G., Béress, L., and Hucho, F. (1981). Radioactive labeling of toxin II from Anemonia sulcata. Anal. Biochem. 113, 277.

Sorokina, Z.A., Chizhmakov, Yu. V., Elyakov, G.B., Kozlovskaya E.P., and Vozhzhova, E.V., (1984). A study of sodium channel inactivation mechanism by a toxin from sea anemone Radianthus macrodactylus and by various chemical agents (in Russian). Fiziol. Zh. 30, 571.

Stone, B.L., and Gray, W.R., (1982). Occurrence of hydroxyproline in a toxin from the marine snail Conus geographus. Arch. Biochem. Biophys. 216, 765.

Tanaka, M., Haniu, M., Yasunobu, K.T., and Norton, T.R. (1977). Amino acid sequence of the Anthopleura xanthogrammica heart stimulant, anthopleurin A. Biochem. 16, 204.

Tesseraux, I., Gulden, M., and Béress, L. (1983). The action of Bolocera toxin on mammalian skeletal muscle. Proc. 5th Europ. Symp. Animal, Plant, and Microbial Toxins (D. Mebs and G. Habermehl, eds). p. 128 (abstract).

Thron, C.D. (1964). Hemolysis by holothurin A, digitonin, and quillaia saponin: estimates of the required cellular lysin uptakes and free lysin concentrations. J. Pharm. Exper. Ther. 145, 194.

Turlapaty, P., Shibata, S., Norton, T.R., and Kashiwagi, M. (1973). A possible mechanism of action of a central

stimulant substance isolated from the sea anemone Stoichactis kenti. Eur. J. Pharmacol. 24, 310.
Ulbricht, K.W. and Schmidtmayer, J. (1981). Modification of sodium channels in myelinated nerve by Anemonia sulcata toxin II. J. Physiol. (Paris) 77, 1103.
Varanda, W., and Finkelstein, A. (1980). Ion and non-electrolyte permeability properties of channels formed in planar lipid bilayer membranes by the cytolytic toxin from the sea anemone, Stoichactis helianthus. J. Membr. Biol. 55, 203.
Vincent, V.P., Balerna, M., Barhanin, J., Fosset, M., and Lazdunski, M. (1980). Binding of sea anemone toxin to receptor sites associated with gating system of sodium channel in synaptic nerve endings in vitro. Proc. Natl. Acad. Sci., USA 77, 1646.
Warashina, A., and Fujita, S. (1983). Effect of sea anemone toxins on the sodium inactivation process in crayfish axons. J. Gen. Physiol. 81, 305.
Warashina, A., Fujita, S., and Satake, M. (1981). Potential-dependent effects of sea anemone toxins and scorpion venom on crayfish giant axon. Pflug. Arch. 391, 273.
Wemmer, D.E., Kumar, N.V., Metrione, R.M., Lazdunski, M., Drobny, G., and Kallenbach, N.R. (1986). NMR analysis and sequence of toxin II from the sea anemone Radianthus paumotensis. Biochem. 25, 6842.
Wunderer, G. (1978). Die disulfidbrucken von Toxin II aus Anemonia sulcata. Hoppe-Seyler's Z. Physiol. Chem. 359, 1193.
Wunderer, G., Fritz, H., Wachter, E., and Machleidt, W. (1976). Amino-acid sequence of a coelenterate toxin: toxin II from Anemonia sulcata. Eur. J. Biochem. 68, 193.
Wunderer, G., and Eulitz, M. (1978). Amino-acid sequence of toxin I from Anemonia sulcata. Eur. J. Biochem. 89, 11.
Zykova, T.A., Vinokurov, L.M., Kozlovskaya, E.P., and Elyakov, G.B. (1986). Amino acid sequence of neurotoxin III from the sea anemone Radianthus macrodactylus (in Russian). Bioorg. Khim 11, 302.

# 21

# THE CHEMISTRY OF CNIDAE

Richard S. Blanquet
Department of Biology
Georgetown University
Washington, D. C. 20057

## SUMMARY

The structural proteins (capsule wall and thread) of several nematocyst types have been shown to contain, as the major constituent, a collagen-like protein linked by disulfide bonds. Proline and hydroxyproline were found in combined amounts (about 25%) that are similar to vertebrate collagens, while glycine levels were somewhat lower (less than 33%). Hydroxylysine has also been reported. When fully discharged, numerous nematocyst types can be solubilized by disulfide reducing agents. Amino acid analyses and electron probe x-ray microanalysis have confirmed the presence of cysteine and high levels of sulfur in the capsule wall and thread. The structural protein of Aiptasia acontial nematocysts has a reported molecular weight of 31,800 daltons. Nematocysts are resistant to digestion by collagenase, elastase and several proteolytic enzymes. The intracapsular contents of some nematocysts have been shown to contain very high levels of calcium, possibly bound to glutamic acid rich proteins. This, in some way, may act to regulate nematocyst discharge. Intracapsular contents are reported here to contain proline hydroxylase activity which may be involved in the post-translational hydroxylation of nematocyst structural protein during development.

## I. INTRODUCTION

Although cnidarians are classified as the earliest

metazoans, they possess intracellular structures, collectively referred to as cnidae, which structurally and functionally represent some of the most complex secretion products yet described. To date, no less than 30 different types have been recognized and broadly characterized into three classes: nematocysts, spirocysts and ptychocysts (Mariscal, 1974, 1984). While differing in certain aspects of their chemistry, morphology and function, they are all similar in basic design and discharge by thread eversion. Though cnidae have been studied for over 200 years, questions regarding basic aspects of their biology, such as the biochemistry of their development and the mechanism of discharge, still remain unanswered despite the wealth of knowledge gained by recent technological advances in instrumentation and analytical techniques.

The basic chemistry of cnidae and comparative analyses of their components such as the capsule wall, thread, armature and intracapsular fluid (exclusive of venom constituents), has received relatively little attention. Almost unstudied are the biochemical events leading to their biosynthesis and assembly during development and maturation.

A major difficulty, in this regard, has been the frustrating inability to isolate suspensions of single nematocyst types in sufficient quantity, free of cellular contamination, for unambiguous assay. Normally, isolation involves disruption of nematocyst-bearing tissue followed by purification attempts designed to separate the nematocysts from cellular debris. Such preparations vary in purity, often contain both discharged and undischarged nematocysts of more than one type, and possibly immature forms whose basic chemistry may differ significantly from fully mature ones. Another difficulty has been the question of whether isolated nematocysts are able to retain low molecular weight components which may escape from the capsule contents during purification. Lubbock and Amos (1981) and Salleo (1984) report that the undischarged capsules of Rhodactis and Pelagia, respectively, are permeable to molecules with molecular weights up to about 700.

An isolation technique which has proved effective in overcoming some of these difficulties is acontial extrusion, in which the forceful expulsion of undischarged nematocysts occurs when acontia are placed in solutions of glycerol or sodium citrate (Yanagita, 1959; Blanquet and Lenhoff, 1966). This procedure results in exceptionally pure nematocyst suspensions but, unfortunately, is effective only with acontia, thereby limiting the types of nematocysts which can be obtained.

Despite these problems, an increasingly clearer picture

of nematocyst chemistry is emerging. Since nematocyst biology ultimately depends on the chemical and physical properties of their constituent molecules, knowledge in these areas will be critical to a fuller understanding of existing contemporary problems. This paper will review the structural chemistry of cnidae as well as intracapsular contents not related to venom function. Nematocyst venoms have been recently reviewed (Blanquet, 1977; Burnett and Calton, 1977; Halstead, 1978) and will also be extensively covered in this symposium.

## II. CHEMISTRY OF THE CAPSULE WALL AND THREAD

Several studies have indicated that the nematocyst capsule wall and thread contain a collagen-like protein (Lenhoff, Kline and Hurley, 1957) linked by disulfide bonds (Blanquet and Lenhoff, 1966; Fishman and Levy, 1967; Mariscal, 1971). Adams (1978), noting the wider range of chemical variability of invertebrate collagens compared to vertebrate collagens, cited the need to define the limits that would include invertebrate proteins under this category compared to the vertebrate "norm." Some general criteria used to define vertebrate collagens include characteristic x-ray diffraction patterns derived from the ordered arrangement of tropocollagen subunits, the occurrence of glycine as approximately one third the imino acids present, and a relatively high combined percentage (about 25%) of the pyrrolidine amino acids, proline and hydroxyproline. These in turn determine the triple helix structure of tropocollagen. Helix formation is believed to require that each collagen chain contain a glycine residue every third site, such that the primary structure can be represented as $(GLY\text{-}X\text{-}Y)_n$. The X and Y positions are usually occupied by proline and hydroxyproline, respectively.

In this light, the term "collagen-like" appears appropriate to describe the nematocyst protein, at this time, since this designation rests almost exclusively on the reported presence of significant amounts of glycine, proline and hydroxyproline. Nematocyst proteins have not been sequenced nor have x-ray diffraction patterns been reported. The birefringence observed in several nematocyst types does demonstrate the presence of ordered parallel arrays of fibrous molecules within the capsule wall. Isolated nematocysts are unaffected by collagenase or elastase and are also resistant to several proteolytic enzymes, perhaps due to the presence of disulfide residues (Blanquet and Lenhoff, 1966; Fishman and Levy, 1967). Nematocysts are also not converted to gelatin when autoclaved or boiled (Lenhoff, Kline and Hurley, 1957).

## A. Occurrence of Proline and Hydroxyproline

Amino acid analyses of nematocysts confirm the presence of proline and hydroxyproline in combined amounts similar to vertebrate collagens (Table 1). In early studies of Hydra nematocyst preparations, (Lenhoff et al., 1957; Lenhoff and Kline, 1958) demonstrated the presence of these imino acids by paper chromatography and stated that they accounted for as much as 22% of the nematocyst protein. Since 80% of the nematocysts analyzed were discharged, they concluded that these amino acids were present in the capsule wall and thread. In a semi-quantitative report on Physalia nematocysts, Lane and Dodge (1958) demonstrated much higher levels of proline in the discharged capsules than in the capsular contents. In an amino acid analysis of nematocysts isolated from the anemone, Metridium, Fishman and Levy (1967) reported values, given as residues per 1,000 residues, to be 69.8 and 143.6 for hydroxyproline and proline, respectively. In a more recent study, Phelan and Blanquet (1985) reported combined proline/hydroxyproline values of 27.9 mole % for Aiptasia mastigophores and 19.4 mole % for the ptychocysts of Pachycerianthus. Another amino acid characteristic of collagens, hydroxylysine, has been reported in small amounts in Aiptasia (Phelan and Blanquet, 1985) and Metridium (Fishman and Levy, 1967) nematocysts. In the above studies, glycine accounted for about 30 mole % in ptychocysts but only 19.2 mole % for Aiptasia mastigophores. Metridium values (20.3%) were similar to those of Aiptasia mastigophores.

With respect to sugar content, vertebrate collagens are glycoproteins containing small amounts of hydroxylysyl residues covalently linked to galacatose or glucosylgalactose (Grant and Jackson, 1976). In basement membrane collagens, smaller quantities of fucose, mannose, hexosamine and sialic acid have been found (Adams, 1978). In contrast, invertebrate collagens are relatively rich in carbohydrate which can also be linked to serine and threonine residues (Adams, 1978). Varying amounts of similar sugars have been reported for nematocysts but the specific type of linkage to the collagen-like protein has not been determined (Blanquet, 1966; Blanquet and Lenhoff, 1966; Goldner, et al., 1969; Phelan and Blanquet, 1985). Phelan and Blanquet (1985) were only able to identify a single sugar, mannose, after mild

Table I. Comparative amino acid profiles of nematocyst structural proteins and vertebrate collagen.

| Amino acid group | *Metridium*[a] | *Aiptasia*[b] | *Pachycerianthus*[b] | Collagen[c] |
|---|---|---|---|---|
| Glycyl | 20.3 | 19.2 | 29.7 | 33.0 |
| Prolyl/ Hydroxyprolyl | 21.3 | 27.9 | 19.4 | 21.7 |
| Cysteinyl | 25.7 | 7.2 | 2.4 | ND |
| Acidic | 15.7 | 10.0 | 10.7 | 12.0 |
| Basic | 7.7 | 5.9 | 4.5 | 8.0 |

[a] From Fishman and Levy (1967)
[b] From Phelan and Blanquet (1985)
[c] From Bailey and Weis-Fogh (1961)
ND - none detected.

hydrolysis of Aiptasia mastigophores.

### B. Occurrence of Sulfur Containing Amino Acids

Early histochemical studies by Hamon (1955) and the solubilization of several nematocyst types in disulfide-reducing agents such as mercaptoethanol, dithiothreitol (DTT) and dithioerythritol (DTE) suggest that nematocyst capsules and their everted threads contain proteins cross-linked by disulfide bridges and that these account for their insolubility and structural integrity (Yanagita, 1959; Blanquet and Lenhoff, 1966; Fishman and Levy, 1967; Mariscal, 1971). Before the discovery of procollagen, it was generally believed that collagens contained no sulfur amino acids. It is now well documented that cysteine residues and disulfide bonds are present within the peptide extensions of some collagen types (type I and II). Collagen types III and IV have also been shown to contain disulfide linkages within the protease-resistant triple helix (Grant and Jasckson, 1976). Invertebrate collagens, like those of some lower vertebrates, contain appreciable amounts of cysteine (Piez and Gross, 1959; McBride and Harrington, 1965; Adams, 1978).

With respect to cnidae, it is unclear as to whether all capsules contain cyteine and disulfide bonds. Mariscal and Lenhoff (1969) and Mariscal (1971) found differing degrees of solubility in DTE in the cnidae of 5 species of coral, 4 species of anemones, 1 species of zoanthid, 1 species of scyphozoan and 1 species of hydroid. The observed resistance to solubilization or differences in solubilization time may have been due, however, to mucus contamination in the smear preparations utilized. As noted by the author, mucus interferes with DTE accessibility to the capsules. Mucus secretions were particularly heavy in coral preparations and nematocysts from these species were found to be the most resistant to solubilization. In addition, the microbasic mastigophores of the coral Fungia were unaffected while the same nematocyst type from the anemones Aiptasia, Boloceroides and Anthopleura dissolved in 1-3 minutes. All spirocysts were found to remain undissolved while, in another study, Phelan and Blanquet (1985) reported that the ptychocysts of the anemone Pachycerianthus were also resistant though amino acid analysis of this nematocyst showed the presence of cysteic acid. It thus remains unclear whether disulfide bonds exist for all the types tested or, if present, are unavailable due to mucus contamination or DTE inaccessibility due to the tertiary structure of the protein itself.

Two other aspects of DTE solubilization deserve special

note. Both Blanquet and Lenhoff (1966) and Mariscal and Lenhoff (1969) reported that only fully discharged nematocysts are capable of solubilization while undischarged or partially discharged ones are not. One notable exception was the microbasic amastigophore of Boloceroides. These observations may reflect the fact that DTE requires access to the interior of the nematocyst capsule to initiate solubilization. In this regard, electron micrographs of many nematocyst types have shown the capsule wall to be composed of an electron-lucent, inner region continuous with the inverted tubule and a more electron-dense outer region. This might suggest that it is the electron-lucent region which contains the disulfide containing protein.

Another interesting observation in the same studies is that the everted threads dissolve at slower rates than the capsule, perhaps reflecting reduced disulfide bond access or content. Reduced sulfur content has been confirmed by amino acid analysis for the everted threads of Aiptasia mastigophores (Phelan and Blanquet, 1985) and by x-ray microanalysis for the thread of Pennaria (Mariscal, 1980).

Several recent studies regarding sulfur metabolism as it might relate to nematocyst maintenance or synthesis deserve mention. Cook (1976) presented preliminary evidence that Hydra which contain algal endosymbionts (zoochlorellae) have an enhanced ability to utilize sulfate as opposed to aposymbiotic or asymbiotic forms. It was suggested that the zoochlorellae might reduce sulfate to a variety of amino acids, as has been shown for other algae, but direct experimental proof was lacking. Auto-radiographic studies using labeled sulfate demonstrate that green hydra incorporate more label into stenoteles than either of the other two Hydra forms and it was postulated that the algae were either supplying some labeled precursor of nematocyst synthesis such as cysteine or enhancing the direct incorporation of sulfate into mucopolysaccharides associated with the capsule wall.

Ohuchida et al. (1981) reported that Hydra japonica treated with 2 mM D,L-c-propargylglycine (PG), a specific inhibitor of an enzyme concerned with L-cysteine biosynthesis, showed a decreased number of desmonemes in their tentacles. Holotrichous and atrichous isorhizas showed smaller drops but stenoteles were unaffected. The decrease in desmosome number was determined to be due to disintegration in the tentacles which was not related to an inhibition of their production or migration from the gastric region. It does not appear, therefore, that PG in this case is inhibiting desmoneme formation, by inhibiting cysteine availability. Isolated nematocysts did not disintegrate on direct exposure to PG.

Finally, Koizumi and Kijima (1980) demonstrated the specific incorporation of S-($^{14}$C)methylglutathione into a nematocyst-rich subcellular fraction of Hydra japonica. They proposed that this molecule reversibly binds to an intermediate macromolecule which is not a catalyzing enzyme. The physiological meaning of this incorporation is unknown but the authors cited the possibility that GSH metabolism may, in some way, be related to the synthesis or maintenance of nematocyst capsules.

## C. Elemental Analysis

A technique which has proved valuable in obtaining comparative analyses of various cnidae subcomponents, such as the capsule wall, tubule and capsule contents, has been electron probe x-ray microanalysis (EPXMA). This technique simultaneously measures total elemental concentrations including components of organic molecules under a highly focused electron beam. Using the stenotele of the hydroid Pennaria, Mariscal (1980) noted that sulfur was the most abundant element in the capsule wall and confirmed the higher sulfur content of the capsule as compared to the thread. Silicon and sodium were also detected in the capsule but in much lower quantities. In some analyses of the thread, however, silicon was the major element. It is interesting to note that the spines were also found to be sulfur-rich but have proved to be insoluble in disulfide reducing agents.

Similarly, Lubbock and coworkers (Lubbock, 1981; Gupta and Hall, 1984) have also shown high sulfur levels in nematocyst structural proteins of Anthopleura and Rhodactis. The undischarged capsule walls of Rhodactis had the highest levels of sulfur (181 mmol $kg^{-1}$ wet wt.). High levels of Na,K and Cl were also reported which were assumed to be bound. The uneverted tubule also had high, but slightly lower levels of sulfur, than the capsule. In addition, high concentrations of Zn, Cu and Na were also found associated with the tubule as opposed to low K and Cl. In undischarged nematocysts, the Cu and Zn were exclusively associated with the tubule. On eversion, the tubule lost most of its Zn but not its Cu (Gupta and Hall, 1984). These authors noted that the capsule and thread contain low levels of Ca and that the high levels reported earlier by Lubbock et al. (1981) may have been due to capsular fluid (rich in Ca) which permeates the capsule wall and the intracapsular spaces between the intricate folds of the uneverted tubule. We have noted in several recent electron microscope studies of maturing

Aiptasia mastigophores (Blake and Chapman, this symposium) that the outer folds of uneverted tubules are locally occupied by small tubular structures which may represent a Ca-rich protein (Fig. 1,2). The function of these structures is unknown but may somehow aid in tubule development or eversion.

## D. Molecular weight and other properties

The ability to solubilize isolated nematocysts with disulfide reducing agents has allowed opportunities for determining the molecular weight and other properties of nematocyst structural proteins. Surprisingly, only a few studies are to be found in this regard. In an early work, Blanquet and Lenhoff (1966) using polyacrylamide gel electrophoresis were able to detect only a single band and proposed that nematocyst capsules of Aiptasia were composed of a single collagen-like protein linked by disulfide bonds. More recently, Phelan and Blanquet (1985) again using solubilized Aiptasia mastigophores isolated a major protein band with a molecular weight of 31,800 daltons. Several other faint bands, however, were also observed. While these may represent minor capsule proteins, they are more likely due to residual toxin components which remain in the capsule despite repeated washing of discharged nematocysts.

The observed changes in volume and shape of nematocysts immediately prior to, during and after discharge have been reported by several investigators (Blanquet, 1966; Robson, 1973; Mariscal, 1974; Salleo, 1984) and suggest some elasticity of the capsule wall. Such properties may be important to the process of nematocyst discharge. Salleo (1984) has suggested that changes in the rigidity of the capsule during discharge may alter the mechanical characteristics of the system and that this limited flexibility may provide part of the force necessary to initiate tubule eversion.

Undischarged nematocysts have also been reported to contain high intracapsular pressure, perhaps as high as 140 atmospheres (Picken and Skaer, 1966; Lubbock and Amos, 1981). In this regard, several aspects of nematocyst function make structural proteins such as the collagens and keratins particularly suitable for nematocyst capsule and tubule. These collagen properties are to a large extent variable through changes in quaternary structure and post-translational modifications (such as hydroxylations and glycosylations) during assembly and permit a degree of flexibility to meet different functional needs. A better

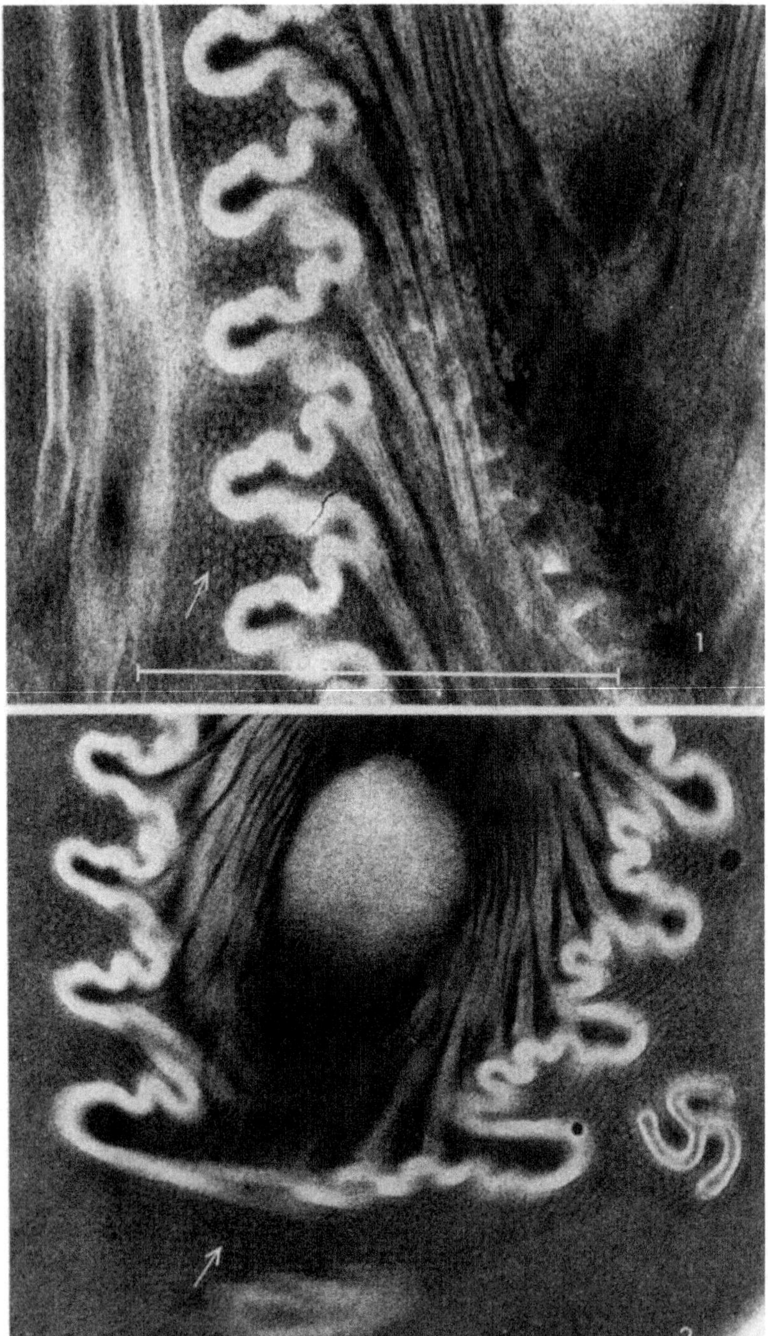
1
2

understanding of the properties of nematocyst structural properties may provide insights into the discharge mechanism.

## III. CHEMISTRY OF THE INTRACAPSULAR CONTENTS

Electron microscopic studies have shown that cnidae are formed within a Golgi derived vesicle (Slautterback, 1963; Carré, 1972; Skaer, 1973; Holstein, 1980). This vesicle, at some time, must contain all the molecules and ions which make up the structural material of the cnidae (capsule, thread and armature), the components of the venom (when present) and other intracapsular components such as the cofactors or enzymes which might be involved with vesicular biochemical modifications relating to activity and assembly. Morphological changes observed during cnidae development suggest a programmed secretion of different components at different times eventually leading to self- or enzyme-assisted assembly within the vesicle. As aptly stated by Gupta and Hall (1984), "What regulates this process of self-assembly has remained a fascinating mystery."

To begin to approach such a problem obviously requires a detailed analysis of the chemical changes within the developing nematocyst and the mature form. To date, however, most studies of capsular contents have been concerned with the identification of components related to venom activity, and no studies have dealt with the possible presence of soluble precursors of the structural proteins or the enzymes which may modify or aggregate these collagen-like molecules.

Another aspect of intracapsular chemistry is how such molecules may relate to the mechanism of discharge. Clearly,

Fig. 1. A longitudinal section through an undischarged microbasic mastigophore of Aiptasia pallida showing a section of the pleated inverted thread with attached spines. Note the circular electron lucent material within the pleates (arrow), some of which seem to have an electron dense core. Bar = 1 micrometer.

Fig. 2. A cross section through an undischarged microbasic mastigophore of Aiptasia pallida. In oblique sections of the inverted thread, the electron lucent material is shown to be organized into filaments which run within the pleats (arrow). Bar = 1 micrometer.

work in this area needs to be expanded.

### A. Elemental Analysis

One of the most striking aspects of the chemistry of cnidae intracapsular contents shown by EPXMA is the extremely high levels of Ca (Mariscal, 1980; Lubbock et al., 1981; Watson and Mariscal, 1984; Gupta and Hall, 1984). Lubbock and coworkers (1981) reported concentrations of Ca in the mature holotrichous isorhizas of Rhodactis and Anthopleura to be 606 and 542 mmol $kg^{-1}$ wet wt. respectively. These levels exceed those of sea water (7 mmol Ca) by a factor of 80. The capsule contents of Rhodactis had surprisingly little Cl, no phosphorus (P), and a low sulfur content (about 50 mmol $kg^{-1}$ wet wt.). It is interesting to note that the sulfur levels of immature nematocysts were about three times this level and could possibly represent soluble structural protein precursors or peptides (see next section). High levels of Ca were not found in developing forms (Lubbock et al., 1981; Gupta and Hall, 1984) where Ca levels approximated that found in the cytoplasm; 9 mmol $kg^{-1}$ wet wt. for Rhodactis and 18 mmol $kg^{-1}$ wet wt. for Anthopleura.

Watson and Mariscal (1984), on the other hand, using a potassium pyroantimonate calcium localization technique found elevated Ca levels at all stages of nematocyst formation in the holotrichous isorhizas from the catch tentacles of Haliplanella luciae. They noted, however, that the same nematocysts from the feeding tentacles lacked Ca but contained high concentrations of phosphorus. The spirocysts lacked both Ca and P but contained small quantities of aluminum. They proposed that for Ca-rich nematocysts the rough endoplasmic reticulum of the nematocyte may function as a Ca sequestering organelle which subsequently concentrates Ca within the Golgi vesicle.

The ability to maintain such high levels of intracapsular Ca may be due to the reported presence of aspartic and glutamic acid-rich proteins (about 80%) whose carboxyl groups may provide specific binding sites (Blanquet, 1968; Lane and Dodge, 1958; Calabrese et al., 1983). Such binding has also been postulated to stabilize the nematocysts and, somehow, to be involved in the discharge mechanism (Blanquet, 1970; Lubbock et al., 1981). Discharge mechanisms vary and a detailed explanation of these is beyond the scope of this review (for discussion see Salleo, 1984; Gupta and Hall, 1984; Mariscal, 1984).

## B. Proline Hydroxylase

In vertebrate collagens, the hydroxylation of proline and lysine residues by hydroxylases is a post-translational event not under direct genetic control. We wish to report here preliminary data on the apparent activity of a proline hydroxylase present in the capsular fluid of the microbasic mastigophores of *Aiptasia pallida* (Cacciabaudo and Blanquet, manuscript in preparation). This enzyme may be involved in the hydroxylation of proline found in the collagen-like protein of this nematocyst during its development.

*Aiptasia* nematocysts were extruded and washed in 1.0 M sodium citrate and finally discharged in 40 mM Tris-HCl buffer (pH 7.6) to obtain the contents (Blanquet, 1968). Collagen substrate labeled with 3,4-$^{3}$H proline was prepared from chick embryo frontal bones and assays for hydroxylase activity were both accomplished as described by Peterkovsky and DiBlasio (1975). Hydroxylation of proline residues results in the release of tritiated water which was determined by normal scintillation counting techniques.

In our studies, the release of tritium was proportional to the nematocyst protein content of the assay system, while the reaction rate was linear (20 to 60 minutes) at chick collagen concentrations of 8.36 to 158 ug protein $ml^{-1}$. Lineweaver-Burke reciprocal plots were also linear giving an apparent $K_m$ of 200 ug protein $ml^{-1}$ and a $V_{max}$ of about 100 cpm $min^{-1}$ (Fig. 3).

The following results are the first to indicate that an enzyme which possibly serves a function related to nematocyst biosynthesis is found within the nematocyst capsule and remains active even after nematocyst maturation. The formation of hydroxyproline in vertebrate collagens is necessary for helix formation and the eventual self-assembly into insoluble collagen fibers. It is tempting to hypothesize that the transformation of a soluble procollagen-like protein to the insoluble nematocyst structural proteins may occur by a similar process. It is of interest to note, that no proline or hydroxyproline was found in the "venom" of *Aiptasia* mastigophores by Phelan and Blanquet (1984) possibly indicating that all soluble structural precursors are absent in mature nematocysts.

Substrate affinity of proline hydroxylase has been shown to depend on several procollagen characteristics such as primary structure, chain length, type of helix formation (if any), and the extent of disulfide bond formation (Kivirikko *et al.*, 1972).

The normal triple helix conformation of vertebrate collagen does not appear to prevent hydroxylation since

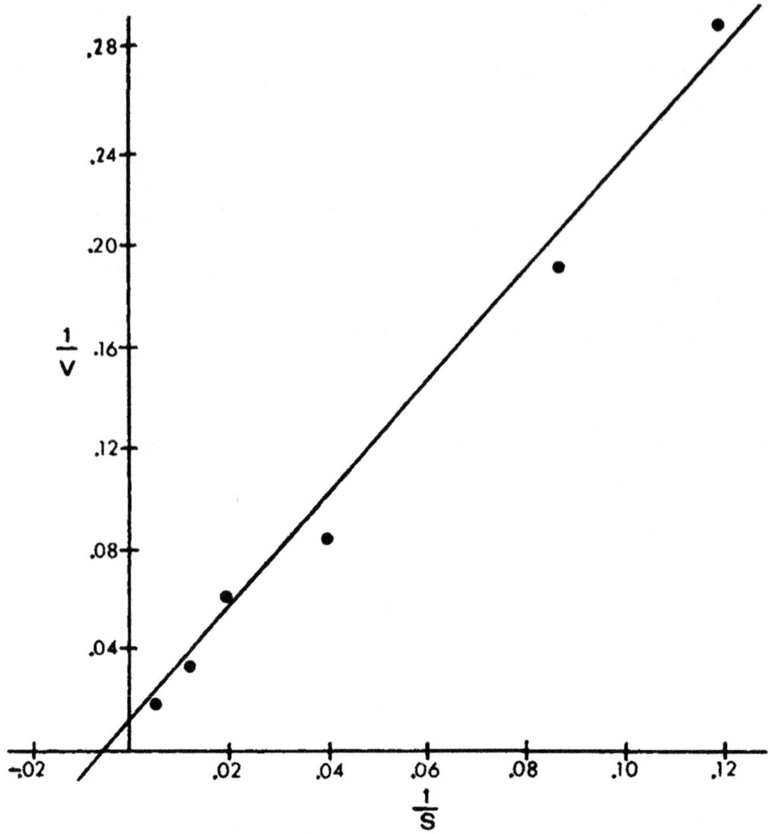


Fig. 3. Lineweaver-Burke plot for the proline hydroxylase activity found in the capsule contents of Aiptasia pallida mastigophores. Tritiated chick procollagen was used as a substrate. Rates were linear, at room temperature, over time periods of 20-60 minutes. Apparent $K_m$ ~200 μg $ml^{-1}$ substrate, $V_m$ = ~100 CPM $min^{-1}$.

results from chick embryos or calf skin (Hutton et al., 1967; Kivirikko et al., 1968) showed no difference in substrate activity between native and denatured preparations. In an interesting study, Fujimoto and Prockop (1968) demonstrated the disulfide-linked, under-hydroxylated collagen of the roundworm Ascaris cuticle to be an unsuitable substrate for chick proline hydroxylase though the denatured form could serve for proline hydroxylation to a significant degree. They suggested that the helical conformation of Ascaris cuticle was conformationally very different from vertebrate collagens and exists as a single polypeptide chain folded back on itself to form an antiparallel collagen type triple helix.

Should the transformation of a soluble procollagen-like nematocyst protein into an insoluble structural protein of the capsule wall and tubule parallel the events which occur during vertebrate collagen biosynthesis, a fruitful line of investigation may be a search within cnidae capsule components for precursor fragments (e.g., procollagen extension peptides) or enzymes involved in post-translational modifications such as procollagen peptidases or UDP-glucosyl transferases.

## III. CONCLUSIONS

It should be pointed out that the data reported here are a composite of relatively few types of the 30 or so different cnidae recognized to date. We still do not know if these data represent a general view of cnidae chemistry or whether more comparative studies will reveal major chemical types as well. Few nematocyst proteins have been isolated and studied in detail and more work in this area needs to be done if we are to fully understand the functioning of these remarkable structures. The presence of proline hydroxylase activity in the intracapsular fluid suggests that nematocyst precursors and the biosynthetic machinery necessary for nematocyst synthesis and assembly may also be present in this subcomponent.

## IV. ACKNOWLEDGMENTS

I acknowledge the work of Ms. Jeanne Cacciabaudo whose experiments on proline hydroxylase is, in part, reported here. I thank Ms. Andrea Blake for preparing the micrographs used in this review and Dr. George B. Chapman who kindly consented to read and criticize this paper in draft.

REFERENCES


Adams, E. 1978. Invertebrate collagens. Science 202: 591-598.

Bailey, K. and T. Weis-Fogh. 1961. Amino acid composition of a new rubber-line protein, resilin. Biochim. Biophys. Acta 48:452-459.

Blanquet, R. S. 1966. The acontia nematocysts of the sea anemone _Aiptasia pallida_: Mechanism of discharge and toxin. Ph.D. Dissertation, Duke University, Dept. of Zoology, Durham, N.C.

Blanquet, R. S. 1968. Properties and composition of the nematocyst toxin of the sea anemone, _Aiptasia pallida_. Comp. Biochem. Physiol. 25:893-902.

Blanquet, R. S. 1970. Ionic effects on discharge of the isolated and _in situ_ nematocysts of the sea anemone _Aiptasia pallida_. Comp. Biochem. Physiol. 35: 451-461.

Blanquet, R. S. 1977. Cnidarian venoms. _In_ A. W. Bernheimer (ed.), _Perspectives in Toxinology_, pp. 149-167. John Wiley and Sons, New York.

Blanquet, R. and H. M. Henhoff. 1966. A disulfide-linked collagenous protein of nematocyst capsules. Science 154:152-153.

Burnett, J. W. and G. J. Calton. 1977. The chemistry and toxinology of some venomous pelagic coelenterates. Toxicon 15:177-196.

Calabrese, L., A. Salleo, M. Alfa and G. LaSpada. 1983. The release of a glutamate rich protein from _Pelagia noctiluca_ nematocysts in the discharge process. 5th Conference of the European Society for Comparative Physiology and Biochemistry. Taormuna, Italy, pp. 53-54.

Carré, D. 1972. Étude du développement des cnidocysts dans le gastrozoide de _Muggiaea kochi_ (Siphonophore calycophore) C. R. Acad. Sci. Paris 275: 1263-1266.

Cook, C. B. 1976. Sulfate utilization in green hydra. _In_ G. O. Mackie (ed.), _Coelenterate Ecology and Behavior_, pp. 415-422. Plenum Press, New York.

Fishman, L. and M. Levy. 1967. Studies on the nematocyst capsule protein from the sea anemone _Metridium marginatum_. Biol. Bull. 133: 464-465.

Fujimoto, D. and D. Prockop. 1968. Denatured collagen from the cuticle of _Ascaris lumbricoides_ as a substrate for protocollagen proline hydroxylase. J. Biol. Chem. 243: 4138-4142.

Goldner, R., J. W. Burnett, J. S. Stone and M. S. Dilaimy. 1969. The chemical composition of sea nettle

nematocysts. Proc. Soc. Exp. Biol. Med. 131: 1386-1388.
Grant, M. E. and D. S. Jackson. 1976. The biosynthesis of procollagen. In P.N. Campbell and W. N. Aldridge (eds.), Essays in Biochemistry, Vol. 12, pp. 77-113. Academic Press, New York.
Gupta, B. L. and T. A. Hall. 1984. Role of high concentrations of Ca, Cu, and Zn in the maturation and discharge in situ of sea anemone nematocysts as shown by x-ray microanalysis of cryosections. In L. Bolis, J. Zadunaisky and R. Gilles (eds.), Toxins, Drugs and Pollutants in Marine Animals, pp. 77-95. Springer-Verlag, Berlin.
Halsted, B. W. 1978. Poisonous and venomous animals of the world, Vol. 1, Chap. IV. Darwin Press, New Jersey.
Hamon, M. 1955. Cytochemical research on coelenterate nematocysts. Nature 176: 357.
Holstein, T. 1980. An ultrastructural and kinematographic study of the nematocyst morphogenesis in selected coelenterate species. In P. Tardent and R. Tardent (eds.), Developmental and Cellular Biology of Coelenterates, pp. 313-330. Elsevier/North Holland, Amsterdam.
Hutton, J. J., A. L. Tappel and S. Udenfriend. 1967. Cofactor and substrate requirements of collagen proline hydroxylase. Arch. Biochem. Biophys. 118: 231-240.
Kivirikko, K. I., H. J. Bright and D. J. Prockop. 1968. Kinetic patterns of protocollagen hydroxylase and further studies on the polypeptide substrate. Biochem. Biophys. Acta 151: 558-567.
Kivirikko, K. I., Y. Kishida, S. Sakakibara and D. J. Prockop. 1972. Hydroxylation of $(X\text{-Pro-Gly})_n$ by protocollagen proline hydroxylase: Effect of chain length, helical conformation and amino acid sequence in the substrate. Biochim. Biophys. Acta 271: 347-356.
Koizumi, O., and H. Kijima. 1980. Specific S-methylglutathione incorporation into a nematocyst-rich fraction of hydra. Biochim. Biophys. Acta 629: 338-348.
Lane, C. E. and E. Dodge. 1958. Toxicity of Physalia nematocysts. Biol. Bull. 115: 219-225.
Lenhoff, H. M., E. S. Kline and R. Hurley. 1957. A hydroxyproline-rich, intracellular, collagen-like protein of Hydra nematocysts. Biochem. Biophys. Acta. 26: 204-205.
Lenhoff, H. M. and E. S. Kline. 1958. High imino acid content of the capsule from Hydra nematocysts. Anat. Rec. 130: 425.
Lubbock, R. and W. B. Amos. 1981. Removal of bound calcium

from nematocyst contents causes discharge. Nature 290: 500-501.

Lubbock, R., B. L. Gupta and T. A. Hall. 1981. Novel role of calcium in exocytosis: Mechanism of nematocyst discharge as shown by x-ray microanalysis. Proc. Nat. Acad. Sci. USA 78: 3624-3628.

Mariscal, R. N. 1971. Effect of a disulfide reducing agent on the nematocyst capsules from some coelenterates with an illustrated key to nematocyst classification. In H. M. Lenhoff, L. Muscatine and L. V. Davis (eds.), Experimental Coelenterate Biology, pp. 157-168. Univ. Hawaii Press, Honolulu.

Mariscal, R. N. 1974. Nematocysts. In L. Muscatine and H. M. Lenhoff (eds.) Coelenterate Biology: Reviews and New Perspectives, pp. 129-178. Academic Press, New York.

Mariscal, R. N. 1980. The elemental composition of nematocysts as determined by x-ray microanalysis. In P. Tardent and R. Tardent (eds.), Developmental and Cellular Biology of Coelenterates, pp. 337-342. Elsevier/North-Holland, Amsterdam.

Mariscal, R. N. 1984. Cnidaria: Cnidae. In J. Bereiter-Hahn, A. G. Matoltsy and K. S. Richards (eds.), Biology of the Integument, Vol. 1 Invertebrates, pp. 57-67. Springer-Verlog, Berlin.

Mariscal, R. N. and H. M. Lenhoff. 1969. Effect of a disulfide reducing agent on coelenterate nematocyst capsules. Experientia 25: 330-331.

McBride, O. W. and W. F. Harrington. 1965. Evidence for disulfide cross-linkages of an invertebrate collagen. J. Biol. Chem. 240: PC4545-4547.

Ohuchida, A., E. Ohtsuka and H. Kijima. 1981. Specific disintegration of nematocysts in Hydra japonica by treatment with propargylglycine. Biol. Bull. 161: 392-401.

Peterkovsky, B. and R. DiBlasio. 1975. Modification of the tritium-release assays for prolyl and lysyl hydroxylases using Dowex-50 colums. Analyt. Biochem. 66: 279-286.

Phelan, M. A. and R. S. Blanquet. 1985. Characterization of nematocyst proteins from the sea anemones Aiptasia pallida and Pachycerianthus torreyi (Cnidaria: Anthozoa). Comp. Biochem. Physiol. 81B: 661-666.

Picken, L. E. R. and R. J. Skaer. 1966. A review of researches on nematocysts. Symp. Zool. Soc. London 16: 19-50.

Piez, K. A. and J. Gross. 1959. The amino acid composition and morphology of some vertebrate and invertebrate collagens. Biochem. Biophys. Acta 34: 24-39.

Robson, E. A. 1973. The discharge of nematocysts in

relation to properties of the capsule. Publ. Seto Marine Lab. 20: 653-673.
Salleo, A. 1984. Discharge mechanism of the nematocysts of Pelagia noctiluca. In L. Bolis, J. Zadunaisky and C. Gilles (eds.), Toxins, Drugs and Pollutants in Marine Animals, pp. 63-76. Springer-Verlag, Berlin.
Skaer, R. J. 1973. The secretion and development of nematocysts in a siphonophore. J. Cell Sci. 13: 371-393.
Slautterback, D. B. 1963. Cytoplasmic microtubules. I. Hydra. J. Cell Biol. 18: 367-388.
Watson, G. M. and R. N. Mariscal. 1984. Calcium cytochemistry of nematocyst development in catch tentacles of the sea anemone Haliplanella luciae (Cnidaria: Anthozoa) and the molecular basis for tube inversion into the capsule. J. Ultrastr. Res. 86: 202-214.
Yanagita, T. M. 1959. Physiological mechanisms of nematocyst responses in the sea anemone. VII. Extrusion of resting cnidae - its possible bearing on the normal nettling response J. Exp. Biol. 36: 478-494.

22

# CHEMISTRY OF HYDRA NEMATOCYSTS[1]

*Jakob Weber*
*Marianne Klug*
*Pierre Tardent*

Zoological Institute
University of Zürich-Irchel
8057 Zürich, Switzerland

## SUMMARY

The development of a novel and rapid procedure for the isolation of pure and undischarged nematocysts from the fresh water cnidarian hydra enabled us to study the chemistry and biochemistry of these complicated secretory products in detail.

The unusually dense nematocysts (d = 1.22 to 1.24 g/ml) are stable in water. Their discharge can be triggered by the addition of disulfide reducing agents.

The water soluble components of a mixture of nematocysts (stenoteles, desmonemes, isorhizas) comprise about 61% of the total dry weight. Beside large amounts of $Mg^{2+}$ and $Ca^{2+}$ mainly proteins and polypeptides are found. Less than 14% of them have a molecular weight larger than 30 000 D; among these large proteins are at least two different toxins which act on nerves or muscles, one of them being strongly hemolytic. The most prominent constituents of the remaining lower molecular weight proteins are polypeptides of about 15 000 D and 3000 D. They may be involved in the discharge mechanisms of the nematocysts rather than in the toxic action of the latter. Most probably they represent intracapsular binding sites for $Ca^{2+}$ and $Mg^{2+}$.

[1]*This project is generously supported by the Swiss National Science Foundation (Grants 3.174.-0.81 and 3.072-0.84).*

## I. INTRODUCTION

The cnidarian nematocysts are often considered as the most complex secretory products produced within an animal cell. Although these fascinating organoids have been studied in detail for a long time, there are many points concerning their synthesis, assembly, and discharge which still are not known. Lately some relevant reports about the mechanisms of triggering and discharge as well as about the toxic action of some selected types of cyst have been published (Mariscal, 1974, 1984; Blanquet, 1977; Burnett and Calton, 1977; Russell, 1984; Salleo, 1984; Gupta and Hall, 1984; Tardent *et al.*, 1985). These documents clearly point to the many still open questions and to the need for developing new techniques for a further characterization of the physical, chemical and biochemical properties of these organoids.

The freshwater hydra, used widely in the laboratory (Lenhoff and Loomis, 1961; Lenhoff, 1983) is easy to grow and to handle and was the first organism in which the development and ultrastructure of nematocysts were investigated in great detail (Slautterback and Fawcett, 1959). The only drawback is the small size of the individual polyp and the difficulties in obtaining sufficient quantities of material for analysis. This is probably one of the main reasons for the fact that most biochemical investigations concerned with cnidarian nematocysts have focused on larger marine cnidarians.

A rather easy and simple method for obtaining fractions of isolated cysts that are free of other cellular residues and that still are capable of discharging has enable us to overcome at least part of the mentioned difficulties.

The present paper describes this isolation procedure and reports a number of biochemical investigations performed on purified preparations of hydra nematocysts.

## II. MATERIAL AND METHODS

### A. Material

*Hydra attenuata* Pall. were mass cultured (Loomis and Lenhoff, 1956) at 14°C and fed with larvae of *Artemia* . Specimens of *Anemonia sulcata* (Anthozoa) were collected in the Mediterranean and kept in artificial sea water. Nematocysts of *Physalia physalis* were kindly supplied by Dr. J.W.Burnett, Baltimore, and additionally purified by a Percoll-centrifugation (see: purification of hydra nematocysts).

## B. Isolation and Purification of Undischarged Nematocysts (see also Weber *et al.*, 1987a)

Samples of whole hydra polyps were washed several times in ice-cold distilled water and, after resting on ice for an hour, quantified volumetrically (Lenhoff, 1983). One $cm^3$ contains about 1000 specimens including their buds. The samples were frozen at -20°C and, not later than 36 hrs, processed as follows. Upon thawing in at least ten volumes of a 50% Percoll solution (Pharmacia Fine Chemicals) in either ice cold water or 300 mM sucrose, the polyps were homogenized by sucking them several times gently through a pasteur pipet. After 30 min on ice, the homogenates were centrifuged for 10 min at 3000 x g (Sorvall RC 5B, 5000 rpm using an SS 34 rotor). The supernatant was discarded. The pellet consisting of a mixture of undischarged stenoteles, desmonemes and isorhizas was carefully resuspended and washed quickly several times by centrifugation (30 sec at 600 x g) in cold water or in the above mentioned sucrose solution. The purified cysts were either processed without delay or stored in a wet or dehydrated state at -20°C.

Isolated cysts discharged normally to nearly 100% within seconds when placed in a solution containing 15 mM dithioerythritol (DTE), 15 mM CaCl and 25 mM Tris HCl (pH 8.8).

## C. Extraction of the Content of Purified Cysts

The mixture of isolated stenoteles, desmonemes and isorhizas was either homogenized in a little distilled water with a small glass mortar and pestle or sonicated until 50-90% of the cysts had disintegrated.

The soluble contents were collected as the supernatant fraction by centrifugation (5 min, 10 000 x g). These crude extracts were stored either in a dessicated or a wet state at -20°C (for months) or at +4°C (for shorter periods).

## D. Analytical Procedures

Proteins were determined as outlined in Table 1. The hemolytic action of crude nematocyst extracts upon human red blood cells were investigated according to the methods described by Klug (1986) and Klug *et al.* (1987).

The paralyzing and lethal effects of extracts were tested by injecting aliquots of the extracts into larvae and adult flies of *Drosophila* according to the methods described by Klug (1986).

Whole nematocysts extract was fractionated by gel filtration on a Sephadex G-50 column (Ø 1 x 48 cm) in 1 M ammonium sulfate (pH 7.5).

The methods for the evaluation of the elemental composition of purified intact nematocysts by x-ray spectral analysis in SEM and for the chemical determination of their accurate Ca and Mg content will be published elsewhere (Weber *et al.*, 1987b).

## III. RESULTS AND DISCUSSION

### A. Properties of Isolated Hydra Nematocysts

Since no simple method for purifying hydra nematocysts was available (Blanquet, 1983) we developed a procedure to isolate intact capsules from hydra. The new method allows to prepare within an hour from a few single polyps or from thousands of animals very clean, intact nematocysts. These nematocysts seem to have preserved most of their native properties and can be used for morphological and functional studies (ultrastructure, mechanism of discharge) as well as for chemical and biochemical analysis of their components (toxins, ions, walls, tubules etc.). Purified nematocysts without any detectable cellular contaminants (Fig. 1, A, B) are obtained in high yield (50-90% recovery). As had also been reported for nematocysts from *Chrysaora* (Burnett and Calton, 1973) the densities of the different types of hydra nematocysts (stenoteles, desmonemes, isorhizas) are all in the narrow range of 1.22 to 1.24 g/ml as determined in urografin-water density gradients (Endean and Rifkin, 1975). Unfortunately the three types can still not be separated from each other.

The walls of all three cyst types are permeable to dyes of molecular weights up to approx. 600 D (neutral red, toluidine blue, methylene blue, janus green B; but not alcian blue 8 GS in water). The metachromatic, cationic dyes which are

*Fig. 1. Purified nematocysts of Hydra attenuata. Bars indicate 20 µm. A. Light micrograph of isolated nematocysts. B. SEM of isolated nematocysts. C-E. Discharge of isolated nematocysts under phase contrast. Discharge was triggered by the addition of 15 mM DTE, 15 mM $CaCl_2$ in 25 mM Tris-HCl (pH 8.8); the nematocysts had been kept in 50% glycerol (optional). Pictures were taken 10, 30 and 60 sec after addition of the reagents. F-H. Discharge of an individual isolated stenotele which had been triggered by the addition of 15 mM DTE, 15 mM $CaCl_2$, 25 mM Tris-HCl (pH 8.8) under the light microscope. Pictures were taken after 10, 20 and 30 sec.*

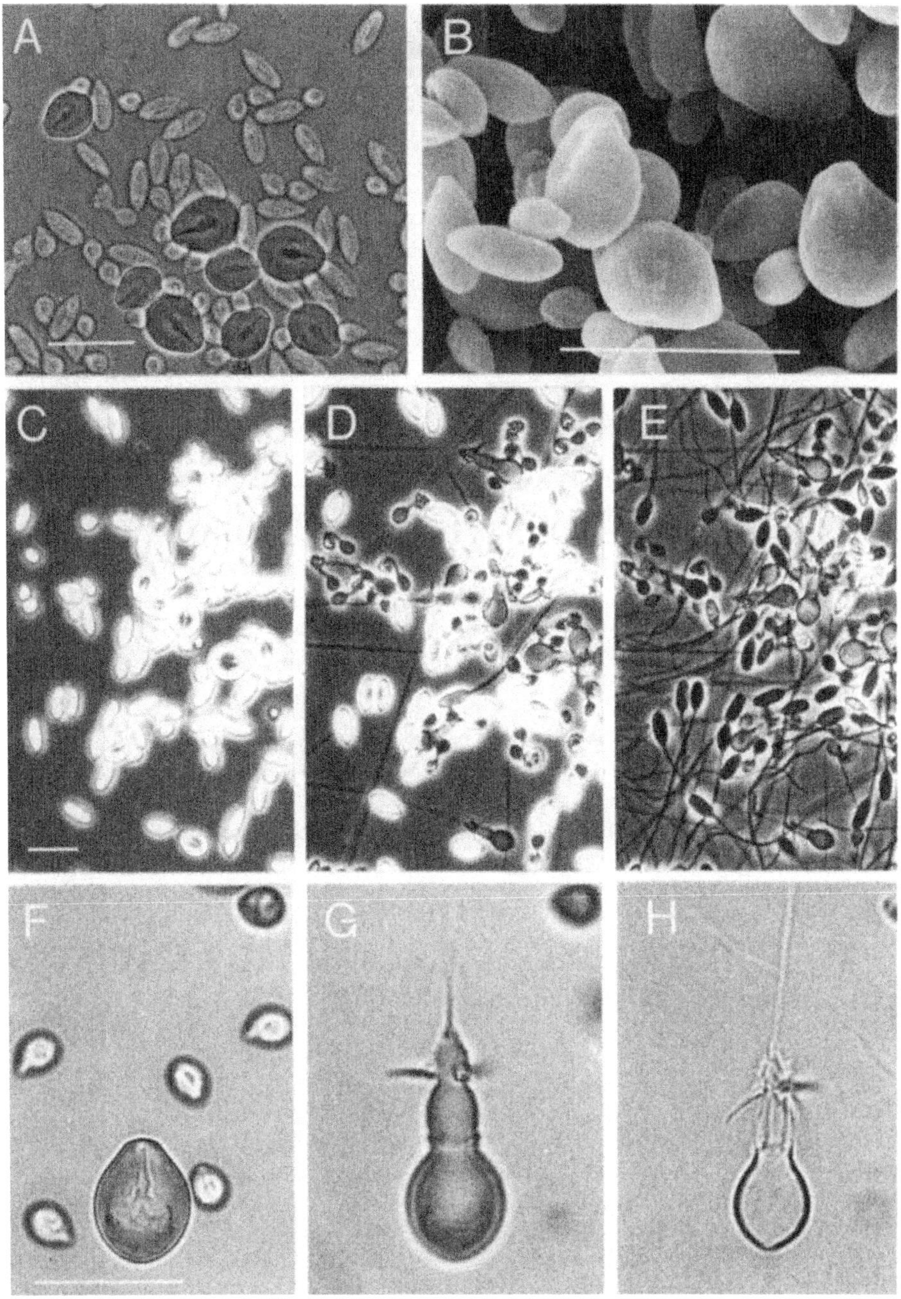
A
B
C
D
E
F
G
H

known to bind cooperatively to polyanionic structures (Cundall *et al.*, 1979) are concentrated manyfold within the nematocysts. Thus, any kind of small molecule up to approx. 600 D may be able to pass readily through the nematocyst wall of hydra, a fact that has been reported also for nematocysts of some marine cnidarians (Gupta and Hall, 1984).

Addition of disulfide reducing agents (DTE, mercaptoethanol) or high concentrations of the chaotropic agent guanidine-SCN causes nematocyst discharge within a few seconds (Fig. 1 C-E, F-H). Under appropriate conditions (cf. legends of Fig. 1) most capsules discharge in the same fashion as it is known for native cysts (see Holstein and Tardent, 1984). However, the time course of the process of discharge is slower (cf. Yanagita and Wada, 1954).

Polyacrylamidegel-electrophoresis of reduced and partially dissolved cysts in the presence of SDS (Fig. 2) reveals a variety of different prominent proteins. This finding stresses the complex nature of these cell organoids at the molecular level. Surprisingly, the gel patterns of nematocysts isolated from different species of hydra and even from different clones of the same species turned out to be remarkably different (Fig. 2, Ha 1 to Hci). Consequently, to prevent conflicting results, it is advisable to restrict chemical investigations to one distinct clone of hydra unless comparative studies are envisaged.

The soluble extracts of the cysts corresponding to 61% of the capsular dry weight (cf Table 1) were separated by centrifugation from the insoluble, structural constituents. By a Coomassie Blue binding assay of the extract (Bradford, 1976) 0.083 mg protein (and 0.178 mg protein in the partially dissolved nematocysts pellet) were determined per mg of dry cyst material. Comparison of the SDS-gel patterns of extract and water insoluble fractions illustrate well the differences exhibited by two fractions and confirms the methodically clean separation of the two (Weber *et al.*, 1987).

In the following chapter some properties of the soluble extract will be characterized.

*Fig. 2. 10% SDS-polyacrylamide gel patterns (Lämmli, 1970) of purified and solubilized nematocysts from various Hydra clones and species after staining with silver (Oakley, 1980). Ha1 to Ha4 are different clones of Hydra attenuata, J4 is a hybrid H. carnea x H. attenuata; Hc, Hm and Hci are H. carnea, H. magnipapillata and H. circumcincta respectively. Nematocysts from twenty polyps were applied to each slot.*

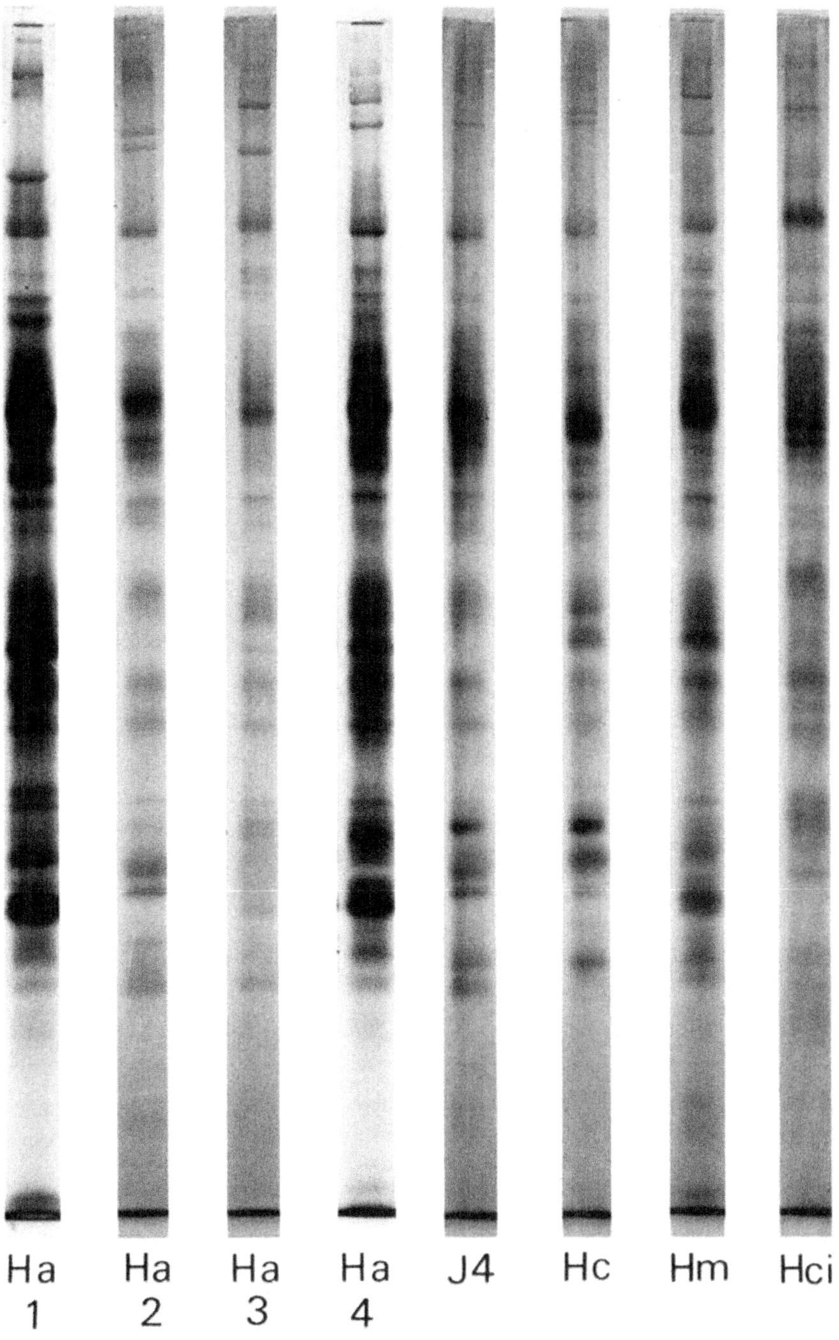
Ha 1
Ha 2
Ha 3
Ha 4
J4
Hc
Hm
Hci

## B. Properties of the Capsular Extract

### 1. Toxicity and inorganic ions

It is to be expected that toxic components and lytic (digestive) enzymes are contained in the soluble nematocyst fraction, as is found in the more extensively studied nematocysts of *Physalia*, *Chironex*, *Chrysaora*, *Aiptasia* etc. (for review see Burnett and Calton, 1977). We were indeed able to detect in hydra cysts at least two different, strongly toxic components which most probably are proteins. They act on nerves and/or muscles (Klug, 1986). One of them causes a general contraction of *Drosophila* larvae after injection of cyst extracts. In addition, the same component exerted an extremely powerful hemolytic activity upon human red blood cells (Fig. 3; Klug, 1986, Klug *et al.*, 1987). The other active component causes rather long term relaxation of insect

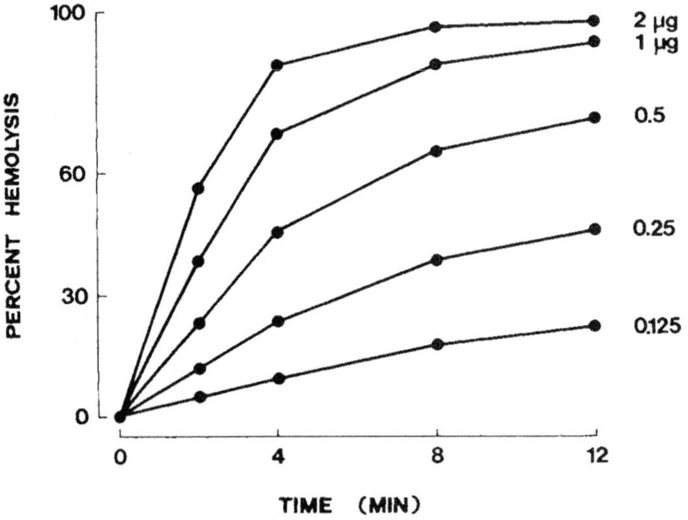


*Fig. 3. Time course of hemolysis of a 1 ml 1% human red blood cell suspension in 120 mM NaCl, 25 mM Na-phosphate, 2 mM EDTA (pH 7.4) at 25°C after the addition of 0.125 to 2.0 µg protein (Bradford, 1976) of a H. attenuata nematocyst extract. Individual samples were withdrawn after the indicated time and assayed for the released hemoglobin (Hessinger and Lenhoff, 1973).*

muscles. Both components are macromolecular and can to some degree be separated by gel filtration on Sephacryl S-300. They both proved to be lethal to adult *Drosophila*. A 50% mortality was obtained with approximately 5 ng protein of the crude extract per fly ($LD_{50}$ of approx. 6 mg/kg).

Purification of these two toxins is difficult due to rapid loss of activities. The hemolysin is particularly labile; it is most stable in distilled water and continuously loses hemolytic (and lethal) activity if the ionic strength is increased or if the pH is changed from neutrality. It is totally inactivated by the addition of low concentrations of detergents (e.g. 0.01% Triton X-100 or SDS in diluted solutions) and by 15% glycerol or ethanol in water. In some properties (chromatographic behaviour on Sephacryl S-300; inhibition by Con A, detergents or lipids; rapid loss of activities in salt solutions) it resembles the hemolysin from *Physalia physalis* which had been isolated by Tamkun and Hessinger (1981).

A potent phospholipase and a very mild protease were found, whereas no chitinase, collagenase or hyaluronidase could be detected (Weber *et al.*, 1987a). In contrast to the venom of nematocysts from the sea anemone *Aiptasia pallida* (Hessinger and Lenhoff, 1976) we have no indications that the phospholipase of hydra cysts supports the action of the hemolysin, since the former is dependent on divalent cations and the hemolysin deploys its maximal activity in the presence of the complexing agent EDTA.

Lubbock (Lubbock *et al.*,1981, Lubbock and Amos, 1981) and Mariscal (1984) detected large amounts of Ca in *in situ* nematocysts obtained from marine cnidarians. Since $Ca^{2+}$ has been claimed to be somehow involved in the discharge process of (at least some) nematocysts (Lubbock and Amos, 1981, Gupta and Hall, 1984) we performed an x-ray spectral analysis in SEM (Fig. 4) on purified hydra nematocysts. It revealed the occurrence of Ca in hydra cysts and of even more Mg (cf. Lane, 1974). The ratio of the divalent cations to each other is constant in all three types of hydra nematocysts (stenoteles, desmonemes, isorhizas). They occur predominantly in the soluble extract of the nematocyst and most are not lost when the undischarged purified cysts are thoroughly washed in water; in spite of the fact that, as mentioned above, molecules of molecular weights up to approx. 600 D are able to pass the capsule wall freely, the cations are retained within the cyst. Consequently $Ca^{2+}$ and $Mg^{2+}$ are probably tightly bound within the nematocysts, possibly to soluble molecules which are too large to pass the cyst's wall. A more detailed, quantitative study by titration revealed an overall Mg and Ca content in hydra nematocysts of 0.80 ± 0.09 and 0.36 ± 0.08 µmole respectively per mg dry capsular material (Weber *et al.*, 1987b).

These values are in good agreement with Lubbock's study (Lubbock *et al.*, 1981) who determined a Ca concentration of 0.5 to 0.6 µmole per mg wet weight in nematocysts from two species of sea anemones. More than 85% of the divalent cations are found in the soluble part of the grinded capsules and are not bound to insolubles structures.

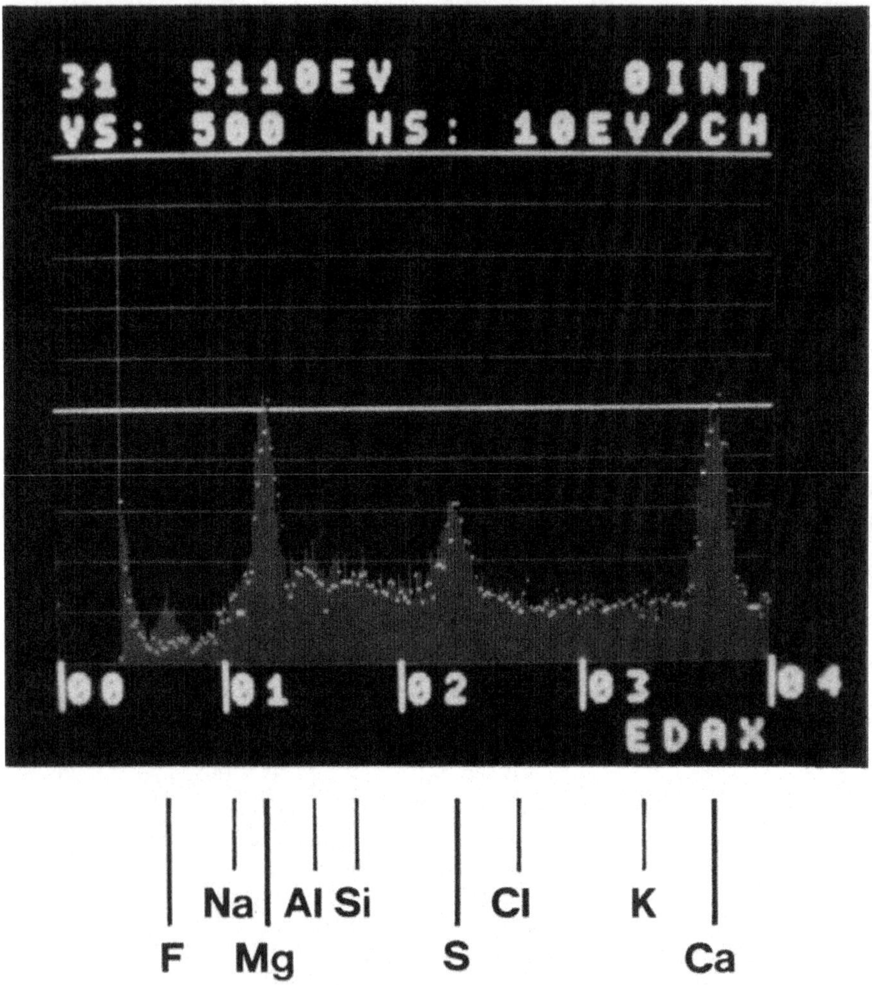


*Fig. 4. x-ray spectral analysis of a single isolated stenotele of H. attenuata in SEM. The figure is adapted from Tardent et al. (1985). The fluorine peak origins from the teflon support. Spectra of desmonemes or isorhizas (not shown) are identical.*

## 2. Protein content of the nematocysts

An intriguing problem is the identity of the nematocysts components to which $Mg^{2+}$ and $Ca^{2+}$ are bound. To preserve electroneutrality the binding molecules are supposed to be highly negatively charged, the total charge concentration being approx. 2.3 µmole per mg dry weight. Possible candidates for these anionic ligands are acidic proteins (Asp/Glu rich or phosphoproteins) and sulfate and carboxylate containing complex carbohydrates or glycoproteins. Since we found in appropriate chemical assays of oxydized (Mg and Ca containing) soluble nematocyst extracts only minor concentrations of sulfate and phosphate, the divalent cations are probably bound to carboxylates (Cl is absent in the EDAX-spectrum of Fig. 4). To compensate for 1.16 µmole Mg plus Ca ions per mg dry weight, at least 0.4 mg Asp or Glu (in polypeptides) per mg crude extract would be needed (at least 85% of the divalent cations are in the soluble extract, which amounts to 61% of the nematocyst dry weight). However only 0.14 mg protein per mg crude extract is detected by the Bradford protein determination method (Bradford, 1976).

We applied several other protein determination methods and found approx. six-fold more protein in the extracts using methods which do not relay upon the presence of particular amino acid side chains, but directly measure the concentration of peptide bonds in the sample, i.e. uv absorbance at short wavelengths (Table 1). Since the absorption coefficients of the amide bonds in proteins depend to some extent on their conformation, one cannot decide whether the results of this study may slightly over- or underestimate the actual protein content of nematocysts.

Thus is appears that proteins are by far the most prominent constituents of nematocysts. It also seems likely that among them are the expected binding sites for the divalent cations. In addition the high protein concentration (roughly estimated over 50% by weight) explains the high buoyant density of the cysts (see above).

## 3. Fractionation of the nematocyst extract

The toxins were difficult to purify because of the tendency of nematocyst extract to aggregate to multimolecular complexes. Only at high ionic strength (at least 1 M ammonium sulfate) could the proteins be separated to some degree by gel filtration (Klug, 1986).

This is not surprising since polyanions that are proposed to bind $Mg^{2+}$ and $Ca^{2+}$ (see above) are expected to interact

TABLE I. Dry weight and protein content of mixtures of purified nematocysts from *Hydra attenuata*. The cysts isolated from 6.6 ± 1.5 ml of packed polyps yielded the following average values ( ± stand. dev.) in mg. Adapted from Weber *et al.*, 1987a.

| | n = | Intact cysts | n = | Soluble extracts after grinding | n = | Insoluble pellet after grinding |
|---|---|---|---|---|---|---|
| Dry weight | | 10.0 | 5 | 6.1 ± 0.4 | 4 | 3.0 ± 0.4 |
| Protein determinations by: | | | | | | |
| Coomassie Blue binding (Bradford 1976) | 4 | 2.7 ± 0.1 | 6 | 0.83 ± 0.11 | | 1.87 (a) |
| Lowry method (Peterson 1977) | | not determined | 5 | 0.62 ± 0.10 | | not determined |
| UV 280/260 nm (Kalckar & Shafran 1947) | | b | 7 | 0.76 ± 0.10 | | b |
| UV 235/280 nm (Whitaker & Granum 1980) | | b | 7 | 0.87 ± 0.10 | | b |
| UV 215/225 nm (Waddell 1956) | | b | 7 | 5.10 ± 0.58 | | b |
| UV 205 nm (Scopes 1974) | | b | 7 | 4.90 ± 0.68 | | b |

n = numbers of separate experiments. (a) calculated as difference between intact cysts and their soluble extracts; b = determination methodically impossible.

electrostatically with cations and cationic domains of macromolecules except at high ionic strength (see e.g. Schwarz and Seelig-Löffler, 1975). The accumulation of some cationic dyes within intact undischarged nematocysts (metachromasia, see above) can also be explained by the presence of these polyanions (Cundall *et al.*,1979).

The elution profile of nematocyst extract from Sephadex G-50 in 1 M ammoniumsulfate (Fig. 5a) indicates that less than 14% of the 205 nm absorbing material elutes at a molecular weight above 30 000 D. This high molecular weight fraction includes all the toxic components and practically every protein band which appears in a silver or Coomassie Blue stained SDS-gel of the extract. On the other hand, two prominent peaks elute at molecular weights of around 15 000 D (named H1, approx. 50% of the total) and of 3000 D (named H2, approx. 30% of the total). Following SDS-gel electrophoresis H1 can only be detected by staining with the cationic dye "stains-all" (Campbell *et al.*, 1983) as a prefront band, whereas H2 remains undetectable. Based on their uv absorbance characteristics both peaks seem to consist of polypeptides with an extremely low content of aromatic side chains. They cooperatively bind the metachromatic dye toluidine blue (forming precipitates at low ionic strength) and are, therefore, favorite candidates for the postulated polyanionic $Mg^{2+}$ and $Ca^{2+}$ binding molecules. They may be composed preferentially of glutamate which, on several occasions has been reported to be by far the most abundant amino acid in different nematocysts (Lane, 1974; Phelan and Blanquet, 1985).

## IV. CONCLUSIONS

Although nematocysts were discovered 150 years ago their chemistry is still rather poorly understood. Research to date has been limited to the cysts of a few cnidarian genera such as the hydrozoan *Physalia*, the cubozoan *Chironex*, the scyphozoans *Chrysaora*, *Pelagia* and *Stomolophus* and to some selected anthozoans (reviewed by Mariscal, 1974; Blanquet, 1977; Burnett and Calton, 1977; Russell, 1984).

The extremely rapid discharge (Holstein and Tardent, 1984) as well as the strongly venomous action of nematocysts (Cleland and Southcott, 1965) has been observed and described frequently and by many authors, but the essential physical and chemical principles governing the recorded phenomena are still more or less obscure.

The comparison of nematocysts from the freshwater hydra

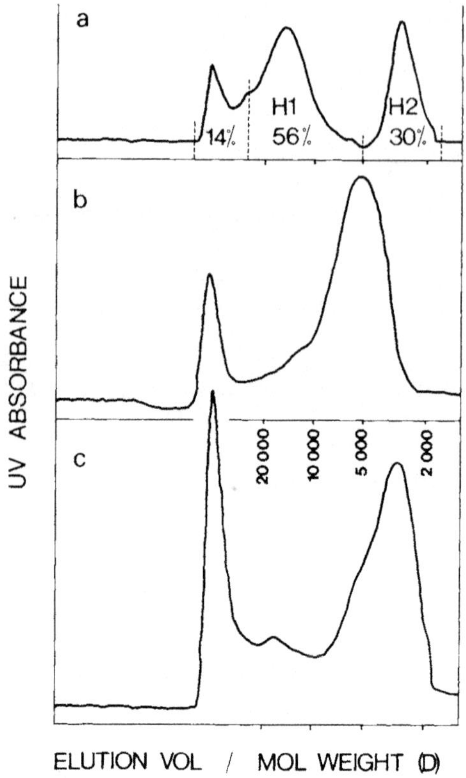


*Fig. 5. Elution profile after chromatographing nematocyst extracts in 1 M $(NH_4)_2SO_4$ (pH 7.5) on a calibrated Sephadex G-50 column. The elution volume, starting with the application of the extract and ending after the elution of one column volume, is plotted versus the uv absorbance (LKB Uvicord S detector). No uv absorbing material eluted later. The figures indicate the molecular weight of the eluted proteins (myoglobin, cytochrome c, aprotinin, insulin alpha chain, bacitracin and azide were taken for calibration).*

a. *Extract of nematocysts from H. attenuata; the indicated protein fractions of the three main peaks were determined spectrophotometrically at 205 nm after pooling.*
b. *Extract of nematocysts (mastigophores) from tentacles of Anemonia sulcata; intact nematocysts had been purified from isolated tentacle ectoderm by the same method as used for Hydra nematocysts.*
c. *Extract of nematocysts (isorhizas) from Physalia physalis.*

with selected marine cnidarians reveals many common characteristics. These include:

- The capability of isolated (cell free) nematocysts to discharge in the absence of any other cellular components (Yanagita and Wada, 1954),
- a buoyant density (presumably due to the high protein content) mostly higher than 1.2 g/ml (Burnett and Calton, 1973),
- the presence of large quantities of divalent cations (in situ roughly one half molar) within the nematocysts (Lubbock and Amos, 1981; Lubbock *et al.*, 1981, Mariscal, 1984, Weber *et al.*, 1987b),
- The occurrence of chemically or physically labile, strongly neurotoxic and lethal proteins, lytic enzymes and possibly pharmacologically active low molecular weight substances and/or specific inhibitors (Burnett and Calton, 1977),
- an unexpectedly large variety of different proteins (of yet unknown function) present in the soluble and insoluble fractions of the cysts, the composition of which may distinctly vary from one clone to another within the same species (Fig. 2), and
- the presence of large amounts of polypeptides which cannot be detected by the routine protein determination methods and which are supposed to be involved in the binding of divalent cations and the process of discharge rather than in toxic actions; as in hydra (Fig. 5a) we also found dominating protein peaks featuring molecular weights lower than 20 000 D among the soluble components of pure nematocysts from *Physalia* and *Chrysaora* (Fig. 5b, c).

An evaluation of these more general characteristics of nematocysts may in future lead to a generally valid structural and mechanistical model for these fascinating cell organoids.

ACKNOWLEDGEMENTS

We would like to thank Dr. R. Stidwill for critically reading and Mrs. B. Werner for typing the manuscript, Mr. U. Jauch for assisting in the x-ray spectral analysis in SEM, Mrs. B. Gyr and Mr. E. Stöhr for their technical assistance and particularly every member of our group in Zürich for many fruitful discussions.

REFERENCES


Blanquet, R.S. 1977. Cnidarian Venoms. *In* A.W.Bernheimer (ed.), *Perspectives in Toxinology*, pp. 149-167. John Wiley and Sons, New York.

Blanquet, R.S. 1983. Isolating undischarged and discharged nematocysts from acontiate sea anemones. *In* H.M.Lenhoff (ed.), *Hydra research methods*, pp. 331-333. Plenum Press, New York.

Bradford, M.M. 1976. A rapid and sensitive method for the quantitation of microgram quantities of protein utilizing the principles of protein-dye binding. Anal. Biochem. 72:248-254.

Burnett, J.W. and G.J. Calton. 1973. The purification of sea nettle nematocyst toxins by gel diffusion. Toxicon 11:243-247.

Burnett, J.W. and G.J. Calton. 1977. Review article. The chemistry and toxicology of some venomous pelagic coelenterates. Toxicon 15:177-196.

Campbell, K.P., D.H. McLennan and O.O.Jorgensen. 1983. Staining of the $Ca^{2+}$ binding proteins, calsequestrin, calmodulin, troponin C and S-100, with the cationic carbocyanine dye "stains-all". J. Biol. Chem. 258:11267-11273.

Cleland, J.B. and R.V. Southcott. 1965. Injuries to man from marine invertebrates in the Australian region. Natl. Health Med. Res. Counc. Spec. Rep. Ser. N° 12.

Cundall, R.B., J.B. Lawton, D. Murray and G.O. Phillips. 1979. The role of cooperativity in determining the binding behavior of hetero-charged polyanions with metachromatic dyes. Makromol. Chem. 180:383-388.

Endean, R. and J. Rifkin. 1975. Isolation of different types of nematocysts from the Cubomedusan *Chironex fleckeri*. Toxicon 13:375-376.

Gupta, B.L. and T.A. Hall. 1984. Role of High Concentrations of Ca, Cu and Zn in the Maturation and Discharge in Situ of Sea Anemone Nematocysts as Shown by x-ray Microanalysis of Cryosections. *In* L. Bolis, J. Zadunaisky and R. Gilles (eds.), *Toxins, Drugs and Pollutants in Marine Animals*. pp. 77-95. Springer Verlag, Heidelberg.

Hessinger, D.A. and H.M. Lenhoff. 1973. Assay and properties of the hemolysis activity of pure venom from the nematocysts of the acontia of the sea anemone *Aiptasia pallida*. Arch. Biochem. Biophys. 159:629-638.

Hessinger, D.A. and H.M.Lenhoff. 1976. Mechanisms of hemolysis induced by nematocysts venom: role of phospholipase A and direct lytic factor. Arch. Biochem. Biophys. 173:603-613.

Holstein, T. and P. Tardent. 1984. An Ultrahigh-Speed Analysis of Exocytosis. Nematocyst Discharge. Science 223:830-833.

Kalckar, H.M. and M. Shafran. 1947. Differential spectrophotometry of the purine compounds by means of specific enzymes. Studies of the enzymes of purine metabolism. J. Biol. Chem, 167:461-475.

Klug, M. 1986. Beitrag zur Kenntnis biologisch aktiver Komponenten isolierter Nematocysten von *Hydra attenuata* Pall. (Hydrozoa, Cnidaria). PhD Thesis, University of Zürich, Switzerland.

Klug, M., J. Weber and P. Tardent. 1987. Assay for direct observation of hemolytic activity of single nematocysts. This volume.

Lämmli, U.K. 1970. Cleavage of structural proteins during the assembly of the head of bacteriophage T 4. Nature 217:680-685.

Lane, Ch.E. 1974. Nematocyst toxins of Coelenterates. *In* H.J. Humm and Ch.E.Lane (eds.), *Bioactive Compounds of the Sea*, pp. 123-137. Marcel Dekker Inc., New York.

Loomis, W.F. and H.M. Lenhoff. 1956. Growth and sexual differentiation of *Hydra* in mass culture. J. Exp. Zool. 132:555-574.

Lenhoff, H.M. 1983. *Hydra Research Methods*, pp. 1-463. Plenum Press, New York.

Lenhoff, H.M. and W.F. Loomis. 1961. *The Biology of Hydra*. pp. 1-467, University of Miami Press, Coral Gables, Florida.

Lubbock, R. and W.B. Amos. 1981. Removal of bound calcium from nematocysts contents causes discharge. Nature 290:500-501.

Lubbock, R., B.L. Gupta and T.A. Hall. 1981. Novel role of calcium in exocytosis: Mechanism of nematocyst discharge as shown by x-ray microanalysis. Proc. Nat. Acad. Sci. USA 78:3624-3628.

Mariscal, R.N. 1974. Nematocysts. *In* L. Muscatine and H.M. Lenhoff (eds.), *Coelenterate Biology*. pp. 129-178. Academic Press, New York.

Mariscal, R.N. 1984. Cnidaria: Cnidae. *In* J. Breiter-Hahn, A.G. Matoltsy, K.S. Richards (eds.), *Biology of the Integument, Vol. 1, Invertebrates*, pp. 57-68. Springer Verlag Berlin, New York.

Oakley, B.R., D.R. Kirsch and N.R. Morris. 1980. A simplified ultrasensitive silver stain for detecting proteins in polyacrylamide gels. Anal. Biochem., 105:361-363.

Peterson, G.L. 1977. A simplification of the protein assay method of Lowry *et al.*, which is more generally applicable. Anal. Biochem. 83:346-356.

Phelan, M.A. and R.S. Blanquet. 1985. Characterization of nematocyst proteins from the sea anemones *Aiptasia pallida* and *Pachycerianthus torreyi* (Cnidaria, Anthozoa). Comp. Biochem. Physiol. B 81:661-666.

Russell, F.E. 1984. Marine Toxins and Venomous and Poisonous Marine Plants and Animals (Invertebrates). *In* J.H.S. Blaxter, F.S. Russell and M. Yonge (eds.), *Advances in Marine Biology*, 21:104-132.

Salleo, A. 1984. Discharge Mechanisms of the nematocysts of *Pelagia noctiluca*. *In* L. Bolis, J. Zadunaisky and R. Gilles (eds.), *Toxins, Drugs and Pollutants in Marine Animals*, pp. 63-76, Springer Verlag, Heidelberg.

Schwarz, G. and A. Seelig-Löffler. 1975. Aggregation of linear biopolymers induced by cooperative binding of ligands. Biochim. Biophys. Acta 379:125-138.

Scopes, R.K. 1974. Measurement of protein by spectrophotometry at 205 nm. Anal. Biochem. 59:277-282.

Slautterback, D.B. and D.W. Fawcett. 1959. The development of the cnidoblast in *Hydra*.An electron microscope study of cell differentiation. J. Biophys. Bicohem. Cytol., 5:441-452.

Tamkun, M.M. and D.A. Hessinger. 1981. Isolation and partial characterization of a hemolytic and toxic protein from the nematocyst venom of the portuguese man-of-war, *Physalia physalis*. Biochim. Biophys. Acta 667:87-98.

Tardent, P. 1987. History and current state of knowledge concerning discharge of cnidae. This volume.

Tardent, P., Th. Holstein, J. Weber and M. Klug. 1985. The morphodynamics and actions of stenotele nematocysts in *Hydra*. Arch. Sc. Genève 38:401-418.

Waddell, W.J. 1956. A simple ultraviolet spectrophotometric method for the determination of protein. J. Lab. Clin. Med. 48:311-314.

Weber, J., Klug, M. and P. Tardent. 1987a. Some physical and chemical properties of purified nematocysts of *Hydra attenuata* Pall. (Hydrozoa, Cnidaria). Comp. Biochem. Physiol. C (in press).

Weber, J., Klug, M. and P. Tardent. 1987b. Detection of high concentrations of Mg and Ca in the nematocysts of various cnidarians. Experientia (in press).

Whitaker, J.R. and P.E. Granum. 1980. An absolute method for protein determination based on difference in absorbance at 235 and 280 nm. Anal. Biochem. 109:156-159.

Yanagita, T.M. and T. Wada. 1954. Effects of trypsin and thioglycollate upon the nematocysts of the sea anemone. Nature (London) 173:171.

23

# NUDIBRANCH NEMATOCYSTS

Paul G. Greenwood

Department of Biological Science
Florida State University
Tallahassee, Florida 32306

## SUMMARY

Most aeolid nudibranchs that feed on cnidarians incorporate nematocysts into the cnidophage cells of their cnidosacs. Recent investigations have provided insights into the mechanisms by which nematocysts are acquired and maintained within nudibranchs. At least one species of nudibranch incorporates late-stage immature nematocysts, which subsequently complete development within the cnidophages. It seems therefore, that nudibranchs may actively maintain nematocysts within their cells. Nematocysts are not digested in cnidophage cells, probably because the membranes of the phagosome break down, and the nematocysts are not exposed to lysosomal enzymes. Cnidophage microfilaments attach to the basal regions of nematocysts and are apparently responsible for the orientation of the nematocysts within the cnidophages. Cnidophage microtubules lie between nematocysts and might hold the nematocysts in their proper orientation. It is thought that nudibranchs retain particular nematocyst types depending on the prey species consumed, and it appears that this selective retention is the result of chemical or physical differences in nematocysts. Nematocysts, even of the same type, might be very different among species of cnidarians.

---

Supported in part by an FSU Psychobiology Traineeship, a Sigma Xi Grant-in-Aid of Research, and an NIH Biotechnology Resource grant to H. Ris for the HVEM facility in Madison, WI.

## I. INTRODUCTION

The mechanisms by which aeolid nudibranchs incorporate cnidarian nematocysts have puzzled biologists for well over a century. Bergh (1862) surveyed several genera of aeolids, which he thought secreted nematocysts. Wright (1863) presented several observations and an experiment suggesting that nudibranch nematocysts came from ingested cnidarians, but he was largely ignored. It was not until Grosvenor's (1903) comprehensive review and confirmation of Wright's experiments that researchers began to accept the idea that aeolids obtained nematocysts from their prey.

Since the early 1900's, several studies on the nudibranch-nematocyst association have served to elucidate the pathway of nematocysts through the nudibranch (Graham, 1938; Kepner, 1943; Edmunds, 1966; Streble, 1968; Kälker and Schmekel, 1976; Conklin and Mariscal, 1977; Moreteau, 1977; Day and Harris, 1978; Cargo and Burnett, 1982). Unfortunately, these studies do not investigate the mechanisms behind the incorporation process. Excellent reviews by Thompson (1960), Edmunds (1966), and Todd (1981) have helped to provide insights into the possible defensive value of nudibranch nematocysts, but they do not address the complex mechanisms controlling nematocyst incorporation. This review focuses on the major problems involved in nematocyst incorporation and maintenance within nudibranchs, and presents several hypotheses on how these problems are overcome.

## II. FUNCTIONS OF NUDIBRANCH NEMATOCYSTS

Nematocysts are generally considered to serve nudibranchs as defensive weapons (Thompson, 1960, 1976; Thompson and Bennett, 1969, 1970; Todd, 1981), though this is often difficult to determine. Edmunds (1966) remarked that it is difficult to separate any defensive effects of nematocysts from the defensive secretions of the nudibranch, or from a general tissue unpalatability, although Allen (1976) found discharged nematocysts in the damaged mouth tissues of a fish that had tried to feed on the aeolid *Hermissenda* (=*Phidiana*) *crassicornis*.

Cuenot (1907) and Glaser (1910) suggested that the defensive benefit of nematocysts to the aeolid is slight, because, in the laboratory, both workers had witnessed predation on aeolids by fish. There are few animal defenses that are 100% effective and it seems premature to conclude, based on a few laboratory observations, that nematocysts are ineffective defensive structures. Harris (1986) witnessed predation by the wrasse Tautogolabrus adspersus on the aeolid Aeolidia papillosa in the field, and he noticed that larger nudibranchs were not eaten. Whether a greater number of nematocysts in the larger nudibranchs contributes to the escape from predation is not known.

Bergh (1862) reported that aeolids use nematocysts as offensive weapons against their cnidarian prey, but there has been only one subsequent report of such an offensive function. Tardy (1964) observed a nudibranch that extended its anterior cerata to its prey, released nematocysts, and then began feeding in the affected area.

Grosvenor (1903), Graham (1938), and Streble (1968) discussed the possible origin of nematocyst-storing structures of aeolids as excretory organs. While it is possible that the concentration of nematocysts at the distal cerata tips evolved as a means of removing a potentially dangerous substance, it appears that nematocysts are now released by nudibranchs only in response to significant external stimulation. Glaser (1910) raised the point that nematocysts probably evolved in the Cnidaria as prey entangling and capturing structures; such structures would provide little defensive benefit to nudibranchs. However, nudibranchs retain particular nematocyst types from their prey species; the retained types are invariably penetrant nematocysts (see below), which often have defensive functions in cnidarians (Mariscal, 1974).

## III. THE PATHWAY OF NEMATOCYSTS AND CERAS MORPHOLOGY

A number of studies of the movements of nematocysts within a nudibranch's digestive system have been conducted (Graham, 1938; Streble, 1968; Rosin, 1970; Kälker and Schmekel, 1976; Conklin and Mariscal, 1977; Greenwood and Mariscal, 1984a). The following description summarizes the movements of nematocysts to the cnidosac, and the structures involved.

Nematocysts are ingested with prey tissues and pass into the digestive diverticulum within the ceras (Fig. 1). A number of unfired nematocysts pass through a narrow ciliated channel into the cnidosac where they are engulfed by the cnidophage cells. The cnidophage cells are large, specialized digestive cells and are extensions of the digestive epithelium. Cnidophages arise from undifferentiated epithelial cells that line the ciliated channel and are pushed distally along a

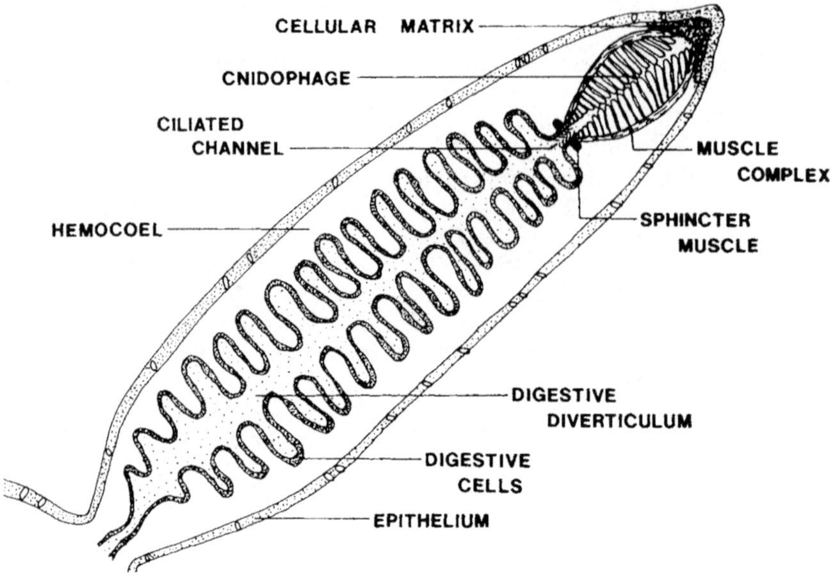


Fig. 1. Illustration of an entire ceras in longitudinal section. The body of the nudibranch is at lower left (from Greenwood and Mariscal, 1984a).

basement membrane by the cells developing behind them. Nematocysts are engulfed by cnidophages in the proximal cnidosac only; distal cnidophages are loaded with nematocysts and are apparently unable to phagocytize more. Nematocysts are held within the cnidophages in groups termed cnidocysts. The nematocysts are almost invariably oriented with their apical ends directed toward the lumen of the cnidosac. Nematocysts are released from the ceras by a contraction of the muscle complex that surrounds the entire cnidosac. Contraction of a sphincter muscle around the ciliated channel prevents the movement of nematocysts from the cnidosac to the diverticulum during cnidosac contraction. Entire cnidophages are ejected through the ceras tip epithelium and the nematocysts discharge following the rupture of the cnidophage membrane.

## IV. PROBLEMS OF ACQUIRING UNFIRED NEMATOCYSTS

How do nudibranchs prevent the discharge of nematocysts during the acquisition process? A considerable amount of interest has been directed towards this problem, and there are differing opinions on whether there is an appreciable discharge of prey nematocysts during nudibranch feeding. Grosvenor (1903) reported that nematocyst discharge was inhibited in some way and he found mostly unfired nematocysts within the nudibranch's body. Graham (1938) and Kälker and Schmekel (1976) also reported that there was little nematocyst discharge during feeding. However, other researchers have found nematocyst discharge to be common during feeding. Harris (1973), Thompson (1976), and I have found discharged nematocysts in the guts and feces of nudibranchs. Discharged nematocysts can represent more than 50% of the total nematocysts in the gut. Yarnall (1972) also reported nematocyst discharge during nudibranch feeding but gave no percentages. While not mentioning it, Moreteau (1977) published photomicrographs of discharged nematocysts in the gut and feces of *Aeolidia papillosa*.

Possibly, different species of nudibranchs manipulate nematocysts in different ways. Nematocysts may discharge more in some nudibranch species than in others, and several hypotheses exist on why nematocyst discharge is limited during feeding. Graham (1938) suggested that copious mucus secretions of the nudibranch inhibited the discharge of nematocysts. Mucus might serve to insulate the nudibranch from discharging nematocysts (Thompson, 1960; Edmunds, 1966), but its value in preventing discharge has not been proven. Conklin and Mariscal (1977) suggested that some sort of acclimation process might be occurring. It is true that some species of anemones that are prey to nudibranchs seem to lose responsiveness to the nudibranch (Conklin and Mariscal, 1977), but other anemones initiate defensive behaviors (Edmunds *et al.*, 1976; Rosin, 1969; Waters, 1973), or immediately contract and are consumed in a contracted state (Glaser, 1910; personal observation).

Yarnall (1972) suggested that a combination of mucus secretion, rapid movement of cnidarian tissues into the nudibranch, and quick digestion of the cnidocyte serves to reduce nematocyst discharge. Graham (1938) had also postulated that rapid intake of tissues prevented discharge. Rapid removal of the cnidocyte from the nematocyst has been reported by Kepner (1943) and Kälker and Schmekel (1976). Such removal would effectively isolate the nematocyst, and Blanquet (1970) showed that isolated nematocysts were generally less likely to discharge than *in situ* nematocysts. But Yarnall (1972) showed that the pH within the gut of *Hermissenda crassicornis* may be as low as pH 1.7; hydrogen ion concentrations this high have been shown to cause nearly 100% discharge even in isolated nematocysts (Blanquet, 1970). However, Graham (1938) found much higher pH levels in the gut of *Aeolidia papillosa*.

Clearly, there are differences in the acidity of nudibranchs' guts, and these differences could account for the disparate observations on nematocyst firing during feeding.

Harris (1971, 1973) reported that several species of hydroid-feeding aeolids suck out the coenosarc of the hydroid and probably obtain few mature nematocysts. Harris speculated that some specialized species of aeolids might incorporate immature nematocysts, which are generally incapable of firing (Lentz and Barrnett, 1961), and allow them to reach functional maturity within the cnidosacs. Todd (1981, 1983) suggested that such a possibility deserved further study. Naville (1926) and Rousseau (1927) also suggested that nudibranchs possessed immature nematocysts, although their reasoning proved to be in error. Naville (1926) supposed that small nematocysts in the cnidosacs of Aeolidiella alderi originated from the cells of the nudibranch and were still growing. Rousseau (1927) reported that he saw developing nematocysts within the cnidosacs of Cuthona aurantiaca but that these originated within the nudibranch.

Recently, Greenwood and Mariscal (1984b) found late-stage immature nematocysts within the cnidosacs of Spurilla neapolitana. We used a histochemical assay specific for late-stage immature nematocysts (see David and Challoner, 1974) and determined that the percentage of immature nematocysts in the cnidosacs of S. neapolitana reduces with starvation (Fig. 2), but that the total number of nematocysts does not change. The reduction of immature nematocysts is attributed to the maturation of nematocysts within the cnidosac. The occurence of mitochondria around the nematocysts and not elsewhere in S. neapolitana cnidophages (Greenwood and Mariscal, 1984a) (Fig. 3) suggests that cnidophages might be supplying energy in the form of ATP to the nematocysts, actively maintaining them in the cnidosac rather than simply holding them there. Unfortunately, little is known about the energetics of nematocyst maturation within the Cnidaria, let alone within a foreign organism. Recently, I have found that ATP concentrations decreased much faster in suspensions of isolated, intact nematocysts than in suspensions of isolated, discharged nematocysts. In addition, the percentage of late-stage immature nematocysts decreases in suspensions of isolated nematocysts that are incubated in solutions of ATP (Greenwood and Mariscal, 1984c; Greenwood, 1987). If isolated nematocysts use ATP for their maintenance, then nudibranchs probably actively maintain nematocysts within their cnidophages.

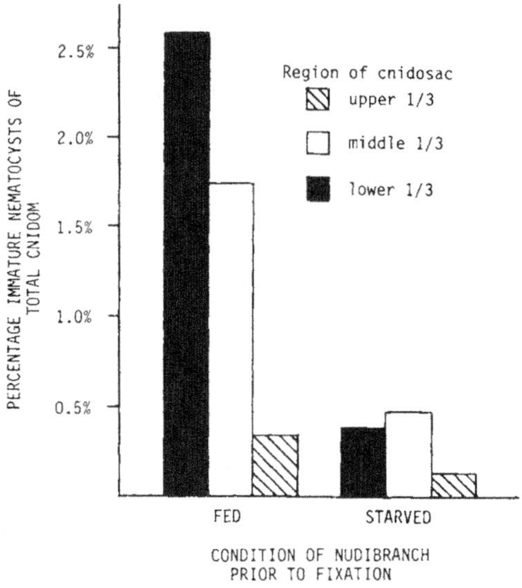


Fig. 2. Results obtained from treating recently fed *Spurilla neapolitana* and a 3-day starved *S. neapolitana* with a substrate of thiolacetic acid and lead nitrate. Note the fivefold decrease in the percentage of immature nematocysts present in the cnidosacs of nudibranchs starved for 3 days prior to fixation (from Greenwood and Mariscal, 1984b).

## V. NEMATOCYST DIGESTIBILITY

Several researchers have studied why nematocysts are not degraded in the nudibranch's digestive diverticula (Glaser, 1910; Graham, 1938; Yarnall, 1972), but few authors have studied why nematocysts are not digested within the cnidophage cells. Intracellular digestion of nematocysts is known to occur in the Cnidaria (see for example Watson and Mariscal, 1983) and in other cnidarian-feeding molluscs (Robertson, 1980), but I have seen few cases of nematocyst degradation within the cnidosacs of *Spurilla neapolitana*. However, Kepner (1943) and Kälker and Schmekel (1976) detected degraded nematocysts within the cnidophages of *Aeolis* (=*Cratena*) *pilata* and *Cratena peregrina* respectively. The intracellular degradation reported by these authors only occurs to those nematocyst types not normally stored within the cnidosacs, so the problem remains as to why certain nematocysts are not digested within the cnidophages.

Jones (1974) discussed three possibilities of how a phagocytized particle could avoid digestion within a host organism. There could be a breakdown of the phagocytic membrane following ingestion, and the ingested particle would not be exposed to lysosomal enzymes. A particle may also avoid digestion by inhibiting lysosomal fusion with the phagocytic membrane. Finally, the ingested particle might simply be resistant to the digestive processes of the cell.

There have been only a few ultrastructural studies on nematocysts within nudibranch cnidosacs (Kälker and Schmekel, 1976; Cargo and Burnett, 1982; Greenwood and Mariscal, 1984a). Greenwood and Mariscal (1984a) found that the phagocytic membranes in the cnidophages of *Spurilla neapolitana* apparently break down shortly after the inclusion of the nematocysts. There are a number of membranous structures within the cnidophages, but it appears that only the original nematocyst membrane remains. The micrographs of Cargo and Burnett (1982) show a membrane surrounding the nematocysts of *Cratena pilata*, and those of Kälker and Schmekel (1976) show a loosely associated membrane around a nematocyst in a cnidophage of *Cratena peregrina*. Kälker and Schmekel considered it to be the phagocytic membrane. The available evidence suggests that the nematocysts of aeolids avoid digestion within the cnidosac because the phagocytic membrane breaks down, or perhaps, because lysosomes are inhibited from fusing with the phagosomes that surround nematocysts.

## VI. NEMATOCYST ORIENTATION

Many researchers have noted that nematocysts in nudibranch cnidophages are oriented with their apical ends directed toward the lumen of the cnidosac (Grosvenor, 1903; Glaser, 1910;

Fig. 3. Two basitrichous isorhiza nematocysts (NE) in a cnidophage cell of *Spurilla neapolitana*. Note the mitochondria (M) near the nematocysts. Bar = 0.5 um.

Fig. 4. Electron micrograph stereo pair of microfilament bundles (arrows) at the base of a microbasic p-mastigophore nematocyst (NE) in a cnidophage of *Spurilla neapolitana*. Bar = 0.5 um.

Fig. 5. Electron micrograph stereo pair of microfilaments (arrows) in a cnidophage of *Spurilla neapolitana*. This cnidophage contained no nematocysts. Bar = 1.0 um.

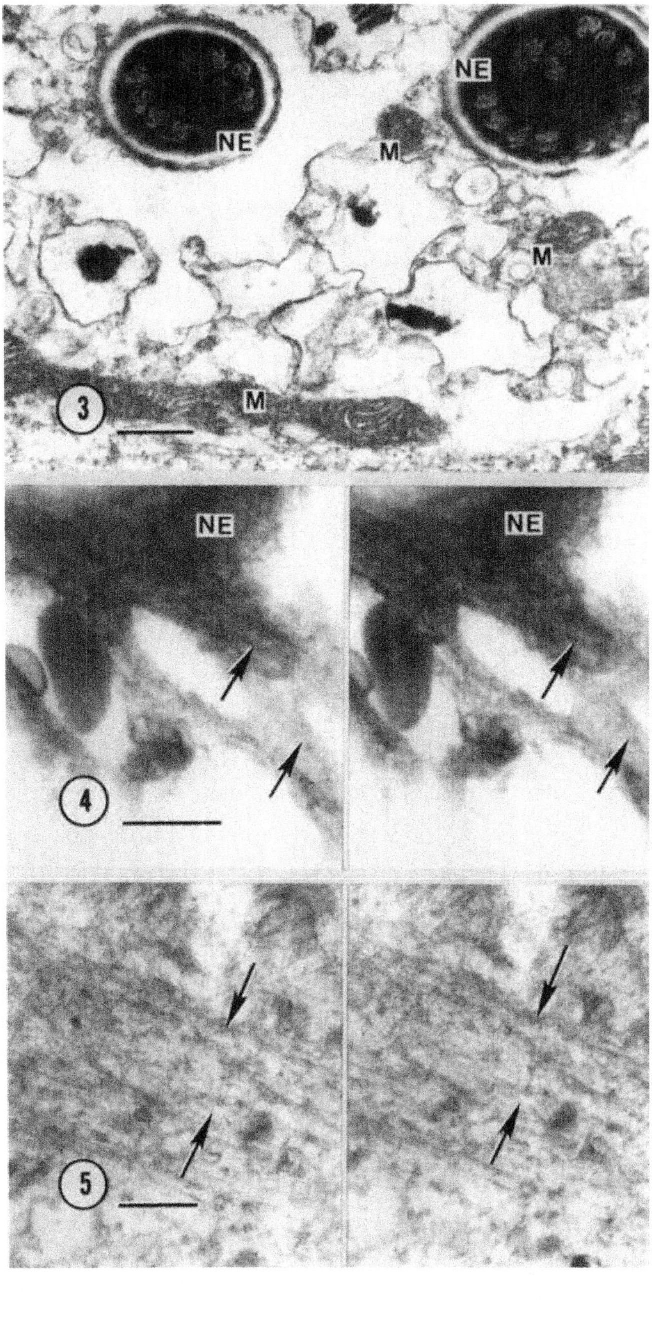
NE
NE
M
M
M
3
NE
NE
4
5

Kepner, 1943; Greenwood, 1983). This orientation is presumed to allow maximum defensive benefit of the nematocysts when they are released from the cnidosac. How this orientation occurs has never been shown. Grosvenor (1903) showed an illustration of nematocysts being engulfed in an oriented position, and Kepner (1943) reported that it was unusual to find a nonoriented nematocyst in the cnidophages. Thus, I suspected that nematocysts are engulfed by the cnidophages in a pre-oriented position, as opposed to being engulfed randomly and then oriented inside the cnidophage.

It is not known if molluscan phagocytic cells recognize a particle as being "self" or "non-self" (Fletcher & Cooper-Willis, 1983). Furthermore, the phagocytosis of a nematocyst by a cnidophage may or may not be triggered by a receptor on the cnidophage recognizing some component of a nematocyst membrane. It has been shown that actin microfilaments are important components in many phagocytic processes (Reaven & Axline, 1973; Pollard, 1981), and they probably serve an active function in cnidophages. I have used conventional and high voltage electron microscopy to study the ultrastructure of the nematocysts in Spurilla neapolitana, and there appear to be filamentous connections between the cnidophage cell membrane and the nematocyst during the incorporation process. There also appear to be bundles of microfilaments at the base of the nematocyst membrane in the cnidophages of the proximal cnidosac (Fig. 4). Cnidophages that have not incorporated nematocysts have numerous filaments running the length of the cell (Fig. 5). The filament bundles have not been investigated using heavy meromyosin so I cannot say definitively that they are actin fibers, but they are about 7-8 nm in diameter and arranged in bundles not unlike the stress fibers of some mammalian cells (see Pollard, 1981).

Rhodamine-labelled phalloidin stains filamentous actin in basal regions of isolated cnidophages (near the basal regions of the nematocysts) and also at the cnidophage surface (Fig. 6). These actin microfilaments are assumed to be the same filaments seen attached to the bases of nematocysts in the electron microscope. Possibly, the basal region of the nematocyst membrane triggers a receptor on the cnidophage membrane to initiate phagocytosis of the nematocyst and the attachment of contractile fibers to the nematocyst. Attachment of microfilaments to only the bases of nematocysts suggests that the protein constituents of nematocyst membranes differ from the apex to the base.

Actin microfilaments probably play the predominant role in nematocyst orientation, but microtubules, which are responsible for some of the nematocyst movements within the cnidocyte, could play a role in maintaining the orientation of nematocysts

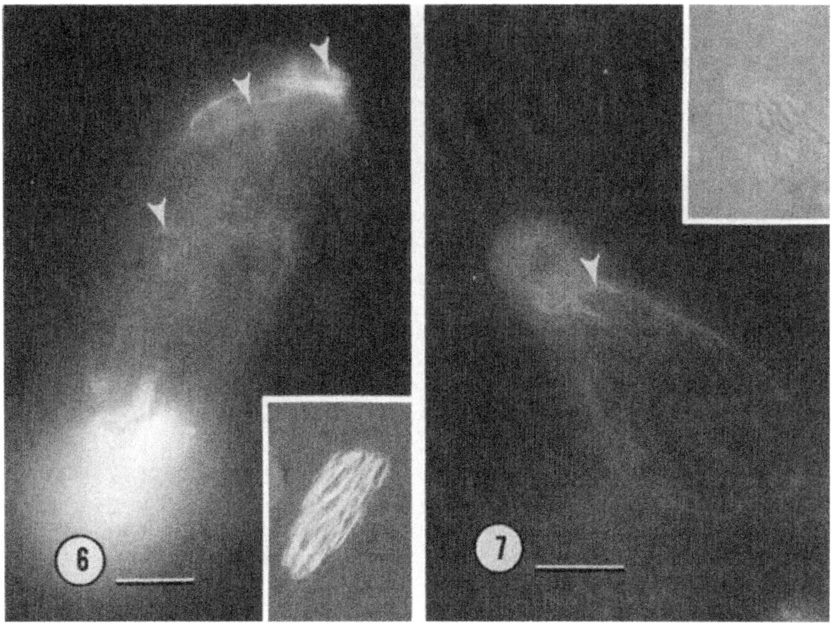


Fig. 6. Rhodamine-labeled phalloidin staining of an isolated cnidophage from Spurilla neapolitana. Fluorescent staining indicates the presence of filamentous actin (arrowheads). Bar = 10 um. Inset: The same isolated cnidophage through transmitted light.

Fig. 7. Fluorescein-labeled anti-tubulin staining of an isolated cnidophage from Spurilla neapolitana. Fluorescent staining indicates the presence of microtubules (arrowheads). Bar = 10 um. Inset: The same isolated cnidophage through transmitted light.

in cnidophages. I used fluorescein-labeled anti-tubulin on isolated cnidophages and found that many of the cnidophages had large numbers of microtubules (Fig. 7). Electron microscopy has revealed that microtubules lie between nematocysts of the cnidophage and are often closely associated with the nematocyst membrane (Greenwood, 1987). Microtubules probably serve to hold the nematocysts in their oriented position within the cnidophage.

## VII. TURNOVER AND RETENTION OF NEMATOCYST TYPES

Grosvenor (1903) reported that the types of nematocysts in the cnidosacs of the aeolid Rizzola (=Cratena) peregrina would change with a change in the prey species consumed. Grosvenor found that nematocysts of the hydroid Eudendrium were almost completely replaced by nematocysts of the hydroid Pennaria after a period of one month. Yarnall (1972) also reported that nematocysts in Hermissenda crassicornis were replaced at a very slow rate. However, Conklin and Mariscal (1977) reported that nematocysts within cnidosacs of Spurilla neapolitana could be replaced in 6 days. Day and Harris (1978) reported even more rapid replacement times; nematocysts within cnidosacs of Coryphella verrucosa were completely replaced in 3 to 4 days.

The rapid replacement times reported by Conklin and Mariscal (1977) and by Day and Harris (1978) indicate that there is a continual replenishing of nematocysts within the cnidosacs. Day and Harris (1978) suggested that there may be a limit to the length of time that a nematocyst remains functional in the cnidosac.

One of the most intriguing aspects of the nudibranch-nematocyst association is the apparent ability of certain nudibranchs to retain particular nematocyst types from a variety of ingested cnida types. Kepner (1943) reported that at least five types of nematocysts were taken into the cnidosacs of the aeolid Aeolis (=Cratena) pilata, but that only one type remained; the others were presumably digested. Table I presents a summary of some of the aeolid species that retain specific nematocyst types, and is intended to illustrate the important features of selective retention.

Thompson and Bennett (1969) reported that the pelagic nudibranchs Glaucus atlanticus and Glaucilla marginata selectively stored only the largest nematocysts of the man-of-war Physalia, even though the nudibranchs also feed on the chondrophores Velella and Porpita. Day and Harris (1978) and Todd (1981) point out that the nudibranchs encountered by Thompson and Bennett might not have selected Physalia nematocysts over those of Velella or Porpita, but that they had simply fed on Physalia most recently. However, the fact remains that Glaucus and Glaucilla that had fed on Physalia still retained only the largest nematocysts of Physalia and digested the smaller types (Thompson and Bennett, 1970).

There is some confusion in the literature regarding nematocyst retention, primarily because the nematocyst types within prey species are often not listed (see Edmunds, 1966; Kalker and Schmekel, 1976). Edmunds (1966) listed several aeolid species that stored more than one nematocyst type, suggesting a lack of selective retention. However, at least one of these species, Catriona aurantia, has since been found to retain particular nematocysts when it feeds on the hydroid Tubularia (see Table I).

Day and Harris (1978) studied a variety of aeolid species and identified the types of nematocysts present in the cnidarian prey. They discovered that many species of aeolids stored specific types of nematocysts within their cnidosacs. They also found that the type of nematocyst retained would often change with a change in the prey species, and that when many nudibranch species fed on the same prey species, they retained the same nematocyst type. I show that the nudibranch Spurilla neapolitana stores basitrichous isorhizas when feeding on Condylactis gigantea, and microbasic mastigophores when feeding on Haliplanella luciae (Greenwood, 1983). These findings are of particular interest because both anemone species possess both types of nematocysts. Cratena pilata also retains different nematocysts from two different hydroids, though both hydroid species possess similar nematocyst types (see Table I).

The mechanisms for the retention of specific nematocyst types are completely unknown. It is difficult to discern the mechanism of nematocyst retention in nudibranchs because, as discussed previously, the function(s) of nudibranch nematocysts remain(s) unclear. Thompson and Bennett (1969, 1970) suggested that the retention of the largest and, presumably, most venomous nematocysts increases the defensive effectiveness of the cnidosacs. If defense of the nudibranch is the primary function of the acquired nematocysts, then selective retention may provide the nudibranch with those nematocysts most effective against a particular predator (Kepner, 1943; Edmunds, 1966). Edmunds (1966) suggested that penetrant nematocysts would be an effective defense against fish, and that other nematocysts, such as desmonemes, might be more effective against crustacean predators. However, this explanation does not account for why different species of nudibranchs retain the same nematocyst type from a given prey species (unless the nudibranchs are subjected to the same predators as the prey species).

The selective retention of nematocysts by nudibranchs is probably based on chemical or physical differences in the nematocysts themselves. Edmunds (1966) hypothesized that nudibranchs capable of selective retention incorporated those nematocysts that were most likely to fire upon release from the cnidosac. Alternatively, the retained nematocysts might be those that are less prone to fire during feeding and incorporation by the nudibranch. The observation that various nudibranch species feeding on the same prey species retain the same nematocyst type, coupled with the observation that a nudibranch may retain one type of nematocyst from one prey species and another type from another prey species (even though both prey species have both types of nematocysts), suggests that nematocysts, even of the same "type," may be very different from one species of cnidarian to the next. Thus, selective retention might be the result of subtle differences in nematocysts that can be distinguished by the nudibranch at the cellular or sub-cellular level.

## VIII. CONCLUSIONS

That nudibranchs are able to take secreted structures from ingested animals from another phylum and use them for their own benefit is unique in the animal kingdom. This review has intended to show that the general features of the nematocyst incorporation process have been known for some time, but that empirical investigations into the mechanisms controlling nematocyst incorporation are long overdue.

Nematocyst incorporation has been termed foreign organellar retention (Taylor, 1974), natural heterotransplantation (of) organelles (Ghilarov, 1983), and symbiosis (Greenwood and Mariscal, 1984a). Terminology aside, recent experiments have indicated that the incorporation process is more complex than was previously supposed. Acquired nematocysts are not simply held in cells, waiting to be fired from the nudibranch, but are incorporated into cnidophage cells in a complex process that leads to their active maintenance. Studies on nematocyst incorporation have routinely focused on characteristics of the nudibranch, but features of the selective retention of nematocysts lead one to conclude that the inherent properties of the nematocyst guide the mechanisms of incorporation. Further experiments into biochemical, physiological, and cellular aspects of the incorporation process might provide interesting insights into nematocyst physiology and into the general mechanisms governing tissue acceptance and rejection.

ACKNOWLEDGEMENTS

Thanks to R. N. Mariscal, G. Schatten, W. R. Ellington, C. Simerly, B. Miller, and T. Fellers for invaluable assistance in my research. Thanks to M. Butler, J. Dalby, J. Martin, and J. Schmidt for excellent critiques of early drafts of this paper. Special thanks to H. Ris and the staff of the HVEM facility.

REFERENCES

Allen, J. K. 1976. Function of nematocysts in eolid nudibranchs. West. Soc. Malac. Ann. Reprt. 9:50 (abst.).

Bergh, M. 1862. On the existence of urticating filaments in the Mollusca. Quart. J. Micro. Sci. 2 (NS):274-277.

Blanquet, R. 1970. Ionic effects on discharge of the isolated and _in situ_ nematocysts of the sea anemone _Aiptasia pallida_: a possible role of calcium. Comp. Biochem. Physiol. 35:451-461.

Cargo, D. G. and J. W. Burnett. 1982. Observations on the ultrastructure and defensive behavior of the cnidosac of _Cratena pilata_. Veliger 24:325-327.

Conklin, E. J. and R. N. Mariscal. 1977. Feeding behavior, ceras structure, and nematocyst storage in the aeolid nudibranch, _Spurilla neapolitana_ (Mollusca). Bull. Mar. Sci. 27:658-667.

Cuenot, L. 1907. L'origine des nematocystes des eolidiens. Arch. Zool. Exp. Gen. 6:73-102.

David, C. N. and D. Challoner. 1974. Distribution of interstitial cells and differentiating nematocytes in nests in _Hydra attenuata_. Am. Zool. 14:537-542.

Day, R. M. and L. G. Harris. 1978. Selection and turnover of coelenterate nematocysts in some aeolid nudibranchs. Veliger 21:104-109.

Edmunds, M. 1966. Protective mechanisms in the Eolidacea (Mollusca: Nudibranchia). J. Linn. Soc. London, Zool. 46:27-71.

Edmunds, M., G. W. Potts, R. C. Swinfen, and V. L. Waters. 1976. Defensive behavior of sea anemones in response to predation by the opisthobranch mollusc _Aeolidia papillosa_ (L.). J. Mar. Biol. Ass. U. K. 56:65-83.

Fletcher, T. C. and C. A. Cooper-Willis. 1983. Cellular defense systems of the Mollusca. _In_ N. Cohen and M. M. Sigel (eds.), _The Reticuloendothelial System: A Comprehensive Treatise_, pp. 141-166. Plenum Press, New York.

Ghilarov. M. S. 1983. Appropriation of functioning organelles of food organisms by phytophagous and predatory opisthobranch molluscs as a specific category of food utilization. Zh. Obshch. Biol. 44:614-620.

Glaser, O. C. 1910. The nematocysts of eolids. J. Exp. Zool. 9:117-142.

Graham, A. 1938. The structure and function of the alimentary canal of aeolid mollucscs, with a discussion on their nematocysts. Trans. Roy. Soc. Edinb. 59:267-307.

Greenwood, P. G. 1983. Ceras ultrastructure and incorporation of immature nematocysts in an aeolid nudibranch. Master's Thesis, Florida State University, 87pp.

Greenwood, P. G. 1987. Nematocyst maintenance and orientation in aeolid nudibranchs. Ph.D. Thesis, Florida State University, 95pp.

Greenwood, P. G. and R. N. Mariscal. 1984a. The utilization of cnidarian nematocysts by aeolid nudibranchs: nematocyst maintenance and release in Spurilla. Tiss. Cell 16:719-730.

Greenwood, P. G. and R. N. Mariscal. 1984b. Immature nematocyst incorporation by the aeolid nudibranch Spurilla neapolitana. Mar. Biol. 80:35-38.

Greenwood, P. G. and R. N. Mariscal. 1984c. Nematocyst maturation in vitro: the effects of ATP on isolated nematocyts. Am. Zool. 24:31A (abst.).

Grosvenor, G. H. 1903. On the nematocysts of aeolids. Proc. Roy. Soc. London 72:462-486.

Harris, L. G. 1971. Nudibranch associations as symbioses. In T. C. Cheng (ed.), Aspects of the Biology of Symbiosis, pp. 77-90. University Park Press, London.

Harris, L. G. 1973. Nudibranch associations. In T. C. Cheng (ed.), Current Topics in Comparative Pathobiology V.II, pp.213-315. Academic Press, New York.

Harris, L. G. 1986. Size-selective predation in a sea anemone, nudibranch, and fish food chain. Veliger 29:38-47.

Jones, T. C. 1974. Macrophages and intracellular parasitism. J. Reticuloendothelial Soc. 15:439-450.

Kälker, H. and L. Schmekel. 1976. Bau und Funktion des Cnidosacks der Aeolidoidea (Gastropoda Nudibranchia). Zoomorphologie 86:41-60.

Kepner, W. A. 1943. The manipulation of the nematocysts of Pennaria tiarella by Aeolis pilata. J. Morph. 73:297-311.

Lentz, T. L. and R. J. Barrnett. 1961. Enzyme histochemistry of Hydra. J. Exp. Zool. 147:125-150.

Mariscal, R. N. 1974. Nematocysts. In L. Muscatine and H. M. Lenhoff (eds.). Coelenterate Biology (Reviews and New Perspectives), pp. 129-178. Academic Press, New York.

Moreteau, J.-C. 1977. Etude sur l'alimentation d'Aeolidia papillosa L. Cah. Biol. Mar. 18:369-384.

Naville, A. 1926. Notes sur les eolidiens. Rev. Suisse Zool. 33:251-289.
Pollard, T. D. 1981. Cytoplasmic contractile proteins. J. Cell Biol. 91:156s-165s.
Reaven, E. P. and S. G. Axline. 1973. Subplasmalemmal microfilaments and microtubules in resting and phagocytizing cultivated macrophages. J. Cell Biol. 59:12-27.
Robertson, R. 1980. Epitonium millecostatum and Coralliophila clathrata: two prosobranch gastropods symbiotic with Indo-Pacific Palythoa (Coelenterata:Zoanthidae). Pac. Sci. 34:1-17.
Rosin, R. 1969. Escape response of the sea anemone Anthopleura nigrescens (Verrill) to its predatory eolid nudibranch Herviella Babba spec. nov. Veliger 12:74-77.
Rosin, R. 1970. On Aeolidoidea, Coelenterata and nematocysts. Argamon 1:43-50.
Rousseau, C. 1927. Les nematocystes des eolidiens. Bull. Soc. Sci. Nat. Ouest France 7:1-28.
Streble, H. 1968. Bau und Bedeutund der Nesselsacke Aeolidia papillosa L., der Breitwarzigen Fadenschnecke (Gastropoda, Opisthobranchia). Zool. Anz. 180:356-372.
Tardy, J. 1964. Comportement predateur de Eolidiella alderi (Mollusque, Nudibranche). Compt. Rend. Acad. Sci. 258:2190.
Taylor, F. J. R. 1974. Implications and extensions of the serial endosymbiosis theory of the origin of the eukaryotes. Taxon 23:229-258.
Thompson, T. E. 1960. Defensive adaptations in opisthobranchs. J. Mar. Biol. Ass. U.K. [N.S.] 39:123-134.
Thompson, T. E. 1976. Nudibranchs. T.F.H. Publications, London. 96 pp.
Thompson, T. E. and I. Bennett. 1969. Physalia nematocysts: utilized by mollusks for defense. Science 166:1532-1533.
Thompson, T. E. and I. Bennett. 1970. Observations on Australian glaucidae (Mollusca:Opisthobranchia). Zool. J. Linn. Soc. Lond 49:187-197.
Todd, C. D. 1981. Ecology of nudibranch molluscs. Oceanogr. Mar. Biol. 19:141-234.
Todd, C. D. 1983. Reproductive and trophic ecology of nudibranch molluscs. In W. D. Russell-Hunter (ed.), The Mollusca, Vol. 6, pp. 225-259. Academic Press, New York.
Waters, V. L. 1973. Food preference of the nudibranch Aeolidia papillosa, and the effect of the defenses of the prey on predation. Veliger 15:174-192.
Watson, G. M. and R. N. Mariscal. 1983. Comparative ultrastructure of catch tentacles and feeding tentacles in the sea anemone Haliplanella. Tiss. Cell 15:939-953.

Wright, T. S. 1863. On the urticating filaments of the Eolidae. Quart. J. Micro. Sci. 3:52-55.

Yarnall, J. L. 1972. The feeding behavior and functional anatomy of the gut in the eolid nudibranchs Hermissenda crassicornis (Eschscholtz, 1831) and Aeolidia papillosa (Linnaeus, 1761). Ph.D. Thesis, Stanford University. 126 pp.

24

# THE CORRELATION BETWEEN NEMATOCYST TYPES AND DIETS IN PELAGIC HYDROZOA[1]

Jennifer E. Purcell

Horn Point Environmental Laboratories
University of Maryland
Cambridge, Maryland 21613

Claudia E. Mills

Friday Harbor Laboratories
University of Washington
Friday Harbor, Washington 98250

## SUMMARY

Among the pelagic hydrozoans, which include the siphonophores, hydromedusae, and velellids (chondrophores), members of some taxonomic subgroups exclusively or primarily consume soft-bodied prey, while others primarily consume hard-bodied (crustacean) prey. The dietary differences are related to the types of nematocysts in the tentacles of these hydrozoans. The predators of crustaceans are the calycophore and physonect siphonophores, Anthomedusae (except the family Pandeidae), Limnomedusae, and Trachymedusae. These groups possess 2 to 5 types of nematocysts, including rhopalonemes (acrophores, anacrophores, and desmonemes) which adhere to and entangle prey, and stenoteles, microbasic euryteles, or microbasic mastigophores, which can penetrate crustacean exoskeleton. Adhesion of nematocyst

[1]Funding for the research on hydromedusae was provided by the Lerner-Gray Fund for Marine Research and a NATO postdoctoral fellowship to J.E.P., a Natural Sciences and Engineering Research Council Strategic Grant #GO871 to G. O. Mackie, and by NSF (Biological Oceanography) Grant #OCE-8214058 to A. R. M. Nowell and C.E.M.

tubules to prey surfaces appears to be important in capture of crustaceans by hydrozoans. Predators of soft-bodied plankton include the cystonect siphonophores, Apolemia uvaria (Physonectae), pandeid Anthomedusae, Leptomedusae, and Narcomedusae. Their cnidoms usually contain one type or one predominant type of nematocyst, primarily isorhizas, mastigophores, or other types often unique to them, that penetrate the soft prey tissues, but they lack surface-adhering nematocyst types.

Herein, we briefly review earlier work on the diets and nematocysts of the Siphonophora, and present new data on the diets and nematocysts of the diverse hydromedusan species from the Strait of Georgia region, Northeast Pacific.

## I. INTRODUCTION

Within the class Hydrozoa, the orders Siphonophora, Chondrophora [now considered to be the Family Velellidae in the Anthomedusae (Bouillon, 1985)], Narcomedusae, and Trachymedusae are holoplanktonic, and many species in the orders Anthomedusae, Leptomedusae, and Limnomedusae have a pelagic medusa stage. The life cycle for medusae in the new order Laingiomedusae (Bouillon, 1978) is not known. The colonial siphonophores and velellids (chondrophores) do not swim while their tentacles are extended to capture prey. Hydromedusae exhibit a diversity of swimming and feeding behaviors (Mills, 1981) and may either fish while they swim, or separate these two activities (Mills, in prep.). For feeding, all of these hydrozoans depend on interception of prey by their nematocyst-laden tentacles. The nematocysts in the tentacles are directly responsible for prey capture, and differences in cnidoms could be expected to result in dietary differences.

Among the cnidarian classes, the greatest number of nematocyst types are found in the class Hydrozoa; 23 of the 30 described types of cnidae occur, and 17 types are unique to this class (reviewed by Mariscal, 1974). Numerous papers describe the cnidoms of many species in this class. Recent reviews, cited below, compile many of these earlier data, and we refer readers to references therein. Cnidoms are very similar among species in each of the siphonophore suborders (Purcell, 1984a; Mackie et al., in press), and in the hydromedusan orders (Bouillon, 1985), with a few exceptions. We limit our present analysis to nematocysts in the tentacles because of their direct relationship to prey capture.

Until recently, little information existed on the diets of pelagic hydrozoans. The natural diets of 25 epipelagic siphonophore species were quantified by Purcell (1981a, b, 1984b). The diets of several hydromedusa species recently have been quantified by Larson (1985) in Saanich Inlet, Vancouver Island, British Columbia, Canada, and by Purcell (unpubl.) in Friday Harbor, Washington and southeastern Vancouver Island. Alvariño (1985) summarized prey captured by numerous siphonophore and medusa species, but virtually all data cited therein were from observations of feeding made in the laboratory. In the present paper, we restrict our analysis to species for which natural diets have been determined, to avoid possible laboratory artifacts.

Only a few studies have examined the effects of nematocysts on prey organisms. Toppe (1909), Ewer (1947), and Tardent and Holstein (1982) studied nematocysts of hydra and found that prey were penetrated by stenoteles and were entangled by desmonemes. Ewer (1947) further noted that the "atrichous" and holotrichous isorhizas of hydra did not discharge on crustacean prey, but were used in locomotion and defense, respectively. Although some discharge of the holotrichous isorhizas was seen against soft-bodied prey organisms, hydra gave no feeding response, and the interaction was interpreted to be defensive.

Purcell (1984a), using scanning electron microscopy to examine the discharged nematocysts of numerous siphonophore species, defined two feeding groups of siphonophores. Species in the suborder Cystonectae eat soft-bodied prey and have only isorhiza nematocysts; physonect and calycophoran siphonophores primarily eat copepods and have 4-5 nematocyst types, including rhopalonemes and stenoteles, organized in batteries. New dietary and nematocyst data presented herein suggest that a parallel situation exists in hydromedusae. Species that primarily eat soft-bodied prey have one type of nematocyst that predominates (usually isorhizas or mastigophores), while most species that eat crustaceans have a mixture of desmonemes and stenoteles. Pelagic hydrozoans that feed on crustaceans have some nematocyst types with tubules specialized to adhere to the hard prey surfaces, while species that feed on soft-bodied prey have nematocyst tubules that penetrate prey.

## II. MATERIALS AND METHODS

Hydromedusae used for dietary analyses during April, 1982 [data cited as Purcell, unpubl. (1982)] and for all nematocyst measurements were collected individually from the docks at the Friday Harbor Laboratories, Washington, USA. Specimens for gut analyses were preserved immediately in 5% formalin. Specimens also were collected using a 0.75 m diameter, 333 μm mesh plankton net in 0.5-3 min tows at 0-5 m depth in Kulleet Bay, on southeast Vancouver Island, British Columbia, Canada during March-June, 1983, April, 1985, or April, 1986 and immediately preserved [cited Purcell, unpubl. (1983), unpubl. (1985), unpubl. (1986)]. Prey were removed from the jellyfish manubria and identified using 8-50x magnification with a dissecting microscope.

Nematocysts in the tentacles of hydromedusae were examined and photographed using a compound microscope. Whole tentacles were placed on a slide along with sea water and $MgCl_2$. When each tentacle was relaxed, a drop of 5% formalin was added to the preparation to prevent ciliary and muscular movement, and a coverslip supported by small plasticene feet was applied. Measurements of undischarged nematocysts were made at 1000x and represent the range of sizes from at least 10 randomly selected nematocysts of each type.

Scanning electron microscopy (SEM) was used to examine nematocyst adhesion and penetration into prey. Medusae were allowed to capture copepods, ctenophores, or other species of medusae in the laboratory, and these prey were quickly retrieved with forceps from the tentacles. In some cases,

TABLE I. Nematocyst types and range of dimensions (µm) in the tentacles of siphonophores (from Purcell, 1984a). Ratios represent the number of species having those nematocyst types / number of species examined in each group

| | Rhopalonemes | Stenoteles | Microbasic mastigophores | Anisorhizas | Isorhizas |
|---|---|---|---|---|---|
| Calycophorae | **12/12**<br>3.0 x 1.5 -<br>23.0 x 3.0 | ---- | **12/12**<br>36.0 x 7.0 -<br>153.0 x 15.0 | **12/12** *<br>12.0 x 2.0 -<br>54.0 x 12.5 | **11/12**<br>7.0 x 5.0 -<br>25.0 x 12.5 |
| Physonectae | **7/8**<br>3.0 x 4.5 -<br>17.5 x 10.0 | **4-5/8**<br>22.5 x 12.5 -<br>109.0 x 29.0 | **2-3/8**<br>112.5 x 20.0 -<br>180.0 x 28.0 | **8/8** *<br>22.6 x 6.0 -<br>72.5 x 12.5 | **P** |
| Cystonectae | ---- | ---- | ---- | ---- | **4/4**<br>10.0 x 10.0 -<br>32.5 x 32.5 |

* numerically predominant

P, present

prey held in forceps were brushed along the tentacles of medusae. The prey specimens were fixed overnight in 5% formalin in sea water and were postfixed for one hour in 2% osmium tetroxide ($OsO_4$) in distilled water. Specimens were dehydrated in a series of 35%, 50%, 70%, and 95% ethanol, followed by two 5-minute rinses in 1, 2 dimethoxypropane. The specimens were critical-point dried using $CO_2$, mounted on stubs with double-stick tape or silver paint, and sputter-coated with gold-palladium. The observations were made using a JEOL JSM-35 scanning electron microscope, operated at 15 KV.

## III. RESULTS

### A. Siphonophores

Data on the nematocysts of siphonophores from Purcell (1984a) are summarized in Table I. The calycophoran and physonect (except Apolemia spp.) siphonophores have one highly-organized nematocyst battery on each of the side branches (tentilla) of the tentacles. In calycophorans, the straight cnidoband contains from 50 homotrichous anisorhizas in small species to 2,000 in large species. At the free end of the battery are found a few isorhizas, and a terminal filament containing small rhopalonemes. At the attached end of the battery are 4 to 30 microbasic mastigophores that are 2 to 4 times longer than the anisorhizas.

In physonects, the coiled cnidoband contains from 150 homotrichous anisorhizas in small species to 20,500 in large species. Along the edge of the cnidoband are 4 to 120 stenoteles or microbasic mastigophores that, again, are much larger than the anisorhizas. The 1 or 2 terminal filaments contain small rhopalonemes.

The nematocysts of cystonect siphonophores are not organized into batteries, but may occur in bands or clusters. Only homotrichous and atrichous isorhizas are found.

Data on the natural diets of siphonophores in Table II were calculated from published data (see Table 3 in Purcell, 1981a and Table 1 in Purcell, 1984b) by adding all percentages of a given prey type among species in a suborder, and then expressing the total for each prey type as a percentage of the summed totals. The compiled data show that 97.3% of the prey of the calycophorans examined were crustaceans, primarily copepods. The diets of physonects consisted mostly of crustaceans (69.3%), including large copepods and larvae of benthic crustaceans, and also included some soft-bodied prey. The diets of cystonect siphonophores exclusively consisted of soft-bodied prey, of which fish larvae predominated (91-100%).

Thus, calycophore and physonect siphonophores have similar cnidoms and similar diets, while cystonects are dramatically different in both respects.

Purcell (1984a) used SEM to examine prey after they were captured by siphonophores but before the prey were ingested. Prey (copepods and chaetognaths) captured by calycophorans and physonects were wrapped and entangled in nematocyst tubules which adhered to the prey surfaces. Nema-

TABLE II. _In situ_ dietary data for siphonophores, as percent of prey items in gut contents (from Purcell, 1981a,b, 1984b)

| | Copepods | Decapod larvae | Clado-cerans | Amphi-pods | Shrimp | Ostra-cods | Molluscs | Chaeto-gnaths | Fish larvae | Gelat. zoopl. | No. of prey/ No. of siphs. examined |
|---|---|---|---|---|---|---|---|---|---|---|---|
| Calycophorae | 87.5 | 0.1 | P | 0.5 | 0.5 | 8.7 | 0.7 | 1.8 | 0.2 | --- | >1,484/360 |
| Physonectae | 39.7 | 6.2 | P | 0.6 | 22.8 | P | 1.1 | 13.2 | 15.2 | 1.2 | 233/117 |
| Cystonectae | --- | --- | --- | --- | --- | --- | 4.1 | 4.9 | 91.0 - 100.0 | --- | >2,174/370 |

P, present

tocyst penetration of the prey was seen only twice. In one case, stenoteles of Nanomia bijuga seemed to have penetrated at the cephalothorax/abdominal joint of a copepod, and in the other case, stenoteles of Forskalia edwardsi penetrated a chaetognath. Cystonect isorhizas penetrated fish larvae, but were not seen to adhere to or penetrate a shrimp (Leander sp.).

Thus, the siphonophores examined appeared to capture small crustacean prey primarily by entangling them in nematocyst tubules. Previously only stenoteles (Toppe, 1909; Tardent and Holstein, 1982) and microbasic mastigophores (Godknecht, 1985) have been shown to penetrate crustacean exoskeleton. In siphonophores, these nematocysts capable of penetration are far outnumbered (20 to 200 times) by other nematocyst types. Few stenoteles were seen penetrating prey of siphonophores, and no microbasic mastigophores had penetrated the prey. Cystonect nematocysts seem unsuitable for adhesion to prey and unable to penetrate crustaceans, perhaps limiting cystonect siphonophores to soft-bodied prey.

## B. Hydromedusae

In Table III, we present new data on the sizes and types of tentacle nematocysts from the hydromedusa species for which we have dietary data, primarily from nearshore waters of Washington State and southern British Columbia. We also include data from Bouillon (1985) who has summarized all nematocyst types known to be present in each of the families of hydromedusae.

Among the Anthomedusae, most species have a combination of desmonemes along with either stenoteles or microbasic euryteles, except species in the family Pandeidae, which lack desmonemes, and have either microbasic euryteles or mastigophores. Bouillon (1980) identified the nematocysts of Indo-Pacific Stomotoca atra as microbasic euryteles, but specimens from the Northeast Pacific examined by Mills (unpubl.) contained microbasic mastigophores (Fig. 1I, 2C, 2D).

The only local species of the Limnomedusae for which we have dietary information, Proboscidactyla flavicirrata, has desmonemes and macrobasic mastigophores. Rees (1979) identified the latter type as macrobasic euryteles, rather than macrobasic mastigophores, in P. flavicirrata. Aglantha digitale, the only local Trachymedusa, has stenoteles and microbasic euryteles [variations in these are discussed in Russell (1940)].

The nematocysts in the tentacles of the Leptomedusae predominantly consist of basitrichous and "atrichous" isorhizas and some microbasic mastigophores. Identification of these small (<15 μm) nematocysts has been difficult using light microscopy, and designations of certain types vary by the author's choice of descriptive names. This appears to be the case for basitrichous isorhizas, microbasic mastigophores, and less often, microbasic euryteles. Russell (1939) identified the nematocysts of Aequorea spp. as basitrichous isorhizas, rather than microbasic mastigophores as we denote them in Table III, and "atrichous" isorhizas. Kubota (1976, 1981) found both microbasic mastigophores and basitrichous isorhizas in tentacles of Obelia sp. medusae (but not both in one specimen). We found only microbasic mastigo-

TABLE III. Nematocyst types and ranges of undischarged dimensions (μm) in the tentacles of hydromedusae (ratios following family names represent the number of species having those nematocysts / number of species examined)

| | Desmonemes | Stenoteles | Microbasic euryteles | Microbasic mastigophores | Isorhizas | References |
|---|---|---|---|---|---|---|
| **Anthomedusae** | | | | | | |
| Fam. Corynidae | **12/13** | **13/13** | **0-1/13** | ---- | **2/13** | Bouillon, 1985 |
| Sarsia eximia | 9.5-12.0 x 4.0-5.5 | 9.5-16.0 x 6.5-10.0 | ---- | ---- | ---- | Russell, 1938 |
| Sarsia tubulosa | 10.8-12.0 x 5.6-6.4 | 14.0-16.0 x 10.4-12.0 | ---- | ---- | ---- | Mills, unpubl. |
| Fam. Rathkeidae | **1/1** | ---- | **1/1** | ---- | ---- | Bouillon, 1985 |
| Rathkea octopunctata | 3.0-5.0 x 2.0-3.0 | ---- | 6.0-10.5 x 2.5-4.0 | ---- | ---- | Russell, 1938 |
| Fam. Bougainvilliidae | **11/11** | ---- | **11/11** | ---- | ---- | Bouillon, 1985 |
| Bougainvillia principis | 4.8-5.2 x 2.4-3.2 | ---- | 6.4-7.2 x 3.6-4.0 | ---- | ---- | Mills, unpubl. |
| Fam. Pandeidae | **1-2/15** | ---- | **14-15/15** | **3-4/15** | **1/15** | Bouillon, 1985 |
| Catablema nodulosa | ---- | 12.0-14.4 x 8.0-9.6 | 8.8-9.6 x 3.6-4.0 | ---- | ---- | Mills, unpubl. |
| Stomotoca atra | ---- | ---- | ---- | 8.4-9.6 x 2.4-3.2 | ---- | Mills, unpubl. |
| Fam. Tubulariidae | **6/6** | **6/6** | **4-5/6** | **0-2/6** | **5/6** | Bouillon, 1985 |
| Hybocodon prolifer | P | P | P | ---- | ---- | Bouillon, 1974 |
| Fam. Polyorchidae | **4/4** | **4/4** | **0-1/4** | ---- | ---- | Bouillon, 1985 |
| Polyorchis penicillatus | 12.0-14.5 x 6.0-7.0 | 15.5-20.5 x 10.0-13.5 | ---- | ---- | ---- | Mackie & Mackie, 1963 |
| Fam. Velellidae | ---- | **2/2** | **1/2** (macro.) | ---- | **2/2** | Bouillon, 1985 |
| Velella velella | ---- | 18.0-21.0 x 14.0-16.0<br>13.0-15.0 x 10.0-11.0 | ----<br>(P, macrobasic) + | ---- | 8-9 x 4.5 | Russell, 1939 |

(continued)

TABLE III (continued)

| | Desmonemes | Stenoteles | Microbasic euryteles | Microbasic mastigophores | Isorhizas | References |
|---|---|---|---|---|---|---|
| **Limnomedusae** | | | | | | |
| Fam. Proboscidactylidae | **4/4** | --- | **2/4** | **4/4** (macro.) | ---- | Bouillon, 1985 |
| *Proboscidactyla flavicirrata* | 5.0-7.0 x 4.0-6.0 | --- | --- | 8.0-12.0 x 5.0-7.2 (macrobasic) | ---- | Hand, 1954; Mills, unpubl. |
| **Trachymedusae** | | | | | | |
| Fam. Rhopalonematidae | ---- | **4-5/5** | **3-5/5** | ---- | **01-/5** | Bouillon, 1985 |
| *Aglantha digitale* | ---- | 11.6-12.0 x 11.2-12.0 | 8.8-9.6 x 5.6 | --- | ---- | Russell, 1940; Mills, unpubl. |
| **Leptomedusae** | | | | | | |
| Fam. Aequoriidae | ---- | --- | --- | **0-3/5** | **5/5** * | Bouillon, 1985 |
| *Aequorea victoria* | ---- | --- | --- | 14.4-16.8 x 3.2 ▵ | 22.4-24.4 x 11.6-12.0 | Mills, unpubl. |
| Fam. Campanulariidae | ---- | --- | --- | **7/7** | **6/7** * | Bouillon, 1985 |
| *Obelia* sp. | ---- | --- | --- | 6.8-8.0 x 2.0-2.4 ▵ | ---- | Mills, unpubl. |
| *Phialidium gregarium* | ---- | --- | --- | 9.6-11.2 x 2.4-2.8 ▵ | ---- | Mills, unpubl. |
| Fam. Mitrocomidae | ---- | --- | --- | **1/4** | **4/4** * | Bouillon, 1985 |
| *Mitrocomella polydiademata* | ---- | --- | --- | 8.8-10.0 x 2.4 ▵ | 12.0-14.4 x 4.8-5.6 | Mills, unpubl. |
| *Mitrocoma cellularia* | ---- | --- | --- | 12.0-14.8 x 3.6-4.2 ▵ | ---- | Mills, unpubl. |
| Fam. Eutimidae | ---- | --- | --- | **5-6/9** | **7/9** * | Bouillon, 1985 |
| *Eutonina indicans* | ---- | --- | --- | 9.2-10.8 x 2.4-2.6 ▵ | ---- | Mills, unpubl. |

(continued)

TABLE III (continued)

| | Desmonemes | Stenoteles | Microbasic euryteles | Microbasic mastigophores | Isorhizas | References |
|---|---|---|---|---|---|---|
| **Narcomedusae** | | | | | | |
| Fam. Aeginidae | ---- | --- | --- | --- | **2/2** ° | Bouillon, 1985 |
| Aegina | ---- | --- | --- | --- | 16.8-20.0 ° | Mills, unpubl. |
| citrea | | | | | 6.4-8.4 ° | |
| Fam. Cuninidae | ---- | --- | --- | --- | **3/3** ° | Bouillon, 1985 |
| Solmissus | ---- | --- | --- | --- | 22.5-25.5 ° | Mills, unpubl. |
| albescens | | | | | 8.4-9.6 ° | |
| Solmissus | ---- | --- | --- | --- | 21.0-23.0 ° | Mackie & |
| marshalli | | | | | 7.0-9.0 ° | Mackie, 1963 |

P present
\+ from Bouillon, 1974
* numerically predominant
△ identified as basitrichous isorhizas in other species of same genera by other authors
° spherical apotrichous isorhizas

phores in Obelia sp. medusa tentacles, in agreement with Östman (1982). Species of Phialidium other than P. gregarium were found to have atrichous or homotrichous isorhizas in addition to microbasic mastigophores (Russell, 1938; Kubota 1978; Östman, 1979).

The Narcomedusae all have only apotrichous isorhizas, usually in two sizes (Mills and Miller, 1984; Bouillon, 1985; Mills, pers. obs.) (Table III, Fig. 1J).

Dietary data show that the anthomedusan species that we have examined primarily consumed crustaceans (75-100%). In addition to the prey types listed in Table IV, the diet of Polyorchis penicillatus contained 14.8% epibenthic crustaceans (gammarid and caprellid amphipods, cumaceans) and 28.7% soft-bodied benthic prey (polychaetes) (Arkett, 1984). Species in the family Pandeidae may primarily eat soft-bodied prey (jellyfish), but few quantitative dietary data are available. We report in situ observations of pandeids repeatedly feeding on other hydromedusae: Stomotoca atra consuming Phialidium gregarium (Mills, Purcell, pers. obs.), and Catablema nodulosa consuming Mitrocomella polydiademata (Mills, pers. obs.).

The limnomedusa Proboscidactyla flavicirrata consumed mostly veligers and tintinnids. Small specimens of the trachymedusa Aglantha digitale consumed large copepods, and tintinnids (Purcell, pers. obs.).

The leptomedusae examined mostly consumed soft-bodied prey, especially larvaceans and invertebrate eggs (euphausiid and copepod). The data of McCormick (1969) for Phialidium gregarium in Yaquina Bay, Oregon, contrast markedly with the data from Purcell (unpubl., 1982) and Larson (1980, 1985) for P. gregarium in the Strait of Georgia region (Table IV). The cause of this discrepancy is not obvious, but raises the point that available prey, and hence the medusa diets, will differ both with location and date. In addition to the prey listed in Table IV, Aequorea victoria also consumed polychaetes in situ (Larson, 1980; Purcell, Mills, pers. obs.). One specimen of Eutonina indicans contained numerous Hybocodon prolifer hydromedusae (Mills, pers. obs.). Data in Zelickman et al. (1969) indicate that the medusa Tiaropsis multicirrata contained 85.5% crustaceans, 5.4% veligers, 1.8% tintinnids, and 7.3% hydromedusae. No data on the nematocysts of T. multicirrata are available.

Narcomedusae may consume exclusively gelatinous zooplankton, but few data exist: one specimen of Aegina citrea at Friday Harbor contained a Mitrocomella polydiademata medusa, Solmissus albescens in the Mediterranean contained Cavalinia pteropods, and a specimen of S. marshalli at Friday Harbor contained a Euphysa medusa (Mills, unpubl.).

In summary, of the species that we examined, the hydromedusae that eat hard-bodied prey include the Anthomedusae (except the family Pandeidae), and probably most Limnomedusae and Trachymedusae. These medusae usually have desmonemes, which adhere to prey, and either stenoteles, which can penetrate crustaceans and soft-bodied prey, or microbasic euryteles, which may serve the same purpose. The hydromedusae that primarily eat soft-bodied prey include the Leptomedusae and pandeid Anthomedusae, having microbasic mastigophores, euryteles, or basitrichous isorhizas, and the Narcomedusae, having apotrichous isorhizas.

TABLE IV. In situ dietary data for hydromedusae, as percent of prey items in gut contents (see additional data in text)

| | Cope-pods | Copepod nauplii | Barnacle larvae | Crab zoeas | Clado-cerans | Veligers | Larva-ceans | Invert. eggs | Tintin-nids | Fish larvae | Gelat. zoopl. | No. of prey/ No. medusae | References |
|---|---|---|---|---|---|---|---|---|---|---|---|---|---|
| **Anthomedusae** | | | | | | | | | | | | | |
| *Sarsia eximia* | 89.7 | 7.7 | ---- | ---- | 2.6 | ---- | ---- | ---- | ---- | ---- | ---- | 39/37 | McCormick, 1969 |
| *Sarsia princeps* | 39 | 18 | 4 | ---- | ---- | <1 | <1 | 39 | <1 | ---- | ---- | 299/11 | Larson, 1985 |
| *Sarsia tubulosa* | 13.0 | 2.5 | 83.9 | ---- | 0.3 | ---- | ---- | ---- | ---- | 0.3 | ---- | 360/26 | Purcell, unpubl.(1982) |
| | 61.9 | 3.2 | 31.7 | ---- | ---- | ---- | ---- | 3.2 | ---- | ---- | ---- | 63/54 | Purcell, unpubl.(1985) |
| *Rathkea octopunctata* | ---- | ---- | 62.5 | ---- | ---- | ---- | 25.0 | ---- | ---- | 12.5 | ---- | 8/8 * | Purcell, unpubl.(1986) |
| *Bougainvillia principis* | 1.2 | 2.4 | 94.0 | 0.6 | ---- | ---- | 1.2 | 0.6 | ---- | ---- | ---- | 168/15 * | Purcell, unpubl.(1982) |
| *Catablema nodulosa* | 1.1 | ---- | 44.4 | ---- | ---- | ---- | 33.3 | ---- | ---- | ---- | 11.1 | 9/8 * | Purcell, unpubl.(1982) |
| *Stomotoca atra* | ---- | ---- | ---- | ---- | ---- | ---- | ---- | ---- | ---- | ---- | M * | --- | Hyman, 1940 |
| *Hybocodon prolifer* | 68.4 | 5.2 | 21.0 | ---- | ---- | ---- | ---- | ---- | 5.2 | ---- | ---- | 19/12 * | Purcell, unpubl.(1982) |
| *Polyorchis penicillatus* | P | 7.3 | 38.0 | 7.3 | ---- | 3.9 | ---- | ---- | ---- | ---- | ---- | >771/>180 | Arkett, 1984 |
| | 55.2 | 1.7 | 0.3 | 1.7 | 0.7 | ---- | 0.7 | 39.7 (1 medusa) | ---- | ---- | ---- | 299/11 | McCormick, 1969 |
| *Velella velella* | 3.0 | ---- | 0.5 | 1.5 | 1.0 | ---- | 11.0 | 78.0 | ---- | 3.5 (eggs) | ---- | ? /38 | Bieri, 1961 |

(continued)

TABLE IV (continued)

| | Cope-pods | Copepod nauplii | Barnacle larvae | Crab zoeas | Clado-cerans | Veligers | Larva-ceans | Invert. eggs | Tintin-nids | Fish larvae | Gelat. zoopl. | No. of prey/ No. medusae | References |
|---|---|---|---|---|---|---|---|---|---|---|---|---|---|
| **Limnomedusae** | | | | | | | | | | | | | |
| _Proboscidactyla flavicirrata_ | <1 | 1 | <1 | ---- | ---- | 67 | 2 | 3 | 25 | ---- | ---- | 259/81 | Larson, 1985 |
| | ---- | ---- | ---- | ---- | ---- | 25 | ---- | ---- | 75 | ---- | ---- | 4/3 * | Purcell, unpubl.(1982) |
| | 57.1 | ---- | ---- | 14.3 | ---- | 28.6 | ---- | ---- | ---- | ---- | ---- | 7/3 | McCormick, 1969 |
| | ---- | 1.0 | 1.0 | ---- | ---- | 27.7 | ---- | 1.0 | 69.3 | ---- | ---- | 101/32 | Purcell, unpubl.(1985) |
| **Trachymedusae** | | | | | | | | | | | | | |
| _Aglantha digitale_ | 5 | 1 | <1 | ---- | ---- | <1 | <1 | 94 | <1 | ---- | ---- | 503/283 | Larson, 1985 |
| | 100 | ---- | ---- | ---- | ---- | ---- | ---- | ---- | ---- | ---- | ---- | 23/6 * | Purcell, unpubl.(1982) |
| | 7.4 | 3.7 | ---- | ---- | ---- | ---- | ---- | ---- | 88.9 | ---- | ---- | 27/3 * | Purcell, unpubl.(1985) |
| | 10.0 | 10.0 | ---- | ---- | ---- | ---- | ---- | 20.0 | 60.0 | ---- | ---- | 12/6 * | Purcell, unpubl.(1986) |
| **Leptomedusae** | | | | | | | | | | | | | |
| _Aequorea victoria_ | ---- | ---- | ---- | ---- | ---- | ---- | ---- | ---- | ---- | ---- | 93.8 | 32/307 | Larson, unpubl.(1980) |
| | 6.6 | P | 2.3 | 1.4 | ---- | ---- | 71.1 | P | P | 5.0 | 12.4 | 16,297/521 | Purcell, unpubl.(1983) |
| _Obelia_ sp. | ---- | ---- | ---- | ---- | ---- | ---- | ---- | ---- | 100 | ---- | ---- | 3/3 * | Purcell, unpubl.(1986) |

(continued)

TABLE IV (continued)

| | Copepods | Copepod nauplii | Barnacle larvae | Crab zoeas | Cladocerans | Veligers | Larvaceans | Invert. eggs | Tintinnids | Fish larvae | Gelat. zoopl. | No. of prey/ No. medusae | References |
|---|---|---|---|---|---|---|---|---|---|---|---|---|---|
| Phialidium gregarium | 79.9 | 0.5 | 1.6 | 0.5 | 3.1 | 0.5 | 6.8 | 7.3 | ---- | ---- | ---- | 192/61 | McCormick, 1969 |
| | 16.3 | ---- | 1.0 | ---- | 8.2 | 1.0 | 15.3 | 45.9 | 5.1 | 6.1 (eggs) | ---- | 98/155 | Larson, unpubl.(1980) |
| Phialidium spp. | <1 | <3 | <1 | ---- | ---- | <1 | 2 | 83 | <1 | ---- | ---- | 4985/768 | Larson, 1985 |
| | 1.8 | ---- | ---- | 1.8 | ---- | ---- | 36.4 | 54.6 | ---- | 1.8 | ---- | 55/24 * | Purcell, unpubl.(1982) |
| Mitrocomella polydiademata | 2 | 4 | <1 | ---- | ---- | <1 | 19 | 75 | <1 | ---- | ---- | 98/23 | Larson, 1985 |
| | 2.3 | ---- | 2.3 | ---- | ---- | ---- | 95.4 | ---- | ---- | M | ---- | 44/18 * | Purcell, unpubl.(1982) |
| Mitrocoma cellularia | <1 | 2 | <1 | ---- | ---- | <1 | <1 | 98 | <1 | M | M | 65/14 | Larson, 1985 |
| | 3.1 | ---- | ---- | ---- | ---- | ---- | 55.4 | 38.5 | ---- | ---- | 3.0 | 65/17 | Purcell, unpubl.(1982) |
| Eutonina indicans | 1.4 | 1.4 | 2.1 | 0.5 | ---- | ---- | 3.2 | 89.5 | ---- | 0.1 | 1.0 | 630/28 | Purcell, unpubl.(1982) |
| **Narcomedusae** | | | | | | | | | | | | | |
| Aegina citrea | ---- | ---- | ---- | ---- | ---- | ---- | ---- | ---- | ---- | ---- | M | 1 obs. * | Mills, unpubl. |
| Solmissus albescens | ---- | ---- | ---- | ---- | ---- | ---- | ---- | ---- | ---- | ---- | M | 10 obs. * | Mills, unpubl. |
| Solmissus marshalli | ---- | ---- | ---- | ---- | ---- | ---- | ---- | ---- | ---- | ---- | M | 1 obs. * | Mills, unpubl. |

* many more specimens were examined that contained no food
M, present (Mills, unpubl.)

The distributions of nematocysts along the tentacles of ten species of hydromedusae are shown in Figure 1. Among species that mostly capture hard-bodied prey, Sarsia tubulosa and Proboscidactyla flavicirrata (Fig. 1A, 1C) have the nematocysts of their tentacles grouped into raised clusters containing two types of nematocysts. This arrangement may be especially effective in capturing crustaceans with complex surfaces. Aglantha digitale does not have raised clusters of nematocysts on its tentacles, and it eats some non-crustacean prey (Table IV).

Species of hydromedusae that eat primarily soft-bodied prey generally have nematocysts of only one type, or one predominant type (Table III). The nematocysts may be scattered regularly along the length of the tentacle (Fig. 1E, 1H), or arranged in a pattern (Fig. 1G, 1I), but rarely occur in raised clusters. Aequorea victoria is unusual in having one nematocyst type on the exumbrellar side of the tentacles, and another type on the subumbrellar side (Fig. 1G). Tentacles without raised nematocyst clusters may be most effective in contacting the broad uniform surfaces of soft-bodied prey.

Tentacles with raised clusters of nematocysts occur throughout the Capitata, which constitutes about half of the Anthomedusae (super-families Tubularioidea, Corynoidea, Zancleopsoidea, Zancleoidea, and Velelloidea) (Bouillon, 1985). Anthomedusae in the Filifera (super-families Pandeoidea, Bougainvillioidea, Rathkeoidea, Hydractinioidea, and Moeridioidea) usually do not have raised nematocyst clusters (except most of the Moeridioidea and incidental species in the other groups). Most, if not all, Limnomedusae have raised nematocyst clusters, frequently in the form of rings, on their tentacles. Leptomedusae, Narcomedusae, Trachymedusae (except the Geryonidae), and Laingiomedusae, have relatively even tentacle surfaces. Their nematocysts may be uniformly distributed, or concentrated along one side of the tentacle or near the tentacle tip.

Examination of the surfaces of prey captured by the anthomedusa Sarsia tubulosa showed some stenoteles to penetrate copepod exoskeleton and others, failing to penetrate, instead wrapping around prey surfaces (Fig. 2A). Although both desmonemes and stenoteles occur together in raised clusters, only desmonemes are found wrapped around the setae of the copepods, and stenoteles are found penetrating larger surfaces (Fig. 2A, 2B). The microbasic mastigophores of the anthomedusa Stomotoca atra, which we believe eats soft-bodied prey (medusae) primarily, were able to penetrate both crustacean exoskeleton (euphausiid larva) and soft-bodied prey (hydromedusa Phialidium gregarium), and also adhered to exoskeleton (Fig. 2C, 2D). Unfortunately, no prey were found in our systematic gut analyses of S. atra. Another pandeid anthomedusa, Catablema nodulosa, consumed crustaceans as well as soft-bodied prey in the field (Table IV). The microbasic mastigophores of the leptomedusa Phialidium gregarium also penetrated and adhered to copepods (Fig. 2E, 2F), even though most gut analyses showed crustaceans to comprise only a small part of its diet in nature (Table IV).

## IV. DISCUSSION

The pelagic hydrozoans fall into two groups, 1) those that primarily eat hard-bodied prey, and 2) those that primarily eat soft-bodied prey (Table V). Species in the first group have rhopalonemes (desmonemes, acrophores, anacrophores), which adhere to the surfaces of hard-bodied prey. Species that eat larger crustaceans (epibenthic species, decapod larvae, shrimps, large copepods) usually have stenoteles (Anthomedusae, physonect siphonophores). Other crustacean-eating species have microbasic euryteles (Anthomedusae) or microbasic mastigophores (calycophoran siphonophores) in addition to rhopalonemes.

In species that primarily eat soft-bodied prey, usually only one type of nematocyst occurs, or one is predominant, and they lack adhesive rhopalonemes. The cystonect siphonophores are the clearest example of this group, having only atrichous or homotrichous isorhizas and eating only soft-bodied prey. Most Leptomedusae have either homotrichous isorhizas or microbasic mastigophores (or basitrichous isorhizas). Their diets include some small crustaceans in addition to soft-bodied prey.

Other species that eat soft-bodied prey have nematocyst types unique to them. The Narcomedusae probably eat only gelatinous zooplankton and have apotrichous isorhizas. These nematocysts presumably penetrate soft-bodied prey, and the large spines on the distal two-thirds of the tubule may aid in anchoring them in soft prey tissues. The physonect Apolemia uvaria has only birhopaloides, which are like isorhizas, but with two small swellings along the tubule. Unlike other physonect siphonophores, the diet of A. uvaria includes many soft-bodied prey (salps, ctenophores, hydromedusae, and chaetognaths; total 20-75.3%), as well as crustaceans (16.7-73.3%).

Species feeding on other medusae do not seem to narcotize or kill these prey upon capture. Prey medusae contract when they are first contacted, but they resume pulsations of the swimming bell during ingestion. In contrast, copepods or fish larvae become quiescent soon after contact with a tentacle and do not subsequently resume movement .

Purcell (1984a) found that the size of copepods eaten by siphonophores increased with increasing size and numbers of nematocysts in the batteries. Comparison of the nematocysts of siphonophores and hydromedusae shows

---

*Fig. 1. Interference contrast micrographs of nematocysts in the tentacles of ten species of hydromedusae. Letters in parentheses following species names indicate Order (a = Anthomedusae, le = Leptomedusae, li = Limnomedusae, n = Narcomedusae, t = Trachymedusae) and predominant prey type (c = crustacean, s = soft-bodied). A. Sarsia tubulosa (a, c). B. Aglantha digitale (t, c). C. Proboscidactyla flavicirrata (li, c). D. Phialidium gregarium (le, s). E. Eutonina indicans (le, s). F. Obelia sp. (le, s). G. Aequorea victoria (le, s). H. Mitrocoma cellularia (le, s). I. Stomotoca atra (a, s). J. Aegina citrea (n, s). Scale bar = 10 μm for all species.*

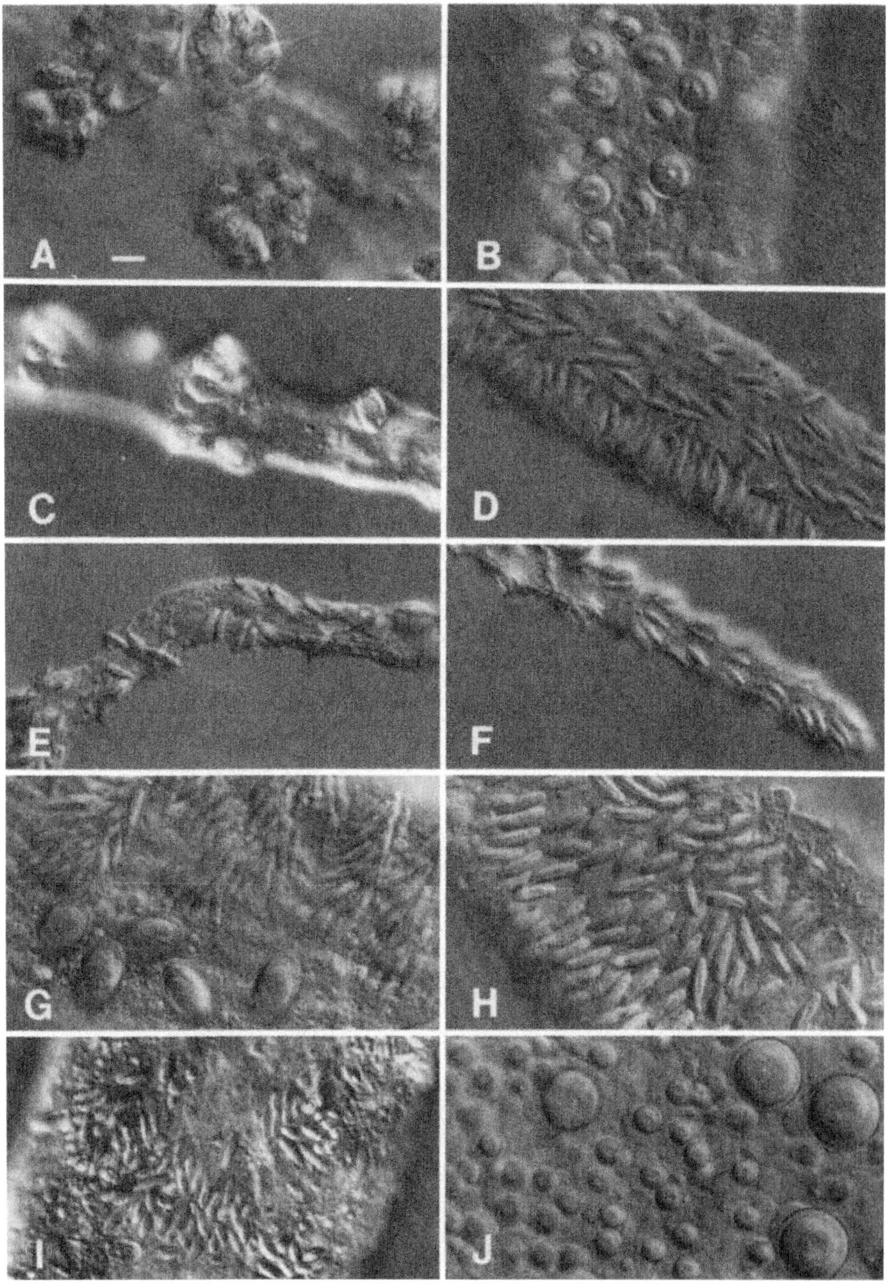


*Figure 1*

that rhopalonemes and isorhizas in the two groups are of similar size. However, stenoteles and microbasic mastigophores of siphonophores are much larger than those of hydromedusae. We estimate the volumes of these nematocysts as in Purcell (1984a) from the formula for the volume of an elipsoid, $4\pi ab^2$, where a = length/2 and b = diameter/2. Stenoteles ranged in volume from 1.84 -17.22 µl in siphonophores, but only 0.21-1.96 µl in hydromedusae, nearly a 9-fold difference. Microbasic mastigophores ranged in volume from 0.68-73.85 µl in siphonophores, but only 0.014-0.090 µl in hydromedusae; 48- to 820-fold differences. Presumably, large nematocysts could contain longer tubules and more venom, and might potentially be more effective in prey capture than small nematocysts.

These differences in nematocyst volume are not reflected in the sizes of copepods captured. Lengths of copepods removed from hydromedusae in the Strait of Georgia region were as follow: Sarsia spp., 1.3 ± 0.7 mm, n = 46; Hybocodon prolifer, 1.0 ± 0.3, n = 12; Aglantha digitale, 1.0 ± 0.5, n = 16; Mitrocoma cellularia, Phialidium gregarium, and Eutonina indicans combined, 0.6 ± 0.2, n = 9; Aequorea victoria, 1.2 ± 0.8, n = 200 (Purcell, unpubl.). Copepods captured by siphonophores ranged in mean cephalothorax length from 0.36 mm to 1.97 mm (Purcell, 1984a). The siphonophore species most common in the Strait of Georgia region, Muggiaea atlantica, consumed copepods 0.36 mm in mean length, substantially smaller than those consumed by the local hydromedusae, even though its nematocysts were generally larger in size [microbasic mastigophores 36 x 6 µm, anisorhizas 15 x 4 µm, rhopalonemes 6 x 4 µm and 6 x 2 µm (Purcell, 1984a)]. We suggest that prey capture primarily by nematocyst adhesion, as in siphonophores, may require larger nematocysts than prey capture that also involves nematocyst penetration, as in hydromedusae.

Fairly consistent trends in cnidom and diet occur among the hydromedusan orders, but the distinctions are not as clear as in the Siphonophora. The nematocysts of medusae that feed on soft-bodied prey (many of which also eat some hard-bodied prey) do penetrate crustacean exoskeleton. Therefore medusae are not limited to soft-bodied prey because of an inability to penetrate crustaceans. An alternative explanation is that soft-bodied feeders lack the adhesive nematocysts that enable ready capture of crustaceans.

---

*Fig. 2. Scanning electron micrographs of hydromedusan nematocysts on prey. A. Stenoteles from Sarsia tubulosa penetrating and adhering to copepod exoskeleon; scale bar = 20 µm. B. Desmonemes from S. tubulosa wrapping around copepod setae; scale bar = 10 µm. C. Microbasic mastigophores from Stomotoca atra penetrating and adhering to crustacean (larval euphausiid) exoskeleton; scale bar = 10 µm. D. Microbasic mastigophores from S. atra penetrating exumbrella of the hydromedusa Phialidium gregarium; scale bar = 5 µm. E. Microbasic mastigophores from P. gregarium penetrating and adhering to copepod exoskeleton; scale bar = 10 µm. F. Microbasic mastigophores from P. gregarium adhering to copepod setae; scale bar = 10 µm.*

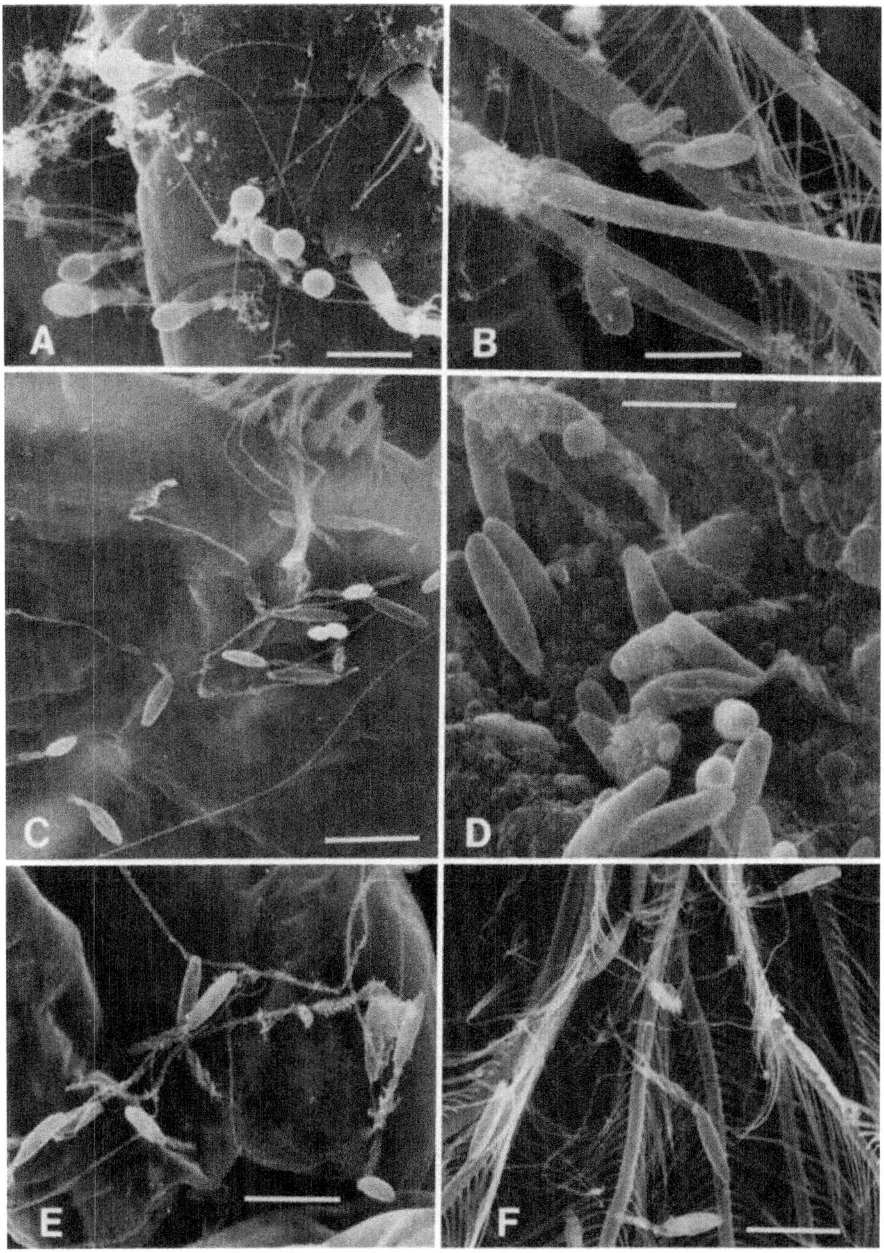


*Figure 2*

TABLE V. Pelagic hydrozoans grouped according to predominant prey type (soft-bodied or crustacean)

| **Soft-bodied prey** | | **Crustacean prey** | |
|---|---|---|---|
| Predator | Predominant nematocyst types | Predator | Predominant nematocyst types |
| Cystonect siphonophores | isorhizas | Calycophore and physonect siphonophores | rhopalonemes<br>microbasic mastigophores or stenoteles<br>anisorhizas * |
| Apolemia uvaria (Physonectae) | birhopaloides | | |
| Family Pandeidae (Anthomedusae) | microbasic euryteles | Anthomedusae | desmonemes<br>stenoteles or euryteles |
| Leptomedusae | microbasic mastigophores<br>isorhizas * | Limnomedusae | desmonemes<br>microbasic euryteles or macrobasic mastigophores |
| Narcomedusae | apotrichous isorhizas | Trachymedusae | stenoteles or microbasic euryteles |

* numerically predominant

Some surface adhesion of nematocyst tubules that failed to penetrate crustaceans was seen in the hydromedusae, but the amount of adhesion was far less than was seen in calycophoran and physonect siphonophores, which discharged thousands of nematocysts with heavily-spined tubules onto prey. The predators of crustaceans all have adhesive rhopalonemes, which are lacking in the predators of soft-bodied plankton. We conclude that adhesion is very important in the capture of crustacean prey.

Differences in the types of nematocysts present probably do not completely explain the dietary differences that exist among the Hydrozoa. It is possible that differences may exist in the stimuli necessary to discharge the nematocysts. Additionally, feeding behaviors differ among hydrozoans, particularly in the hydromedusae, and these may affect the types of prey captured. Morphological differences among species and in ontogeny also are great, and undoubtedly influence prey capture. Diets will change with location and time, due to variation in prey availability. We stress that most of the dietary data presented here are from specimens of many species collected concurrently, and that true differences in diet exist among pelagic hydrozoan species.

## V. CONCLUSIONS

There are consistent trends in nematocyst types and in diets among the orders of hydromedusae and among the suborders of siphonophores. Species that mostly eat crustaceans generally have rhopalonemes and other nematocyst types that adhere to the prey surfaces, and stenoteles that can penetrate the prey exoskeleton. Species that mostly eat soft-bodied prey lack adhesive rhopaloneme nematocysts, and usually have a simple cnidom of one or two nematocyst types, both of which penetrate soft-bodied organisms. Surface adhesion of the nematocyst tubules, and the entanglement of prey is critical in the capture of crustaceans, but adhesion may be ineffective in the capture of soft-bodied organisms.

## ACKNOWLEDGMENTS

We thank the Friday Harbor Laboratories and the University of Victoria for the generous use of their facilities and equipment.

## REFERENCES


Alvariño, A. 1985. Predation in the plankton realm, mainly with reference to fish larvae. Inv. Mar. CICIMAR Vol. 2:1-122.

Arkett, S. A. 1984. Diel vertical migration and feeding behavior of a demersal hydromedusan (Polyorchis penicillatus). Can. J. Fish. Aquat. Sci. 41:1837-1843.

Bieri, R. 1961. Post-larval food of the pelagic coelenterate, Velella lata. Pac. Sci. 15:553-556.

Bouillon, J. 1974. Description de Teissiera milleporoides, nouveau genre et nouvelle espèce de Zancleidae des Seychelles (Hydrozoaires; Athécates-Anthoméduses), avec une révision des Hydroïdes "Pteronematoidea". Cah. Biol. Mar. 15:113-154.

Bouillon, J. 1978. Hydroméduses de la mer de Bismarck (Papouasie, Nouvelle-Guinée). Partie II. Limnomédusa, Narcomédusa, Trachymédusa et Laingiomédusa (sous-classe nov.). Cah. Biol. Mar. 19:473-483.

Bouillon, J. 1980. Hydroméduses de la mer de Bismarck (Papouasie, Nouvelle-Guinée). Partie III: Anthomédusae-Filifera (Hydrozoa - Cnidaria). Cah. Biol. Mar. 21:307-344.

Bouillon, J. 1985. Essai de classification des Hydropolypes-Hydroméduses (Hydrozoa-Cnidaria). Indo-Malayan Zool. 1:29-243.

Ewer, R. F. 1947. On the functions and mode of action of the nematocysts of Hydra. Proc. Zool. Soc. Lond. 117: 365-376.

Godknect, A. 1985. Die Entladung der B-Rhabdoiden der Tentakel von Anemonia sulcata Pennant (Cnidaria, Anthozoa). Diploma Thesis, Univ. Zurich.

Hand, C. 1954. Three Pacific species of "Lar" (including a new species), their hosts, medusae, and relationships. (Coelenterata, Hydrozoa). Pac. Sci. 8:51-67.

Hyman, L. H. 1940. The Invertebrates: Protozoa through Ctenophora. McGraw-Hill, New York. 726 pp.

Kubota, S. 1976. Notes on the nematocysts of Japanese hydroids, I. J. Fac. Sci. Hokkaido Univ. Ser. VI, Zool. 20:230-243.

Kubota, S. 1978. The life-history of Clytia edwardsi (Hydrozoa; Campanulariidae) in Hokkaido, Japan. J. Fac. Sci. Hokkaido Univ. Ser. VI, Zool. 21:317-354.

Kubota, S. 1981. Life-history and taxonomy of an Obelia species (Hydrozoa; Campanulariidae) in Hokkaido, Japan. J. Fac. Sci. Hokkaido Univ. Ser. VI, Zool. 22:379-399.

Larson, R. J. 1980. Diet, daily ration and clearance rates for the hydromedusae Aequorea victoria and Phialidium gregarium. (17 pp.) Zooplankton Ecology-Oceanography 533 class report, Friday Harbor Laboratories Library, University of Washington.

Larson, R. J. 1985. Trophic ecology of gelatinous predators (Cnidaria and Ctenophora) in Saanich Inlet, Vancouver Is., B.C., Canada. Ph.D. Thesis, Univ. of Victoria., Canada. 253 pp.

Mackie, G. O., and G. V. Mackie. 1963. Systematic and biological notes on living hydromedusae from Puget Sound. Natl. Mus. Canada (Contrib. Zool.) Bull. 199:63-84.
Mackie, G. O., P. R. Pugh, and J. E. Purcell. 1987. Siphonophore biology. Adv. Mar. Biol. in press.
Mariscal, R. N. 1974. Nematocysts. Pp. 129-178 in Coelenterate Biology: Reviews and New Perspectives. L. Muscatine & H. Lenhoff, eds. Academic Press, New York.
McCormick, J. M. 1969. Trophic relationships of hydromedusae in Yaquina Bay, Oregon. Northwest Sci. 43:207-214.
Mills, C. E. 1981. Diversity of swimming behaviors in hydromedusae as related to feeding and utilization of space. Mar. Biol. 64:185-189.
Mills, C. E. and R. L. Miller. 1984. Ingestion of a medusa (Aegina citrea) by the nematocyst-containing ctenophore Haeckelia rubra (formerly Euchlora rubra): phylogenetic implications. Mar. Biol. 78:215-221.
Östman, C. 1979. Nematocysts in the Phialidium medusae of Clytia hemisphaerica (Hydrozoa, Campanulariidae) studied by light and scanning electron microscopy. Zoon 7:125-142.
Östman, C. 1982. Nematocysts and taxonomy in Laomedea, Gonothyraea and Obelia (Hydrozoa, Campanulariidae). Zool. Scripta 11:227-241.
Purcell, J. E. 1981a. Dietary composition and diel feeding patterns of epipelagic siphonophores. Mar. Biol. 65:83-90.
Purcell, J. E. 1981b. Feeding ecology of Rhizophysa eysenhardti, a siphonophore predator of fish larvae. Limnol. Oceanogr. 26:424-432.
Purcell, J. E. 1984a. The functions of nematocysts in prey capture by epipelagic siphonophores (Coelenterata, Hydrozoa). Biol. Bull. 166:310-327.
Purcell, J. E. 1984b. Predation on fish larvae by Physalia physalis, the Portuguese man of war. Mar. Ecol. Prog. Ser. 19:189-191.
Rees, J. T. 1979. Growth stages of the medusa of Proboscidactyla flavicirrata Brandt, 1835 (Hydrozoa: Limnomedusae). Can. J. Zool. 57:551-557.
Russell, F. S. 1938. On the nematocysts of Hydromedusae. J. Mar. Biol. Ass. U.K. 23:145-165.
Russell, F. S. 1939. On the nematocysts of Hydromedusae. II. J. Mar. Biol. Ass. U.K. 23:347-359.
Russell, F. S. 1940. On the nematocysts of Hydromedusae. III. J. Mar. Biol. Ass. U.K. 24:515-531.
Tardent, P. and T. Holstein. 1982. Morphology and morphodynamics of the stenotele nematocyst of Hydra attenuata Pall. (Hydrozoa, Cnidaria). Cell Tissue Res. 224:269-290.
Toppe, O. 1909. Über die Wirkungsweise der Nesselkapseln von Hydra. Zool. Anz. 33:798-805.
Zelickman, E. A., V. I. Gelfand, and M. A. Shifrin. 1969. Growth, reproduction and nutrition of some Barents Sea hydromedusae in natural aggregations. Mar. Biol. 4:167-173.

# 25

# IMPORTANCE OF NEMATOCYSTS TO ACTINIAN TAXONOMY

Daphne Gail Fautin

Department of Invertebrate Zoology
California Academy of Sciences
San Francisco, California 94118-9961

## SUMMARY

Despite the fact that actinians possess only six of the approximately 25 types of nematocysts currently recognized, or less than a third the variety of the Hydrozoa, cnidae play a far greater role in anthozoan than hydrozoan taxonomy. Indeed, actinian families are defined in part by cnidom, and an inventory of cnida types and sizes is an indispensable component of contemporary actinian species descriptions, although it is demonstrably not the case that each species has a unique array (types and/ or sizes) of cnidae. Moreover, it is now clear that various biological factors can affect the categories of cnidae possessed by an individual actinian, so it may be impossible to define an absolute cnidom for all species.

Data on nematocysts have been used in taxonomy mainly qualitatively, but there is an increasing tendency to treat them quantitatively. However, methodologies for obtaining and handling such data have never been made explicit. Among matters to be resolved -- aside from the most fundamental of a classification system acceptable to all scientists -- are minimum number of animals and capsules from which measurements are taken, statistical presentation of the data, meaning of statistical comparisons between samples, and replicability. Whether a uniform protocol should be developed rests ultimately upon determining the level of significance that should be attached to cnidae, qualitatively or quantitatively.

## I. INTRODUCTION

Approximately 25 types of nematocysts are currently recognized, in addition to spirocysts and ptychocysts (these and the following figures are from Mariscal, 1974, with modifications from Rifkin and Endean, 1983). Both non-nematocyst cnidae are confined to the Anthozoa, as are macrobasic and microbasic amastigophores. Of the additional four categories of nematocysts possessed by anthozoans, atrichous and holotrichous isorhizas occur in all classes, while basitrichous isorhizas are shared with the Hydrozoa, and microbasic mastigophores with Hydrozoa and Cubozoa. Scyphozoans and cubozoans each possess four types of cnidae -- not the same array, and none peculiarly. Therefore, all but the two exclusively anthozoan types of nematocysts occur among members of class Hydrozoa, 17 of them uniquely.

In light of these numbers, it may seem curious that cnidae have played a far greater role in anthozoan than in hydrozoan taxonomy. They have been particularly important for actinians, most families of which are defined in part and several primarily by cnidom (Carlgren, 1949). Weill (1934*b*) found study of the actinian cnidom particularly onerous because tentacles, body wall, pedal disc, pharynx, mesenterial filaments, and various specialized structures such as acontia, pseudotentacles, and acrorhagi each have a different array of nematocysts. Therefore, in contrast to other cnidarians, all these tissues must be assayed to establish the cnidom of a species.

An implicit assumption -- indeed, a conviction -- in using cnidae for taxonomy is that they are useful at some level of classification, but many of Weill's (1934*b*) concerns about how exogenous and endogenous factors may influence distribution and size of nematocysts have yet to be addressed. Moreover, despite the current convention of including nematocyst data in every species description (which implies that such may even distinguish species), procedures for obtaining and handling these data have not been committed to paper. Rather, this unwritten lore has been handed down in a sort of apprenticeship from professor to student.

I shall put into the context of issues raised by Weill (1934*a*, *b*) some of the information that has accumulated during the half century since he wrote, and examine critically some philosophical considerations in using nematocysts as taxonomic characters for sea anemones. This subject is hardly new, constituting as it does the second of Weill's two volumes containing results of 11 years of study, but many matters have yet to be fully explored, much less resolved.

## II. HOW CNIDAE CAME TO BE USED IN ACTINIAN TAXONOMY

In 1900, Carlgren proposed that cnida size is of systematic value, further claiming that no description of an actinian is complete without such data. That this opinion was not immediately accepted probably had more to do with the inadequacy of the several classification schemes for cnidae that had been proposed before Weill's (1934*a*) (summarized by him and by Carlgren, 1940) than with philosophical skepticism. None of the earlier taxonomies of cnidae recognized more than four categories, and, indeed, Carlgren's own early measurements were either for "thin-walled" and "thick-walled" capsules (e.g. 1900) or for categories to which he attached no names at all (e.g. 1928).

Stephenson (1920, 1921, 1922) did not mention cnidae in his work "On the classification of Actiniaria." By 1928, he agreed with Carlgren about the value of nematocysts to anemone systematics. In his 1929 paper, devoted to exploring the limits of this character, Stephenson concluded that size is of taxonomic significance in only some tissues of some species. Although he later (1935) provided size data for cnidae of the British actinians, he resisted applying Weill's scheme, yet, significantly, abandoned his use of "penicilli" and "spirulae" that Weill had criticized.

Weill (1934*a*) defined 19 categories of nematocysts, but his hierarchical scheme, that recognizes differences within classes of basically similar capsules, did not entirely satisfy even its creator, as he pointed out instances of intermediates. It is widely recognized that Weill's classification does not accommodate the variability of cnidae, so that those of similar form differing in details of shaft length, tubule spine pattern, capsule shape, etc. are assigned the same name. The advent of electron microscopy allowed resolution of detail that showed, for example, the tubule of some "atrichs" are, in fact, holotrichous, the spines being too small to be seen clearly with all but the most sophisticated light microscopy (Cutress, 1955; Bigger, 1976, 1982). Some such issues are addressed by Schmidt's (1972*a*, *b*) system, but it has not displaced Weill's terminology, which has been tinkered with and debated almost since it was proposed (e.g. Carlgren, 1940; Cutress, 1955; Hand, 1961; Mackie and Mackie, 1963; Conklin, Bigger, and Mariscal, 1977).

Although data on size of nematocysts were accumulating well before there was an adequate nomenclature for cnidae (Stephenson, 1935), Weill's (1934*b*) work bolstered Carlgren's long-standing advocacy of cnidae as a taxonomic character by providing terms to attach to the size ranges. Carlgren's early adoption of Weill's terminology (e.g. 1938), followed by two papers (1940, 1945) simply listing cnidae distributions and sizes in 78 species of anthozoans, mostly sea anemones, were precedent-setting in the practice of actinian taxonomy. Indeed, his procedure rapidly gained the status as convention for the field (see Hand, 1955*a*).

## III. HOW CNIDAE ARE USED IN ACTINIAN TAXONOMY

Initially, published size ranges gave no indication of how many capsules contributed to the figure, nor how many animals had been sampled.

Carlgren's (1940) criticism of an earlier worker for omitting to mention number of individuals examined rings hollow since he did not do so himself. Hand (1955*a*,*b*, 1956) followed the practice of Stephenson (1929) in providing both the number of animals from which cnidae were obtained and the number of capsules measured. However, whereas Stephenson (1929) provided data on only the "majority-form" cnida for each tissue, Hand (1955*a*, *b*, 1956) dealt with all types present. The number of cnidae Hand measured in the various tissues is rather similar, so there is no way of knowing whether all capsules were equally common. While not stating the number of cnidae measured, Sebens and Laakso (1978) indicated whether each type was abundant, common, or rare. I initially specified both number of animals examined for a particular tissue and number of cnidae of each type measured (e.g. Dunn, 1981). However, because I sometimes find a particular variety of cnida in smears from only some individuals of a species, I now list the number of animals in which I found each type of cnida as a fraction of the total examined (e.g. Dunn, 1982).

Stephenson (1929) was also apparently the first to provide, in addition to the range, a mean length and width for each variety of nematocyst. Yet nowhere has a protocol been specified for gathering samples of nematocyst measurements appropriate to the type of analysis to be done. Methods sections of most taxonomic papers concentrate on physical preparation of specimens (e.g. Stephenson, 1928, 1929; Schmidt, 1972*a*), or terminology of cnidae (e.g. Schmidt, 1972*b*). Hand (1955*a*), exceptional in describing how his data were actually obtained, advocated measuring only the larger and smaller capsules to obtain the range. I cautioned that to avoid unwittingly lumping cnidae of more than one size class, after establishing the low and high end of an apparent range, it is necessary to measure capsules of intermediate sizes to bridge the gap (Dunn, 1981).

## IV. QUANTITATIVE VERSUS QUALITATIVE DATA

Implicit in the use of cnidae for taxonomy is the assumption that each taxon has a characteristic size and/or distribution of nematocysts. Actual data reveal that this is not always so. Some closely related species of actinians that share a cnidom differ in size and/or distribution of cnidae. However, others do not, contrary to the findings of Mackie and Mackie (1963) in hydromedusae.

There are no consistent, significant qualitative or quantitative differences in nematocysts of four species of *Epiactis* found along the Pacific coast of North America (Fautin and Chia, 1986). However, the 12 species of tropical actinians, belonging to three families, dealt with in my monograph (Dunn, 1981) are separable on the basis of cnidae. All have a cnidom of spirocysts, basitrichs, and microbasic *p*-mastigophores, which is the most common in the Actiniaria. Certain species, however, lack microbasic *p*-mastigophores in the column. The column basitrichs of some species fall within a single size class, whereas as many as three can be distinguished for others. Similar qualitative and quantitative differences exist in the cnidae of other tissues as well. None of the species is distinguished solely on the basis of cnidae; those data simply correlate with other characters such as morphology and ecology. Similarly, Jaworski (1938) recognized that geographical races of an actinian differed in cnidae, but this was neither a necessary nor a sufficient condition for distinguishing the races.

Weill (1934*b*) posed two questions about the taxonomic value of nematocysts: 1) Is the cnidom of a particular species constant in face of variables such as age, sex, segment of life cycle, nutritional state, and external conditions?; and 2) Is the cnidom identical among all individuals of a species when they are in identical circumstances? Note that these concern usefulness of the *cnidom* to coelenterate taxonomy. The original definition of this much-misused term is: *l'ensemble des nématocystes que présente un Cnidaire à un moment donné de son existence* (Weill, 1934*b*, page 351; italics his) -- in other words, simply an inventory of an organism's cnidae, without regard to abundance, size, or distribution of the capsules. Although studies have not been designed explicitly to address these issues, there are sufficient data in the natural history and taxonomic literature to conclude that neither question can be answered in the affirmative.

Carlgren (1929) established that catch tentacles possess a different array of nematocysts than ordinary tentacles, so individuals of a species with catch tentacles differ in cnidom from those lacking them (Carlgren, 1949). Holotrichs characterize mature catch tentacles (Watson and Mariscal, 1983), which are used in intraspecific aggression (Purcell, 1977). Acrorhagi, structures with a similar function in certain actiniids (Francis, 1973), are also defined in part by possession of holotrichs. At least some individuals of *Epiactis* have holotrichs in their outer tentacles (Fautin and Chia, 1986). I suspect, from their position and by analogy, that they serve an aggressive function in *Epiactis* as well. Position within a clone affects the number of acrorhagi that a specimen of *Anthopleura elegantissima* has; this is correlated with the extent to which the anemone is involved in intraspecific agonism (Francis, 1976). Thus holotrichs seem responsive at least to external conditions, so systematic weight should not be attached to their presence/absence.

Insufficient attention has been paid to age, sex, developmental stage, and other such attributes of individuals to ascertain how extensively they affect cnidom. Planulae of *Anthopleura elegantissima* contain more types of nematocysts than are found in the adult (Chia and Koss, 1983). Young individuals of anemone species characterized by acrorhagi may lack them, and thereby the "atrichs" (=holotrichs) that are typically confined to them (Carlgren, 1949). Polyps and medusae of certain scyphozoans may not have the same arrays of cnidae, since both additions and deletions can occur during strobilation (Calder, 1977). Hydrozoan polyps and medusae of the same species may also differ in cnidom (e.g. Rees, 1983).

As to Weill's second issue, I found that samples of a particular tissue from different individuals living under apparently identical conditions may not contain the same array of cnidae. If, for example, in examining eight individuals of a species, I found microbasic *p*-mastigophores in the actinopharynx of only five, I would list 5/8 beside that type of cnida from that tissue. The significance of this figure is unclear to me. Does it imply that only 62.5% of the individuals in that population possess microbasic *p*-mastigophores in their actinopharynx? Or are chances of finding them only 62.5% because the cnidae are rare or patchily distributed? This phenomenon is not confined to the Anthozoa: Calder (1977) found that some ephyrae of *Aurelia* (a scyphozoan) have a third type of cnida absent in most individuals. Additional information on function and occurrence of cnidae may resolve the issue of whether different significance should be attached to finding a type/size of cnida not reported, and not finding a type/size that has been reported. Perhaps, as more data of this sort are gathered, it will become clear that certain types of cnidae, regardless of position in the body, typically are more variable in occurrence and/or size than others. Alternatively, cnidae of all types may be more variable in particular tissues. Thus it is not possible to define absolutely a single cnidom for all species.

Carlgren (1940, page 5), in rebuttal to Weill, whose concern was strictly with cnidom and who claimed that nematocyst size lacks taxonomic importance, asserted "all information as to the size of the cnidae is nevertheless valuable, especially for distinguishing species." However, he provided no support for this statement, nor guidelines as to how quantitative data should be analyzed and what comparisons are appropriate. One of the attractive attributes of cnidae as a taxonomic character is that they presumably are recognizable only when mature, and do not grow thereafter. Operationally, there appears to be sufficient consistency among individuals of a species that a size range for each type of cnida from each tissue can be compiled. However, cnida size may vary with age and/or size of anemone (Dunn and Bakus, 1977; Dunn, 1981), as also may be true of hydromedusae (e.g. Rees, 1983), but the correlation is neither invariant nor regular (Stephenson, 1929).

Range is the statistic Carlgren chose, for unspecified reasons, to apply to his cnidae size data. It can be a useful descriptor for taxonomic purposes, an indicator of the variability encountered within a species. Random sampling is not required for range data, nor is range affected by the actual underlying distribution of the character under consideration. There are, however, no statistics for comparing ranges. The least ambiguous case is non-overlap, but significance ascribed to any overlap is a matter of experience and judgment, often based on how many observations comprise the statistic and degree of coincidence. At best, it is semi-quantitative.

From the limited data available, I conclude that the variability in methodology and uncertainty about the significance of information precludes use of cnidae as taxonomic characters at all but the grossest level (i.e. substantially non-overlapping ranges, or presence of unusual types) for species determinations. Indeed, I know of no instance of an actinian species being recognized solely on the basis of cnidae.

Tim Blanchard studied in my laboratory acontia cnidae of *Diadumene cincta* from San Francisco Bay to evaluate comparative efficiency and usefulness of alternative ways of gathering and analyzing data on nematocysts for taxonomic purposes. He measured randomly in 10 animals length of 100 undischarged capsules of each of the two types present -- basitrichs and microbasic *p*-mastigophores. Measurements from each successive group of 10 cnidae were tallied separately. A "true" range was also established by purposeful search for the largest and smallest of each type of cnida in the 10 anemones.

Basitrich length range was 10-21 μm, with a strongly defined mode of 17 μm, which was also the mode in eight of the anemones. A grand mean/standard deviation were not computed, but for individuals the mean was about 17, and the standard deviation varied from 6 to 12% of the mean. The common range (that portion of the size range common to all samples) of 15-18 μm contained 87% of the nematocysts sampled (varying from 82% to 94% in the 10 individuals), and distributions were fairly consistent among specimens.

By contrast, the range of microbasic *p*-mastigophores was broad (40-64 μm) and the common range, which contained only 28% of the capsules measured, was narrower than for basitrichs (50-52 μm). The mode of 50 μm was only somewhat higher than peaks of approximately equal magnitude at 48, 52, and 55 μm; only two individuals had modes of 50 μm; and distributions were poorly correlated between specimens. The mean for individual anemones, which varied between about 46 and 53 μm, was always very near the modal length for that animal, and standard deviations varied between 4 and 22% of the mean.

For basitrichs, the "true" range for the specimen was ultimately obtained by random sampling in six of the 10 specimens, whereas for microbasic *p*-mastigophores it was found in eight. The full size range for the species, as established by purposeful search, was not obtained by random sampling of any specimen, and in only one individual was the full "true" range characteristic of the species procured by purposeful search, that for basitrichs. Some individuals in which the standard deviation of basitrichs was small had highly variable microbasic *p*-mastigophores, and vice versa, whereas in others both were large or small. Capsules at the extremes of the range (for the individual) were encountered anywhere from the third to the tenth (final) set of 10, with an average of seven sets (60-70 nematocysts). About an equal number produced the full range for each set of 10 cnidae from each of the 10 animals. The mean size of 100 nematocysts was more variable within each of the 10 anemones than it was for 10 cnidae from each anemone.

## V. HOW DATA ON CNIDAE MAY BE USED IN TAXONOMY

Ultimate resolution of whether cnidae have taxonomic value lies in whether they provide significant information of use to identification. Just as for other taxonomic characters, descriptions should ideally contain data from several individuals encompassing a range of variability to which the species is subject. However, there is still virtually no information on the extent to which cnidae vary quantiatively or qualitatively within a species geographically, through time (which includes ontogenetically, with age, seasonally, and even evolutionarily), or with environment. Such data are tedious to gather and may ultimately prove inconclusive.

Qualitative information may be at least as revealing for taxonomic purposes as quantitative. Stephenson (1929), arriving at a similar conclusion, proposed that some differences probably characterize genera, others species. Adshead, Mackie, and Paetkau (1963) were able to identify species of *Hydra* largely on the basis of qualitatively distinctive nematocysts. Campbell (1983) also advocated use of such features as relative width of capsule and coiling of nematocyst shaft to identify hydras to species. Perhaps more attention should be paid in actinians to features such as relative length of shaft, extent of tubule spination, uniformity of spine length, etc., which are almost infinitely variable. Obviously such features are best observed in the discharged state. Not only is this impossible with preserved material, which is often all that is available, but poorly understood physiological differences among cnidae make some types virtually impossible to discharge even in fresh preparations. Another operational obstacle to accurate qualitative characterization is optics. Increasing reliance is placed on electron microscopy, which is not universally available, nor is it practical in certain situations such as the field.

It is unlikely that any nomenclatorial system could express adequately the extreme variability in possibly significant features of cnidae. A way of avoiding debate about terms, and inconsistencies in the way they are used that can make comparisons among taxonomists difficult, is provision of a figure for each type of cnida measured. To convey maximum information, these should not be diagrammatic (e.g. Hand, 1955*a*) or even tracings of photographs (e.g. Dunn, 1981), but should be detailed drawings (e.g. Schmidt, 1972*b*) or, best of all, photomicrographs (e.g. Rifkin and Endean, 1983; Fautin and Chia, 1986).

In light of the dearth of knowledge about patchiness of cnidae, site of synthesis, and migration within the animal, it is premature to propose a universal method for gathering quantitative data, or even to conclude whether one is possible. However, each taxonomist must make explicit at least the number of animals and cnidae used, and how the data were gathered: whether randomly or by purposeful search for those of particular sorts and sizes. Jaworski (1938), for example, measured 100 capsules, from which he obtained both range and mean, but did not mention whether his sample was random. For identification of specimens of hydras, as opposed to species diagnoses, Campbell (1983) advocated obtaining an average length and width from measurement of five capsules. He omitted mention of how those five should be selected. A mean based on measurements not obtained randomly has limited value; but any random method risks missing important information, and a range compiled from a random sample of the small number of cnidae typically specified in taxonomic papers is unlikely to be representative of the species. Most statistical tests require random samples, and normalcy, assumed by many, has yet to be demonstrated for cnidae.

Obtaining a random sample has some practical difficulties beyond the issue of how randomness is to be assured. At the high magnifications necessary for adequate resolution, a very thin smear must be made so that most capsules are oriented in the plane of focus, and thereby each one to be sampled is amenable to measurement. If fresh material is studied, since the conventional practice is to measure only undischarged capsules (thereby permitting comparison with preserved specimens), there is a finite risk that those left undischarged, and therefore in the pool from which the sample is drawn, are immature, defective, or somehow atypical in other respects.

It is clear from Blanchard's data and those of Thomason and Brown (1986) on corals that generalizations about statistical distributions of cnidae are currently untenable. Even distributions of the same type of cnidae from different tissues may vary. For example, Blanchard (unpublished) found that microbasic *p*-mastigophores in tentacles of *Diadumene cincta* lack a common range: at the extremes, those of one individual were 40-51 μm long, while those of another were 56-62 μm. A feature such as shape of the distribution curve might serve as a character itself, which could also provide insight into appropriate sampling strategy. I infer that for cnidae with a very steep distribution curve, encountering the smallest and largest capsules is less likely by random sampling than for cnidae characterized by a flat, broad curve. Thus, counterintuitively, nematocyst types with a very narrow range probably require a larger random sample to ascertain the full scope than do those with a broader range. Establishing a range, if desired for statistical purposes or for comparison with data that have been accumulating throughout this century, clearly is more efficient through deliberate search than by means of random sampling. Although size information appears to accumulate about equally rapidly through random sampling whether measuring 100 cnidae in one animal or 10 in each of 10 individuals, it is important to obtain a composite figure from numerous individuals, since no one usually spans the entire range (Dunn, 1981; Blanchard's data).

It may go without saying that contamination must be avoided, but even Weill (1934*b*) made the point, especially for actinians, in which the anatomy is sufficiently complex, and the array of cnidae present in the various tissues so different, that it is all too common. As a result, it can be tempting to discount rare capsules as exogenous, thereby missing unusual or rare varieties.

## VI. TAXONOMY VERSUS SYSTEMATICS

Carlgren (1949) defined all member families and genera of the Ptychodactiaria, Corallimorpharia, and Actiniaria in his catalog. All for which data are available possess spirocysts and basitrichs. That constitutes the cnidom of the Actinodendronidae, the anemones with the most potent sting to humans (Hansen and Halstead, 1971). Those with three sorts of cnidae (half the 38 actinian families recognized by Carlgren, in addition to Liponematidae -- see Dunn and Bakus, 1977, and Actinoscyphiidae -- see Riemann-Zürneck, 1978) invariably have, also, microbasic *p*-mastigophores. Beyond this level, the six families with four types all contain those three, and two each have microbasic *b*-mastigophores, microbasic amastigophores, and macrobasic amastigophores. Knowledge of the functional significance of especially the last three types of cnidae might provide leads about whether their presence is more likely to reflect ecology or phylogeny. Nor is position within the body, which may provide clues to function, random: some types occur most commonly in particular tissues, such as spirocysts in tentacles, and microbasic *p*-mastigophores in mesenterial filaments and acontia.

Schmidt (1972*a*, page 422) stated that because cnidae of different anthozoan orders are so variable "they are of little taxonomic and phylogenetical value." This does not distinguish between their utility in identifying specimens of anemones, and their potential for differentiating higher taxa and establishing evolutionary links among them. Indeed, nematocysts are used to define some acontiate families. That distribution of distinctive types of cnidae correlates with other features in defining families supports their use at higher levels. At the highest of all, as Mariscal (1984) pointed out, the cnidom of anthozoans differs fundamentally from those of the other classes, bespeaking an essential dichotomy that is apparent also in life history and additional features. Knowledge of how function correlates with structure is particularly critical to use of cnidae in phylogenetic analysis, where it might provide perspective on which morphological features are mechanically necessary, which are responsive to physical or biotic environment, and which are indicative of phylogeny/ancestry. The ideal would be, as with any taxonomic character, if cnidae were stable intraspecifically, but sufficiently changeable through evolutionary time to reflect phylogeny.

## VII. CONCLUSIONS

Why is it that knowledge about size and distribution of cnidae is so much more complete in the Actiniaria and Corallimorpharia than in any other cnidarian order? Is such analysis inappropriate for members of the other three cnidarian classes? In 1972, Calder (page 213) could state: "Nematocysts . . . of most scyphozoan species have yet to be examined and described," believing that "nematocyst differences may be of value in resolving certain taxonomic problems in the Scyphozoa, despite the impoverished cnidome of the class." In fact, variety in cnidom may be irrelevant to the issue, or, counterintuitively, promote the use of cnidae in taxonomy.

Werner (1965) correlated richness of cnidom with complexity in other morphological features, as well as behavior. Thus, hydrozoans exhibit sufficient diversity that sound taxonomic determinations can be made without considering cnidae, although Mackie and Mackie (1963, page 63) believed it "probable that each species [of hydrozoan] could, if necessary, be distinguished solely by its nematocysts." Analogously, cnidae have been almost totally ignored for taxonomic purposes in scleractinians, where skeletal characters are of paramount importance for a variety of historical, practical, and biological reasons (Lang, 1984). In her appeal for use of a wider array of features, particularly those of the living animal, Lang (1984) cited only three papers using cnidom for coral taxonomy. Thomason showed in his poster for this meeting that the nematocysts of some congeneric species of corals do differ. Anthozoans that form colonies and skeletal elements, such as scleractinians and octocorals, manifestly possess a wealth of morphological features relative to the non-skeletalized, solitary sea anemones, whose body plan is uniform, with few exceptions that are thereby easily identifiable.

This dearth of features, combined with relative taxonomic richness (about 800 species; Carlgren, 1949), has made anemone systematists desperate for characters. An appealing attribute of cnidae is their being one of the only measurable morphological entities in animals whose very dimensions are highly changeable. Carlgren's practice of using cnidae measurements as a character has become so conventional that taxonomists must, by dint of precedent and peer pressure, provide such data to compare with those published for previously described species. Whether such data are truly taxonomically useful, especially in light of the considerable time necessary to gather the figures, has hardly been questioned. Thus, actinian taxonomists may have ascribed such prominence to cnidae by default rather than through any demonstrated utility.

## ACKNOWLEDGMENTS

I thank Tim Blanchard for gathering nematocyst measurements, Barbara Weitbrecht for discussions and bibliographic assistance, Lisbeth Francis and George Mackie for lively conversations on this topic, and Kerry Irons and R. W. Buddemeier for critical reading of an earlier version of this manuscript.

## REFERENCES


Adshead, P.C., G.O. Mackie, and P. Paetkau. 1963. On the hydras of Alberta and the Northwest Territories. Bull. Natl. Mus. Canada 199:1-13.

Bigger, C.H. 1976. The acrorhagial response in *Anthopleura krebsi*: intraspecific and interspecific recognition. *In* G.O. Mackie (ed.), *Coelenterate Biology and Behavior*, pp. 127-136. Plenum Press, New York.

Bigger, C.H. 1982. The cellular basis of the aggressive acrorhagial response of sea anemones. J. Morph. 173:259-278.

Calder, D.R. 1972. Nematocysts of the medusa stage of *Rhopilema verrilli* (Scyphozoa, Rhizostomeae). Trans. Amer. Micros. Soc. 91(2):213-216.

Calder, D.R. 1977. Nematocysts of the ephyra stages of *Aurelia*, *Chrysaora*, *Cyanea*, and *Rhopilema* (Cnidaria, Scyphozoa). Trans. Amer. Micros. Soc. 96(1):13-19.

Campbell, R.D. 1983. Identifying hydra species. *In* H.M. Lenhoff (ed.), *Hydra: Research Methods*, pp. 19-28. Plenum Press, New York and London.

Carlgren, O. 1900. Ostafrikanische Actinien. Mitt. Naturh. Mus. Hamburg 17:1-124.

Carlgren, O. 1928. Actiniaria der deutschen Tiefsee-Expedition. Wissen. Ergeb. Deutschen Tiefsee-Exped. 22(4):125-266.

Carlgren, O. 1938. South African Actiniaria and Zoantharia. K. Svenska Vetenskapsakad. Handl., ser. 3, 17(3):1-148.

Carlgren, O. 1940. A contribution to the knowledge of the structure and distribution of the cnidae in the Anthozoa. K. Fysiogr. Sallsk. Handl. n.f. 51(3):1-62.

Carlgren, O. 1945. Further contributions to the knowledge of the cnidom in the Anthozoa especially in the Actiniaria. K. Fysiogr. Sallsk. Handl. n.f. 56(9):1-24.

Carlgren, O. 1949. A survey of the Ptychodactiaria, Corallimorpharia and Actiniaria. K. Svenska Vetenskapsakad. Handl., ser. 4, 1(1):1-121.

Chia, F.-S. and R. Koss. 1983. On the fine structure of the nematocysts in the planula larva of a sea anemone, *Anthopleura elegantissima*. Biol. Bull. (Taiwan Natl. Norm. Univ.) 12:12-24.

Conklin, E.J., C.H. Bigger, and R.N. Mariscal. 1977. The formation and taxonomic status of the microbasic q-mastigophore nematocyst of sea anemones. Biol. Bull. 152:159-168.

Cutress, C.E. 1955. An interpretation of the structure and distribution of cnidae in Anthozoa. Syst. Zool. 4(3):120-137.

Dunn, D.F. 1981. The clownfish sea anemones: Stichodactylidae (Coelenterata: Actiniaria) and other sea anemones symbiotic with pomacentrid fishes. Trans. Amer. Phil. Soc. 71(1):1-115.

Dunn, D.F. 1982. *Paraphelliactis pabista*, a new species of hormathiid sea anemone from abyssal northeastern Pacific waters (Coelenterata: Actiniaria). Syesis 15:51-56.

Dunn, D.F. and G.J. Bakus. 1977. Redescription and ecology of *Liponema brevicornis* (McMurrich, 1893), with definition of the family Liponematidae (Coelenterata, Actiniaria). Astarte 10:77-85.

Fautin, D.G. and F.-S. Chia. 1986. Revision of sea anemone genus *Epiactis* (Coelenterata: Actiniaria) on the Pacific coast of North America, with descriptions of two new brooding species. Can. J. Zool. 64:1665-1674.

Francis, L. 1973. Intraspecific aggression and its effect on the distribution of *Anthopleura elegantissima* and some related sea anemones. Biol. Bull. 144:73-92.

Francis, L. 1976. Social organization within clones of the sea anemone *Anthopleura elegantissima*. Biol. Bull. 150:361-376.

Hand, C. 1955*a*. The sea anemones of central California part I. The corallimorpharian and athenarian anemones. Wasmann J. Biol. 12(3) (for 1954): 345-375.

Hand, C. 1955*b*. The sea anemones of central California part II. The endomyarian and mesomyarian anemones. Wasmann J. Biol. 13(1):37-99.

Hand, C. 1956. The sea anemones of central California part III. The acontiarian anemones. Wasmann J. Biol. 13(2)(for 1955):189-251.

Hand, C. 1961. Present state of nematocyst research: types, structure and function. *In* H.M. Lenhoff and W.F. Loomis (eds.), *The Biology of Hydra and of Some Other Coelenterates: 1961*, pp. 187-202. Univ. Miami Press, Coral Gables, Florida.

Hansen, P.A. and B.W. Halstead. 1971. The venomous sea anemone *Actinodendron plumosum* Haddon of South Vietnam. Micronesica 7(1-2):123-136.

Jaworski, E. 1938. Untersuchungen über Rassenbildung bei Anthozoen. Thalassia 3(1):1-57.

Lang, J.C. 1984. Whatever works: the variable importance of skeletal and of non-skeletal characters in scleractinian taxonomy. Palaeont. Amer. 54(Proc. 4th Intl. Symp. Fossil Cnidaria):18-44.

Mackie, G.O. and G.V. Mackie. 1963. Systematic and biological notes on living hydromedusae from Puget Sound. Bull. Natl. Mus. Canada 199: 63-83.

Mariscal, R.N. 1974. Nematocysts. *In* L. Muscatine and H.M. Lenhoff (eds.), *Coelenterate Biology: Reviews and New Perspectives*, pp. 129-178 (chap. 3). Academic Press, New York and other cities.

Mariscal, R.N. 1984. Cnidaria: cnidae. *In* J. Bereiter-Hahn, A.G. Matoltsy, and K.S. Richards (eds.), *Biology of the Integument*, Vol. 1 (invertebrates), pp. 57-68 (chap. 6). Springer-Verlag, Berlin and other cities.

Picken, L.E.R. and R.J. Skaer. 1966. A review of researches on nematocysts. *In* W.J. Rees (ed.), *The Cnidaria and their Evolution* (Symp. Zool. Soc. London 16), pp. 19-50. Academic Press, London and New York.

Purcell, J.E. 1977. Aggressive function and induced development of catch tentacles in the sea anemone *Metridium senile* (Coelenterata, Actiniaria). Biol. Bull. 153:355-368.

Riemann-Zürneck, K. 1978. Tiefsee-Aktinien der Familie Actinoscyphiidae aus dem Nordatlantik (Actiniaria, Mesomyaria). Zool. Scripta 7:145-153.

Rifkin, J. and R. Endean. 1983. The structure and function of the nematocysts of *Chironex fleckeri* Southcott, 1956. Cell Tissue Res. 233:563-577.

Schmidt, G.H. 1982. Replacement of discharged cnidae in the tentacles of *Anemonia sulcata*. J. Mar. Biol. Assn. U. K. 62:685-691.

Schmidt, H. 1972*a*. Die Nesselkapseln der Anthozoen und ihre Bedeutung für die phylogenetische Systematik. Helgo. Meeresunters. 23:422-458.

Schmidt, H. 1972*b*. Prodromus zu einer Monographie der mediterranen Aktinien. Zoologica 121:1-146.

Sebens, K.P. and G. Laakso. 1978. The genus *Tealia* (Anthozoa: Actiniaria) in the waters of the San Juan Archipelago and the Olympic Peninsula. Wasmann J. Biol. 35(2)(for 1977):152-168.

Stephenson, T.A. 1920. On the classification of Actiniaria. Part I. Forms with acontia and forms with a mesogloeal sphincter. Quart. J. Micros. Sci. n.s. 64(256):425-574.

Stephenson, T.A. 1921. On the classification of Actiniaria. Part II. Consideration of the whole group and its relationships, with special reference to forms not treated in Part I. Quart. J. Micros. Sci. n.s. 65(260):493-576.

Stephenson, T.A. 1922. On the classification of Actiniaria. Part III. Definitions connected with the forms dealt with in Part II. Quart. J. Micros. Sci. n.s. 66(262):247-319.

Stephenson, T.A. 1928. *The British Sea Anemones*, vol. I. Ray Society, London.

Stephenson, T.A. 1929. On the nematocysts of sea anemones. J. Mar. Biol. Assn. U. K. 16(1):173-201.

Stephenson, T.A. 1935. *The British Sea Anemones*, vol. II. Ray Society, London.

Thomason, J.C. and B.E. Brown. 1986. The cnidom: an index of aggressive proficiency in scleractinian corals. Coral Reefs 5(2):93-101.

Watson, G.M. and R.N. Mariscal. 1983. The development of a sea anemone tentacle specialized for aggression: morphogenesis and regression of the catch tentacle of *Haliplanella luciae* (Cnidaria, Anthozoa). Biol. Bull. 164(3):506-517.

Weill, R. 1934a. Contribution a l'étude des cnidaires et de leurs nématocystes. I. Recherches sur les nématocystes. Trav. Sta. Zool. Wimereux 10:1-347.

Weill, R. 1934b. Contribution a l'étude des cnidaires et de leurs nématocystes. II. Valeur taxonomique du cnidome. Trav. Sta. Zool. Wimereux 11:348-700.

Werner, B. 1965. Die Nesselkapseln der Cnidaria, mit besonderer Berücksichtigung der Hydroida I. Klassifikation und Bedeutung für die Systematik und Evolution. Helgol. Wissen. Meeres. 12:1-39.

# 26

# NEMATOCYSTS AS TAXONOMIC CRITERIA WITHIN THE FAMILY CAMPANULARIIDAE, HYDROZOA

Carina Östman

Department of Zoology
Uppsala University, Box 561,
S-751 22 Uppsala, Sweden.

## SUMMARY

Nematocysts of the hydroid family Campanulariidae were examined with light and scanning electron microscopy, and taxonomically useful differences were shown. Distinctions in the cnidoms of *Gonothyraea hyalina* and *Gonothyraea loveni* for example, strengthen the validity of the two hydroids. SEM investigations revealed that the previously assigned microbasic mastigophores of Campanulariidae lack the enlarged basal part of the tubule characteristic of these nematocysts. A new sub-category of nematocysts, the 'pseudo-microbasic b-mastigophores', is proposed. The presence of a morphologically very similar nematocyst type within the pseudo-microbasic b-mastigophores in all species of the sub-families Campanulariinae and Clytiinae examined so far, and in some species of the sub-family Obeliinae might indicate affinity between the three subfamilies and the genera assigned to them. Detailed descriptions of the pseudo-microbasic b-mastigophores of *Campanularia hincksi* and *Rhizocaulus verticillatus* are given.

## 1. INTRODUCTION

In the thecate hydroid family Campanulariidae species definition solely on the basis of differences in skeletal morphology has often caused taxonomic problems. Hence it is useful to establish additional criteria to define species limits, and the use of nematocysts for hydrozoan classification has lately been emphasized (Östman, 1979 and subsequent papers; Picard, summaries in Bouillon, 1984, 1985; Watson, 1985; Gravier-Bonnet, 1987).

This study is part of a taxonomic investigation of campanulariid hydroids based mainly on light and scanning electron microscopic (SEM) observations on nematocysts. In the Campanulariidae two main categories of nematocysts have been identified, the microbasic b-mastigophores and the atrichous isorhizas. The classification system adopted is that of Weill (1934), somewhat modified by Carlgren (1940), Cutress (1955) and Mariscal (1974). SEM investigations of campanulariid nematocysts identified as microbasic mastigophores (Weill, 1934) or as microbasic b-mastigophores (Itô & Inoue, 1962; Cornelius, 1975; and Kubota, 1978) have shown, however, that these nematocysts do not fulfil the criteria for the category of microbasic mastigophores. Hence a new sub-category, 'pseudo-microbasic b-mastigophores' is proposed here (following discussion with Prof. Jean Bouillon). In previous papers (Östman, *op. cit.*) I have referred to the 'microbasic b-mastigophores' of campanulariids as b-rhabdoids following the terminology of Schmidt (1974), pending revision of nematocyst nomenclature.

A small pseudo-microbasic b-mastigophore nematocyst with spindle-shaped capsule and axially discharged tubule has been identified in all campanulariid hydroids examined (Figs 1-3). Previously it had been described and referred to as 'A-type' (Östman, *op. cit.*). In all other campanulariid hydroids examined (except in *Gonothyraea loveni* (Allman, 1859)), one or more additional pseudo-microbasic b-mastigophore or isorhizous nematocysts were found. As far as known,some are species-specific, whereas others are characteristic of genera or sub-families (Östman, 1987). These specific nematocysts occurred throughout the colonies except on the tentacles of the hydranth. Usually, they were most abundant in the gonangia, stolons and hydrocauli. In most species they also formed a ring around the hypostome below the tentacle bases (Figs 4-7). Detailed descriptions of the specific pseudo-microbasic b-mastigophores of *Campanularia hincksi* Alder, 1856, and *Rhizocaulus verticillatus* (Linnaeus, 1785) are given.

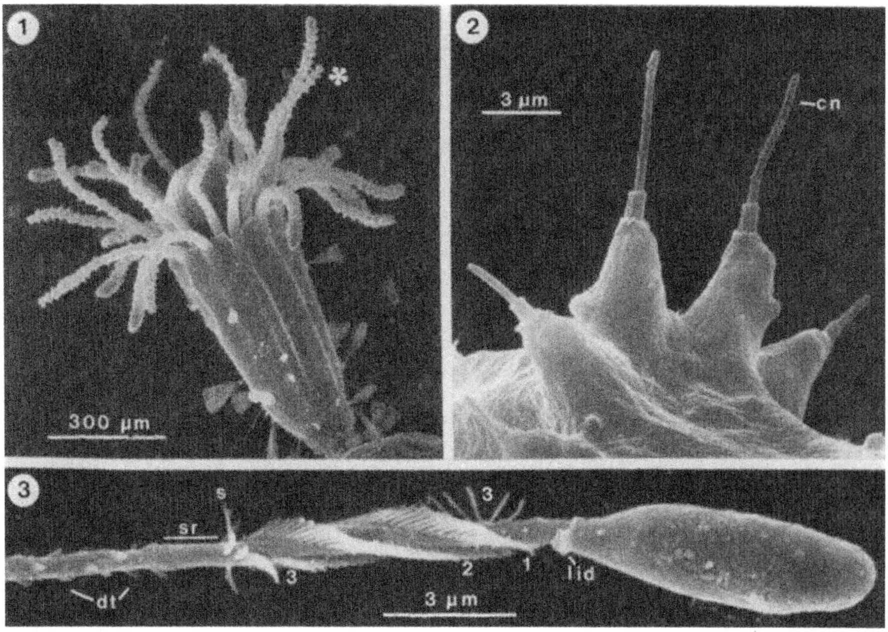


Figs 1-3. Scanning electron photomicrographs. -1. Hydranth of *Gonothyraea loveni*. -2. A higher magnification of the area represented by the star in figure 1 showing cnidocils (*cn*) from pseudo-microbasic b-mastigophore 'A-type' nematocysts. -3. Discharged pseudo-microbasic b-mastigophore 'A-type' nematocyst.

*dt*, distal tubule; *s*, some proximal spines no adhered to the others; *sr*, spineless region; *1, 2, 3*, separate strands of long, paired spines forming the right handed triple helix in the proximal armature.

The relation between the deeper water and northern boreal *Gonothyraea hyalina* Hincks, 1866, and the littoral and more southerly *G. loveni* has been discussed. Many hydroid workers have regarded *G. loveni* and *G. hyalina* conspecific under the name *G. loveni*, and considered the morphological distinctions to be phenotypic (Hummelinck, 1936; Naumov, 1960, 1969; Cornelius, 1982; for others see Calder, 1970). The present investigation, however, has suggested strongly that the cnidoms of *G. hyalina* and *G. loveni* are distinct, strengthening arguments for the validity of the two species.

Observations reported here further strengthen the possible phylogenetic value of the nematocysts discussed in previous papers (Östman, 1987; Östman, *et al.*, 1987).

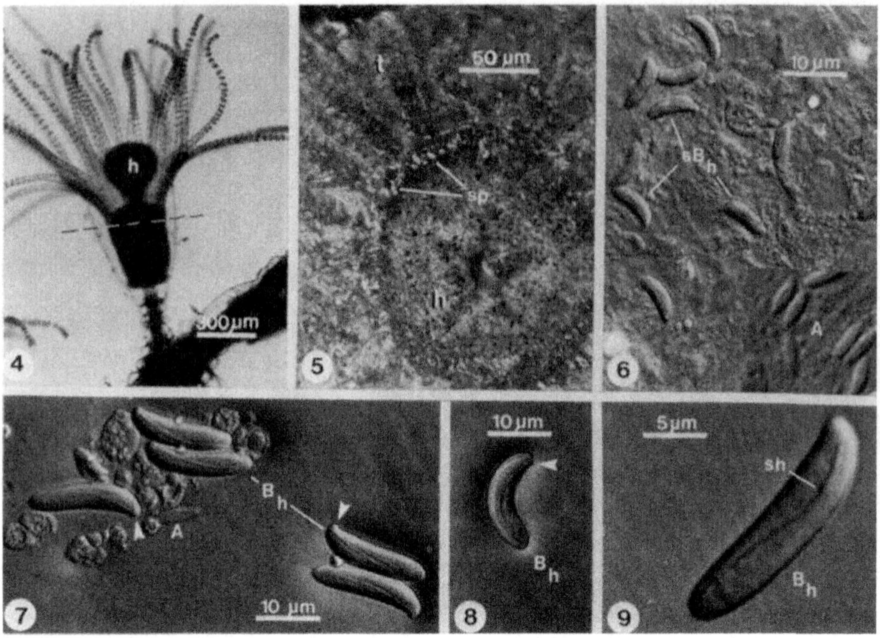


Fig. 4. Hydranth of *Laomedea flexuosa.* The broken line indicates where to cut to obtain a squash preparation of the upper part of the hydranth. Fig. 5. Upper part of hydranth of *Clytia hemisphaerica* spread out on a slide showing specific pseudo-microbasic b-mastigophore nematocysts *(sp)* forming a ring around the hypostome (*h*). Figs 6-9. Interference contrast photomicrographs of pseudo-microbasic b-mastigophores. -6. 's$B_h$-type' nematocysts from a hydranth of *Gonothyraea hyalina.* 7-9. '$B_h$-type' nematocysts of *Campanularia hincksi.*

*A*, 'A-type' nematocysts; $B_h$, '$B_h$-type' nematocysts; *h*, hypostome; $sB_h$, '$sB_h$-type' nematocysts; *sh*, shaft. *sp*, specific nematocysts; *t*, tentacles.

## 2. MATERIALS AND METHODS

Campanulariid hydroids were collected from Norway near Bergen, 60°23' N, 5°20' E, and near Tromsø, 69°42' N, 19°00' E; in the Gullmar Fjord, Bohuslän, on the west coast of Sweden, 58°16' N, 11°26' E; and in England at Plymouth, Devon, 50°23' N, 4°10' W, and Hunstanton, Norfolk, 52°57' N 0°30' E. The hydroids were either collected by hand along the shore, or gathered by skin diving or dredging. The nematocysts were examined and photographed under interference contrast microscopy. Measurements of undischarged nematocysts capsules from mostly fresh material were made to an accuracy of 5 $\mu m$. Whenever possible, at least ten length and width measurements from different parts of different colonies were made on each nematocyst type of each species. Squash preparations for nematocyst studies using both light and scanning electron microscope were made as described elsewhere (Östman, 1987). Detailed nematocyst descriptions were made from the following species:

(1) *Campanularia hincksi*, on *Lophelia prolifera* and *Eunicella verrucosa*, 100-125 m depth, from Brattholmen reef, near Bergen, Norway, 10, 20 June, 1983; and on stones and shells of *Modiolus modiolus*, 20-25 m depth, at Vattlestraumen, near Bergen, Norway, 7 June, 1984; on shells, 20-30 m depth, at Löken in the Gullmar Fjord, Bohuslän, Sweden, 14 July, 1983.

(2) *Rhizocaulus verticillatus*, substrate not recorded, 70 m depth, 20 km south of Plymouth, England, 26 July, 1983; from intertidal pools (after stranding) at Holme-next-the-Sea, near Hunstanton, Norfolk, England, 16-25 October, 1983; on shells and stones, 20-30 m depth, at Löken, in the Gullmar Fjord, Bohuslän, Sweden, 11 July, 1984, 26 July, 1985, 14 July, 1986; on old *Chlamys islandica* shells, 25-60 m depth, Kval Sundet, near Tromsø, Norway, 19 June, 2 July, 1984.

(3) *Gonothyraea hyalina*, on shells of *Modiolus modiolus*, 20-25 m depth, at Vattlestraumen, near Bergen, Norway, 20 June, 1983, and 7 June, 1984.

(4) *Gonothyraea loveni*, intertidal estuary of Ådlands-straumen, near Bergen, Norway, 9 June, 1984; on *Fucus vesiculosus* and *Ascophyllum nodosum*, 0-0.5 m depth, Långgap, in the Gullmar Fjord, Sweden, June, 1977, 1978, 1983, 1984.

## 3. RESULTS

### A. *General observations*

The small nematocyst, identical with the previously described 'microbasic b-mastigophore' or b-rhabdoid, 'A-type' (Östman 1979*a*, *b*, 1982), was abundant in all species examined (Table I, Fig. 3). It is here referred to as a pseudo-microbasic b-mastigophore. Nematocysts morphologically similar to the previously recorded 'microbasic b-mastigophore' or b-rhabdoid, 'B-type' (Östman, 1979, 1983) were also present in *C. hincksi* and *R. verticillatus*. These nematocysts are here termed pseudo-microbasic b-mastigophores, '$B_h$-'and '$B_v$-types' (*h* and *v* represent the first letter in the species names *hincksi* and *verticillatus*). In *G. hyalina*, but not in *G. loveni*, a similar but smaller pseudo-microbasic b-mastigophore was found, referred to as '$sB_h$-type' (*s* stands for small, *h* for *hyalina*). These three

TABLE I. Measurements in μm of undischarged pseudo-microbasic b-mastigophore, 'A-type' nematocysts of campanulariid hydroids from the north east Atlantic

| | No. of colonies | Length range | Length $\bar{x}$±S D (n) | Breadth range | Breadth $\bar{x}$±S D (n) | Range of length:breadth ratio |
|---|---|---|---|---|---|---|
| *Campanularia hincksi* (Be,N) | | | | | | |
| from tentacles | 6 | 7.0-9.0 | 8.0±0.4(71) | 2.0-2.5 | 2.0±0.3(71) | 2.8-4.3 |
| from hydrocaulus | 6 | 7.0-8.5 | 7.5±0.5(63) | 2.0-3.0 | 2.0±0.3(63) | 2.8-4.0 |
| *Gonothyraea hyalina* (Be,N) | | | | | | |
| from tentacles | 4 | 6.0-8.5 | 7.0±0.5(130) | 2.0-2.5 | 2.0±0.2(130) | 2.8-4.3 |
| from hydrocaulus | 4 | 6.0-8.5 | 7.0±0.4(65) | 2.0-2.5 | 2.0±0.2(65) | 2.8-4.0 |
| *Rhizocaulus verticillatus*[a] (Tr,N) | | | | | | |
| from hydranth | 4 | 6.0-8.0 | 7.5±0.5(33) | 1.5-2.5 | 2.0±0.2(33) | 2.8-4.7 |
| from hydrocaulus | 4 | 6.0-8.0 | 7.0±0.4(65) | 1.5-2.5 | 2.0±0.2(65) | 2.6-4.7 |
| *Rhizocaulus verticillatus* (Gu,S) | | | | | | |
| from hydranth | 6 | 7.0-8.5 | 7.5±0.6(85) | 1.5-2.5 | 2.5±0.4(85) | 2.8-5.6 |
| from stolon | 6 | 6.0-8.0 | 7.0±0.7(69) | 2.0-2.5 | 2.0±0.3(69) | 2.8-4.8 |
| *Rhizocaulus verticillatus* (No,E) | | | | | | |
| from hydranth | 5 | 6.0-10.0 | 7.5±0.7(72) | 2.0-2.5 | 2.1±0.2(72) | 2.8-3.5 |
| from hydrocaulus | 5 | 6.0-8.0 | 7.0±0.6(49) | 2.0-2.5 | 2.0±0.2(49) | 3.0-3.5 |
| *Rhizocaulus verticillatus* (Pl,E) | | | | | | |
| from hydranth | 1 | 6.0-7.0 | 6.5±0.3(22) | 2.0-2.5 | 2.5±0.3(22) | 2.6-3.5 |
| from stolon | 1 | 6.0-7.0 | 6.7±0.4(11) | 2.0-2.5 | 2.0±0.2(11) | 2.6-3.5 |

$\bar{x}$ = mean value; S D = standard deviation; n = No. of nematocysts; Be,N = Bergen, Norway; Tr,N = Tromsø, Norway; Gu,S = The Gullmar Fjord, Sweden; No,E = Norfolk, England; Pl,E = Plymouth, England.

[a] Preserved in 70% ethanol.

TABLE II. Measurements in μm of undischarged pseudo-microbasic b-mastigophore, 'B-type' nematocysts of campanulariid hydroids from the north east Atlantic

| | No. of colonies | Length range | $\bar{x}$±S D (n) | Breadth range | $\bar{x}$±S D (n) | range of length:breadth ratio |
|---|---|---|---|---|---|---|
| *Campanularia hincksi* (Be,N) | | | | | | |
| from hydranth | 5 | 15.0-19.0 | 17.0±0.9(56) | 4.0-5.0 | 4.5±0.3(56) | 3.5-4.6 |
| from hydrocaulus | 10 | 15.0-22.0 | 17.0±0.7(76) | 3.5-5.0 | 4.0±0.3(76) | 3.6-4.9 |
| from male gonangium | 2 | 16.0-18.5 | 17.5±0.6(79) | 4.0-4.5 | 4.0±0.1(79) | 3.5-4.6 |
| from female gonangium | 1 | 15.0-17.0 | 16.0±0.7(20) | 3.5-4.5 | 4.0±0.3(20) | 3.6-4.6 |
| *Campanularia hincksi* (Gu,S) | | | | | | |
| from hydranth | 1 | 14.5-18.0 | 17.0±0.8(16) | 4.0-4.0 | 4.0±0.0(16) | 3.5-4.3 |
| from hydrocaulus | 1 | 14.5-18.0 | 17.0±0.7(16) | 4.0-4.5 | 4.0±0.2(16) | 3.5-4.3 |
| *Gonothyraea hyalina* (Be,N) | | | | | | |
| from hydranth | 4 | 8.0-10.0 | 8.5±0.5(123) | 2.0-3.0 | 2.5±0.2(121) | 2.8-5.0 |
| from hydrocaulus | 4 | 7.0-9.0 | 8.5±0.5(75) | 2.0-3.0 | 2.5±0.3(75) | 2.8-4.5 |
| *Rhizocaulus verticillatus*[a] (Tr,N) | | | | | | |
| from hydranth | 4 | 14.0-17.0 | 16.0±0.7(76) | 3.5-5.0 | 4.5±0.3(76) | 3.3-4.6 |
| from hydrocaulus | 4 | 14.0-17.5 | 16.0±0.7(100) | 3.5-5.0 | 4.0±0.3(100) | 3.3-4.6 |
| from gonangium | 1 | 15.5-17.5 | 16.0±0.5(23) | 3.5-5.0 | 4.0±0.3(23) | 3.3-4.6 |
| *Rhizocaulus verticillatus* (Gu,S) | | | | | | |
| from hydranth | 7 | 12.0-17.5 | 15.5±0.7(106) | 3.0-5.0 | 4.0±0.6(106) | 3.1-5.0 |
| from stolon | 7 | 16.5-18.0 | 16.5±1.2(110) | 3.0-5.0 | 4.5±0.5(110) | 3.5-5.0 |
| from gonangium | 1 | 14.5-18.0 | 16.0±0.8(23) | 3.0-4.5 | 4.0±0.3(23) | 3.3-4.0 |
| *Rhizocaulus verticillatus* (No,E) | | | | | | |
| from hydranth | 5 | 12.5-16.0 | 14.5±0.8(71) | 3.0-5.0 | 4.0±0.3(71) | 3.1-4.3 |
| from hydrocaulus or stolon | 5 | 12.0-16.0 | 14.0±1.2(78) | 3.0-5.0 | 4.0±0.5(78) | 3.1-5.0 |
| *Rhizocaulus verticillatus* (Pl,E) | | | | | | |
| from colony | 2 | 13.0-16.0 | 15.0±0.8(22) | 3.0-4.5 | 4.0±0.4(22) | 3.1-4.7 |
| from hydranth | 2 | 13.5-15.5 | 14.5±0.7(28) | 3.0-4.5 | 3.5±0.4(28) | 3.3-4.7 |
| from hydrocaulus or stolon | 2 | 12.5-16.5 | 14.5±1.0(38) | 3.0-4.0 | 3.5±0.4(38) | 3.2-4.8 |

$\bar{x}$ = mean value; S D = standard deviation; n = No. of nematocysts; Be,N = Bergen, Norway; Gu,S = The Gullmar Fjord, Sweden; Tr,N = Tromsø, Norway; No,E = Norfolk, England; Pl,E = Plymouth, England.

[a] Preseved in 70% ethanol

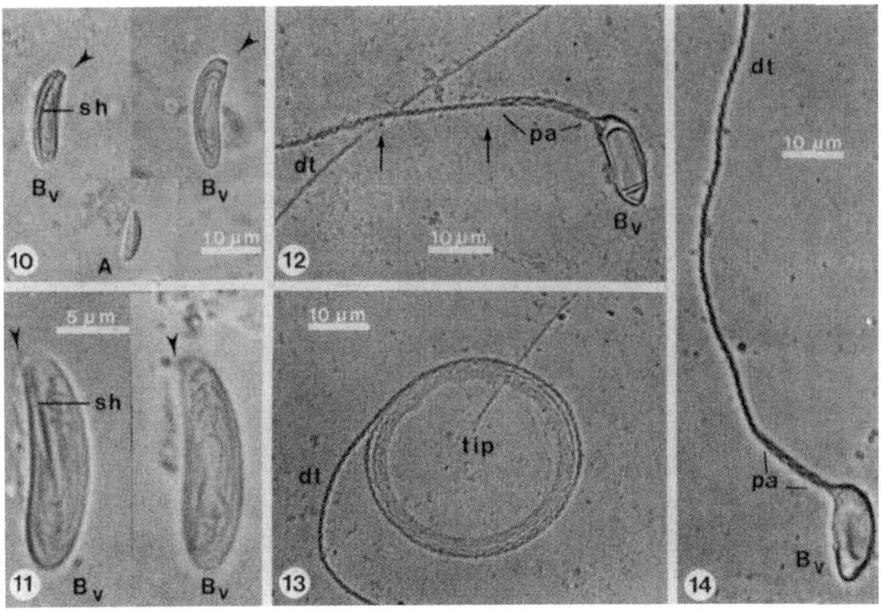


Figs 10-14. Interference contrast photomicrographs of the pseudo-microbasic b-mastigophore '$B_V$-type' nematocyst of *Rhizocaulus verticillatus*. -10, 11. Undischarged nematocysts ($B_V$) ; note the shaft (*sh*) and the pattern of the longitudinal coils inside the capsules when differently focused; *arrowheads* mark the protruding tips. Inset (Fig. 10): 'A-type' nematocyst (*A*) for comparison. -12, 14. Discharged nematocysts; note the prominent, proximal armature (*pa*) and the helical pattern on the distal tubule (*dt*). Next to the proximal armature (Fig. 12, between *arrows*) the diameter of the tubule is narrower than that of the region slightly distally. -13. Coiled distal tubule (*dt*); note the helical pattern and compare the diameter with that of the tip from another '$B_V$-type' nematocyst.

'B-type' nematocysts occurred throughout the colonies except on the tentacles.

1. The large, elongate, bean-shaped pseudo-microbasic b-mastigophores, the '$B_h$-' and '$B_V$-type' of *C. hincksi* and *R. verticillatus* (Figs 7-26).

a. Light microscopy (LM). The 'B-type' nematocysts quite outnumber the smaller 'A-type' in the gonangia, pedicel bases, stolons and in the polysiphonic stems of *R. verticillatus*. The capsules (length 12.0-22.0 $\mu m$, breadth 3.0-5.0 $\mu m$, Table, II) are mostly bean-shaped but in

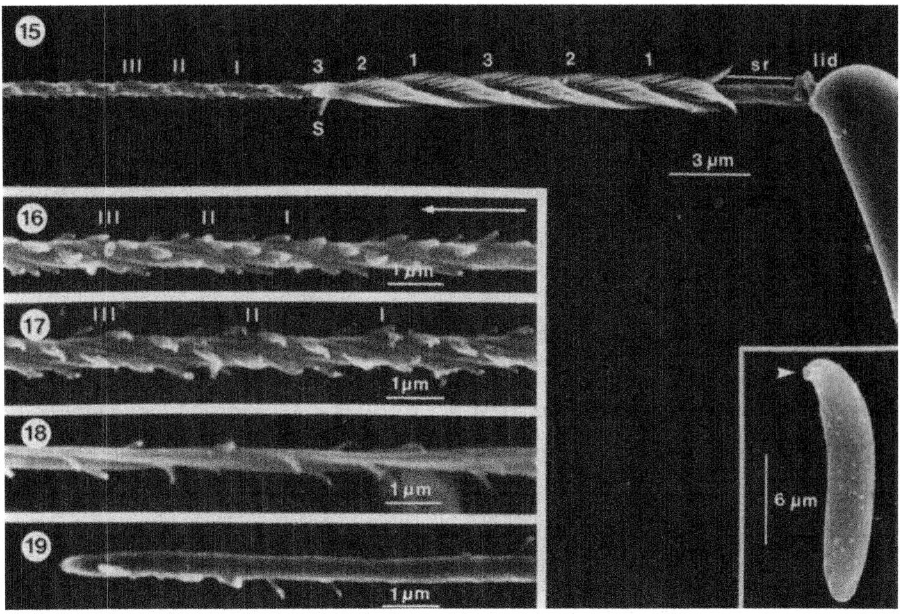


Figs 15-19. Scanning electron photomicrographs of the pseudo-microbasic b-mastigophore '$B_V$-type' nematocyst of *R. verticillatus*. -15. Discharged nematocyst with prominent, proximal armature. The tubule roughly forms a right angle with the long axis of the capsule. Inset: undischarged capsule; note the protruding tip (*arrowhead*). -16-19. Parts and tip of distal tubule. The shortest spines are next to the proximal armature (Fig. 16) and the longest spines are close to the tip (Figs 18, 19). The distances between the spines are increasing towards the end of the tubule (Figs 18, 19).

*arrow* points distally; short spineless region; *1, 2, 3*, three rows of long, paired spines; *I, II, III*, three rows of minute, paired spines on the distal tubule.

*C. hincksi* a few are markedly curved (Figs 7-11). The tubule is longitudinally coiled and its proximal part is a slightly S-shaped or bowed rod inside the capsules. When discharged the capsule diminished somewhat in size. The tubule everts from a small tip protruding from the concave side of the capsule at roughly a right angle with the long axis of the capsule (Fig. 12). The proximal armature of the everted tubule, with four to seven coils of spines, is the same length or slightly longer than the

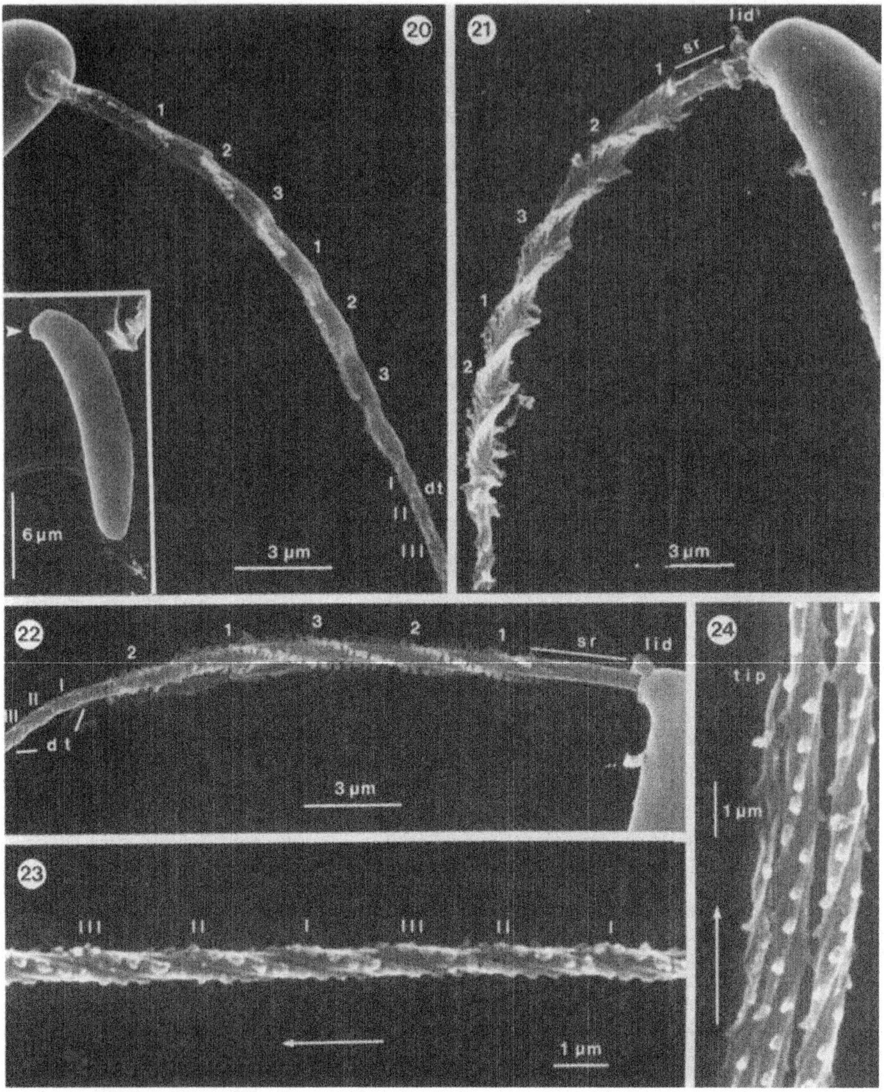


Figs 20-24. Scanning electron photomicrographs of the pseudo-microbasic b-mastigophore 'B$_h$-type' nematocyst of *Campanularia hincksi*. -20-22. Discharged nematocysts. The tubule roughly forms a right angle with the long axis of the capsule. In figures 20, 22 the proximal spines in the strands marked *1, 2, 3* are

capsule (Fig. 12). The S-shaped portion of the inverted tubule may correspond to the longer proximal armature. The distal tubule is wide and taper slightly and uniformly from base to tip. The tubule is often coiled with the tip innermost and a right-handed, helical pattern can be seen throughout its length (Figs 12-14). The length of the tubule is up to 510 $\mu m$ in *C. hincksi* and up to 560 $\mu m$ in *R. verticillatus.*

b. Scanning electron microscopy (SEM). The protruding tip, from which the tubule discharges, and the apical flaps ("lid") are shown in figure 15. The tubule tapers slightly, if at all, from base to tip. It has no distinct basal enlargement abruptly narrowing to the distal tubule (Figs 15-19). However, proximally the tubule is heavily armed with strands of close-set, paired spines forming a dextral triple helix of four to seven coils (Fig. 15). Generally the lengths of the spines vary from being around one and a half times to twice as long as the diameter of the tubule. In some of the '$B_h$-type' of *C. hincksi* the proximal spines are very short and less prominent (Figs 20-22). Close to the capsule and mostly distal to the proximal armature the tubule is spineless for a short region (*sr*, Figs 15, 21, 25). The remaining distal tubule is armed with three rows of minute, paired, hook-shaped spines forming a dextral triple helix (Figs 16-19). Also nematocysts, with tiny proximal spines, are similarly armed with these distal spines (Figs 22-24). The size of the distal spines and their pattern are almost uniform throughout the whole tubule. However, towards the tip the diameter of the tubule decreases markedly, and the size of the spines and their distance between them are slightly increased (Figs 18, 19).

---

shorter compared with those in figure 21. The spines on the distal tubule (*dt*) are however clearly seen. Inset (Fig. 20): undischarged capsule; note the protruding tip (*arrowhead*). -23. Part of distal tubule next to the proximal armature. -24. parts of coiled distal tubule including tip.

*Arrows*, point distally, *dt*, distal tubule; *sr*, short spineless region; *1, 2, 3*, three strands of long, paired spines in the proximal armature; *I, II, III*, three rows of minute, paired spines on the distal tubule.

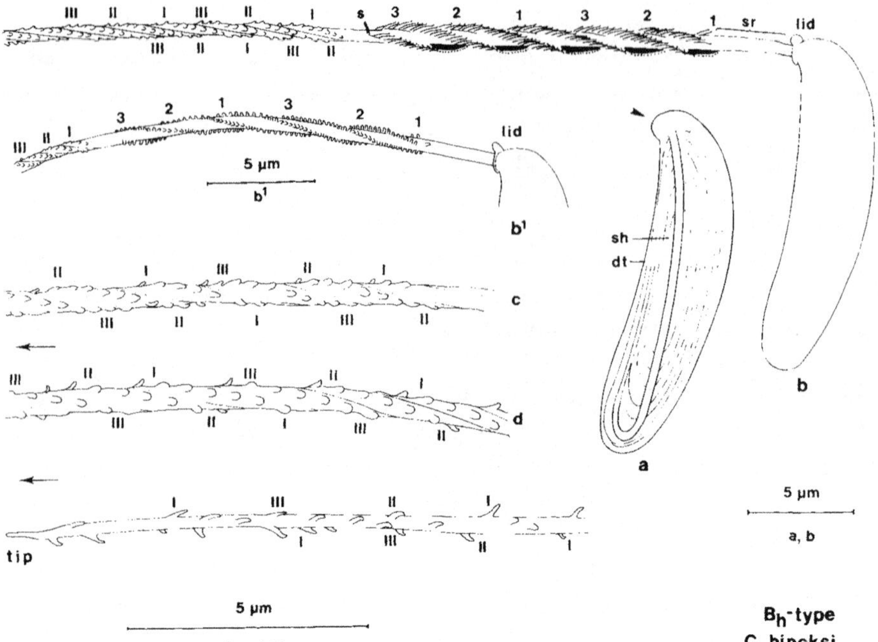


Fig. 25

Figs 25, 26. Summary of the pseudo-microbasic b-mastigophore 'B-type' nematocysts of *Campanularia hincksi* ($B_h$) and *Rhizocaulus verticillatus* ('$B_v$).

*a*, undischarged capsule; *arrows*, point distally; *arrowheads*, point to the protruding tip; *b, b'* discharged capsule showing the proximal armature, note the minute spines in the proximal armature in *b'* of *C. hincksi*; *c, d, e, f,* and *tip*, segments of the distal tubule or its tip; *dt*, coils of distal tubule; *s*, some proximal spines not adhered to the other; *sh*, shaft; *sr*, spineless region; *1, 2, 3*, separate strands of long, paired spines forming the right-handed triple helix in the proximal armature; *I, II, III*, rows of minute spines, mostly paired, hook-shaped, forming an extended, right-handed triple helix.

2. The small bean-shaped, pseudo-microbasic b-mastigophore, the '$sB_h$-type' of *G. hyalina.*

These nematocysts (length 7.0-10.0 $\mu m$, Table II) appear in the hydranths, hydrocauli and stolons, but are not as numerous as the 'A-type'. The morphology and pattern of the capsule, tubule and spines are similar to those of the larger '$B_h$-' and '$B_v$-types' of *C. hincksi* and *R. verticillatus* (Fig. 6).

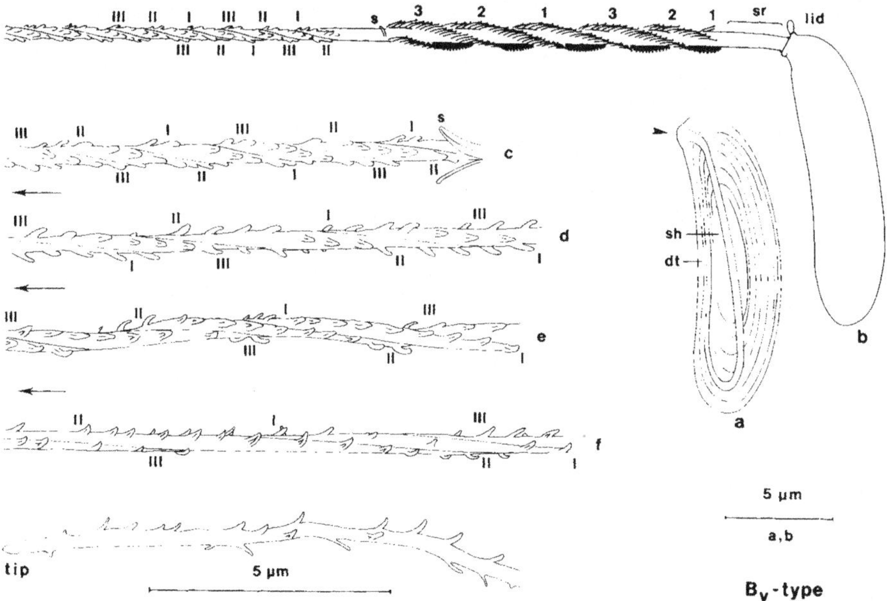


Fig. 26

$B_V$-type
R. verticillatus

## 4. DISCUSSION

SEM investigations of 'microbasic b-mastigophore' nematocysts within the Campanulariidae revealed that their tubules taper almost unnoticeably from base to tip (Figs 3, 15, 22). The distinct proximal enlargement of the tubules, characteristic of the microbasic mastigophores (Weill, 1934), is thus lacking here. The proximal armature of the tubule, however, forms an enlargement of spines clearly distinguishable from the remaining spined tubule (Figs 3, 25, 26). These nematocysts might thus better correspond to the category heterotrichous anisorhizas. But under LM it might be impossible to discern whether these nematocysts have an enlarged basal portion of the tubule or just well developed proximal spines, and to base LM nematocyst identification on ultrastructural characteristics seems impractical. Besides, observations made under both SEM and LM on probably immature 'microbasic b-mastigophores' within the Campanulariidae have revealed that until the spines have developed they sometimes have a basal enlargement of the tubule (Östman, 1982). Therefore, I propose a new sub-category termed pseudo-microbasic b-mastigophores for these and similar nematocysts. This sub-category can include all cnidae having a proximal armature that is distinct from the remaining spined tubule, but under SEM lacking a clearly visible enlarged basal portion of the tubule. This proposal, however, has the drawback that it affects the definition given by Weill (1934) of the main group heteronemes.

The large pseudo-microbasic b-mastigophore '$B_h$-' and '$B_v$-types' of *Campanularia hincksi* and *Rhizocaulus verticillatus* (Figs 25-26), were similar in morphology to the large 'microbasic b-mastigophore' or b-rhabdoid 'B-type' nematocyst earlier discribed by Östman (1979 and subsequent papers). The presence of the 'B-type' nematocysts in all species examined within the two sub-families Campanulariinae and Clytiinae may indicate an affinity between them (Östman, 1987; for the division of Campanulariidae into sub-families see Cornelius, 1982). In the third subfamily Obeliinae of Campanulariidae the 'B-type' nematocyst has been identified in both *Gonothyraea hyalina* and *Laomedea exigua* Gegenbaur, 1857 (in Cornelius & Östman, 1987) which might further strengthen the phylogenetic value of the 'B-type' nematocysts (see also Östman, 1987). Size differences between the 'B-type' nematocysts is probably of little phylogenetic value. Small 'B-type' nematocysts were identified in Mediterranean hydroids of Campanulariinae and Clytiinae (Östman, *et al.*, 1987). Slight size differences were also found between the English and the Scandinavian populations of the '$B_v$-type' nematocysts of *R. verticillatus* (Table II).

In the Obeliinae other types of pseudo-microbasic b-mastigophores, slightly different in capsule shape and spine pattern from the 'B-type', have been identified in *Laomedea flexuosa*, Alder, 1856, *Obelia longissima* (Pallas, 1766), and *Obelia bidentata*, Clarke, 1875 (Östman, 1982, 1987). In each genus within the Campanulariidae one or more specific pseudo-microbasic b-mastigophores have thus been identified, strengthening the affinity between the three subfamilies of Campanulariidae. A similar pattern for the isorhizous nematocysts recognized within the Campanulariidae has not been found (Östman, 1979*b*, 1982, 1987).

Further, the small pseudo-microbasic b-mastigophore '$sB_h$-type' of *G. hyalina* was not identified in *G. loveni*. Observed cnidom differences between *G. hyalina* and *G. loveni* strengthen the arguments for the validity of the two hydroids.

## 5. ACKNOWLEDGEMENTS

I thank Dr. Paul F. S. Cornelius for help collecting most of the material and for discussing and criticizing the manuscript, Mrs. Vibeke Nilsson for drawing the figures and Mr. Gary Wife for assistance with the SEM work. The work was carried out at the Institute of Marine Biology, University of Bergen, Blomsterdalen, Espegrend, Norway, at the Field Station of the Norfolk Naturalists' Trust, Holme-next-the-Sea, near Hunstanton, Norfolk, England, and at the Kristineberg Marine Station Fiskebäckskil, near Lysekil, Sweden. Some of the material was collected by the r. v. *Sepia* of the Marine Biological Association of the United Kingdom, Plymouth, Devon, England. Financial support was obtained from the Royal Swedish Academy of Science, the Foundation of Lars Hjertas Minne and the Foundation of Magnus Bergvall.

## 6. REFERENCES

Bouillon, J. 1984. Révision de la famille des Phialuciidae (Kramp, 1955) (Leptomedusae, Hydrozoa, Cnidaria) avec un essai de classification des Thecatae - Leptomedusae. Indo-Malayan Zool. 1: 1-24.

Bouillon, J. 1985. Essai de classification des hydropolypes - hydroméduses (Hydrozoa - Cnidaria). Indo-Malayan Zool. 2: 29-243.

Calder, D. R. 1970. Thecate hydroids from the shelfwaters of northern Canada. J . Fish. Res. Bd, Canada 27: 1501-1547.

Carlgren, 0. 1940. A contribution to the knowledge of the structure and distribution of the cnidae in the Anthozoa. Acta Univ. lund. (2) 36 (3): 1-62.

Cornelius, P. F. S. 1975. The hydroid species of *Obelia* (Coelenterata, Hydrozoa: Campanulariidae), with notes on the medusa stage. Bull. Br. Mus. nat. Hist. (Zool.) 28: 249-293.

Cornelius, P. F. S. 1982. Hydroids and medusae of the family Campanulariidae recorded from the eastern North Atlantic, with a world synopsis of genera. Bull. Br. Mus. nat. Hist. (Zool.) 42: 37-148.

Cornelius, P. F. S. & Östman, C. 1987. Redescription of *Laomedea exigua* Sars, a hydroid new to Scandinavia, with comments on its nematocysts, life-cycle and feeding movements. Zool. Scr. 16: 1-8.

Cutress, C. E. 1955. An interpretation of the structure and distribution of cnidae in Anthozoa. Syst. Zool. 4: 120-137.

Gravier-Bonnet, N. 1987. Nematocysts as discriminators of thecate hydroids in light microscopy. *In* Bouillon, J., Boero., Cicogna, F., and Cornelius, P. F. S. (eds.), *Modern trends in systematics, ecology and evolution of hydroids and hydromedusae.* Oxford University Press, Oxford. (In press).

Hincks, T. 1866. On new British Hydroida. Ann. Mag. nat. Hist. (3) 18: 269-299.

Hummelinck, P. W. 1936. Hydropoliepen. *Flora Fauna Zuiderzee* (Suppl.): 41-64.

Itô, T. & Inoue, K. 1962. Systematic studies on the nematocysts of Cnidaria. I. Nematocysts of Gymnoblastea and Calyptoblastea. Mem. Ehime. Univ. (2, B) 4: 445-460.

Kubota, S. 1978. The life-history of *Clytia edwardsi* (Hydrozoa; Campanulariidae) in Hokkaido. Japan. J. Fac. Sci. Hokkaido. Univ. Ser. (Vl-Zool.). 21: 317-354.

Mariscal, R. N. 1974. Nematocysts. *In* Muscatine L. and Lenhoff, H. M. (eds.), *Coelenterate biology*, pp. 129-178. Academic Press, New York, San Francisco & London.

Naumov, D. V. 1960. Gidroidy i gidromedduzy morskikh, solonovatvodnykh i presnovodnykh basseinov S.S.S.R. Fauna S.S.S.R. 70: 1-626.

Naumov, D. V. 1969. Hydroids and hydromedusae of the U.S.S.R. Fauna S.S.S.R. 70: 1-660. Israel Program for Scientific Translations, cat. no. 5108.

Östman, C. 1979 a. Two types of nematocysts in Campanulariidae (Cnidaria, Hydrozoa ) studied by light and scanning electron microscopy. Zool. Scr. 8: 5-12.

Östman, C. 1979 b. Nematocysts in the *Phialidium* medusae of *Clytia hemisphaerica* (Hydrozoa, Campanulariidae) studied by light and scanning electron microscopy. Zoon, Uppsala 7: 125-142

Östman, C. 1982. Nematocysts and taxonomy in *Laomedea, Gonothyraea* and *Obelia* ( Hydrozoa, Campanulariidae). Zool. Scr. 11: 227-241.

Östman, C. 1983. Taxonomy of Scandinavian hydroids (Cnidaria, Campanulariidae): A study based on nematocyst morphology and isoenzymes. Acta Univ. upsaliensis 672: 1-22.

Östman, C. 1987, New Techniques and old problems in hydrozoan systematics. *In* Bouillon, J., Boero, F., Cicogna, F., and Cornelius, P. F. S. (eds.), *Modern trends in the systematics, ecology and evolution of hydroids and hydromedusae.* Oxford University Press, Oxford. (In press).

Östman, C., Piraino, S., and Roca, I., 1987. Nematocyst comparissons between some Mediterranean and Scandinavian campanulariids

(Cnidaria, Hydrozoa). *In* Bouillon, J., Boero, F., Cicogna, F., and Cornelius, P. F. S. (eds.), *Modern trends in the systematicics, ecology and evolution of hydroids and hydromedusae.* Oxford University Press, Oxford. (In press).

Schmidt, H. 1974. On evolution in the Anthozoa. *In* Great Barrier Reef Committe, 1974 (eds.), *Proceedings of the second international symposium on coral reefs*, 1: 533-560. Brisbane.

Watson, J. E. 1985. The genus *Eudendrium* (Hydrozoa: Hydroida) from Australia. Proc. R. Soc. Victoria 97: 179-221.

Weill, R. 1934. Contribution à l'étude des cnidaires et de luer nématocyst. II. Trav. Stn. zool. Wimereux 10 & 11. 1-701.

27

# FUNCTIONAL ANATOMY OF NEMATOCYTE INNERVATION IN BATTERY CELL COMPLEXES OF THE <u>HYDRA</u> TENTACLE

L.A. Hufnagel

Department of Microbiology
University of Rhode Island
Kingston, R.I. 02881

G. Kass-Simon

Department of Zoology
University of Rhode Island
Kingston, R.I. 02881

## SUMMARY

The ectodermal layer of the distal tentacle of hydra is composed exclusively of battery cell complexes, each consisting of a battery cell enveloping separately nested nematocytes and one or two neurons. To better understand the behavioral influence exerted on nematocytes, we have investigated the anatomy of the complex surrounding, and including, the basal ends of the nematocytes. In this region nematocytes, nerve cells and myonemes are in contact. Each nematocyte terminates proximally in one or more spatulate processes, which insert into cup-like, lateral extensions of the myonemes, with which they form desmosomal junctions. Nearby lies the cell body of a ganglion cell, which is in extensive contact with the cell body of a sensory cell. Close to desmosomal regions, broad neurites filled with dense-cored vesicles are common and the following types of putative synapses occur: (1) from ganglion cell bodies onto (a) sensory cell bodies, (b) nematocytes, (c) battery cell bodies, and (d) myonemes; and (2) from unknown neurites onto ganglion cell bodies. These observations support the idea that the ganglion cell modulates the functions of all cells of a single battery cell complex. We also present the first ultrastructural evidence that the ganglion cell of one battery cell complex may influence the functions of sensory cells of other battery cell complexes.

## I. INTRODUCTION

The ectoderm of the tentacles of the freshwater polyp, hydra, is constructed of many similar structural units, the battery cell complexes, which appear to function both autonomously and in concert with other, neighboring complexes (Hufnagel et al, 1985). Battery cell complexes each contain a cluster of nematocysts of three different types, the stenoteles, desmonemes and isorhizas. These dischargeable organelles (organoids), contained in nematocytes, respond to mechanical and chemical stimulation.

Discharge can be induced or inhibited by appropriate external stimulation. This control is selective for nematocyst type (Ewer, 1947, Smith et al, 1974; Ruch and Cook, 1964; Kass-Simon, this volume; 1986; Grosvenor, 1986).

Ultrastructural evidence suggests that nematocyst discharge is under nervous control: Lentz and Barrnett (1965) and Slautterback (1967) reported that nerve processes lie close to, or end on, nematocytes and Westfall et al. (1971) showed that neuro-nematocyte synapses occur in hydra.

In our previous ultrastructural analysis of battery cell complexes (Hufnagel et al, 1985), we noted that nerve cell bodies and nerve processes were most numerous in sections which also contained desmosomal junctions. Synapse-like structures also were particularly abundant in these regions. We are now extending these findings by focusing on these synapses especially those between the ganglion cell and nematocytes. Our observations are consistent with a role for the nervous system of hydra in influencing nematocyst discharge, and for the ganglion cell in modulating the function of all cells of the battery cell complex. We also report the first ultrastructural observation of a tract of neurites extending between and contacting adjacent battery cell complexes.

## II. MATERIALS AND METHODS

Young, Swiss *H. attenuata*, grown in modified BVC solution and fed *Artemia* nauplii, were starved for 24 hr and some were relaxed with menthol just prior to fixation. The methods used to fix, embed and section hydra were as described in detail previously (Hufnagel _et al_, 1985). Sequences of sections having prominent desmosomes and myonemal processes were photographed as montages, at sufficient magnification to resolve synapses. We use the term "synapse" to identify structures which closely resemble structures identified as synapses by Westfall (Westfall _et al_., 1971; Westfall and Kinnaman, (1984). It remains to be demonstrated that these structures are in fact specialized sites of chemical transmission.

## III. RESULTS

Within the battery cell complexes of the distal tentacle, nematocytes usually terminate basally in more than one process; some fit into cup-like lateral extensions of the myonemes, while others extend directly to the mesoglea (Hufnagel et al, 1985). We were particularly interested in the former because nerve cell bodies and neurites frequently were found nearby. Sequential sagittal sections cut roughly parallel to the tentacle axis were analyzed (Fig. 1). In the most basal sections each nematocyte process was surrounded exclusively by the myoneme and its extension. The association with the myonemal extension was by means of a desmosomal junction. In more apical sections containing desmosomes, neurites and nerve cell bodies were often encountered (Figs. 2, 3). Thus, the desmosomes were useful guides to associations of nerves with nematocytes.

Neurites were identified by their narrow diameter, content of microtubules, and electron-lucent or electron-dense vesicles (Figs 3, 4). Close associations of two or three neurites were common near desmosomes (Fig. 4). Broad neurites filled with large numbers of electron-dense or electron-lucent vesicles were also common in this region (Fig. 4, 5), whereas they were seldom observed far from desmosomal junctions.

In each battery cell complex of the distal tentacle, we always found one internally situated neuron, which we previously designated the Type I neuron (Hufnagel _et al_, 1985). Because of its similarity to the ganglion cell described by Westfall (1973), we will refer to it as such. (c.f. Westfall and Kinnamon, 1984). The ganglion cell was always in proximity to desmosomes (Fig. 2, 6). Ganglion cells had an internal ciliary complex (Fig. 2, 6) and a typical neuronal nucleus of non-uniform electron density

(Fig. 6). A second type of neuron, which resembled the sensory cell of Westfall and Kinnamon, (1984), was observed in some battery cell complexes. This neuron rested basally on the ganglion cell body (Fig. 14) and extended apically to the surface of the tentacle (see Hufnagel et al, 1985, Fig. 21). Following Westfall (c.f. Westfall and Kinnamon, 1984), we will call this neuron a sensory cell (type II neuron; Hufnagel et al, 1985).

Because the ganglion cell is centrally located relative to nematocytes, myonemes and sensory cells and occurs in all battery cell complexes, we sought to determine its synaptic relationships with other cells of the complex, especially nematocytes. Following the criteria established by Westfall et al (1971), synapses were identified by the presence of electron-lucent or electron-dense vesicles close to thickenings of the plasma membrane of the nerve cell. In some cases, the post-synaptic membrane was also thickened, and in some cases both pre- and post-synaptic membrane thickenings were not apparent but the vesicles were closely attached to the nerve cell plasma membrane by electron-dense connections.

Fig. 1. Basal, sagittal section through nematocyte processes. Nematocytes (Ne) terminate basally in one or more spatulate processes, which insert into cup-like lateral extensions of myonemes (My). Nematocytes and myonemes are joined by desmosomal junctions (D).

Fig. 2. Section through desmosomal junctions (D), containing nearby ganglion cell body ($N^G$). An internal ciliary complex (arrow) is identified by microvilli projecting into an internal vacuole. A cilium, if present, is out of the plane of the section.

Fig. 3. Neurite (N) in close proximity to a desmosomal junction (D) between a nematocyte (Ne) and myonemal process (My). Note in this neurite the longitudinally sectioned microtubules and electron-lucent vesicles (arrow).

Fig. 4. Several neurites (N), in serial contact, near a desmosome (D). Note that these neurites contain electron-dense vesicles. Ne-nematocyte; My-myoneme.

Fig. 5. Neurite swelling filled with electron-dense vesicles (arrow), in proximity to desmosomal junctions (D).

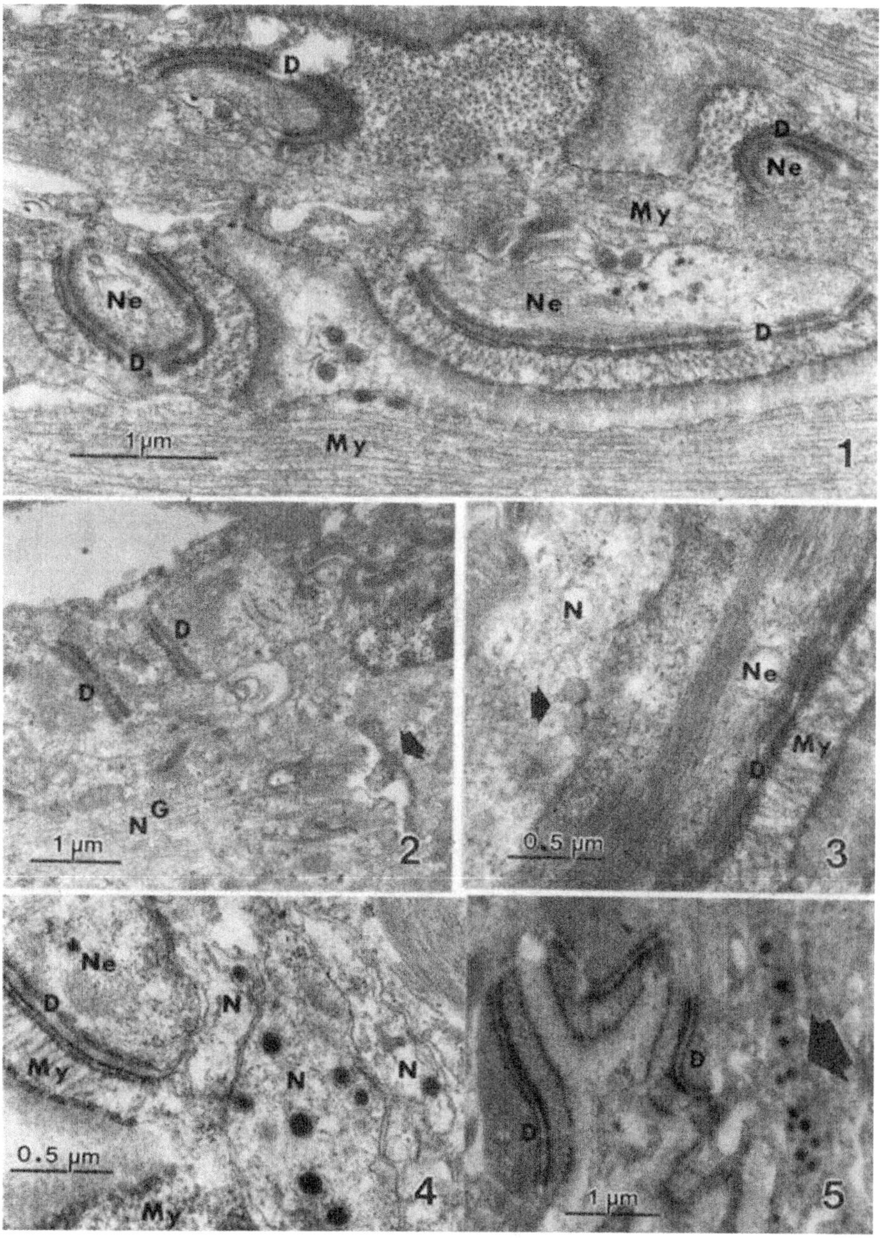
D
D
Ne
My
Ne
Ne
D
D
1 μm
My
1
D
D
N
G
1 μm
2
N
Ne
My
D
0.5 μm
3
Ne
D
N
My
N
N
0.5 μm
My
4
D
D
1 μm
5

Synapses of ganglion cell bodies onto cell bodies of sensory cells (Fig. 7) and nematocytes (Fig. 8) were encountered. Ganglion cells appeared also to synapse directly onto the battery cell body (Fig. 11), and onto myonemes, (Figs. 9, 10). Synapses involving neurites of unknown origin were frequently observed. These occurred between neurites and (1) nematocytes (Fig. 12), (2) battery cells, (3) the ganglion cell body, and (4) between two different neurites (Fig. 13). In general, synapses onto nematocytes exhibited both pre- and post-synaptic membrane thickenings, and the thickened membranes were more extensive than with other types of synapses (Figs. 8, 12). To summarize, the synapses we observed fell into two classes: (1) synapses of ganglion cells onto other cells and (2) synapses of unknown neurites onto other cells.

In a series of sections cut almost parallel to the tentacle axis, a major neurite was observed to extend from a sensory cell of one battery cell complex, to a neighboring battery cell complex. The sensory neurite of the first complex was found to extend parallel to and come into contact with an oppositely directed neurite of a ganglion cell from the second complex (Fig. 14). A third neurite, from the ganglion cell of the first complex, also extended toward the

Fig. 6. Ganglion cell ($N^G$; Type I neuron), typically close to desmosomal junctions (D). Note in ganglion cell an internal ciliary complex (arrow) and typical neuronal nucleus (non-uniform electron density). My-myonemes. (Adapted from Hufnagel et al, 1985, Fig. 12).

Fig. 7. Synapse (S) of ganglion cell ($N^G$) onto sensory cell ($N^S$). A dense cored vesicle is present and the presynaptic membrane (arrow) appears to be more electron dense than the postsynaptic membrane.

Fig. 8. Synapse (S) of ganglion cell ($N^G$) onto nematocyte (Ne). Clear vesicles (somewhat collapsed) are present at the presynaptic membrane. Arrow-internal ciliary complex.

Fig. 9. Synapse (S) of ganglion cell ($N^G$) onto myoneme (My). Pre-and post-synaptic membranes are only slightly thickened. Me-mesoglea.

Fig. 10. Similar to Fig. 9. Note dense bridges linking an electron-dense vesicle to the neuronal plasma membrane, and a vesicle (V) in the post-synaptic cytoplasm near neuro-myonemal synapse.

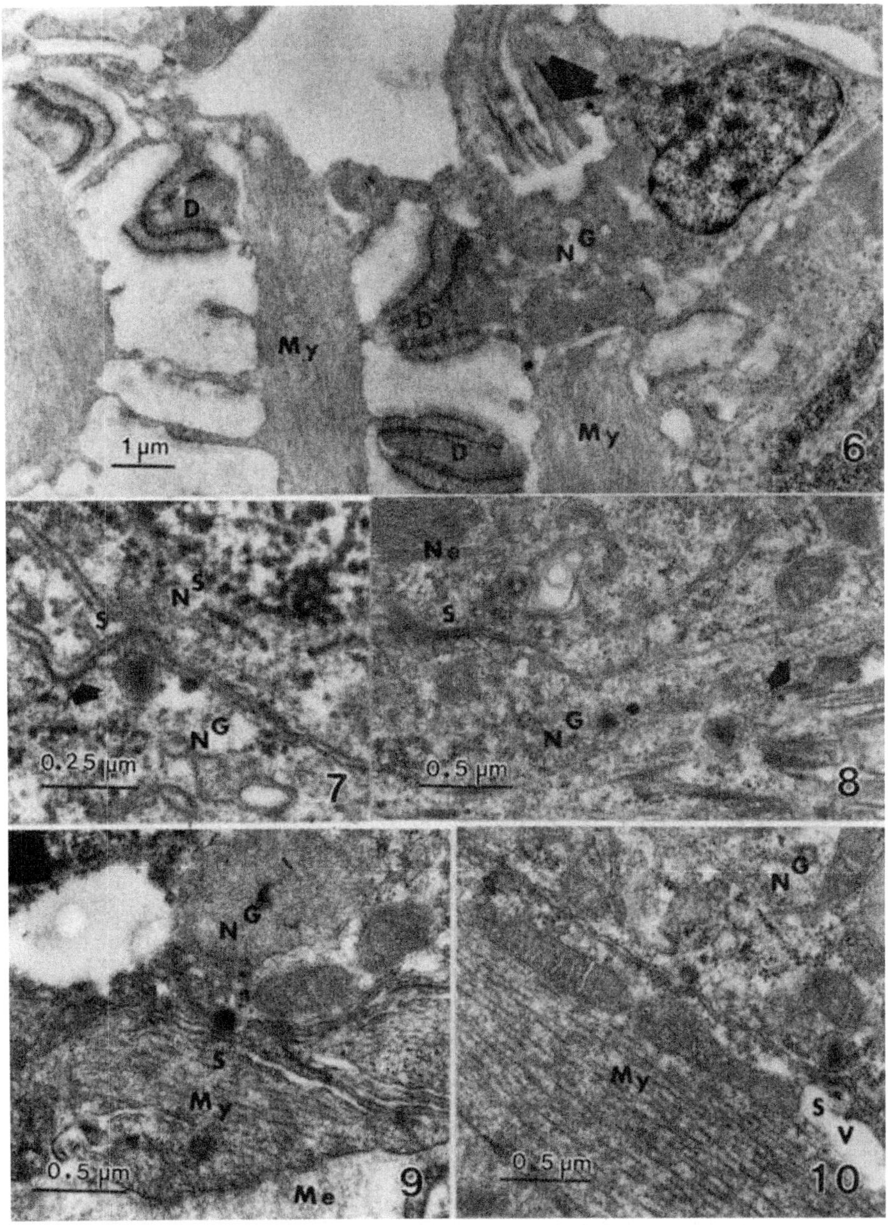
D
D
D
My
My
N G
1 µm
6
N S
S
N G
0.25 µm
7
Ne
S
N G
0.5 µm
8
N G
S
My
0.5 µm
Me
9
N G
My
S
V
0.5 µm
10

second complex. The neurites appeared to comprise a nerve tract linking the adjacent battery cell complexes. In other words, we observed that battery cell complexes of different tentacle rings (Hufnagel *et al*, 1985) are in contact via nerve cell processes. We also noted a possible synapse from the ganglion cell neurite onto the sensory cell neurite (Fig. 14, arrow). This synapse was located near the septate junction between the two adjacent battery cells.

## IV. DISCUSSION

By focusing on the distal tentacle of *H. attenuata*, our observations on synapses extend and confirm those in the tentacle, hypostome, peduncle and basal disc of *H. littoralis* by Westfall et al (1971), Westfall (1973) and Westfall and Kinnamon (1984). Westfall has also examined hydra's distal tentacle and in this volume reports observations which substantiate our own. Our observations suggest that the ganglion cell may influence the functions of all cells in its

Fig. 11. Synapse (S) of ganglion cell ($N^G$) onto battery cell body (BC), identified by numerous small vesicles (V). Arrow-rootlet structures of the internal ciliary complex of the ganglion cell.

Fig. 12. Synapse(S) between neurite (N) of unknown origin and nematocyte (Ne). Both pre-and post-synaptic membranes are thickened and electron-lucent vesicles are present.

Fig. 13. Synapse (S) between two neurites (N).

Fig. 14. Montage of photographs from a single section and tracings based on two serial sections, showing neurites connecting two battery cell complexes. Complexes can be distinguished from one another by the presence of septate junctions (SJ) between the different battery cells. A neurite from the sensory cell ($S_1$) of one complex contacts an oppositely projecting neurite from the ganglion cell of a second complex ($G_2$). The ganglion cell of the first complex is also present ($G_1$). Tracings of dense cored vesicles (arrow) indicate the site of a possible synaptic junction between the neurites. Note its close proximity to the septate junction (SJ) joining the two complexes. For an enlargement of this region in another section of this series, see Fig. 28 in Hufnagel *et al* (1985).

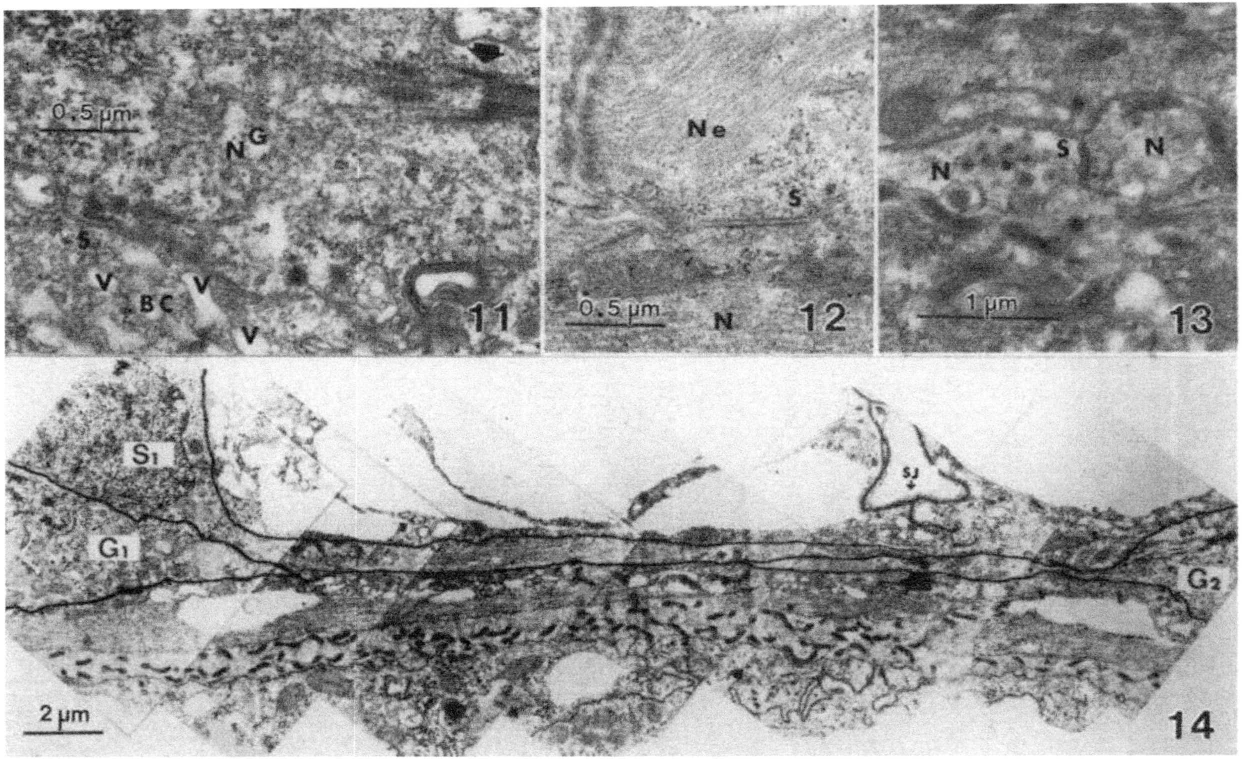
0.5 µm
N
G
S
V
BC
V
V
11
Ne
S
0.5 µm
N
12
N
S
N
1 µm
13
S1
G1
SJ
G2
2 µm
14

own battery cell complex, as well as some cells of nearby complexes. This is consistent with the possible role of the ganglion cell as an inhibitor of nematocyst discharge, proposed by Kass-Simon in this volume.

We also report the connection of neighboring battery cell complexes of the tentacle via neurites grouped into tracts or bundles. We believe this observation, plus the ultrastructural evidence for a synapse from the ganglion cell of one complex to the sensory cell of the other complex, constitute the first ultrastructural evidence for neuronal contact and possible synapsing between complexes. Yu et al. (1985) demonstrated contact between different complexes via neurons, using light microscopic immunocytochemistry, but to demonstrate functional connectivity via synapses, ultrastructural evidence is required, in addition to behavioral, neurophysiological and histochemical evidence.

## V. ACKNOWLEDGEMENTS

We are very grateful to Mary K. Lyon for preparation of material from which these observations were made. We also thank Ken Davignon for preparation of diagrams and photographic work, and Sue Haynes for photographic assistance. Helpful discussions with Jane Westfall and David Hessinger are also gratefully acknowledged.

## REFERENCES

Ewer, R.F. 1947. On the functions and mode of action of the nematocysts of Hydra. Proc. Zool. Soc. London 117:305-376.

Grosvenor, W. 1986. Control of nematocyst discharge in Hydra attenuata during locomotion and feeding. M.S. Thesis, University of Rhode Island.

Hufnagel, L.A., G. Kass-Simon and M.K. Lyon. 1985. Functional organization of battery cell complexes in tentacles of Hydra attenuata. J. Morphol. 184: 323-341.

Ruch, R.J. and C.B. Cook. 1984. Nematocyst inactivation during feeding in Hydra littoralis. J. Exp. Biol. 111:31-42.

Slautterback, D.B. 1967. The cnidoblast-musculoepithelial cell complex in the tentacles of hydra. Z. Zellforsch. Mikrosk. Anat. 79:296-318.

Smith, S., J. Oshida, and H. Bode. 1974. Inhibition of Nematocyst Discharge in Hydra Fed to Repletion. Biol. Bull. 147:186-202.

Westfall, J.A., S. Yamataka and P.D. Enos. 1971. Ultrastructural evidence of polarized synapses in the nerve net of Hydra. J. Cell Biol. 51:318-323.

Westfall, J.A. 1973. Ultrastructural evidence for a granule-containing sensory-motor-interneuron in Hydra littoralis. J. Ultr. Res. 42:268-282.

Westfall, J.A. and J.C. Kinnamon. 1978. A second sensory-motor-interneuron with neurosecretory granules in Hydra. J. Neurocytol. 7:365-379.

Westfall, J.A. and J.C. Kinnamon. 1984. Perioral synaptic connections and their possible role in the feeding behavior of Hydra. Tissue and cell 16:355-365.

Yu, S.-M., J.A. Westfall and J.F. Dunne. 1985. Light and electron microscopic localization of a monoclonal antibody in neurons in situ in the head region of Hydra. J. Morphol. 184:183-193.

# 28

# TOWARDS A NEUROETHOLOGY OF NEMATOCYST DISCHARGE IN THE TENTACLES OF HYDRA

G. Kass-Simon

Department of Zoology
University of Rhode Island
Kingston, RI 02881

## SUMMARY

All ectodermal cells of the distal tentacle of Hydra attenuata are epitheliomuscular cells which encase three types of nematocytes, a ganglionic nerve cell and sometimes a sensory nerve. Together they form the battery cell complex.

Ewer's early experiments and our more recent studies indicate that mechanical and chemical excitation of stenoteles and desmonemes and their chemical inhibition by high molecular weight food molecules are local, but may or may not involve the nerves of the battery cell complex. Mechanical excitation, and chemical inhibition by food substances of atrichous isorhizas appears to be local, but their inhibition during locomotion may involve inhibitory synapses either from the battery's ganglion cell onto the nematocytes and/or onto the sensory cells. The postulated mechano-receptor function of the ganglion cell also suggests that its synapses with nematocytes are inhibitory. Since specific and directed excitation does not require the presence of a synapse, but specific inhibition does, these inhibitory synapses may have been among the earliest synapses to appear during the evolution of the nervous system.

## I. INTRODUCTION

If "ethology" means the study of the behavior of animals in their natural habitat (Jaynes, 1969) and if "neuroethology" refers to the description of how nerves control that behavior, then, it is possible to talk of the neuroethology of nematocyst discharge and mean the description of the neuronal mechanisms controlling the behavior of nematocysts in their "natural habitat". In hydra's tentacles, the natural habitat of nematocysts is the outside world and the battery cell complex.

## II. THE BATTERY CELL COMPLEX

The battery cell complex is the morphological/functional unit of hydra's tentacular ectoderm; all other ectodermal cells exist within it (Hufnagel, et al., 1985). It is responsible for all the functions performed by the tentacular ectoderm: sensory perception, impulse conduction, tentacle contraction and bending, capture and killing of prey and adhesion to surfaces. It consists of a ganglion cell, one or two stenoteles surrounded by a complement of isorhizas and desmonemes, and usually, but not always, a sensory cell (Fig. 1). In the ectodermal battery cells around the mouth, unlike those in the distal tentacle, there are sometimes two sensory cells, and that these may make reciprocal synapses with each

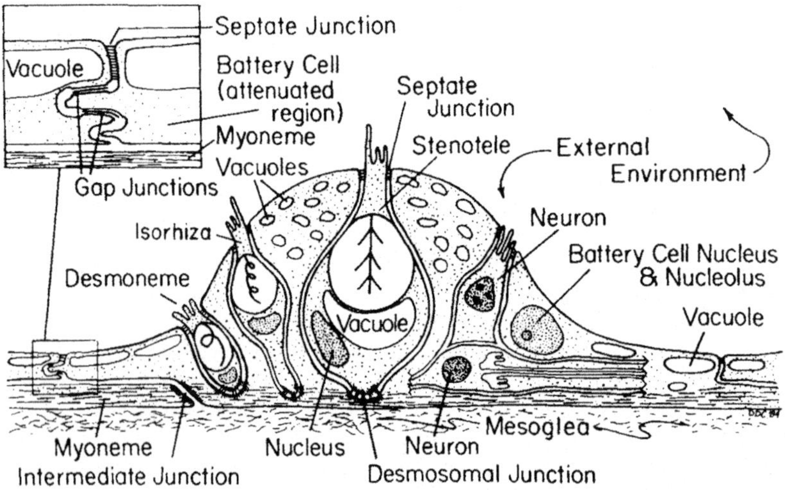


Fig. 1. Schematic representation of the battery cell complex. The three types of nematocytes, a sensory cell and a ganglion cell (with its cilium) are imbedded in the battery cell. Attenuated regions of adjacent battery cells are coupled by gap junctions and connected by septate junctions. (Modified from Hufnagel, et al., 1985.)

other. There are also reciprocal synapses between ganglion cells and between ganglion cells and sensory cells. Ganglion cells and sensory cells also make synapses onto battery cell myonemes (Westfall and Kinnamon, 1978, 1984; Kinnamon and Westfall, 1982.) In the distal tentacle, the body of the ganglion cell synapses onto nematocytes, sensory cell bodies, battery cell bodies and myonemes. We have also observed neurites of unidentified cells synapsing onto ganglion cells, nematocytes, and battery cell bodies, and have observed a neurite of a single sensory cell entering a second battery cell complex and coming into contact with the neurite of its ganglion cell (Hufnagel, et al., 1985; Hufnagel and Kass-Simon, this volume.) Westfall (this volume) finds that ganglion cells send processes to nematocyst batteries in different battery cells as well as to more than one nematocyte within a single battery cell. Lentz and Barrnett (1961) have identified cholinesterase activity at both the nematocyte and in the nematocyst capsule. These anatomical and histochemical relationships imply neuronal control of nematocysts discharge and place certain constraints on our interpretation of their behavioral physiology.

## III. EXCITATION OF STENOTELES AND DESMONEMES

Early studies on nematocyst behavior were interpreted to mean that nematocysts were, in Parker's terminology, independent effectors (Parker, 1919). Thus, Ewer (1947) following Pantin's studies on anemones (1942), showed that stenoteles and desmonemes in isolated tentacles could be discharged in by mechanical and chemical stimuli. By touching the isolated tentacles with a glass capillary, Ewer found that only 3 out of 14 tentacles discharged stenoteles and desmonemes, but when the glass was coated with food extract, the discharge ratio was 10 out of 14. Ewer also found that in the presence of food extract, desmonemes required stronger mechanical stimulation than did stenoteles. Food extract alone did not cause discharge. Since Ewer's experiments were done on small areas of isolated tentacles and since the differences in stenotele and desmoneme response to mechanical stimuli could be interpreted as differences in threshold, Ewer concluded, as did Pantin, that stenoteles and desmonemes responded directly and independently to mechanical and chemical stimuli--the mechanical stimulus being necessary and sufficient, the chemical stimulus lowering the discharge threshold. That is, stenoteles and desmonemes were considered to be independent effectors.

Although there is evidence that in anemones nematocyst discharge from the tentacles may be enhanced by mechanical stimulation to the column (Conklin and Mariscal, 1976), and although in hydra, the possibility of nervous excitation cannot be discounted on anatomical grounds, apart from recent studies indicating enhancement of nematocyst discharge by food extracts (Ruch and Cook, 1984), no further behavioral studies have been done, so that the idea of their direct and independent excitation in hydra may, for the present, remain unaltered.

## IV. CHEMICAL INHIBITION OF STENOTELES AND DESMONEMES

The question of independence has also been raised with respect to the inhibition of nematocyst discharge. When hydra are fed to repletion, they stop killing Artemia (Burnett, Lentz, and Warren, 1960). Smith, Oschida and Bode (1974) suggested that metabolites from the ingested prey were responsible for inhibiting nematocyst discharge. Artemia homogenate injected into the gut significantly reduced the number of Artemia subsequently caught and killed by an animal.

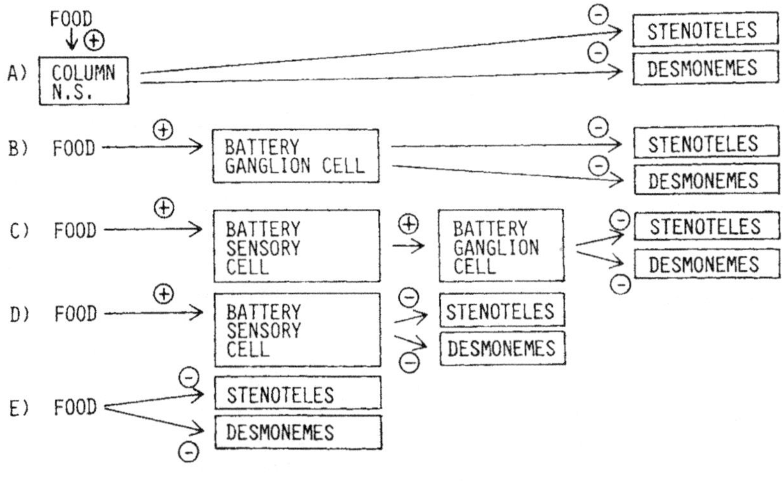


Fig. 2. Possible connections that would cause inhibition of stenotele and desmoneme discharge by food substances. See text for further explanation.

In general, five control mechanisms for this inhibition are possible: (1) Food might excite the column's nervous system, which in turn would inhibit stenoteles and desmonemes (Fig. 2a). (2) Food might excite the ganglion cells of the tentacle either directly (Fig. 2b), or (3) through the sensory cell of the battery cell complex (Fig. 2c). The ganglion cell would in turn inhibit the nematocytes. (4) Alternatively, food might excite the sensory cell which would in turn inhibit the nematocytes; or, finally, (5) food might act directly on either the nematocyte or the nematocyst to inhibit discharge. In all cases, during chemical inhibition, if the ganglion cell were involved as a final path to the nematocytes, its synapses with them must be inhibitory.

It is possible to determine if the nerves of the gut or body wall are necessary for inhibiting stenoteles and desmonemes. If an individual tentacle is isolated and Artemia homogenate carefully pipetted onto it, the segment of the tentacle which was exposed to homogenate no longer captures or kills Artemia (Grosvenor, 1986; Grosvenor and Kass-Simon, in prep.) (Fig. 3). In six tentacles whose bases

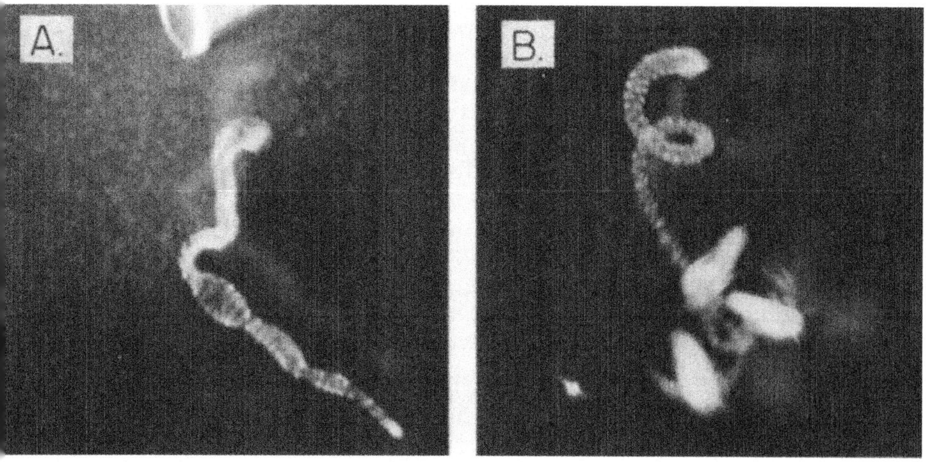


Fig. 3. Local inhibition of stenotele and desmoneme discharge by food extracts. (A) Food extract is pipetted onto the base of an isolated tentacle (cloudy region). (B) When the tentacle is transferred to fresh culture medium, Artemia are only captured and killed by the section of tentacle which was untouched by homogenate.

had been covered with Artemia extract, the number of Artemia killed by the bases was reduced to 6% (of the total number of Artemia which made contact with the tentacle) from their control value of 75%, while the reduction for the tips of such tentacles (not covered with extract) was only 75% from their control rate of 82%. The differences between the killing values for tentacle bases and tips is significant at p=0.003 (Wilcoxin Rank Sums).

It is therefore possible to say that the nerves of the hypostome, gut and body column are not necessary for the chemical inhibition of stenoteles or desmonemes in the tentacle. Further, if the ganglion or sensory cells of the tentacle are involved, then their effectiveness in causing this inhibition along the length of the tentacle is at least limited to within a given region. So far, anatomically, there are synapses from sensory cells onto nematocytes (Westfall, this volume) from ganglion cells onto nematocytes (Hufnagel, et al.,1985) and from ganglion cells onto other ganglion cells (Westfall, this volume.) Anatomical evidence therefore cannot yet be used to differentiate between the inhibitory pathways represented by Figures 2b and 2c.

Such differentiation can only be made by determining where the receptor sites for the inhibitory molecule(s) lie and tracing the afferent paths to the nematocytes experimentally. To do this one can begin to look for the receptor sites for the chemical inhibitor by isolating the inhibitory substance and finding where it acts on the battery cell complex. If Artemia homogenate is fractionated by allowing it to flow through a Sephadex chromatography column, and the different fractions tested for their effects on the feeding behavior, it is seen that the different molecular weight fractions have differential effects on the various aspects of feeding behavior (Grosvenor and Kass-Simon, 1979 and in prep.; Grosvenor, 1986). When hydra are immersed in each of the fractions, the high molecular weight fractions significantly reduce the number of Artemia killed, while the lower molecular weight fractions cause a significant increase in the amount of mouth opening and tentacle activity. (These latter fractions come off the gel column in the same region as reduced glutathione.)

## V. INHIBITION OF NEMATOCYST DISCHARGE DURING LOCOMOTION

Hydra's propensity to perambulate over glass surfaces, often in response to a distant light stimulus, has been well known since Trembley (1744). The best description of this behavior is given by Ewer (1947). Locomotion begins with an extension and bending of the body column from the attached base; followed by the settling of the tentacles onto the substrate. The base detaches from the substrate while the body contracts, causing the animal to 'stand on its head'. The body then extends, base first, until it touches the substrate again; the tentacles are released from their attachment and the animal stands itself upright. During this sequence, according to Ewer, the tentacles attach to the substrate with their atrichous isorhizas, although we have also observed that all three types of nematocysts are discharged during locomotion. In five animals, the adhering tentacles (average 6 tentacles/hydra) discharged 12.2 $\pm$ 9.3 stenoteles. 46.6 $\pm$ 7.0 desmonemes and 90.2 $\pm$ 23 isorhizas.

Food extracts differentially affect various aspects of hydra's behavior (Grosvenor and Kass-Simon, 1979). With respect to nematocysts, Ewer found that by stroking the tentacles of sedentary hydra (in ordinary water) with a silica coated glass rod, 16 out of 20 animals fired their atrichous isorhizas. Only one out of 20 responded if the glass rod was also coated with food extract. Again, since the response occurred directly at the site where the rod had made contact with the tentacle, Ewer concluded that both the excitatory response of the isorhizas to the hard surface and the inhibitory response to food extract were direct and without intervention of the nervous system. Nonetheless, these findings do not preclude neuronal inhibition of atrichous isorhizas (Fig. 4a, b). It also remains to be shown whether the fractions of Artemia homogenate which inhibit isorhiza discharge are the same as those which inhibit stenotele and desmoneme discharge.

Neuronal inhibition of atrichous isorhizas can, however, be demonstrated in hydra whose tentacles are making contact with the substrate during locomotion. If the unattached tentacle of a locomoting animal is stroked with a glass rod, the number of atrichous isorhizas that are discharged is significantly reduced. Only 2 out of 8 locomoting animals responded to the stroking of their free tentacles while among sedentary hydra, 8 out of 12 responded. Although these differences did not prove significant because of the small sample size, the differences in the number of atrichous isorhizas that were discharged by the responding tentacles were significant. For the two locomoting hydra that responded, each only fired 1 isorhiza per tentacle, while an average of 8 isorhizas per tentacle (8.38 $\pm$ 5.4) were fired by the by the sedentary hydra. The difference is significant at $p=0.02$ (Wilcoxin Rank Sums). Ewer also noted that there were seldom any isorhizas discharged from "walking" hydra.

Therefore, the mechanical stimulation which results in excitation of atrichous isorhizas in the adhering tentacles also results ultimately in their inhibition in the non-adhering tentacle. Again, a number of pathways, which are consistant with the anatomical evidence of the battery cell complex, are possible; several are given in Figure 4c, d, e, f. Thus, the hard substrate might excite the isohizas in the attaching tentacle either directly or through a sensory cell (Fig. 4c, d). In the attaching tentacles, the sensory cell would either be directly excitatory to the isorhizas (Fig. 4d) or would excite an intervening ganglion cell which would then excite the isorhizas (Fig. 4e, f). At the same time, the excitation would be transmitted via the body or hypostomal nerves to the ganglion cells of the free tentacle which would then inhibit isorhiza discharge in the free tentacle either directly (Fig. 4e) or by inhibiting the sensory cells of the free tentacle (Fig. 4f). If only one ganglion cell in each battery cell complex made contact with an isorhiza, it is not likely, though not impossible, that its synapses onto that cell would be both excitatory and inhibitory. Therefore, if we wish to postulate that ganglion

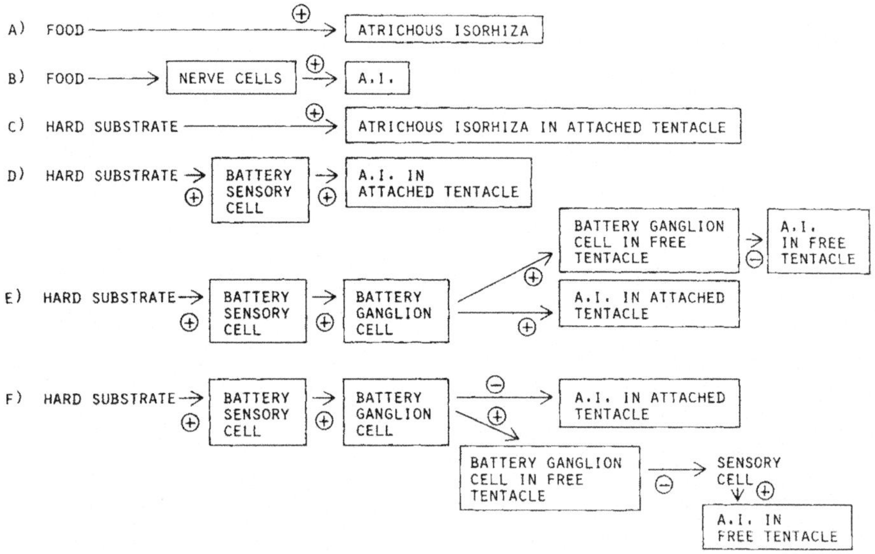


Fig. 4. Connections that would result in the simultaneous excitation of artrichous isorhizas in attaching tentacles and their inhibition in free tentacles of locomoting *Hydra*. See text for further explanation.

cells excite isorhizas, then we should probably postulate that more than one ganglion cell synapses with a given isorhiza. (The tract of multiple neurites found in the battery cell complex (Hufnagel and Kass-Simon, this volume) makes this, anatomically, at least, a distinct possibility.) But, if only a single ganglion cell synapsed onto each isorhiza, it would not make much sense for it to make excitatory contact with an isorhiza at the same time that it were inhibiting its response to mechanical stimulation via the sensory cell (Fig. 4f). Here too, its synapses onto both cells would probably be inhibitory.

Ultrastructural evidence also suggests that the synapses of the ganglion cell onto the nematocyte are inhibitory. We have recently investigated the ultrastructural relationships of the ganglion cell with respect to its possible proprioceptive function (Kass-Simon, Hufnagel et al., in prep.) and have found that the cilium in contracted tentacles can be bent. If this resulted in an excitation of the ganglion cell, since the ganglion cell synapses directly onto nematocytes, then those synapses must not be excitatory, otherwise, every time a tentacle contracted, the nematocysts would tend to be discharged. Clearly, it is more likely that these synapses are inhibitory; in this way discharge would be prevented whenever the tentacle contracted. If the synapses from the ganglion cell onto the nematocyte are inhibitory, then excitation of the atrichous isorhizas could occur either directly through the nematocyte itself or through the sensory cell (Fig. 4c, d). The finding that the ganglion cell makes synapses not only onto nematocytes, but also onto the sensory cell (Hufnagel and Kass-Simon, this volume) would support the notion that the ganglion cell might be able to inhibit the sensory cell in the free tentacle (Fig. 4f).

Evidence for the distant inhibition of nematocyst discharge by intervening nerves has been described for sea anemones (Davenport, Ross and Sutton, 1961; Ross and Sutton, 1964). During swimming, Stomphia's feeding nematocysts are inhibited and, in Calliactus, the threshold for firing attaching nematocysts is raised in animals that are already attached to their preferred shell substrates.

The idea that the ganglion cell makes inhibitory synapses with nematocytes and possibly with the sensory cell, must, of course, be experimentally tested. But it is nonetheless possible to speculate that if these synapses are inhibitory, they may have been among the very first to appear during the evolution of the nervous system. Over short distances, specific and directed excitation does not require a synapse -- electrical coupling between cells and organelles (or organoids) suffices. But, specific and delineated inhibition, even over such short distances as exist within the battery cell complex, demands the presence of a synapse.

## REFERENCES

Burnett, A.L., T. Lentz, and M. Warren. 1960. The Nematocysts of Hydra (Part I): The Question of Control of the Nematocyst Discharge Reaction by Fully Fed Hydra. Ann. Soc. Roy. Zool. Belg. 90: 247-267.

Conklin, M.H. and R.N. Mariscal. 1976. In G.O. Mackie (ed.), Coelenterate Ecology and Behavior. Plenum Press, New York. pp. 549-558.

Davenport, D., D.M. Ross, and L. Sutton. 1961. The Remote Control of Nematocyst Discharge in the Attachment of Calliactis parasitica to the Shells of Hermit Crabs. Vie Milieu 12: 197-209.

Ewer, R.F. 1947. On the Functions and Mode of Action of the Nematocysts of Hydra. Proc. Zool. Soc. London 117: 305-376.

Grosvenor, W. 1986. Control of Nematocyst Discharge in Hydra attenuata during locomotion and feeding. M.S. Thesis, University of Rhode Island.

Grosvenor, W. and G. Kass-Simon. 1979. Differential Responses of the Contraction Burst System in Hydra to Gel Chromotography Fractions of Artemia Homogenates. Amer.

Grosvenor, W. and G. Kass-Simon, Inhibition of Nematocyst Discharge I: Effects of Artemia Homogenate. In Prep.

Hufnagel, L.A., G. Kass-Simon, and M.K. Lyon. 1985. Functional Organization of the Battery Cell Complexes in Tentacles of Hydra attenuata. J. Morphol. 176: 323-341.

Hufnagel, L.A., G. Kass-Simon, 1987. Functional Anatomy of Nematocyst Discharge in the Hydra Tentacle. This volume.

Jaynes, J. 1969. The Historical Origins of "Ethology" and "Comparative Psychology". Anim. Behav. 17: 601-606.

Kass-Simon, G., L. Hufnagel and M. Lyon. Ultrastructural analysis of proprioception in Hydra. In prep.

Kinnamon, J.C. and J.A. Westfall. 1982. Types of neurones and synaptic connections at hypostome-tentacle functions in Hydra. J. Morph. 173: 119-128.

Lentz, T.L. and R.I. Barnett. 1961. Enzyme Histochemistry of Hydra. J. Exp. Zool. 147: 125-149.

Pantin, C.A.F. 1942. The Excitation of Nematocysts. J. Exp. Biol. 19: 294-310.

Parker, G.H. 1919. The Elementary Nervous System. Lippincott, Philadelphia.

Parker, G.H. and M.A. Van Alstyne. 1932. The Control and Discharge of Nematocysts, Especially Metridium and Physalia. J. Exp. Zool. 63: 329-344.
Ross, D.M. and L. Sutton. 1964. Inhibition of the Swimming Response by Food and of Nematocyst Discharge during Swimming in the Sea Anemone, Stomphia coccinea. J. Exp. Biol. 41: 751-757
Ruch, R.J and C.B. Cook, 1984. Nematocyst inactivation during feeding in H. littoralis. J. Exp. Biol. 111: 31-42.
Trembley, A. 1744. In S.G. Lenhoff and H.M. Lenhoff (eds.), Hydra and the Birth of Experimental Biology. 1986. The Boxwood Press, Pacific Grove, Ca.
Smith, S., J. Oshida, and H. Bode. 1974. Inhibition of Nematocyst Discharge in Hydra Fed to Repletion. Biol. Bull. 147: 186-202.
Westfall, J.A. 1986. Neuronematocyte Synapses and Possible Pathways Controlling Discharge of Nematocysts in Hydra. This volume.
Westfall, J.A. and J.C. Kinnamon. 1978. A Second Sensory Motor Interneuron With Neurosecretory Granules in Hydra. J. Neurocytol. 7: 365-379.
Westfall, J.A. and J.C. Kinnamon. 1984. Perioral Synaptic Connections and their Possible Role in Feeding Behavior. Tissue and Cell 16: 355-365.

# 29

# DIRECT OBSERVATION OF HEMOLYTIC ACTIVITY ASSOCIATED WITH SINGLE NEMATOCYSTS[1]

*Marianne Klug*
*Jakob Weber*
*Pierre Tardent*

Zoological Institute
University of Zürich-Irchel
8057 Zürich, Switzerland

SUMMARY

Hemolytic activity of some kinds of individual nematocysts can be tested in the light microscope by discharging the capsules in a dense red blood cell suspension. We developed this assay for the nematocysts of the fresh water polyp *Hydra attenuata* and applied it to nematocysts of other cnidarians (*Hydra fusca*, *Hydra circumcincta*, *Podocoryne carnea*, *Physalia physalis*, *Chrysaora quinquecirrha*, *Cassiopea andromeda*, *Anemonia sulcata*, *Actinia equina* and *Calliactis parasitica*). This test enables one to detect not only the hemolytic activity of individual nematocysts, but also their mode of toxin release.

## I. INTRODUCTION

In several cnidarians hemolytic activity of the nematocyst venom has been described. An extract prepared from isolated nematocysts was usually used for detection of this activity (Hessinger and Lenhoff, 1973; Mariscal, 1974; Blanquet, 1977;

[1]*This work was supported by the Swiss National Science Foundation (Grant Nr. 3.072-0.84)*

Burnett and Calton, 1977). In most cases the nematocyst preparations contained more than one type of capsule; so, it is not clear which type of cysts contained the hemolytic activity (Mariscal, 1974). Similarly, a strong hemolytic activity was detected in extracts of isolated nematocysts of *Hydra attenuata* (Klug, 1986; Weber *et al.*, 1987). This extract was obtained from a mixture of all four types of nematocysts found in hydra, because physical separation of the different capsules was not successful. We developed a simple method that allowed direct observation of hemolytic capacity in individual nematocysts. By this means we could demonstrate that stenoteles as well as holotrichous isorhizas lyse red blood cells (Klug, 1986). We applied this test to three species of hydra and to seven species of marine cnidarians, amongst them *Physalia physalis* and *Chrysaora quinquecirrha*, whose hemolytic activities already have been described (Tamkun and Hessinger, 1981; Burnett and Calton, 1977).

## II. MATERIAL AND METHODS

*Podocoryne carnea* (Hydrozoa), *Anemonia sulcata*, *Calliactis parasitica* and *Actinia equina* (Anthozoa) were collected from the Mediterranean Sea, and *Cassiopea andromeda* (Scyphozoa) from the Red Sea. Their nematocysts and those of *Hydra attenuata*, *Hydra circumcincta* and *Hydra fusca* (Hydrozoa) were isolated according to the method described by Weber *et al.* (1987). Isolated and freeze-dried nematocysts of *Physalia physalis* (Hydrozoa) were kindly provided by J.W. Burnett (Baltimore); another batch was purchased from Sigma Chemicals. Isolated and freeze-dried nematocysts of *Chrysaora quinquecirrha* (Scyphozoa) were also from Sigma Chemicals.

Fresh human red blood cells were washed four times in 120 mM NaCl, 25 mM $NaH_2PO_4$, 2 mM EDTA (pH 7.4), and stored as a 50% suspension at 4°C. Before use the red blood cells were diluted to a final concentration of 10-20% in the above buffer; this buffer was found to be suitable for assaying the hemolytic activity of crude extract from nematocysts of *H. attenuata*. For the tests, a drop of the red blood cell suspension was mixed with isolated nematocysts on a slide, then covered and observed under a photomicroscope for the appearance of areas of lysed cells after discharge of nematocysts. To increase the rate of nematocyst discharge, 5 µl of 0.1 M dithioerythritol (DTE) was added (cf. Weber *et al.*, 1987). To better identify the capsules among the red blood cells we sometimes stained the capsules with brilliant cresyl blue (0.01%). When indicated,

3 mM $CaCl_2$ were added to the red blood cell suspension to exclude a possible inhibitory effect of EDTA (Hessinger and Lenhoff, 1976).

The nematocysts of the marine cnidarians were identified according to Mariscal (1974).

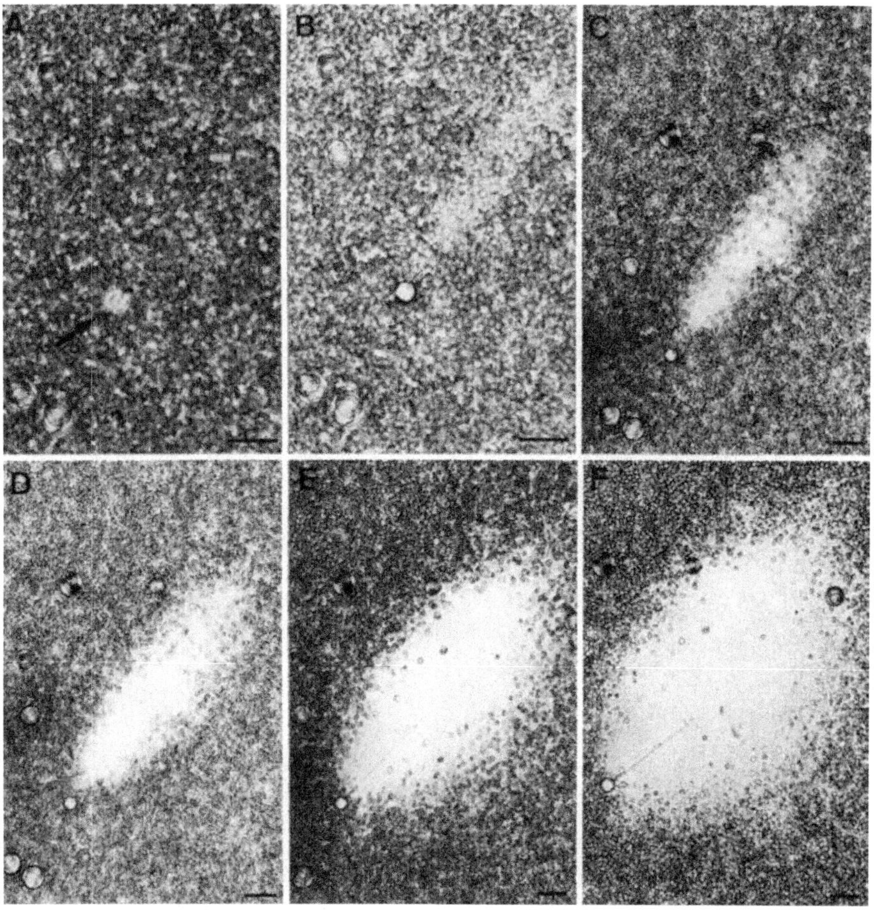


*Fig. 1. Sequences of the hemolysis produced by a discharged eurytele of Physalia physalis. A. Undischarged eurytele among the red blood cells. B. The same eurytele 2 sec after discharge: lysis of the cells occurs immediately. C. 5 sec after discharge. D. 10 sec after discharge. E. 15 sec after discharge. F. 30 sec after discharge (bars: 50 μm).*

TABLE I. Hemolytic activity of nematocysts. The tissues from which the nematocysts were isolated are indicated in parentheses.

| Origin of nematocysts | hemolysis | no hemolysis |
|---|---|---|
| *Hydra attenuata* (whole polyp) | stenotele<br>holotrichous isorhiza | desmoneme |
| *Hydra circumcincta* (whole polyp) | stenotele<br>holotrichous isorhiza | desmoneme |
| *Hydra fusca* (whole polyp) | stenotele<br>holotrichous isorhiza | desmoneme |
| *Podocoryne carnea* (whole polyp) | eurytele | desmoneme |
| *Physalia physalis* | eurytele | anisorhiza |
| *Chrysaora quinquecirrha* | eurytele<br>anisorhiza | isorhiza |
| *Cassiopea andromeda* (whole polyp) | eurytele | isorhiza |
| *Anemonia sulcata* (tentacle) | -- | mastigophore |
| *Anemonia sulcata* (acrorhagi) | -- | mastigophore<br>isorhiza |
| *Calliactis parasitica* (acontial filaments) | -- | mastigophore |
| *Actinia equina* (tentacle) | -- | mastigophore |

## III. RESULTS AND DISCUSSION

Hemolysis was usually observed within the first few seconds after discharge of a nematocyst (Fig. 1). Delayed lysis was not observed. The hemolytic activities of the various nematocysts are summarized in Table I.

Stenoteles and holotrichous isorhizas of all three fresh water polyps have a strong hemolytic activity (Klug, 1986;

comparable to the activity of *Physalia physalis* euryteles shown in Fig. 1). This could not be established for the atrichous isorhizas, since this type of capsules is rather rare and could not be identified reliably among the red blood cells.

Desmonemes of all species investigated showed no hemolytic activity. Nevertheless, they may contain hemolytic active components which can not be detected by this assay, since their contents are not released upon nematocyst discharge (Burnett, 1960).

All euryteles lysed red blood cells (Table I). It is worth mentioning that in both samples of *Physalia physalis* hemolytic eurytele-like nematocysts could be detected (Fig. 1 and 2). Other authors (Lane and Dodge, 1958; Hulet *et al.*, 1974; Burnett and Calton, 1977) have only described the more abundant anisorhizas in the Portuguese man-of-war (Fig. 2A). We do not know whether the presence of euryteles are a special quality of our samples of *Physalia physalis*.

Although Tamkun and Hessinger (1981) purified a powerful hemolysin from *Physalia* anisorhizas, we never observed lysed red blood cells around the discharged small or large anisorhizas. In addition, if a *Physalia* nematocyst venom (acc. to Tamkun and Hessinger, 1981) is partially purified by gel filtration and analysed on SDS-gels, none of Hessingers physalitoxin associated bands could be detected (Weber, unpublished). Thus our two batches of *Physalia* nematocysts seem, in some respects, to be different from Hessingers.

In the case of *Chrysaora quinquecirrha* we only considered the three types illustrated on Figure 2 (E-G) for the test, even though more than three types of nematocysts have been described (Sutton and Burnett, 1969). In contrast to the anisorhizas of *Physalia physalis* the anisorhizas of *Chrysaora quinquecirrha* lysed red blood cells (Fig. 2C and 2I). In the tested anthozoans we found no hemolytically active nematocysts. The addition of $CaCl_2$ to the red blood cell buffer did not affect this result. Ferland and Lebez (1974) described a hemolytically active component in the extract of nematocysts of *Actinia equina*. It is possible that a hemolytic component exists which we could not detect under our experimental conditions; it may also be that not the nematocysts themselves but the spirocysts or some tissue components of the anthozoans contain a hemolytically active substance.

## IV. CONCLUSIONS

With the exception of the euryteles from *Chrysaora quinquecirrha*, which lysed red blood cells only at the tip of the everted tubule, all other hemolytically active nematocysts

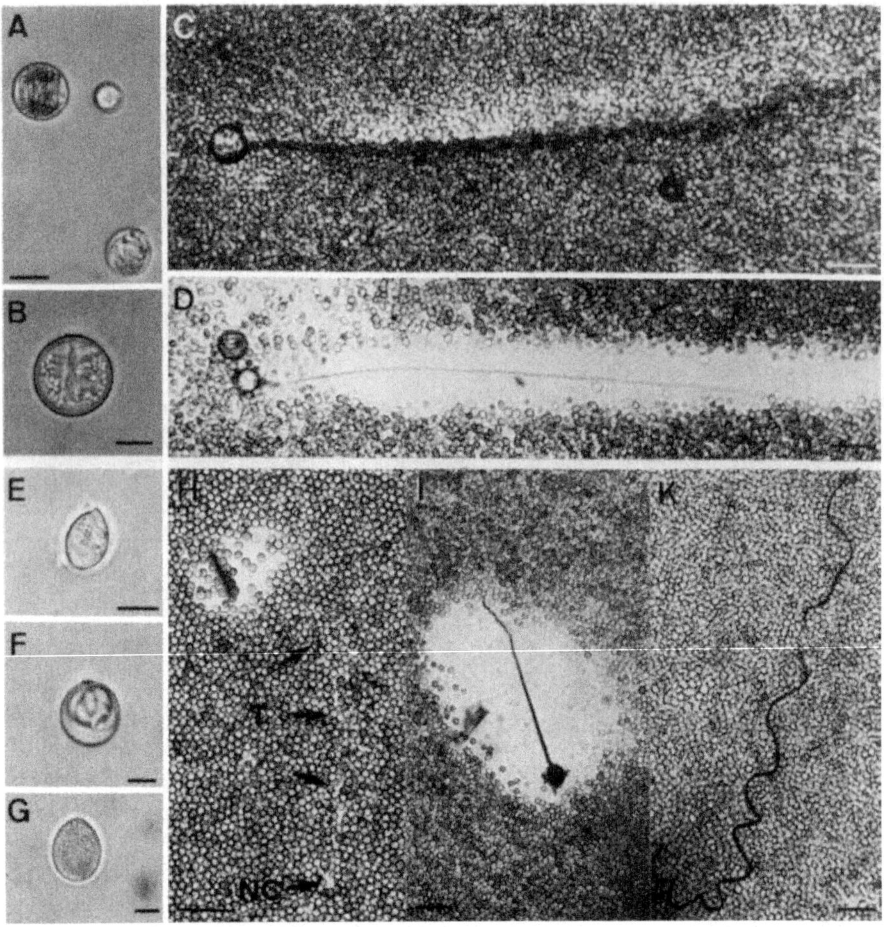


*Fig. 2. Isolated nematocysts of Physalia (A-D) and Chrysaora (E-K) and their hemolytic capacity. A. Anisorhizas of varying sizes (bar: 20 μm), B. Eurytele (bar: 10 μm), C. Discharged anisorhiza. No hemolysis occurs, some of the red blood cells adhere to the barbed tubule (bar: 50 μm), D. Discharged eurytele. Hemolysis occurs along the tubule (bar: 50 μm), E. Eurytele (bar: 10 μm), F. Anisorhiza (bar: 10 μm), G. Isorhiza (bar: 10 μm), H. Discharged eurytele. Hemolysis occurs only at the tip of the everted tubule. Tubule (T) and the capsule of the nematocyst (NC) are marked with arrows (bar: 50 μm). I. Discharged anisorhiza, stained with brilliant cresyl blue. Hemolysis occurs even though the tubule is not fully everted (bar: 50 μm). K. Tubule of a discharged isorhiza, stained with brilliant cresyl blue, no hemolysis occurs (bar: 50 μm).*

produced a halo along the everted tubule (Fig. 1 and 2). Even if the tubule was not fully everted, for example the anisorhiza on Figure 2I, hemolysis nevertheless occurred. Thus we conclude that in most cases a complete eversion of the tubule is not necessary for the release of biologically active components (see also Tardent *et al.*, 1985).

Although we did not optimise the conditions of this method for each individual species tested, it allowed us to analyze the hemolytic capacity of several different nematocysts and the mode of toxin release. This method may be a simple tool for identifying hemolytic components in nematocysts and for comparative analysis of the hemolytic potency of different types of nematocysts in a particular species.

## ACKNOWLEDGEMENT

We thank Drs. Marianne Bienz and Robert Stidwill for carefully reading the manuscript and Dr. J.W. Burnett for the generous gift of nematocysts.

## REFERENCES

Blanquet, R.S. 1977. Cnidarian venoms. *In* A.W. Bernheimer (ed.) *Perspectives in Toxinology*, pp. 149-167. John Wiley and Sons, New York.

Burnett, A.L. 1960. The maturation of nematocysts in *Hydra*. Ann. Soc. Roy. Belg. 90:269-280.

Burnett, J.W. and G.J. Calton. 1977. Review article: The chemistry and toxicology of some venomous pelagic coelenterates. Toxicon 15:177-196.

Ferlan, I. and D. Lebez. 1974. Equinatoxin, a lethal protein from *Actinia equina*. I. Purification and characterization Toxicon 12:57-61.

Hessinger, D.H. and H.M. Lenhoff. 1973. Assay and properties of the hemolysis activity of pure venom from the nematocysts of the acontia of the sea anemone *Aiptasia pallida*. Arch. Biochem. Biophys. 159:629-638.

Hessinger, D.H. and H.M.Lenhoff. 1976. Membrane structure and function. Mechanism of hemolysis induced by nematocyst venom: Roles of phospholipase A and direct lytic factor. Arch. Biochem. Biophys. 173:603-613.

Hulet, W.H., J.L. Belleme, G. Musil and Ch.E. Lane. 1974. Ultrastructure of *Physalia* nematocysts. *In* H.J.Humm and Ch.E. Lane (eds.), *Bioactive Compounds from the Sea*, pp. 99-114, Marcel Dekker, New York.

Klug, M. 1986. Beitrag zur Kenntnis biologisch aktiver Komponenten isolierter Nematocysten von *Hydra attenuata*. Dissertation, Universität Zürich.

Lane, Ch.E. and E. Dodge. 1958. The toxicity of *Physalia* nematocysts. Biol. Bull. 115:219-226.

Mariscal, R.N. 1974. Nematocysts. *In* H.M.Lenhoff and L. Muscatine (eds.) *Coelenterate biology, reviews and new perspectives*, pp. 129-178, Academic Press, New York.

Russell, F.E. 1984. Marine Toxins and Venomous and Poisonous Marine Plants and Animals (Invertebrates). *In* J.H.S. Blaxter, F.S. Russell and M. Yonge (eds.), *Advances in Marine Biology*, 21:104-132.

Sutton, J.S. and J.W. Burnett. 1969. A light and electron microscopic study of nematocytes of *Chrysaora quinquecirrha*. J. Ultrastruct. Res. 28:214-234.

Tamkun, M.M. and D.H. Hessinger. 1981. Isolation and partial characterization of a hemolytic and toxic protein from the nematocyst venom of the portuguese man-of-war *Physalia physalis*. Biochem. Biophys. Acta 667:87-98.

Tardent, P., T. Holstein, J. Weber and M. Klug. 1985. The morphodynamics and actions of stenotele nematocysts in *Hydra*. Arch. Sc. Genève 38:401-418.

Weber, J., M. Klug and P. Tardent. 1987. Chemistry of *Hydra* nematocysts (this volume).

# 30

# DYNAMICS OF RELEASE OF FREE CALCIUM DURING THE DISCHARGE OF HOLOTRICHOUS ISORHIZA OF NEMATOCYSTS OF *PELAGIA NOCTILUCA*.

A. Salleo
G. La Spada
M.G. Denaro
G. Falzea

Institute of General Physiology
University of Messina, Italy

## SUMMARY

To observe the release of free $Ca^{2+}$ during the discharge of holotrichous isorhiza nematocysts from *Pelagia noctiluca* the light emitted by aequorin was recorded when a suspension of isolated resting nematocysts was added to a medium containing both aequorin and the discharging agent $SCN^-$. The light recordings had two components: a sudden spike-shaped one, and a delayed one having a rise phase that began 31 (±6.5) sec after contact of nematocysts with $SCN^-$ and a peak at 102 ( ± 20) sec, followed by an exponential decay that did not depend on exaustion of aequorin. During the evolution of the entire light signal less than 10% nematocysts discharged. This shape of the light signal was not peculiar response to $SCN^-$, because it was quite similar when the capsules were discharged by thioglycolate. Finally, the signal was significantly reduced when the discharge rate was slowed with $La^{3+}$. These data suggest that free $Ca^{2+}$ is released from the discharging nematocysts and that this occurs before the eversion of the tubule.

## INTRODUCTION

Calcium has been proposed (i) to aggregate protein molecules in the capsule fluid (Lubbock and Amos, 1981; Lubbock *et al.*, 1981; Gupta and Hall, 1984;), (ii) to maintain the apical flaps in the closed position and (iii) to stabilize $Ca^{2+}$ binding subunits that, in turn, mantain the everted tubule in a folded and coiled state ( Blanquet, 1970; Watson and Mariscal, 1985). The discharging effect of chaotropic anions, with the order of effectiveness of the Hofmeister series ( Salleo *et al.*, 1984 a, b) suggests that dissociation of proteins is involved in the discharge mechanism (Sawyer and Puckridge, 1973). Moreover, not only exogenous $Ca^{2+}$ added to the medium inhibits the discharge induced by chaotropic anions, but also other di- and polyvalent cations such as $Ba^{2+}$, $Ru^{3+}$, $Th^{4+}$ (Salleo, 1984; Salleo *et al.*, 1984) have an even stronger effect.

$Ca^{2+}$ has been shown in nematocysts by X-ray microanalysis (Lubbock *et al.* , 1981; Gupta and Hall, 1984; Mariscal, 1984) or by histochemical methods (Johnson and Lenhoff, 1958; Watson and Mariscal, 1984; 1985). It is not currently known whether during nematocyst discharge $Ca^{2+}$ is released (i) from protein molecules in the capsule fluid or (ii) from sealing structures. In either cases, free $Ca^{2+}$ is expected to be released into the medium. In the first case, treating a population of isolated holotricous isorhiza nematocysts of *Pelagia noctiluca* with $SCN^-$ which does not induce simultaneous discharge and only causes about 50% discharge by 15 min (Salleo *et al.* , 1983), most $Ca^{2+}$ is expected to be released in the medium with the capsule fluid as the discharge proceeds. In the second case the release of free $Ca^{2+}$ is expected to precede the discharge. Since the release of $Ca^{2+}$ is controversial (Watson and Mariscal, 1985) and since no data are available about the kinetics of $Ca^{2+}$ release during discharge, the purpose of the present investigation is to study the release of free $Ca^{2+}$ from $SCN^-$-stimulated nematocysts by measuring the light emitted from aequorin ( Shimomura *et al.* , 1963). Aequorin has an advantage: its high molecular weight (Blinks *et al.* , 1976, 1978) ensures that it can only be activated by released $Ca^{2+}$ since it cannot penetrate the nematocysts wall.

## MATERIALS AND METHODS

The nematocysts were extruded and isolated from tentacles of *Pelagia noctiluca* immediately after capture and stored (Salleo *et al.* , 1983, 1984 a).

To ascertain whether a detectable amount of free $Ca^{2+}$ is released during discharge, a preliminary test was performed with murexide to measure free $Ca^{2+}$ (Onhishi and Ebashi, 1963). A 2 - ml suspension of isolated nematocysts, whose concentration had been previously determined in a Thoma - Zeiss chamber (mean value : $608/mm^3$ ) was induced to discharge by adding NaSCN at a final concentration of 500 *mM* (Salleo *et al.* , 1984 b). After 60 min the suspension was centrifuged to remove the capsules. The supernatant containing the released contents of about 80% of nematocysts was centrifuged (3,000 x g) for 45 min in a Microconcentrator Centricon 10 in order to remove protein molecules having a molecular weight higher than 10,000 daltons. A 1-ml sample of filtrate was added to 9-ml of a solution containing 112.5 μ*M* murexide in 10 *mM* Tris-HCL pH 6.8. Although $SCN^-$ does not reduce $Ca^{2+}$ activity, (Dani *et al.*, 1983) $SCN^-$ was added to control solution at the same final concentration as to test solution. The absorbance difference obtained using a Varian Cary 219 between the isosbestic point (507 nm) and 540 nm was used as an index to compare the $Ca^{2+}$ concentration of test solution to that of control solutions.

Since detectable amounts of free $Ca^{2+}$ ( see below ) were found in the released contents using murexide, the experiments with aequorin were performed. Aequorin (Sigma, type III) was dissolved in 4 ml of distilled water. Before use 0.1 ml of aequorin solution was added to 0.3 ml of a solution containing 833 *mM* $SCN^-$ , 333 μ*M* $MgCl_2$ and 20 *mM* Tris-HCl at pH 7.4 (Azzi and Chance, 1969). The cuvette containing this reaction medium was introduced in the chamber of a Luminometer (LKB model 1250) equipped with a recorder (LKB model 2210). Then 200 μl of nematocysts suspension were added suddenly to the reaction medium and the light emission was recorded. Control solutions containing either 10, 20 or 100 μ*M* $CaCl_2$ were added to the same medium. It is worth nothing that $SCN^-$ reduced the light yield, as does Iodide (Blinks *et al.*, 1976). Thus, this anion was always present at the same concentration in the reaction medium either for test solutions or for control ones. Moreover, since the sensitivity of aequorin dissolved in distilled water decays quite rapidly (Blinks *et al.*, 1976) all measurements performed with each vial were completed within 5 hours. Reference curve were repeated in each experimental session. Comparative tests revealed that aequorin did not alter the discharge rate of nematocysts. The medium that had contained nematocysts was tested separately in order to ensure that it did not emit light comparable to that observed during the experiments with nematocysts.

In order to ascertain whether free $Ca^{2+}$ release is peculiar to discharge induced by $SCN^-$ we used also 25 or 50 *mM* Na thioglycolate at pH 8.2 (Blanquet and Lenhoff, 1966) as a discharging agent. It was observed that 25 *mM* thioglycolate induces a discharge rate similar to that of 500 *mM* $SCN^-$ without dissolving the discharged capsules (Salleo *et al.*, 1986). Known amounts of $Ca^{2+}$ were added as a control. During each light emission measurement the discharge rate of nematocysts of the same batch was observed under a microscope.

In the final experiment the inhibitory effect of Lanthanum on discharge as well as on $Ca^{2+}$ release was tested. This cation is believed to bind more strongly to the same sites as $Ca^{2+}$ (Weiss, 1974; Curtis *et al.* , 1986). Nematocysts were suspended in either 2, 5, or 25 *mM* $LaCl_3$ for 5 min before applying 500 *mM* $SCN^-$ as a discharging agent. A 25 *mM* concentration, which inhibited completely the discharge (see below) was chosen for subsequent tests. Since such a concentration could trigger a high luminescence in the aequorin molecule, the nematocysts previously incubated in 25 *mM* $La^{3+}$ were repeatedly washed in distilled water and centrifuged so that the final concentration of $La^{3+}$ in the medium was lower than 2 *μM*. This concentration could not be detected with the amplification factor generally employed. The discharge rate of lanthanized and washed nematocysts was significantly slowed with respect to control. Therefore, they were then examined for aequorin light emission during $SCN^-$ induced discharge as described above. This test was repeated at 15, 30 and 60 min after washing.

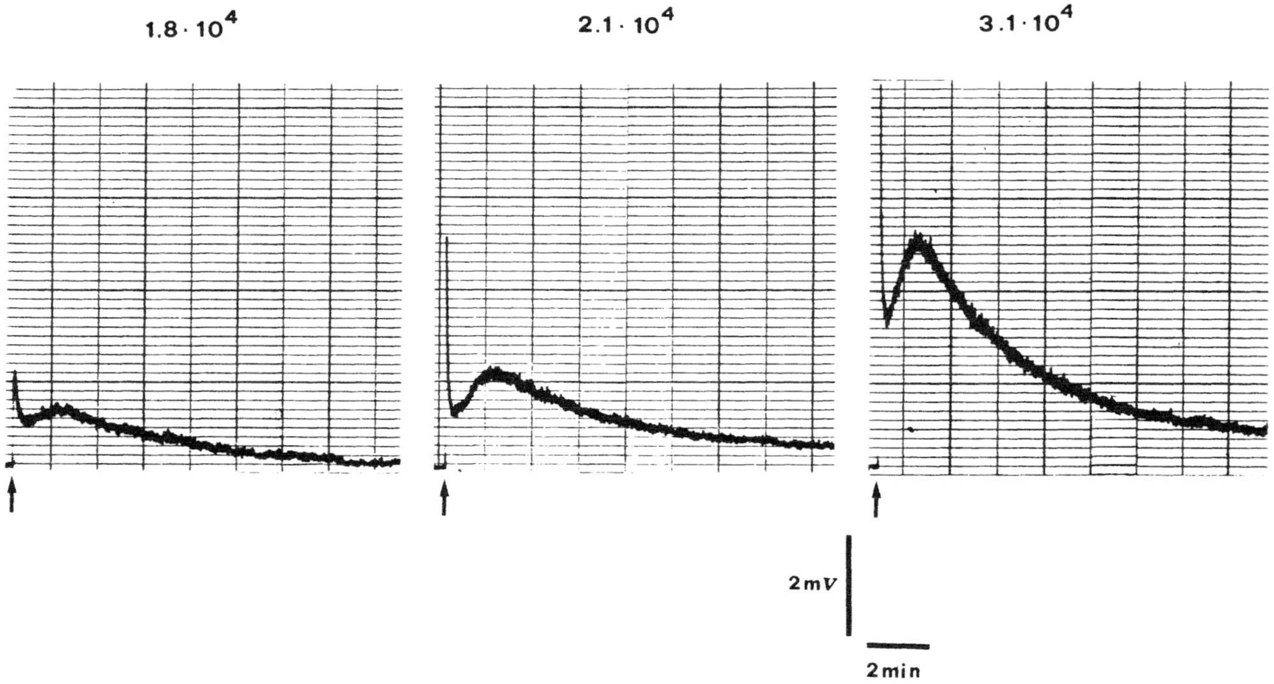


Fig. 1. Light emission during discharge induced by $SCN^-$.

## RESULTS

The preliminary experiment performed with murexide revealed that the discharge of 486.4 (± 128) nematocysts/$mm^3$ releases in the medium an average amount of 675 (± 236) moles of free $Ca^{2+}$. Therefore, it could be predicted that, even with lower concentrations of nematocysts as well as with a lesser percentage of discharged capsules, free $Ca^{2+}$ release could be detected with aequorin.

Typical light emission signals obtained during the discharge are shown in Figure 1. As soon as the nematocysts dropped into the reaction medium, an early light emission was observed, that evolved with a sharp peak and a rapid decay. The time interval between the introduction of the resting nematocysts into the cuvette and the onset of this signal could not be measured with the equipment employed, but it was certainly of < 1 sec. Such a first phase was followed by a slower one that began 31.1 (± 6.5) sec after contact with $SCN^-$ and reached a maximum after 102 (±19.9) sec (Tab.1). Then the light emission decayed exponentially without reaching the baseline during 15 min.

The decay of light emission which follows the slower component was not found to depend on exaustion of aequorin. In fact, new responses were observed by adding repetitively nematocysts to the reaction medium.

We ensured that light emission was due to discharge and not to the mere presence of resting nematocysts in the cuvette. When nematocysts were introduced in a reaction medium where no $SCN^-$ had been added, no signal was detected, although in the absence of $SCN^-$ the light yield of aequorin is much higher. Therefore, the light emission could be interpreted as being dependent on discharge. This was confirmed also by the fact that the signal amplitude increases with the concentration of nematocysts and shows that the light emission depends on discharge (Figure 1).

In Figure 2 the response of aequorin and the discharge rate are compared. During the evolution of both rapid or slow signal phases less than 10% of the nematocysts had discharged. Then, while the signal decayed, the discharge continued so that by 15 min the discharged nematocysts exceeded 50%.

TABLE I. Onset and peak of the slow component of light emission signal after contact of nematocysts with the discharging agent.

| | Onset (sec) | Peak (sec) |
|---|---|---|
| | 30 | 90 |
| | 35 | 135 |
| | 35 | 115 |
| | 35 | 100 |
| | 20 | 85 |
| | 20 | 70 |
| | 35 | 95 |
| | 35 | 120 |
| | 35 | 110 |
| Mean | 31.1 | 102 |
| S.D. | ±6.5 | ±19.9 |

The time course of light emission signal produced when thioglycolate was used as a discharging agent in place of $SCN^-$ was similar to that observed with the latter, although its amplitude was much greater because thioglycolate does not impair the light yield of aequorin (Figure 3). Thioglycolate at 50 *mM* produced a larger signal as well as a faster discharge rate than did a 25 *mM* concentration.

The incubation in $La^{3+}$ produced a significant inhibition of $SCN^-$ induced discharge (Figure 4). $La^{3+}$ at 25 *mM* inhibited discharge completely. This effect was not abolished by the removal of $La^{3+}$ from the medium, although it was significantly reduced. Moreover, the inhibitory effect gradually disappeared so that, while the discharge rate appeared to be slower with respect to control 15 min after washing, by 60 min it had became quite normal (Tab.2).

The light emission observed with lanthanized nematocysts was significantly reduced with respect to the control. In Figure 5 it is shown the light signal at different time intervals from the removal of $La^{3+}$. The rapid phase of the signal in-

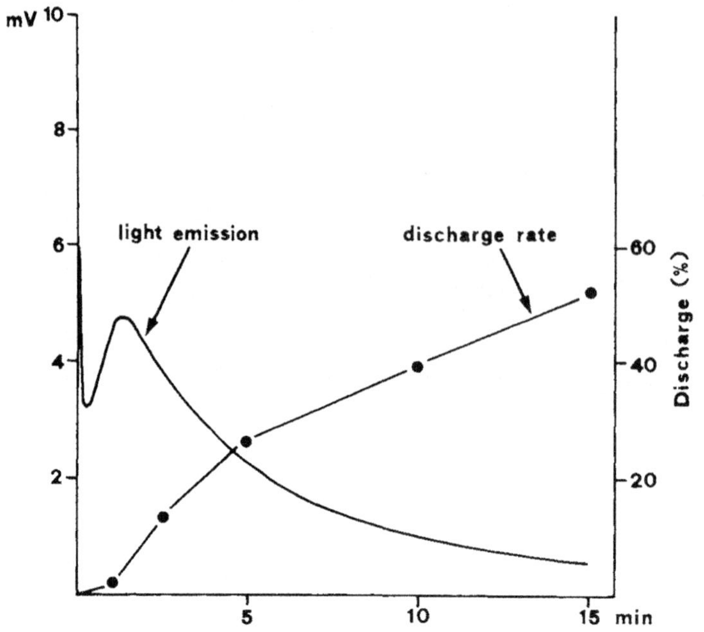


Fig. 2. Light emission and discharge rate. Note that light emission tends to subside during the beggining of discharge.

creased gradually with time, so that by 60 min it did not differ significantly from that of untreated nematocysts. The slow phase increased to a lesser extent.

A quantitative interpretation of the response to aequorin in terms of [$Ca^{2+}$] has not been attempted (Blinks *et al.*, 1978) owing to the following limitations: (i) the kinetics of aequorin - $Ca^{2+}$ reaction is complicated by its sensitivity to mixing, temperature, and rapidity of changes in [$Ca^{2+}$]; (ii) the relation between [$Ca^{2+}$] and light emission varies with resting [$Ca^{2+}$], as suggested by Blinks *et al.* (1976). Nevertheless, it is worthy to note that the amplitude of both phases of light signal increases with the concentration of nematocysts. The order of magnitude of released $Ca^{2+}$ estimated by comparing the experimental signal to that of control solution did not differ substantially from that observed with murexide (0.9

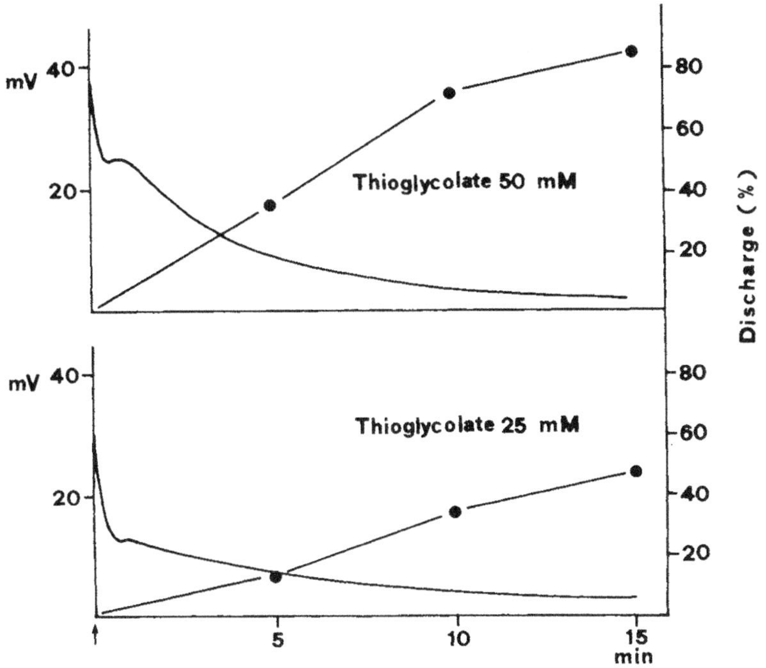


Fig. 3. Light emission during discharge induced by thio-glicolate.

picamole and 1.4 picamole/discharged nematocyst respectively). It should be stressed that the above quantities reflect free $Ca^{2+}$ released during discharge and not the total calcium content of nematocysts.

DISCUSSION

The results of the present investigation reveal that free $Ca^{2+}$ is released by discharging nematocyst of *Pelagia noctiluca*. Since most $Ca^{2+}$ appears in the medium while only 10% nematocysts have discharged, it is reasonable to infer that $Ca^{2+}$ release precedes the eversion of the tubule. Such a result is in agreement with the possible role of calcium proposed by Watson and Mariscal (1985), although they were unable to find calcium surrounding the discharging nematocysts. In fact whether

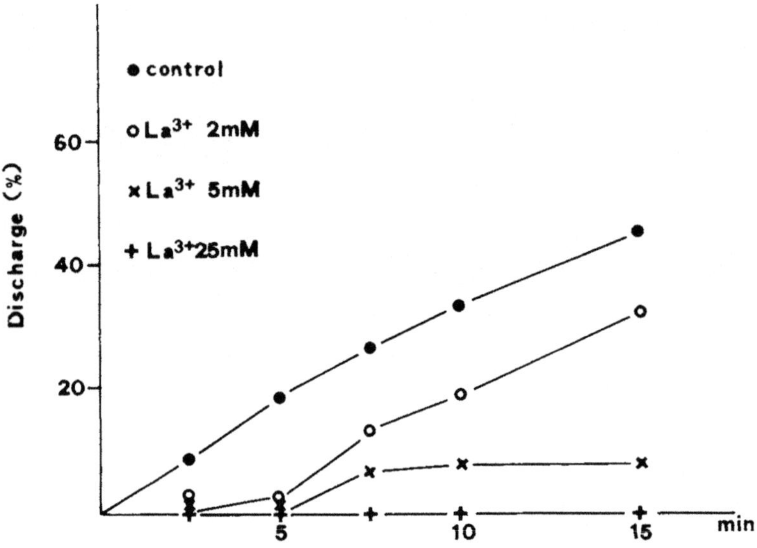


Fig. 4. Discharge rate of lanthanized nematocysts.

calcium is bound either to the seams of the tip and/or to the subunits placed on both surfaces of the tubule in pre-discharged nematocysts, it should be released prior to the eversion of the tubule. Lubbock *et al.* (1981) observed the presence of calcium in the fluid surrounding discharging nematocysts. If $Ca^{2+}$ is unbound from proteins of the capsule fluid, although some $Ca^{2+}$ could diffuse outward through the wall prior to discharge, thus giving rise to a light signal, a part of it is expected to remain in the capsule fluid. Therefore, the concentration of $Ca^{2+}$ in the medium should increase further as an

TABLE II. Discharge rate of untreated and lanthanized nematocysts at different time intervals after washing.

| Minutes | 2.5 | 5 | 7.5 | 10 | 15 |
|---|---|---|---|---|---|
| | | | Discharged nematocysts (%) | | |
| Untreated | 8.66 | 19.29 | 26.82 | 33.85 | 46.11 |
| S.D. | (±7.6) | (±7.4) | (±4 ) | (±9 ) | (±6.4) |
| Lanthanized | | | | | |
| 15 min | 3.76 | 7.63* | 12.05* | 17.66* | 20.6** |
| S.D. | (±3.21) | (±6.02) | (±7.45) | (±13.3) | (±11.4) |
| 30 min | 3.66 | 8.7* | 12.93* | 20 | 33.26 |
| S.D. | (±2.03) | (±3.11) | (±5.55) | (±9.7) | (±15.72) |
| 60 min | 7.13 | 14.46 | 28. | 32.36 | 40.03 |
| S.D. | (±5.6) | (±5.95) | (±13.8) | (±10 ) | (±8.5) |

Experimental vs.control
* $P < 0.05$
** $P < 0.01$

increasing number of nematocysts eject the capsule the capsule fluid.

The early rapid phase of light emission that started as soon as the nematocysts were dropped into the reaction medium could be interpreted to be the results of a higher affinity of $Ca^{2+}$ for aequorin than for the nematocysts proteins. On the other hand, when the reaction medium did not contain $SCN^-$ no response to aequorin was observed; thus, this interpretation does not seem to hold. It is reasonable to propose that the earlier phase of light signal is due to free $Ca^{2+}$ unbound from some structure near the surface as a consequence of contact with $SCN^-$. The slower phase of the light signal evolves in less than 2 min. Then $Ca^{2+}$ release probably subsides.

When the discharge is induced by thioglycolate free $Ca^{2+}$ is released with the same time course. This discharging agent acts by reducing disulfide bridges in a collagenous protein which is a main component of the capsule wall (Blanquet and Lenhoff, 1966; Phelan and Blanquet, 1985). Such a mechanism of discharge is completely different from that of $SCN^-$. Whether by reducing disulfide bridges or by interacting with hydrogen bonds (Sawyer and Puckridge, 1973), both discharging agents supposedly share the common effect of dissociating protein molecules. In this case not only free $Ca^{2+}$ is released, but also a swelling of the capsule is expected to precede the discharge (Holstein and Tardent, 1984; Salleo *et al.* , 1986); Tardent and Holstein, 1982).

Finally, it has been demonstrated that $La^{3+}$ inhibits the $SCN^-$ induces discharge, and that in lanthanized and washed nematocysts the changes in discharge rate are paralleled by those in light signal. This result confirms that free calcium release is a relevant part in the discharge mechanism.

In conclusion, the results of the present investigation confirm that calcium plays a role in maintaining the nematocysts in the resting state. Moreover, they suggest that in the holotrichous isorhiza nematocysts of *Pelagia noctiluca* the Ca-binding sites that release free $Ca^{2+}$ during the discharge are likely to be located in the operculum and/or in the tubule rather than in the capsule fluid. Nevertheless, since total calcium has not been determined in the nematocysts of *Pelagia noctiluca* it cannot be excluded that an additional amount of calcium is kept bound to proteins during discharge, and does not participate to the discharge process.

## REFERENCES

Azzi, A. and B. Chance. 1969. Energized state of mitochondria Life - Time and ATP equivalence. Biochem. Biophys. Acta 189:141-151.

Blanquet, R. 1970. Ionic effect on discharge of the isolated and *in situ* nematocysts of the sea anemone *Aiptasia pallida*: a possible role of calcium. Comp. Biochem. Physiol.35:451-461.

Blanquet, R. and H. M. Lenhoff. 1966. A disulfide-linked collagenous protein of nematocyst capsules. Science 154:152-153.

Fig. 5. Light emission during discharge of lanthanized nematocysts. A: untreated nematocysts; B: 15 min after washing in distilled water; C: 30 min; D: 60 min. in the central diagram are shown the respective rates of discharge.

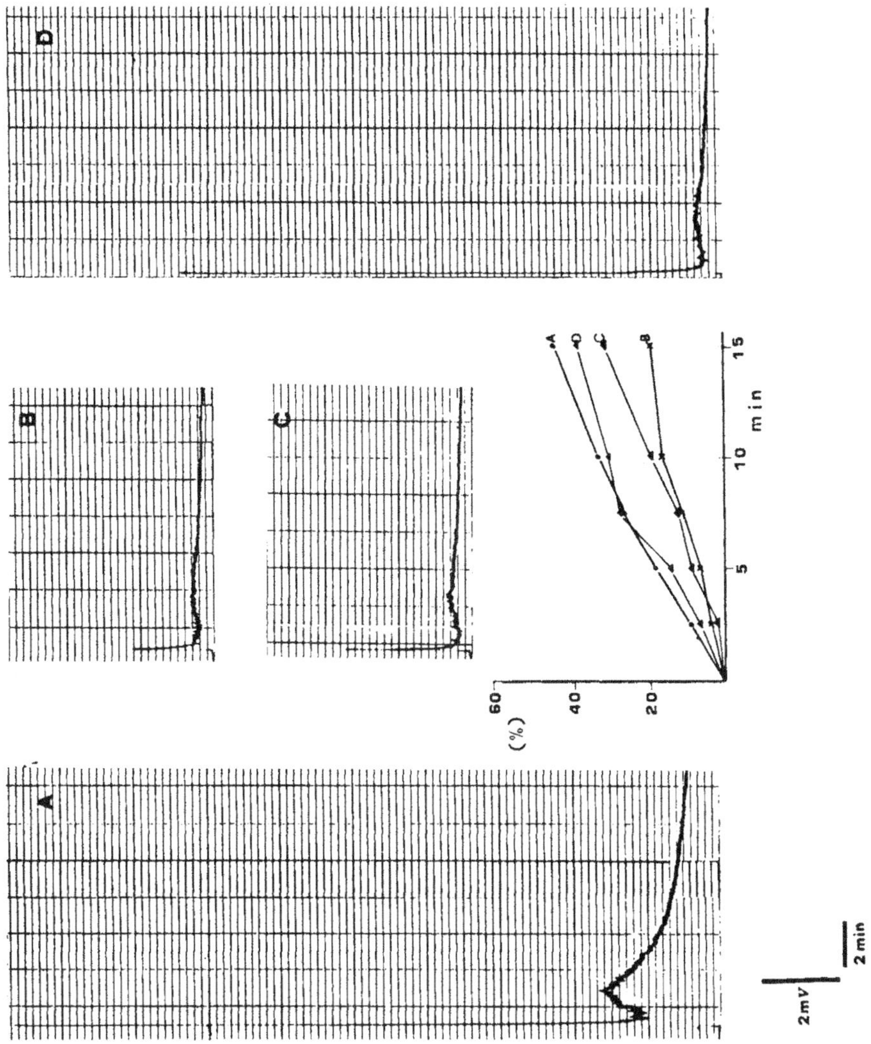
A
B
C
D
(%)
60
40
20
5
10
15
min
2mV
2 min

Blinks, J. R., F. G. Prendergast, and D. G. Allen. 1976. Photoproteins as biological Calcium indicators.Pharmacol. Rev.28:2-87.

Blinks, J. R., R. Rudel, and S. R. Taylor. 1978. Calcium transient in isolated amphibian skeletal muscle fibres: detection with aequorin.J.Physiol.277:291-323.

Curtis, M. J., D. M. J. Quastel,and D.A.Saint. 1986. Lanthanum as a surrogate for Calcium in transmitter release at mouse motor nerve terminals. J.Physiol. 373:243-260.

Dani, J. A., J. A. Sanchez,and B. Hill. 1983. Na channel gating Ca electrode response. J.Gen.Physiol. 81:255-281.

Gupta, V. L. and T. A. Hall. 1984. Role of high concentrations of Ca, Cu and Zn, in the maturation and discharge *in situ* of sea anemone nematocysts as shown by X-ray microanalysis of cryosection. In:L.Bolis, J. Zadunaisky and R. Gilles. (eds.), *Toxins, Drugs and Pollutants in marine animals*,pp. 77-95. Springer-Verlag,New York.

Holstein, T. and P. Tardent. 1984. An ultrahigh-speed analysis of exocytosis: nematocysts discharge. Science 223:830-833.

Johnson F. B. and H. M. Lenhoff. 1958. Histochemical study of purified *Hydra* nematocysts. J.Histochem.Cytochem. 6:394

Lubbock, R. and W. B. Amos. 1981. Removal of bound calcium from nematocysts causes discharge. Nature 290:500-501.

Lubbock, R., B. L. Gupta,and T. A. Hall. 1981. Novel role of calcium in exocytosis: mechanism of nematocysts discharge as shown by X-ray microanalysis. Proc.Natl.Acad.Sci. USA 78:3624-3628.

Mariscal,R. N. 1984. Cnidaria: Cnidae. In:J.Bereiter-Hann, A. G. Maltotsy and K. S. Richards (eds.) *The Biology of Integuments* pp.57-68. Springer-Verlag,New York.

Onishi T. and S. Ebashi. 1963. Spectrophotometrical measurement of istantaneous calcium binding of the relaxing factor of muscle. J.Biochem. 54:506-511.

Phelan M. A. and R. Blanquet. 1985. Characterization of nematocysts proteins from the sea anemone *Aiptasia pallida* and *Pachycerianthus torreyi*. Comp.Biochem.Physiol. 81:661-666.

Salleo, A. 1984. Discharge mechanism of the nematocysts of *Pelagia noctiluca*. In: L.Bolis, J. Zadunaiskj and R. Gilles. (eds.), *Toxins, Drugs and Pollutants in marine animals*.pp. 63-76. Springer-Verlag,New York.

Salleo, A., G. La Spada,and M. Alfa. 1983. Blockage of trypsin induced discharge of nematocysts of *Pelagia noctiluca* by $Ca^{2+}$ Mol.Physiol.3:89-97.

Salleo, A., G. la Spada, M. G. Denaro,and G. Falzea. 1984*a*. Discharging effect of anions and inhibitory effect of divalent cations on isolated nematocysts of *Pelagia noctiluca*. Mol.Physiol. 5:23-34.

Salleo, A., G. La Spada, M. G. Denaro,and G. Falzea. 1984*b*. Discharging effectiveness of lyotropic anions on nematocysts of *Pelagia noctiluca*. Mol.Physiol. 6:19-26.

Salleo, A., G. La Spada, M. G. Denaro,and G.Falzea. 1986. Effect produced by $SCN^-$ and thioglycolate on isolated nematocysts of *Pelagia noctiluca*. Cell.Mol.Biol.32(6):661-666.

Sawyer, H. W. and J. Puckridge. 1973. The dissociation of proteins by chaotropic salts.J.Biol.Chem. 248:8429-8433.

Shimomura, O., F. H. Johnson,and Y. Saiga. 1963. Microdetermination of Calcium by aequorin luminescence. Science 140: 1339-1340.

Tardent, P. and T. Holstein. 1982. Morphology and morphodynamics of the stenothele nematocysts of *Hydra attenuata*. (Cnidaria). Cell.Tissue Res. 224:269-290.

Watson, G. M. and R. N. Mariscal. 1984. Calcium cytochemistry of nematocyst development in catch tentacles of the sea anemone *Haliplanella luciae* (Cnidaria: anthozoa) and the molecular basis for tube inversion into the capsule. J. Ultrastruct.Res. 86:202-214.

Watson, G. M. and R. N. Mariscal. 1985. Ultrastructure of nematocysts discharge in catch tentacles of the sea anemone *Haliplanella luciae* (Cnidaria). Tissue and Cell 17: 199-213.

Weiss, G. B. 1974. Cellular pharmacology of lanthanum. Pharmacol. Rev. 14:343-354.

# 31

# POLYMERIZED ACTIN IN THE APICAL REGION OF HYDRA NEMATOCYTES[1]

Robert P. Stidwill
Thomas G. Honegger
Pierre Tardent

Department of Zoology
University of Zürich
Zürich, Switzerland

SUMMARY

The occurrence of polymerized actin in the apical part of hydra nematocytes is demonstrated by staining with the F-actin specific probe rhodamine phalloidin. Decoration of the filaments with heavy meromyosin reveals a uniform polarity. Detergent treated tentacles observed in the scanning electron microscope allow the localization of actin filament bundles connected to other cytoskeletal elements. In the transmission electron microscope up to eight hundred strictly hexagonally bundled actin filaments can be counted on cross sections through the microvillar structures supporting the cnidocil.

Possible functions of the actin filament bundles as part of a complex cytoskeleton are discussed.

## I. INTRODUCTION

The occurrence of actin in vertebrate and invertebrate nonmuscle cells is well established (Goldman, et al., 1975). Its crucial role has been demonstrated (i) in actomyosin-based events of cell motility and cell shape changes (Weeds, 1982) (ii) in actin-based shape changes regulated by assembly,

[1]Supported by Grant No. 3.072-0.84 of the Swiss National Science Foundation.

disassembly (Tilney, et al., 1973) or rearrangement of actin filaments (Burridge, 1981) and (iii) as the major constituent of scaffolding cytoskeletal elements with static functions. This latter property has been particularly well investigated in the microvilli of intestinal epithelial cells (see Mooseker, 1985) and in the stereocilia of the vertebrate inner ear hair cells (DeRosier, et al., 1980). In the latter case it has been shown that displacement of the actin filament bundles results in the rapid opening of ionic channels. Mechanical stimulation of the cytoskeleton is thus transduced into electrical signals (Hudspeth, 1985).

The cytoskeleton of cnidarian nematocytes may serve as an invertebrate model for cytoskeletal structures with stimulus-transducing functions. In the present study we have analyzed the arrangement and polarity of actin filaments in the apical part of hydra nematocytes by means of fluorescence microscopy, SEM and TEM. We have concentrated on the class of largest nematocytes, the stenoteles. Our results however most probably hold true for all four types of nematocytes in hydra.

## II. MATERIAL AND METHODS

Animals of the species *Hydra attenuata* were cultured as described by Loomis and Lenhoff (1956).

Fluorescence microscopy: Staining with rhodamine coupled phalloidin was carried out according to the method of Wood and Novak (1982). Observations were made on a ZEISS photomicroscope equipped with epifluorescence illumination.

TEM: Specimens were fixed in 0.05M cacodylate buffer pH 7.4 containing 1% glutaraldehyde and 0.2% tannic acid. After postfixation in $OsO_4$ (1% in 0.1M phosphate buffer pH 6.0) the objects were embedded in Spurr/Epon (Honegger, 1983). Ultrathin sections were inspected in a Siemens ELMISKOP 102. For the study using heavy meromyosin (HMM) whole animals were incubated in a permeabilizing buffer (modified from Matsudaira and Burgess, 1982) containing 75mM KCl, 5mM $MgCl_2$, 1mM EGTA (Ethyleneglycol-bis-(β-aminoethylether)- N,N,N',N'-tetraacetic Acid), 1mM DTE (Dithioerythritol), 10mM Imidazole, pH 6.9 with 1% Triton-X-100 for 10min before being exposed to HMM (1.2mg/ml) in the same buffer without the detergent for another 10min. The objects were then fixed and processed for TEM.

SEM: Specimens were fixed and postfixed as described for the TEM study except that no tannic acid was added. After critical point drying and sputter coating with gold or gold/palladium, objects were studied in a Cambridge S-4

Stereoscan. Animals were pretreated with an extracting buffer slightly modified from Jeffery and Meier (1983) containing 10mM PIPES, 300mM Sucrose, 100mM KCl, 1mM EGTA, 4M glycerol and 0.25% Triton-X-100 at a pH of 6.8.

Rhodamine coupled phalloidin was kindly provided by Dr. T. Wieland, Max Planck Institut, Heidelberg FRG.

All chemicals including HMM were purchased from SIGMA.

## III. RESULTS

### A. Phalloidin Staining

After incubation of whole animals with rhodamine coupled phalloidin the pattern of polymerized actin in nematocytes can be seen in the fluorescence microscope (Fig. 1). The approximately nine rod-like structures around the cnidocil are stained. They can be detected in all four types of nematocytes although the intensity of fluorescence and thus apparently also the amount of polymerized actin present is higher in stenoteles than in the other two nematocyte types.

### B. SEM

Treatment of animals with a buffer which stabilizes actin filaments and contains the detergent Triton X-100, removes the membranes and, depending on the duration of the treatment, some of the cellular material in the vicinity of the capsule.

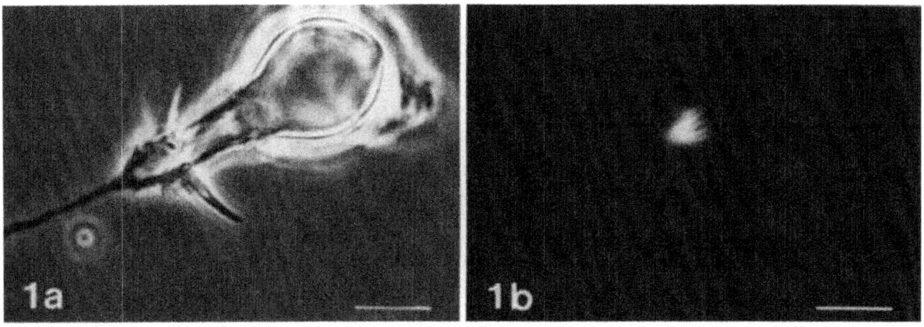


Fig. 1: Isolated discharged stenotele stained with rhodamine coupled phalloidin. a.) phase contrast photograph. b.) the identical cell as it appears with epifluorescence illumination. (Bars = 10µm).

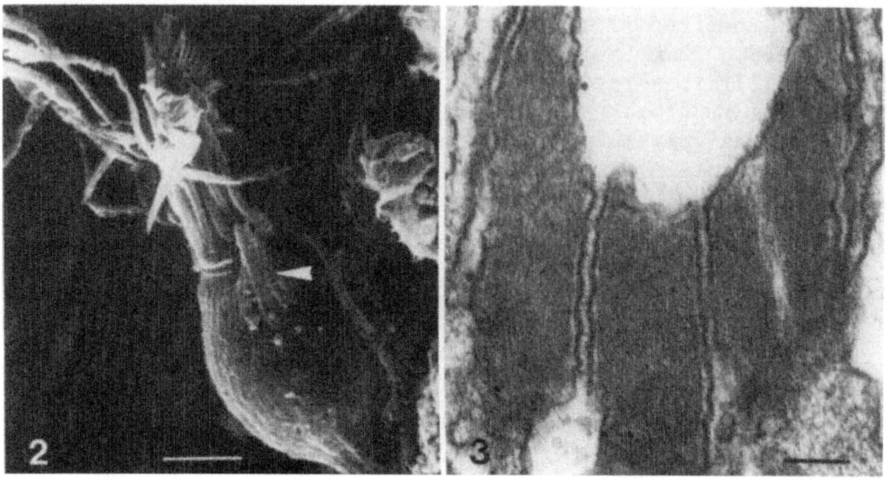

Fig. 2: Scanning electron micrograph of a stenotele after extraction with a detergent containing buffer (see methods). Arrowhead indicates the actin filament bundles. (Bar = 7.5μm).

Fig. 3: TEM of longitudinal section through the microvillar region showing parallel strands of filaments. (Bar = 0.25μm).

Thereby a very complex basket-like structure around the nematocysts becomes visible in the scanning electron microscope (Fig. 2). Elements corresponding to the bundles of actin filaments seen in the fluorescence microscope can easily be identified at the apical part of the capsules in contact with other cytoskeletal structures.

## C. TEM

In longitudinal sections through the microvillar region highly ordered parallel strands of actin filaments can be seen (Fig. 3). They appear to be bundled in a strictly hexagonal paracristalline order in cross sections (Fig. 4), where up to eight hundred individual filaments can be counted at the largest diameter.

Incubation of detergent-treated tentacles with HMM results in the characteristic arrowhead decoration of the actin filaments (Fig. 5). The pointed ends of the decorated filaments are located at the proximal parts of the microvilli, i.e. pointing from the apex of the cell to its base.

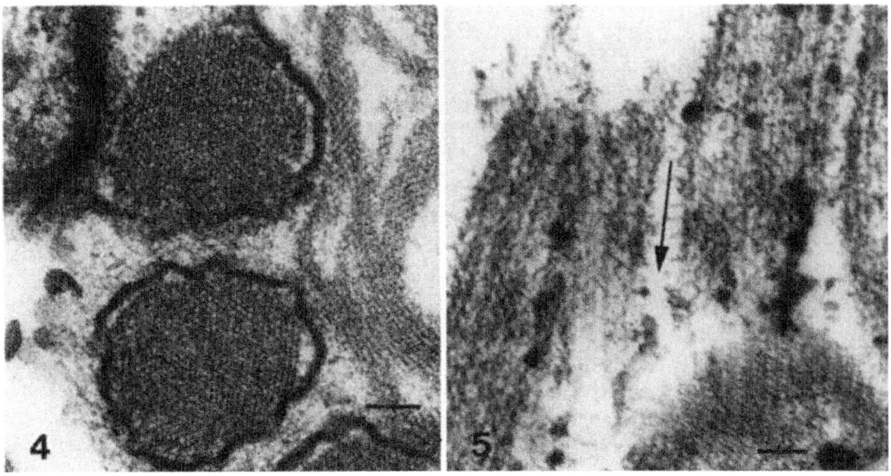


Fig. 4: TEM of cross section through the microvillar region. Note the hexagonal packing of the filaments. (Bar = 0.1µm).

Fig. 5: Decoration of the actin filaments with heavy meromyosin. The characteristic arrowheads all point to the interior of the cell revealing a uniform polarity of the filaments. The arrow indicates filament polarity. (Bar = 0.1µm).

## IV. DISCUSSION

While actin in its polymerized and unpolymerized form is found in virtually every cell in the animal kingdom, the occurrence of bundled actin filaments with static functions is generally limited to the core filaments of microvillus like structures.

In hydra "supporting rods of the cnidocil" were described by various authors (Hyman, 1940; Chapman and Tilney, 1959). Wood and Novak (1982) first demonstrated the presence of polymerized actin in these elements by staining with the F-actin specific probe NBD-phallacidin. Our experiments with rhodamine coupled phalloidin and our ultrastructural studies confirm and extend the findings of these authors.

Unlike in the microvilli of intestinal epithelial cells (Mooseker and Tilney, 1975) or in those of sea urchin eggs at fertilization (Burgess and Schroeder, 1977), where less than twenty actin filaments are found per microvillus, the actin containing structures in hydra nematocytes comprise up to eight hundred bundled filaments. This number is, to our knowledge, exceeded only by the stereocilia in the inner ear hair cells of certain vertebrates, e.g. the alligator lizard, where more than three thousand filaments were counted (DeRosier, et al., 1980; Tilney, et al., 1980).

Actin filament bundles in the nematocytes of hydra may be part of a cytoskeletal structure with a similar role as the elements of the inner ear hair cells, namely that of a transducer of mechanical stimuli from the cnidocil (a mechanoreceptor) to the cyst. The cytoskeletal complex may also be responsible for triggering the discharge of the latter. Additionally it quite possibly can modulate the incoming stimuli thus acting as a control element of the discharge mechanism.

At first examination many ultrastructural similarities between the elements in hydra nematocytes and the stereocilia of the inner ear may be seen. It is therefore very tempting to draw functional analogies between the two organelles, especially since in the nematocytes connections between the actin filament bundles and the cnidocil apparatus on the one hand and the cytoskeletal complex around the capsule on the other hand may speak in favour of a stimulus transducing function of the actin bundles. These speculations however are certainly premature because there are some rather crucial differences between the two structures. Apart from the different number of actin filaments, the bundling patterns of the filaments appears to be different. While the filaments in the stereocilia are bundled in a "festooning" pattern (Tilney, et al., 1983), the ones in the nematocytes of hydra seem, in our opinion, to be packed in a strictly hexagonal paracristalline fashion, whereas Slautterback (1967) saw a rather festooning pattern. Thickness and angle of the sections as well as the fixation protocol may be responsible for these contradictory findings. Only X-ray diffraction studies will answer the question about the steric arrangement of these structures.

The stereocilia have been found to be able to bend, probably due to the "festooning" bundling pattern (Tilney, et al., 1983). It seems rather unlikely to us that the bundling fashion of the nematocyte actin filaments would allow bending to a considerable degree, although we have no experimental evidence so far to support this hypothesis.

Beside their dimension and their binding capacity of phallo-toxins, the typical interaction of the filaments with heavy meromyosin is the third and clear proof that they are indeed composed of actin. The repeated arrowheads observed after decoration with HMM uniformly point to the cell interior, indicating that the in vitro fast growing ends of the filaments (the barbed ends) are located at the distal tips of the microvillar structures, a polarity which has been found nearly dogmatically in every so far investigated actin filament containing structure.

Investigations of the biochemical identity of the actin bundling protein(s) and its (their) properties will hopefully enable us to comment more precisely on the *in vivo* behaviour and possible function of the actin filament bundles in hydra nematocytes.

## ACKNOWLEDGEMENTS

We thank Dr. T. Wieland (Heidelberg) for his generous gift of rhodamine-phalloidin, Ms. Bea Gyr and Mr. Elmar Stöhr for excellent technical assistance, and Mr. Urs Jauch for advice in SEM preparation.

## REFERENCES

Burgess, D.R. and T.E. Schroeder. 1977. Polarized bundles of actin filaments within microvilli of fertilized sea urchin eggs. J. Cell Biol. 74: 1032-1037.

Burridge, K. 1981. Are stress fibers contractile? Nature 294: 691-692.

Chapman, G.B. and L.G. Tilney. 1959. Cytological studies of the nematocysts of hydra I. Desmonemes, isorhizas, cnidocil, and supporting structures. J. Biophys. and Biochem. Cytol. 5: 69-78.

DeRosier, D.J., L.G. Tilney, and E.H. Egelman. 1980. Actin in the inner ear: the remarkable structure of the stereocilium. Nature 287: 291-296.

Goldman, R.D., E. Lazarides, R. Pollack, and K. Weber. 1975. The distribution of actin in non-muscle cells. Exp. Cell Res. 90: 333-344.

Honegger, T.G. 1983. Ultrastructural and experimental investigations of sperm-egg interactions in fertilization of *Hydra carnea*. Roux's Arch. Dev. Biol. 192: 13-20.

Hudspeth, A.J. 1985. The cellular basis of hearing: The biophysics of hair cells. Science 230: 745-752.

Hyman, L.H. 1940. The Invertebrates: Protozoa through Ctenophora. McGraw-Hill Book Company, New York. pp 1-726.

Jeffery, W.R. and S. Meier. 1983. A yellow crescent cytoskeletal domain in ascidian eggs and its role in early development. Dev. Biol. 96: 125-143.

Loomis, W.F. and H.M. Lenhoff. 1956. Growth and sexual differentiation of hydra in mass culture. J. Exp. Zool. 132: 555-573.

Matsudaira, P.T. and D.R. Burgess. 1982. Organization of the cross-filaments in the intestinal microvilli. J. Cell Biol. 92: 657-664.

Mooseker, M.S. 1985. Organization, chemistry, and assembly of the cytoskeletal apparatus of the intestinal brush border. In: G.E. Palade (ed.), Annual Review of Cell Biology. Vol 1, pp 209-241.

Mooseker, M.S. and L.G. Tilney. 1975. The organization of the actin filament-membrane complex: filament polarity and membrane attachment in the microvilli of intestinal epithelial cells. J. Cell Biol. 67: 725-743.

Slautterback, D.B. 1967. The cnidoblast-musculoepithelial cell complex in the tentacles of hydra. Z. Zellforsch. 79: 296-318.

Tilney, L.G., S. Hatano, H. Ishikawa, and M.S. Mooseker. 1973. The polymerization of actin: its role in the generation of the acrosomal process of certain echinoderm sperm. J. Cell Biol. 59: 109-126.

Tilney, L.G., D.J. DeRosier, and M.J. Mulroy. 1980. The organization of actin filaments in the stereocilia of cochlear hair cells. J. Cell Biol. 86: 244-259.

Tilney, L.G., E.H. Egelman, D.J. DeRosier, and J.C. Saunders. 1983. Actin filaments, stereocilia and hair cells of the bird cochlea II. Packing of the actin filaments in the stereocilia and in the cuticular plate and what happens to the organization when the the stereocilia are bent. J. Cell Biol. 96: 822-834.

Weeds,A. 1982. Actin-binding proteins - regulators of cell architecture and motility. Nature 296: 811-816.

Wood, R.L. and P.L. Novak. 1982. The anchoring of nematocysts and nematocytes in the tentacles of hydra. J. Ultrastruct. Res. 81:104-116.

# 32

# THE ALLOMETRY OF NEMATOCYSTS[1]

J.C. Thomason

Department of Zoology
University of Newcastle upon Tyne
Newcastle upon Tyne
Tyne & Wear
U.K.
NE1 7RU

## SUMMARY

Data on the size (length and width) of nematocysts was obtained from the extensive literature on cnidarian taxonomy. The data covered 230 anthozoan and 87 meduzoan species and represented the measurement of 21,000 to 210,000 nematocysts. From these data additional variables - surface area, volume and a packing index (number of nematocysts $mm^{-2}$) - were computed. An allometric analysis quantified the relationship between surface area and volume ($SA = 6{\cdot}7\ VOL^{0{\cdot}67}$) as well as the relationship between surface area/volume ratio and the packing of the nematocysts ($SA/VOL = 0{\cdot}0094\ [\text{number of nematocysts } mm^{-2}]^{0{\cdot}048}$). The slope of the line for surface area on volume suggested that nematocysts maintain a low surface area to volume ratio, and this, in accordance with an osmotic hypothesis of nematocyst discharge, would maintain the high capsular forces required for effective discharge. In order to reduce the latency period between stimulus reception and discharge it is suggested that nematocysts could become smaller and not increase their surface area to volume ratio. The detection of outliers by residuals analysis indicated that these data might

[1]Funded by a grant from the Science and Engineering Research Council of the United Kingdom.

represent nematocysts that discharge by the release of tension within the capsule. It was concluded that the size and shape of nematocysts are at least partially the result of an optimum resolution of three constraints, i.e.the maintenance of a low surface area/volume ratio, the geometrical requirements of packing and the reduction of the latency period.

## I. INTRODUCTION

Owing to the usefulness of cnidom characteristics as an identification aid in certain cnidarian taxa there is a considerable wealth of data pertaining to the type, distribution and size of nematocysts. It was felt that further use could be made of the data by making an allometric analysis to find some clues about the constraints on nematocyst size and shape, especially in terms of the physico-chemical processes of discharge (sensu Lubbock et al., 1981) and the geometry of packing of the nematocysts.

## II. METHODS

### A. Sources and Coverage of the Data

The thirty-four papers used to obtain the data are listed in Table I. A total of 230 anthozoan species and 87 meduzoan species were covered in the data, providing 2157 sets of data. These data represent between 21,000 and 210,000 measurements of nematocysts.

### B. Encoding of the Data

Species, family and order were encoded for the data. Length, width and type of the nematocysts within the cnidom of each species were also encoded. Where a size range was given by a author for length and/or width, the median value was used. In order to reduce entry time of the data into the computer, length and width values (in µm) were rounded to the nearest whole integer. Nematocyst type was encoded as 'missing' when this was uncertain or not recorded in the literature.

TABLE I. A summary of the thirty-four papers used in the present study to obtain the raw data. The taxonomic subdivisions give a rough indication of the coverage of the papers.

| Taxon | Author(s) |
|---|---|
| Anthozoa | Parry (1951, 1952); Correa (1973); Stephenson (1929, 1935); Carlgren (1937, 1940, 1945, 1950); Weill (1934); Manuel (1977); den Hartog (1977a, 1977b, 1977c, 1980); Thomason & Brown (1986) |
| Scyphozoa & Cubozoa | Burnett (1971a, 1977b); Calder (1983); Calder & Peters (1975); Papenfuss (1936) |
| Hydrozoa | Kubota (1978); Hashimoto (1981); Kustor (1977); Ostman (1979); Nagao (1973); Yamada & Konno (1973); Purcell (1984); Russell (1938, 1939, 1940); Hesthagen (1971); Cairns (1983) |

C. Analysis of the Data

The raw data was stored and processed on an Amdahl mainframe computer operating on the Michigan Terminal System. All subsequent analyses and computations were performed using the program $SPSS^X$ (release 2:1). For a complete description of the program see the User's Guide (SPSS Inc., 1983).

From each pair of length (L) and width (W) values the surface area (SA) and volume (VOL) of the nematocyst were computed thus:

$$SA = 2\pi (W/2)\ L - W) + 4\pi (W/2)^2$$

$$VOL = 4/3\pi (W/2)^3 + \pi(W/2)^2\ (L - W).$$

Thus, if a nematocyst is spherical, L = W, hence:

$$SA = 4\pi (W/2)^2.$$

Likewise,

$$VOL = 4/3\pi (W/2)^3.$$

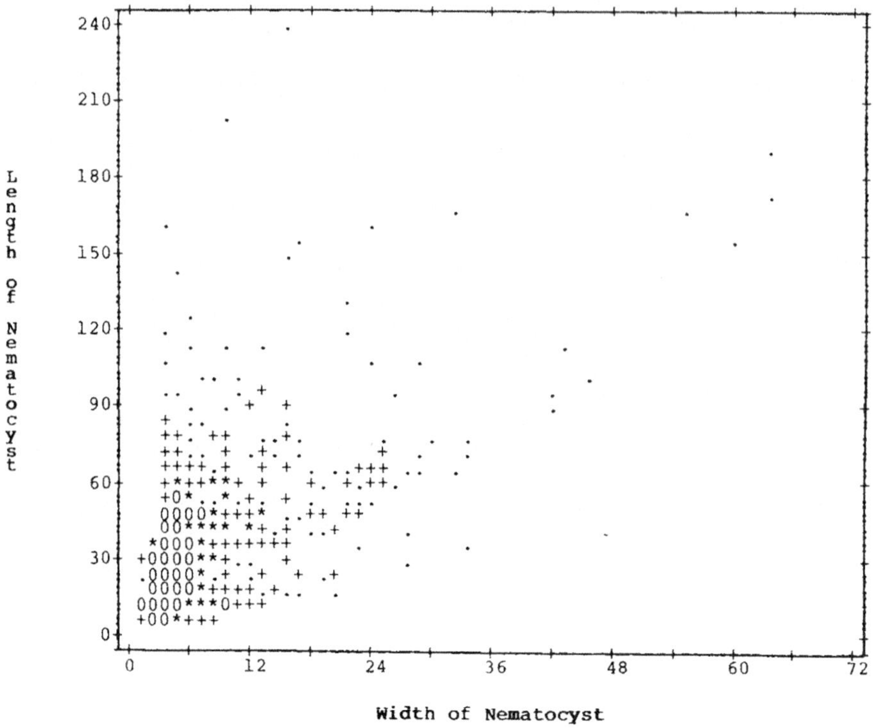

Length of Nematocyst
240
210
180
150
120
90
60
30
0
0
12
24
36
48
60
72
Width of Nematocyst


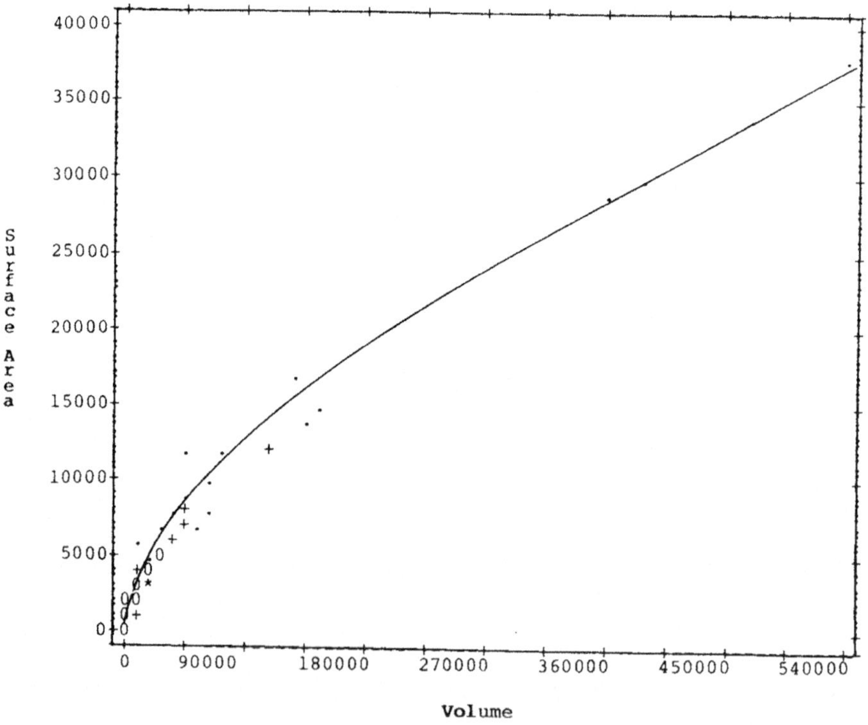

Surface Area
40000
35000
30000
25000
20000
15000
10000
5000
0
0
90000
180000
270000
360000
450000
540000
Volume

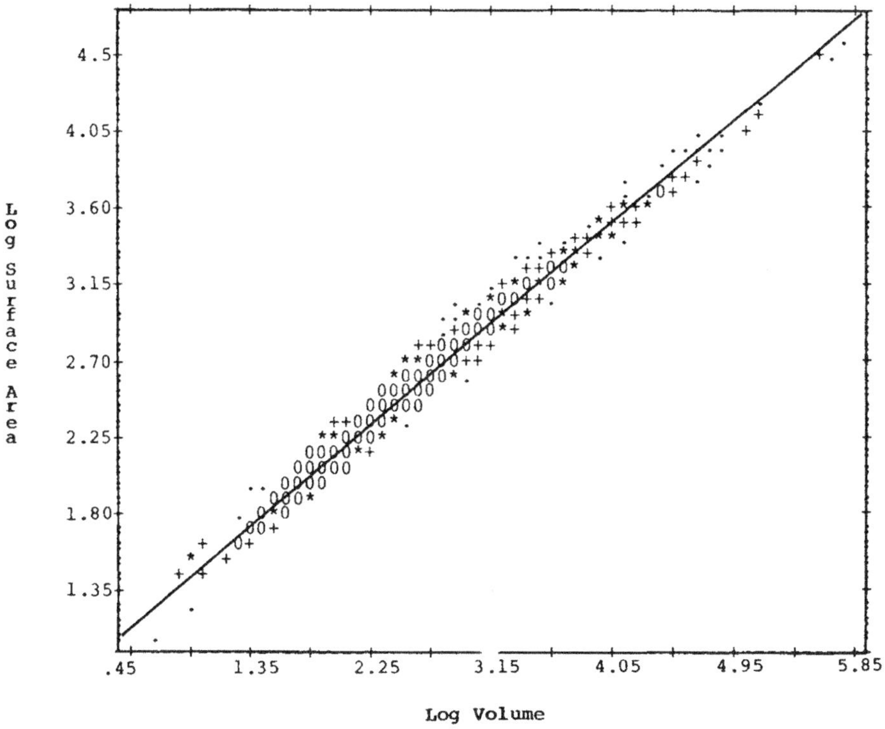


Fig. 1c.

Values obtained from these computations were used for the allometric analysis. Regression equations (intercepts, a and regression coefficients, b), squares of the Pearson product - moment correlation ($R^2$), and the residuals statistic, h (leverage coefficient) were calculated, with forced entry of the variables into the regression equation (see SPSS inc., 1983).

Fig. 1. (a) Density scattergram of length (um) and width (um) data for nematocysts used in the present study. (b) Density scattergram of surface area ($um^2$) and volume ($um^3$) of nematocysts. (c) Density scattergram of log-transformed surface area and volume data. $R^2 = 0{\cdot}98$ (p 0·05), $Y = 7{\cdot}17 X^{0{\cdot}66}$. For all graphs, '·' = 1-4, '+' = 5-9. '*' = 10-14 and '0' = >15 points plotted

## III. RESULTS

The raw data (length and width values) are presented in Figure 1a. The points are scattered, hence the low correlation coefficient of $R^2 = 0{\cdot}3648$, $p < 0{\cdot}05$. The computed surface areas (SA) and volumes (VOL) have a curvilinear relationship (Fig. 1b, $R^2 = 0{\cdot}8451$, $p < 0{\cdot}05$). Log transformation of the SA and VOL data yields a linear relationship (Fig. 1c, $R^2 = 0{\cdot}9803$, $p < 0{\cdot}05$). Since the calculated regression line for this relationship, $Y = 0{\cdot}8554\ X^{0{\cdot}658}$, provides a slope, $\underline{b} = 0{\cdot}658$, that is theoretically impossible (i.e. $< 0{\cdot}667$), the residual statistic, $\underline{h}$ was computed. Excluding data with $\underline{h}$ values greater than $n/\overline{4}$ (n = 2157; see Hoaglin & Welsch, 1978) gave a regression line of $Y = 6{\cdot}7066\ X^{0{\cdot}6702}$, with only 127 pairs (5·8%) of the data from the original line excluded. This line ($R^2 = 0{\cdot}9702$, $p < 0{\cdot}05$) is significantly different from the original fit ($p < 0{\cdot}05$) and is assumed to be the better approximation of the relationship between the two variables, SA and VOL. Thus, $SA = 6{\cdot}7066\ VOL^{0{\cdot}6702}$.

The excluded data (i.e. where $\underline{h} > 4/n$) was analysed and plotted (Fig. 2) in order to identify any possible relationships within this subset of the data. Like the original raw data, the points are well scattered, $R^2 = 0{\cdot}3788$, $p < 0{\cdot}05$. Further investigation revealed that of the 127 excluded pairs of data, 105 were anthozoan and 22 meduzoan. This reflects the original proportions contributed by each group to the raw data. Conversely, of the meduzoan data all but one excluded pair were from hydrozoan sources, representing 14 species of siphonophore, and the bottom-dwelling trachymedusan, *Tesserogastria musculosa* Beyer. Of the anthozoan data, excluded data were obtained from the Actiniaria, Ceriantharia, Corallimorpharia, Zoanthidea and Scleractinia. No patterns within these groups were found, except in the Ceriantharia, Corallimorpharia and Zoanthidea. In the Ceriantharia, excluded values (n = 5) were from measurements of atrichs. The exclusions from the Zoanthidea (n = 30) were all from species in the genus *Palythoa* and were measurements of holotrichs. The exclusions from the Corallimorpharia (n = 31) represent data from all 9 species used in the present study, covering all of the extant corallimorpharian families.

The original regression of all the data was further analysed by taxonomic group and nematocyst type. A regression line was calculated for each of these subsets. For the data divided by taxon, the line calculated for the Ptychodactiaria is significantly different ($p < 0{\cdot}001$, Fig. 3a) from all the other lines. This is owing to the few data points used in the calcualtion and their similarity in width (all width values = 4µm). The data for this group was not excluded in the residuals

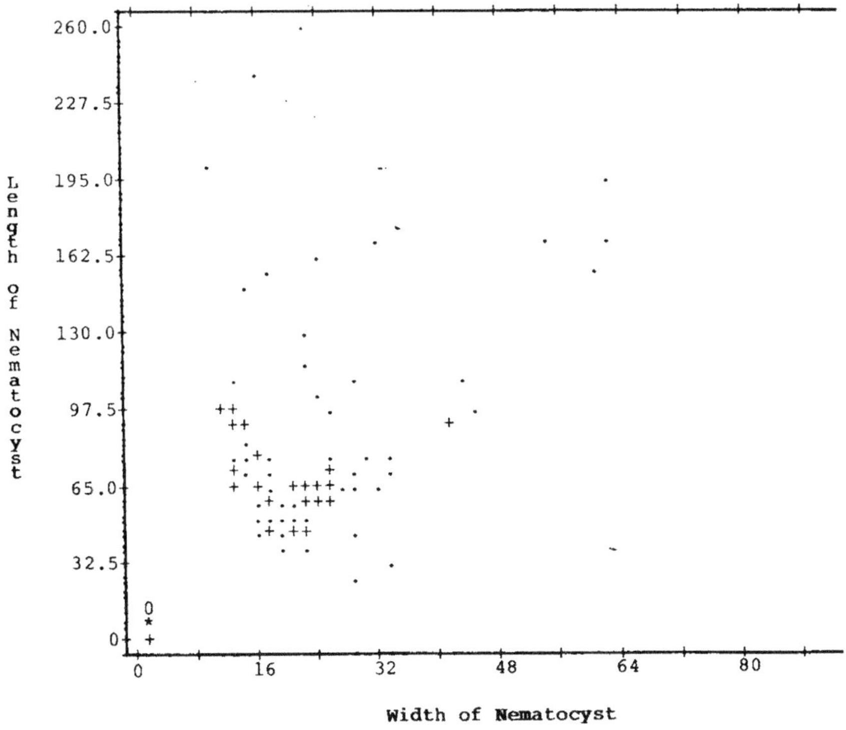


Fig. 2. density scattergram of length (μm) and width (μm) data that gave a leverage coefficient (h) n/4. '·' = 1-4, '+' = 5-9, '*' = 10-14, and '0' = >15 points plotted.

analysis. Whilst the zoanthids provided a slope (b) significantly different from the other groups ($p < 0{\cdot}05$, Fig. 3a), this regression line is theoretically impossible (i.e. $b < 0{\cdot}6667$), though a large number of points were excluded in the residuals analysis. Similarly, whilst the b value obtained for the euryteles (Fig. 3b) is also theoretically impossible, this may reflect the small number of data found for this nematocyst type. The difference in b values between the three nematocyst types of eurytele, desmoneme and stenotele and the four types of mastigophore (Fig. 3b) is probably indicative of the divergence from sphericity in the mastigophores.

Although some subjective indication of the constraints of packing requirements upon nematocyst size can be obtained from the raw data (Fig. 1a) by assessing the deviation from Y = X, it can be quantified by relating SA/VOL ratio to the number of nematocysts $mm^{-2}$. The packing index, number of nematocysts $mm^{-2}$, was defined and calcualted thus:

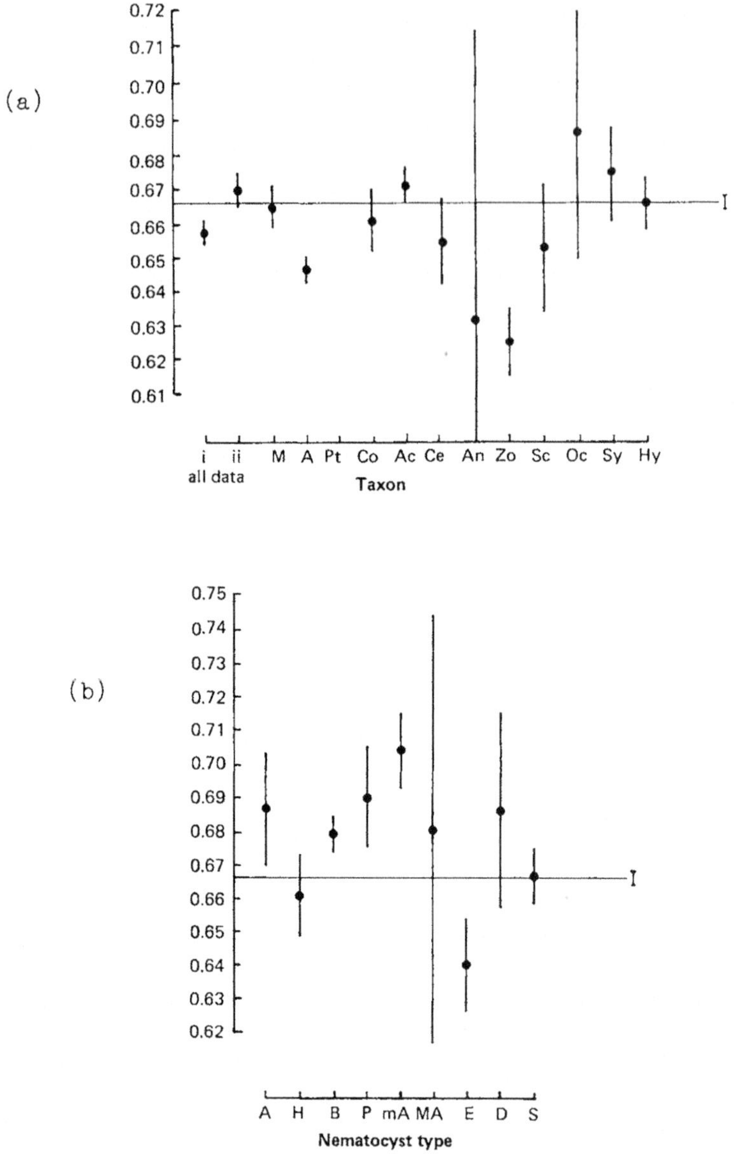


Fig. 3. (a) 95% confidence intervals for b, from the regression of surface area on volume for various taxa. I = 0·6667. All data (i) = both anthozoan and meduzoan data included, all data (ii) = data excluded if it gave a leverage coefficient (h) n/4. M = Meduzoa, A = Anthozoa, Pt = Ptychodactiaria, Co = Corallimorpharia, Ac = Actiniaria, Ce = Ceri-

$$\text{number of nematocysts mm}^{-2} = 1{,}000{,}000/\pi W^2$$

The index was rounded to the nearest whole integer. This equation assumes that a close packing arrangement is applicable. This is certainly the case in many sea anemones, particularly in the acontia and tentacles. A spaced (loose) packing arrangement cannot be quantified, except by observation of that particular tissue, as the distances between the nematocysts could infinitely variable. The results were linearized by the log transformation of both variables (Fig. 4). The fact that a regression line could be fitted, and that the correlation coefficient is high ($R^2 = 0{\cdot}99$, $p < 0.05$) is taken as evidence that the assumption of close packing is valid. Thus, as the SA/VOL ratio increases so does the number of nematocysts per unit area, according to the equation, $Y = 0{\cdot}00938X^{0{\cdot}4794}$.

## IV. DISCUSSION

These results clearly demonstrate that nematocysts are, in general, with $SA = 6{\cdot}17\ VOL^{0{\cdot}67}$ approaching the smallest possible surface area (SA) to volume (VOL) ratio (i.e. $SA = 4.84\ VOL^{0{\cdot}67}$), which is that of the sphere. Deviation from sphericity (i.e. $a > 4.84$) can be explained by requirements to maximize the number of nematocysts per unit area. These two factors are in conflict, as SA/VOL ratio decreases so does the number of nematocysts per unit area according to the equation $Y = 0{\cdot}0094\ X^{0{\cdot}048}$, where Y = SA/VOL and X = nematocysts $mm^{-2}$. Presumably each species, in relation to its niche, and hence the use(s) of its cnidom, has adapted in terms of the best possible solution within the confines of these two constraints. Flexibility within these constraints is obtained by varying the structure of the discharged tubule. The many different types of nematocyst recognised by current workers on the basis of tubule structure is evidence for this hypothesis.

antharia, An = Antipatharia, Zo = Zoanthidea, Sc = Scleractinia, Sy = Scyphozoa, Hy = Hydrozoa, Oc = Octocorallia. (b) 95% confidence intervals for *b*, from the regression of surface area on volume for each nematocyst type. I = 0·6667. A = atrich, H = holotrich, B = B-mastigophore, P = P-mastigophore, mA = microbasic amastigophore, MA = macrobasic amastigophore, E = eurytele, D = desmoneme, S = stenotele.

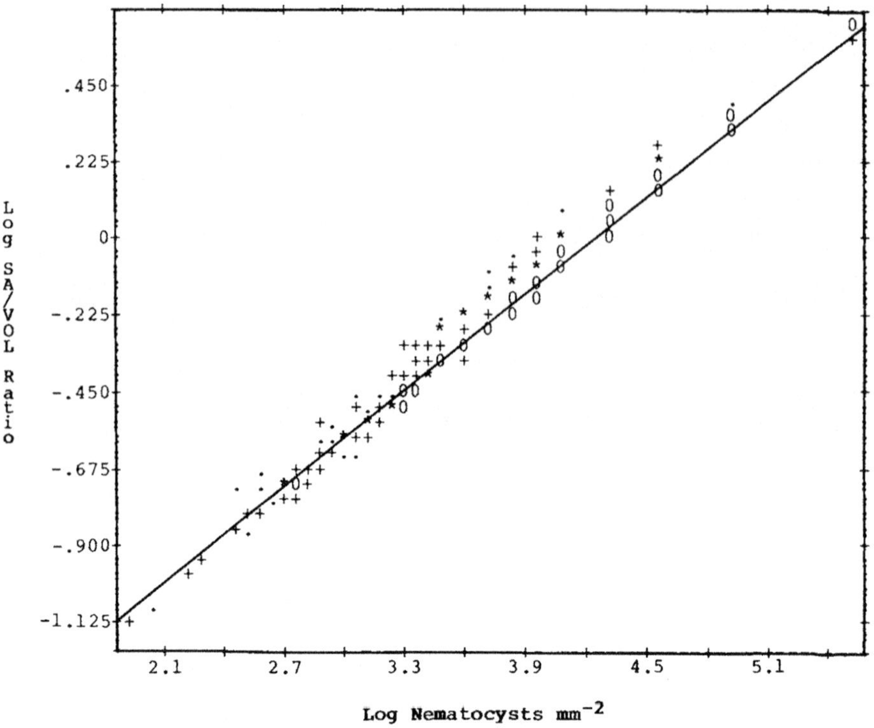


Fig. 4. Density scatterplot of log-transformed surface area/volume (SA/VOL) ratio and the packing index, nematocysts $mm^{-2}$. '.' = 1-4, '+' = 5-9. '*' = 9-14, and '0' = >15 points plotted. $R^2 = 0{\cdot}99$ ($p < 0{\cdot}05$), $Y = 0{\cdot}01\ X^{0{\cdot}48}$.

Considering SA/VOL ratio, this study shows that nematocysts can be divided into two groups, with the majority (94·2%) maintaining a small SA/VOL ratio. The maintenance of a small SA/VOL ratio by the majority of nematocysts supports the osmotic hypothesis of nematocyst discharge proposed by Lubbock _et al._ (1981), _viz_ a rapid influx of solvent into the capsule is owing to an increase in intracapsular osmotic pressure caused by the dissociation of 'macromolecules' within the capsule. Thus, a large volume of 'macromolecule' is maintained, whilst keeping the SA as small as possible hence the pressure per unit area on the capsule wall maximised. On the other hand, it could be argued that the rapidty of solvent influx would be facillitated by a large SA and a high intracapsular concentration of solute, thus the maintenance of a small SA/VOL ratio would not favor an osmotic hypothesis. However, the rapidity of solvent influx affects only the latency period between stimulus receipt and tubule eversion and not the amount of force developed in the capsule wall. In addition, assuming a constant concentration of solute ( i.e. 'macromolecule') then a reduction in nematocyst VOL would also facillitate solvent influx as, proportionally, the SA is increased. Thus, for nematocysts to reduce the latency period between stimulus reception and discharge there should be a decrease in size and to maximize the forces developed in the capsule the SA/VOL ratio should be minimized. Hence, both of these objectives can be achieved without conflict.

The detection of 14 siphonophores as outliers, by the residuals analysis, suggests that a different mechanism of discharge is involved in the nematocysts of these species. This has been suggested recently by Carré (1980). This hypothesis proposes that discharge is caused by the release of tension within the capsule. In this case, therefore, SA should be maximized in preference to VOL. A combination of the two proposed discharge mechanisms, i.e. tension and osmosis, could also be envisaged and supported by the results in the present study. Holstein & Tardent (1984), working with *Hydra* stenoteles, also suggested a combination of these two factors as a possible mechanism for nematocyst discharge. This hypothesis might also explain the presence of outliers detected in the groups Zoanthidea, Ceriantharia and Corallimorpharia, though, as yet there is no experimental evidence to support this suggestion.

Smith (1980) has criticized many allometric studies for the uncritical use of the methodology. Specifically, uncritical and exclusive use of the log-log transformation, the use of substituted measures (i.e. the use of length where volume would be more applicable), and the widespread use of *post hoc* interpretations were berated. The methodology in the present study avoids these particular pitfalls. Transformations were not arbitrarily imposed, but only used if this improved the fit of a particular regression line. In addition to Pearson product -moment correlation coefficients, residual statistics were examined, SA and VOL were used instead of just length and width, and an *a priori* slope (i.e. $b = 0 \cdot 67$) was used a test for the observed regressions. Therefore, it can be concluded that the size and shape of a nematocyst are governed by the physico-chemical demands for effective discharge, i.e a small SA/VOL ratio, which is in conflict with the maximizing of the packing density of the nematocysts, and that to reduce the response time for discharge, capsule size should be minimized. Thus, the size and shape of nematocysts are, at least partially, an optimum resolution of three constraints. Detection of outliers suggests that these nematocysts may discharge by the release of intrinsic tension forces within the capsule, though this needs further examination.

ACKNOWLEDGEMENTS

Many thanks to Dr. Barbara Brown for her support and to Dr. David Hessinger for helping to considerably improve this paper.

REFERENCES

Burnett, J.W. 1971a. An ultrastructural study of the nematocysts of the polyp of Chrysaora quinquecirrha. Chesapeake Sci. 12:225-230

Burnett, J.W. 1971b. An electron microscopic study of two nematocytes in the tentacle of Cyanea capillata. Chesapeake Sci. 12:67-71.

Cairns S.D. 1983. A generic revision of the Stylasterina (Coelenterata, Hydrozoa): 1. Description of the genera. Bull. Mar. Sci. 33:427-508.

Calder D.R. 1983. Nematocysts of stages in the life cycle of Stomopholus meleagris, with keys to scyphistomae and ephyrae of some Western Atlantic Scyphozoa. Can. J. Zool: 1185-1192.

Calder D.R. and E.C. Peters. 1975. Nematocysts of Chiropsalmus quadrumanus with comments on the systematic status of the Cubomedusae. Helgol. Wiss. Meeresunters. 27:364-369.

Carlgren, O. 1937. Ceriantharia and Zoantharia. Rep. Great Barrier Reef Exped. 5:177-207.

Carlgren, O. 1940. A contribution to the knowledge of structure and distribution of cnidae in the Anthozoa. Lunds. Univ. Arsskr. Avd. 2. 36:1-62.

Carlgren, O. 1945. Further contributions to the knowledge of the cnidom in the Anthozoa, especially in the Actiniaria. Lunds. Univ. Arsskr. Avd. 2. 41:1-24.

Carlgren, O. 1950. Actiniaria and Corallimorpharia. Rep. Great Barrier Reef Exped. 5:427-457.

Carré, D. 1980. Hypothèse sur le mechanisme de l'evagination due filament urticant des cnidocystes. Eur. J. Biol. 20: 265-271.

Corea, D.D. 1973. On the sea-anemone Actinoporous elegans Duchassaing. Publ. Seto Mar. Biol. Lab. 20:157-164.

den Hartog, J.C. 1977a. The marginal tentacles of Rhodactis sanctithomae (Corallimorpharia) and the sweeper tentacles of Montastrea cavernosa (Scleractinia), their cnidom and function. Proc. 3rd Int. Coral Reef Symp. Miami 1:463-470.

den Hartog, J.C. 1977b. Notes on the little known sea anemone Cataphellia brodicii, new record, and on the allied Hormathia coronata and Paraphellia expansa (Actiniaria, Hormathidae). Neth. J. Zool. 23:237-244.

den Hartog, J.C. 1977c. Descriptions of two Ceriantharia from the Caribbean region, Pachycerianthus curacaoensis n.sp. and Arachnanthus nocturnus n.sp. with a discussion of the cnidom and of the classification of the Ceriantharia. Zool. Meded., Leiden. 51:211-242.

den Hartog, J.C. 1980. Caribbean shallow water Corallimorpharia. Zool. Verh., Leiden. (176):1-83.

Hashimoto, H. 1981. The freshwater hydroid Astrohydra japonica, new genus, new species. Annot. Zool. Jap. 54: 207-2212.

Hesthagen, I.H. 1971. On the biology of the bottom-dwelling Trachymedusa Tesserogastria musculosa Beyer. Norw. J. Zool. 19:1-19.

Holstein, T. and P. Tardent. 1984. An ultrahigh-speed analysis of exocytosis: nematocyst discharge. Science 223:830-833.

Kubota, S. 1978. The life history of Clytia edwardsi (Hydrozoa, Campanulariidae) in Hokkaido, Japan. J. Fac. Sci. Hokkaido Univ. Ser. VI. Zool. 21:317-354.

Lubbock, R., B.L. Gupta and T.A. Hall. 1981. Novel role of calcium in exocytosis: mechanism of nematocyst discharge as shown by X-ray microanalysis. Proc. Nat. Acad. Sci. U.S.A. 78:3624-3628.

Manuel, R.L. 1977. A redescription of Edwardsia beautempsi and E. timida (Actiniaria: Edwardsidae). Cah. Biol. Mar. 18:483-497.

Ostman, C. 1979. Two types of nematocyst in Campanulariidae (cnidaria, Hydrozoa) studied by light and scanning elctron microscopy. Zool. Scr. 8:5-12.

Parry, G. 1951. The Actiniaria of New Zealand, pt 1. Rec. Cant. Mus. 6:83-119.

Parry, G. 1952. The Actiniaria of New Zealand, pt 2. Rec. Cant. Mus. 6:121-141.

Purcell, J.E. 1984. The functions of nematocysts in prey capture by epeipelagic siphonophores (Coelenterata, Hydrozoa). Biol. Bull. 166:310-327.

Russell, F.S. 1938. On the nematocysts of the Hydromedusae. J. Mar. Biol. Assoc. U.K. 23:145-165.

Russell, F.S. 1939. On the nematocysts of the Hydromedusae, II. J. Mar. Biol. Assoc. U.K. 23:347-359.

Russell, F.S. 1940. On the nematocysts of the Hydromedusae, III. J. Mar. Biol. Assoc. U.K. 24:515-523.

Smith, R.J. 1980. Rethinking allometry. J. Theor. Biol. 87:97-111.

SPSS inc. 1983. SPSS$^{X}$ User's Guide. McGraw-Hill Book Co. London. 806p.

Stephenson, T.A. 1929. On the nematocysts of sea anemones. J. Mar. Biol. Assoc. U.K. 16:173-200.

Stephenson, T.A. 1935. The British Sea Anemones, Vol. 2. Ray Soc. London. 426p.

Thomason J.C. and B.E. Brown. 1986. The cnidom: an index of interspecific aggressive proficiency in scleractinian corals. Coral Reefs 5:93-101.

Weill, R. 1934. Contribution a l'étude des cnidaires et de leurs nématocystes. II. Valeur taxonomique du cnidome. Trav. Stn. Zool. Wimereux 11:351-701.

Yamada, M. and K. Konno. 1973. Polyp and medusa of the hydroid Sphaerocoryne multitentaculata (Warren) from Japan. Publ. Seto. Mar. Biol. Lab. 20:103-109.

# Index

G



H



I

## S

## W



## X



## Z

Printed in Dunstable, United Kingdom

91338422R00346